Improving Almost Anything

Improving Almost Anything

Ideas and Essays

Revised Edition

George Box
and Friends

A JOHN WILEY & SONS, INC., PUBLICATION

Library of Congress Cataloging-in-Publication Data:

Box, George E. P.
  Improving Almost Anything: ideas and essays / George E. P. Box – Rev. ed.
    p. cm.
  Rev. ed. of: Box on quality and discovery. c2000.
  Includes bibliographical references and index.
  ISBN-13: 978-0-471-72755-2 (pbk : acid-free paper)
  ISBN-10: 0-471-72755-5 (pbk : acid-free paper)
  1. Mathematical statistics. 2. Quality control – Statistical methods. 3. Experimental design.
  4. Control theory. I. Box, George E. P. Box on quality and discovery. II. Title.

QA276.B678 2006
658.5'62015195 – dc22

                                        2005055153

10 9 8 7 6 5 4 3 2 1

To Claire,
my never failing source of encouragement and inspiration

Contents

Special Note: Most of the chapters in this book are written for the general reader involved in improving quality and improving processes. The chapters characterized below by a star (☆) will be of most interest to those concerned with the derivation of new techniques or the detailed comparison of existing methods. It is believed however that the summaries and conclusions in these chapters are of general interest. The source of the material is indicated in the title page of each chapter.

Some of the articles in this book were written as research and preparation for the second edition of *Statistics for Experimenters* by Box, Hunter, and Hunter. There is therefore occasional duplication of material in the two books.

Foreword

Some time has passed since the publication of *The Collected Works of George E. P. Box*, Volumes I and II, under the editorship of George Tiao (1985). These have been active years in the life of the gentleman for whom these early volumes were produced. His continuing originality, and verve, provide us today with a cornucopia of new writings.

The first printing of this book was called *Box on Quality and Discovery: With Design, Control, and Robustness*. George has now revised it and aptly changed its name. At George Tiao's urging he extracted, and arranged some of his publications in logical order to form the chapters of this book. We now have vignettes of his thoughts, reminiscences of earlier days, and renewed insights. The book is divided into six parts. The first of these are directed to managers, another to those who would apply experimental design, another for those who study and work in experimental design, a group encouraging the flirtation between control engineering and statistics, a group devoted to variance reduction and the concept of robustness and, finally, some of his songs.

In addressing the concerns of managers, Box recognizes that they are more likely to be interested in prescriptions for problem solving and prevention rather than in descriptions of specific statistical tools. These chapters thus emphasize the application of the scientific method. They advise the manager to view their processes as ongoing iterative learning exercises. An early requirement is to open avenues of communication with those who run each process and then to listen. Success further requires the willingness to make changes based on the information gathered—read "experiment." Analogous the Deming's "Plan, Do, Check, Act" is Box's refrain "Conjecture, Design, Experiment, and Analyze," which I much prefer, particularly the advice that analysis has as its primary objective the creation of new conjectures. That's Box's message to managers, workers, students, scientists, and statistical academics alike.

The second part is titled "Design of Experiments for Quality". When the journal *Quality Engineering* was started, Box was asked if he would contribute a "column" to each issue. The result was a series of popular short articles describing the uses of statistics, principally experimental design, adapted to the pursuit of *quality* products and processes. His audience were those who were anxious to employ tools that might lead to improved quality of products and processes, and sequential experimentation received special emphasis. All the papers are terse and clear, and give new meaning to the acronym KISS (Keep It Simple, Statistician). Two additional articles are included in this group, one discusses factorial design dispersion effects and the second the need for hands-on practice in the teaching of statistics.

The third collection, "Sequential Investigation and Discovery," has the consulting statistician in mind. Here *discovery* joins with *quality*, the theme in Box's mind for this entire volume of works. The emphasis of the first few chapters is on the necessary alternation between inference and deduction, estimation and hypothesis testing. Box argues that, in applying the scientific method, planning for data that place inferences in jeopardy is every bit as important as increasing the power of a test. "Surprise!" is every bit as important as "Gotcha!" This point is driven home with Box's description of a detective gathering evidence to find a thief contrasted with that of a prosecutor who must prove the guilt of the accused. The final chapters discuss fractional factorial and response surface designs and offer Box an opportunity to reminisce on the early events that stimulated his interest. Clearly, close rapport between experimenter and statistician was the essential element.

The fourth group is titled "Control." In his introduction he recalls the events that brought him to this important arena. This work vastly changes the field of Statistical Process Control (SPC). Historically, SPC evokes the application of Shewhart or similar "control" charts. Their purpose is to monitor processes assumed to be reasonably stable. When unusual events are discovered, they stimulate searches for assignable causes that may lead to process improvement. The several chapters in this group describe both theory and practice leading to feedback control appropriate for the SPC environment. A simple graphical technique (recently named Box–Jenkins charts to honor their originators) is described for adjusting a process to acquire minimum variability about target. The idea of appropriately adjusting processes to target is new to most quality professionals and much education lies ahead. SPC now has two mutually supporting techniques, one for monitoring and a second for adjustment. Many of the papers on which these chapters were based were precursors to the book by George Box and Alberto Luceño, *Statistical Control by Monitoring and Feedback Adjustment* (Wiley, 1997).

The final group is titled "Variance Reduction and Robustness." The subject of split plot designs allows Box to bring into focus Dr. Genichi Taguchi's contributions to the application of statistics in industry. Split plot designs and Taguchi's inner and outer array designs have identical structure. Taguchi employed these designs to determine conditions providing "robust" manufactured product and process environments. In the early 1950s, Box was the first person to employ the word robust in describing the influence of nonnormality and other departures from assumptions upon the usefulness of the t and F tests. The origins of the statistical uses of the word "robust" are reviewed in Box's commentary. From the papers in this section we find that robustness is a function of specifications, variability caused by measurement, the process, the environment, the transmission of error, the choice of metric, and the search for an optimum or best conditions.

As one reads this book, fundamental ideas emerge, blend, and reemerge. In contrasting the themes of inference versus testing, or mathematical exactitude versus statistical robustness, one is reminded of the horns and strings in an orchestra. Each component has its separate role and yet their fullest expression occurs when they are combined. Additional themes of *quality* and *discovery* provide resonance and balance. The pursuit of quality requires meeting standards and positive feedback leading to improvement. Discovery requires planning for information-laden data and welcoming the phenomenon of surprise. And to pursue musical analogy further, technical flourishes occur, as in the work on projected factorial designs, Cuscore charts, and the selection of sampling intervals for discrete feedback control. Alphabetic optimality of experimental designs adds just the touch of required dissonance.

All the computations in this book can be done with the statistical language R (R development Core Team, 2004), available at CRAN (http://cran.R-project.org). There is as well commercial software, such as the SCA Statistical System and SAS Jump, which some readers willl find easier to use.

In their totality, the papers on which this book is based form a collection of master works. They are unique. Read an opening paragraph and you know the name of the composer. Their full expression cannot be captured in a few words. Several have refrains so simple you are sure you thought of them first. "All models are wrong, some models are useful." "When Murphy speaks—listen." And who can forget robust statistics, happenstance data, and lurking variables? On occasion, these works contain elements of considerable complexity and yet they are all suffused with a unifying message. We learn that statistics is not mathematics but possesses its own philosophy, that statistics and the scientific method are together entwined, that collaborative work provides a statistician's greatest rewards, and that statistics well taught easily captures the imagination of the student, worker, manager, and scientist.

I met George Box in January 1953 when, at the invitation of Miss Gertrude Cox, he accepted a one-year appointment to the Institute of Statistics in Raleigh, North Carolina. He became my major professor and to this day I wonder at my good fortune.

J. STUART HUNTER
Professor Emeritus, Princeton University

Co-Authors and Friends of George Box

Chapter

Søren Bisgaard, Eugene M. Isenberg Professor of Technology Management, Eugene M. Isenberg School of Management, University of Massachusetts, Amherst, Massachusetts — B5

Conrad A. Fung, Consultant, Brookfield, Wisconsin and adjunct faculty member of the College of Engineering, University of Wisconsin, Madison, Wisconsin — E2 and 3

Stephen P. Jones, Boeing Company, Seattle, Washington — E7 and 8

Tim Kramer, Hewlett-Packard, Battle Ground, Washington — D5

Alberton Luceño, Professor, E.T.S. de Ingenieros de Caminos, University of Cantabria, Santander, Cantabria, Spain — D6, 7, 8, and 10

Patrick Y. Liu, Sr. Staff Engineer, Kohler Company, Kohler, Wisconsin — C1

Dr. R. Daniel Meyer, Pfizer, Inc., Groton, Connecticut — B11 and C7

Dr. José G. Ramírez, W.L. Gore & Associates, Inc., Newark, Delaware — D11

D. M. Steinberg, Department of Statistics and Operations Research, Tel Aviv University, Tel Aviv, Israel

John Tyssedal, Associate Professor, The Norwegian University of Science and Technology, Trondheim, Norway — C8

My Professional Life

By George Box

My professional life has been profoundly influenced by many generous and talented people. It was forever changed when J. Stuart Hunter, by a happy conspiracy with Miss Gertrude Cox and Frank Grubbs of ARO, brought me in 1953 to the United States from England. I came on leave of absence from my industrial job at ICI and spent a wonderful year with the folks at the Institute of Statistics at North Carolina's State College. Stu and I worked together on RSM, and other aspects of experimental design, and became close friends.

After my return to England, I spent three more years with ICI further developing response surface methods and Evolutionary Operation and working with the physical chemists on kinetic models. The study of mechanistic models such as these allowed a deeper understanding of our processes and led to interesting work on nonlinear estimation and the numerical solution of simultaneous differential equations.

In 1956 John Tukey and Sam Wilks invited me to head up the Statistical Techniques Research Group at Princeton. Martin Wilk was deputy director and our long-term visitors included Henry Scheffé, Martine Beale, Norman Draper, Don Behnken, Colin Mallows, H. L. (Curly) Lucas, and G. S. Watson. Some of the permanent members—in particular, Stu Hunter, Merve Muller, and John Tukey—spent part of their time with the group and part working in industry. The result was a never-ending series of lively discussions from which we all greatly benefited. The arrival of Gwilym Jenkins marked another turning point in my life and began a long and happy collaboration on time series, forecasting, and control.

During my earlier visit to the United States, in 1953 I had met Cuthbert Daniel. He took me on my first consulting trip and I was, and remained, fascinated by this wise and witty man so full of valuable and original ideas. Sometime in 1958, Stu, Cuthbert, and I decided that a new journal was needed to meet the needs of applied statisticians. We believed that balance might be maintained if it was jointly managed by ASQ and ASA but we needed $10,000 to get it started. The Chemical Division of ASQ came up with $5000 and the three of us got the other $5000 by organizing and teaching a short course for industry. The name *Technometrics* was suggested by R. A. Fisher. The journal would never have got off the ground without the talent and enthusiasm of its founding editor, J. Stuart Hunter, under whose guidance it quickly became widely read and respected.

In 1960, I was invited to start a new statistics department at the University of Wisconsin–Madison and Stu Hunter and Norman Draper both came to help. I had almost no previous teaching experience but among many other duties I found myself responsible for a course on the advanced theory of statistics. I had seven students. By great good fortune these included George Tiao, Bill Hunter, and Sam Wu. I made up the course as I went along, and although as a student I had been thoroughly indoctrinated with Neyman–Pearson theory, my attempts to make sense of statistical inference became more and more Bayesian with every passing week. George Tiao was my bellwether and, whenever he looked worried, I knew I should look again at what I had just written on the blackboard. Our subsequent work together began a lifelong friendship and, eventually, a book on Bayesian methods.

The beer and statistics seminar got started very soon after I arrived in Madison. Students, mostly from statistics and engineering, met, originally in my basement, every Monday night. For more than 30 years, the ever-changing group discussed a great variety of problems brought by engineers, chemists, biologists, and administrators. They came from university departments, from industry, and from local government. I believe we helped them and they certainly educated us. One of the enthusiastic Monday-nighters was Bill Hunter, whom I had known as an engineering student at Princeton. With the help of Olaf Hougen, Bill and I initiated joint projects with the Chemical Engineering Department on automatic optimization and on model building techniques. It was not long after that, that Stu, Bill, and I started our book *Statistics for Experimenters*.

George Tiao and Bill Hunter became faculty members, who both profoundly influenced the development of the Statistics Department at Madison. Norman Draper has been another long-term pillar of the department. He and I worked together for many years; and our collaboration resulted in books on evolutionary operation and on response surface methods.

The department prospered and attracted many talented faculty. In particular, I remember debates with Steve Stigler whose office was next to mine and who liked to tease me about my Bayesian proclivities. I continue to learn from his fascinating historical discoveries and to enjoy his sense of humor. (He once told me, with only a trace of a grin, that my sawtooth diagram representing the interplay between induction and deduction had recently been discovered carved on the wall of a cave and dated about 3000 B.C.)

Beginning in the 1970s George Tiao and I made several visits to Spain to teach short courses on the design of experiments and on time series. The moving spirits who brought us there were Daniel Peña from Madrid and the late Albert Prat from Barcelona. This was the start of a long and continuing association with a beautiful country and with two enterprising and generous friends.

Bill Hunter and I became interested in "Quality" some sixteen or so years ago. We had, for a long time, worked together on statistical methods, in particular experimental design, for the improvement of industrial processes; but Bill quickly saw that quality ideas should also be employed to improve the social environment. He instigated and taught a new course in Quality Improvement held in the evening with enthusiastic participants from city and state governments, health care, local banks, and industry. With Bill's help, Quality Improvement projects were begun in the city department of motor vehicles, in the police force, in garbage collection, and so forth. As a consequence, the University set up the Center for Quality and Productivity Improvement (CQPI) in the College of Engineering with Bill as its first director. Its mission was to teach and conduct research on the many aspects of quality. These initiatives continued to flourish under the wise direction of Søren Bisgaard and they remain a lasting tribute to Bill's memory. Members of the Center and our many visitors produced over 180 publications (many of the papers that appear in this volume were initially CQPI reports) describing research on many different aspects of quality. Also, we taught a large number of short courses for industry emphasizing experimental design.

One of the Center's most enthusiastic supporters was Bill Hill then of Allied Signal, who a very long time ago was my graduate student. He has always provided an important link between academia and industry and recently, under the banner of the "Six Sigma" movement has helped top management appreciate the enormous potential of modern process improvement methods. These techniques, appropriately organized and applied at all levels in his company, have met with outstanding success.

One of the more recent visitors to CQPI was Alberto Luceño, a professor from the University of Cantabria. We have worked together particularly on feedback adjustment methods appropriate in the context of SPC. This has resulted in my again visiting Spain for many months and in Alberto's coming to Madison for similar visits. Our happy companionship eventually led to the publication of the book on control.

I cannot end without reference to some earlier happenings. I have told elsewhere (see De Groot, 1987) how in WWII I accidentally became an amateur statistician and how as the result of a series of lucky circumstances, the first professional statistician I ever met was R. A. Fisher. He invited me, an unknown and ignorant army staff sergeant, to his house and with great patience and kindness spent the whole day with me, not only solving my problem but also working with me on the calculations.

After this I had no doubt that I wanted to be a statistician and, with the help of a post-war program similar to the G.I. Bill, went to study at University College London. I was welcomed there with great kindness by Egon Pearson and I remember with particular affection the generosity and humor of my thesis supervisor, H. O. Hartley.

My greatest debt is to the late Professor George A. Barnard, who for over 50 years was my mentor and cherished friend; without his early and continued encouragement my life would have been very different and much less interesting.

Not least for all the fun we have had together, I must express my gratitude to my co-authors—Søren Bisgaard, Conrad Fung, Stephen Jones, Tim Kramer, Alberto Luceño, Patric Liu, Dan Meyer, José Ramirez, David Steinberg, and John Tyssedal.

Special thanks are due to Murat Kulahci, Lan Zhang, and Ernesto Barrios for their invaluable help in preparing the manuscript and the diagrams for this book, and to Lisa Van Horn who produced the completed book with cheerful thoroughness and dispatch.

PART A

Some Thoughts on Process and Quality Improvement

CHAPTER A.0

Introduction

In day to day encounters with industry, government departments, hospitals, universities, airlines, restaurants, and indeed every sort of organization, three things seem equally perplexing:

- that the level of quality is frequently poor
- that this situation is tolerated by society
- that most of the problems seem to be elementary, yet impossible to solve.

The way forward was clearly indicated by the profound insights of W. Edwards Deming, encapsulated in his fourteen points. But perhaps nothing is less common than common sense, and putting into effect needed changes is often extremely difficult.

The following pieces written at widely different times discuss some of the many issues and are intended to assist process change.

Improving Almost Anything: Ideas and Essays, Revised Edition. By George Box and Friends
Copyright © 2006 John Wiley & Sons, Inc.

CHAPTER A.1

Good Quality Costs Less? How Come?

At the heart of the process improvement revolution is an idea expressed some 400 years ago by Sir Francis Bacon. He was Lord Chancellor of England, a distinguished philosopher of science, and man of such stature as to be credited by some with writing Shakespeare's plays.

He said "Knowledge itself is power." The application of that profound statement is this: To the extent that we know more about our process, our product, our customers' needs, and about all the operations we perform—manufacturing, billing, invoicing, dispatching, and so forth— and *only to that extent* can we do a better job, make a better product, and so please our customer more.

So process improvement and good quality is produced by knowing more about what we are doing.

Now, perhaps we could learn more about what we are doing if we had more people and spent more money, but the good news is that we don't need to take that route. We have three important resources we can draw on. All of them are free and we can put them together to continuously generate the new information we need to get ahead and stay ahead. These ideas can be applied to improve every industry, every government department, every hospital, and every university; but most of the time they are underutilized or not used at all.

The three resources relate to the following: (1) all human beings are creative; (2) the operation of any system generates information on how it can be improved; and (3) experimental design can increase the efficiency of experimentation many times over. Let's talk about each of these in turn.

From Box, G. E. P. (1990–1991), *Quality Engineering*, 3(1), 85–90. Copyright © 1990 by Marcel Dekker, Inc.

Improving Almost Anything: Ideas and Essays, Revised Edition. By George Box and Friends
Copyright © 2006 John Wiley & Sons, Inc.

EVERY PERSON IS CREATIVE

The characteristic that most distinguishes humankind from the rest of the animal kingdom is *creativity*. Just as "fish gotta swim" and "birds gotta fly" every human being possesses creativity and feels the need to use it. If you had been able to look at horses in a field 20,000 years ago and you looked at their present descendants they would be doing about the same things then as now. But this would not be true for human beings, who would have found ways to clothe and shelter themselves, to get clean drinking water, to converse with each other, to write, to transmit messages, . . . and so on and so on, seemingly without end.

Now, not all people have the same degree of technical sophistication, anymore than they have the same height. Indeed, like height, this characteristic will have a hump-shaped frequency distribution as in Figure 1a. In the past only a selected managerial and scientific elite were recognized as licensed to put their creativity to use. These were a small number of people supposed to be in the right-hand tail of this distribution. Now look at Figure 1b. This is a frequency distribution of problems

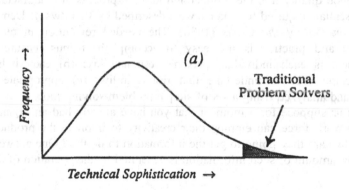

(a)

**Traditional
Problem Solvers**

Technical Sophistication →

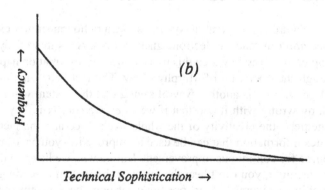

(b)

Technical Sophistication →

Figure 1. (a) Frequency distribution of workers with a given degree of technical sophistication. (b) Frequency distribution of problems requiring a given degree of technical sophistication for their solution.

that might beset some organization, classified by the degree of technical sophistication needed to solve them. This is a Pareto-like distribution with a large number of problems not requiring a high degree of technical sophistication and the frequency falling off as the problems become more challenging. If you look at these two diagrams together, you see that, under the old system, we threw away the creativity of an enormous number of people. You do not need advanced qualifications to figure out how to ensure that the right screw is delivered to a work station, or the right hospital test results are available when a patient comes in for an examination. And yet such problems not only occur, but *persist*, in very many organizations.

So why then was this potentially vast problem-solving resource not used? It was because, as Dr. Deming said, the people closest to the system often had no expectation that it could be better, or any understanding of how to make it better, and because they believed they were powerless to change it.

To remedy that situation required a radically different management philosophy in which the old idea of a quality control department acting as a quality *policeman* to perform the (hopeless) task of inspecting out bad quality was replaced by the concept of the whole work force acting as quality *detectives* to discover new ways of building good quality into the product and into the process of manufacture.

The essentials required for change were described by W. Edwards Deming in his classic book *Out of the Crisis* (1986). The needed revolution in management philosophy and practice is not easy to accomplish it must ensure that: (a) improvement is each individual person's responsibility; (b) each individual is suitably empowered to undertake that responsibility; (c) appropriate data are collected and analyzed using a set of simple problem-solving tools.

But let us suppose for a moment that you have already had an organization in which the work force can employ their creativity to improve the product and the process. How are they going to get the information to do this? One answer is that a tremendous amount of such information is generated by the operation of the system itself.

EVERY SYSTEM GENERATES INFORMATION

As illustrated in Figure 2, an operating system is like a radio transmitter except that it transmits *information* instead of electromagnetic waves. As is more fully explained in the next chapter, one way in which this transmission of information happens (Box, 1989a) is through the operation of Murphy's Law. The fact that "anything that can go wrong will go wrong" is another way of saying that the system will *tell* us when there's something wrong with it and that if we *listen* we can fix it. If you put these two things together—the creativity of the whole work force and the fact that every system generates information that can be used to improve it—you have an extremely powerful resource for continuous improvement. But how *do* we listen to the process? Like a radio transmitter, you can't hear the message the system is sending out unless you have suitable receivers. For an operating system these receivers are simple devices for collecting and analyzing data—flow charts, check sheets, Pareto diagrams, fishbone charts, graphs, Shewhart charts and so forth. You will find most

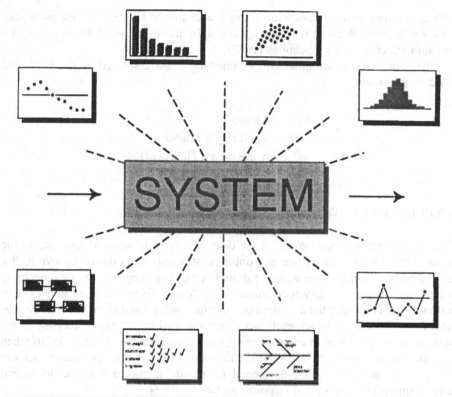

Figure 2. The system transmits information received and interpreted by (reading clockwise from upper left) a run chart, a Pareto chart, a scatter diagram, a histogram, a Shewhart chart, a fishbone chart, a check sheet, and a flow diagram.

of them described and illustrated with real examples in the wonderful book by Ishikawa (1982). Such tools are complemented by elementary experimental design.

Process operators, nurses in hospitals, workers in city government can all learn to use these tools. They can be used, for example, to improve a canning process in a factory, the distribution of medications in a hospital, and the issue of driver's licenses in the Department of Motor Vehicles. You can read about some of these latter applications in a paper that appeared in *Quality Progress* by Box, Joiner, Rohan, and Sensenbrenner (1989) and also, for example, Box and Bisgaard (1987).

The benefits provided by worker participation are twofold. Quality is improved because of the finding and fixing of a very large number of problems, but also, and perhaps equally important, morale is improved. It is enormously satisfactory to be allowed to be creative, and frustrating to be treated merely as a pair of hands. The bird in the cage, once it has overcome its initial disbelief, will find it wonderful to be allowed to fly.

It has been said that more than 85% of quality problems come from the system itself and that "therefore" only management can solve them. But by setting up problem-solving teams led and encouraged by management, the work force, in effect, becomes part of management and is available to help solve these problems. One excellent book about the *team* approach, by means of which this idea may be put into effect, is that by Scholtes (1988).

The idea is nicely summarized in something I saw displayed by the UAW and Ford management:

Tell me—I'll forget

Show me—I may remember

Involve me and I'll understand

EXPERIMENTAL DESIGN

The quality improvement tools so far discussed provide ways of listening to the system in its *normal operation* and doing what it tells us to do to improve it. But engineers and scientific management should be tackling deeper questions concerning what would happen if they tried something *different*. To find this out they need to experiment. The recognized method of experimentation used to be the *"one-factor-at-a-time"* method in which each factor was changed in turn while keeping all the rest constant. That way of experimenting became outdated in the early 1920s when Ronald Fisher discovered much more efficient methods of experimentation factorial designs. These were further developed to include fractional designs, orthogonal arrays split plot designs, and response surface methods.

In Chapter B.1 of this book, I discuss a very simple factorial design used by Hellstrand (1989) at SKF for the improvement of the design of a bearing. In this experiment, which was one of a series saving many millions of dollars, three factors—heat treatment, outer ring osculation, and case design—were tested each at two levels. This experiment resulted in a fivefold increase in bearing life, and this factorial design required only a total of eight runs!

Now, for most organizations not only is it true that insufficient attention is given to experimentation, but often the experimentation that is done is done extremely inefficiently. Sometimes not even the one-factor-at-a-time method is used but just *"pick and try."* The extraordinary truth is, that 80 years after Fisher invented modern experimental design, it is still not widely taught in schools of engineering and science in our universities. Industry must help academia to remedy this distressing situation.

Notice that all the things I've talked about—the *creativity of the whole work force*, the *information continually generated by an operating process*, the running of experiments according to the *principles of statistical experimental design*—do not, of themselves, cost anything. The resources are there but are largely unused. They do not require the hiring of more people or the purchasing of more equipment.

Their use does however require *profound reorganization* of management and extensive training that *does* involve considerable expenditure of time and money. But just as teaching a man to fish can provide him with food for the rest of his life, so reorganization and training can set in place a system of continuous improvement that *never ends*.

THE ESSENTIAL MISSING INGREDIENT FOR THE SIX SIGMA REVOLUTION

In the past, although statistical tools had been used in American industry, its successes had rarely permeated the "Bottom Up" transmission of information. The successful projects that came to light were assumed to be isolated cases requiring expensive and specialized talent.

The situation in Japan was different. After World War II, as part of an initiative to help Japanese industry get back on its feet, Drs. Edwards Deming and Joseph Juran lectured in Japan about quality and process improvement. These lectures were attended by some of the most *senior* Japanese executives and they took them very seriously.

Thirty years later the effects were felt here – the excellence of imported Japanese automobiles and almost every other kind of mechanical and electronic device threatened many of our industries with extinction. After Dr. Deming made his television presentation, "If they can, why can't we?" great interest was aroused in American industry. But Deming refused to discuss his ideas with anyone but top executives.

So at last, an authoritative statistical scientist got to talk with higher management in this country and to explain what was wrong with their policies and how they could change. Once the ice was broken such interchanges become more common and eventually produced remarkable results.

Now, if you set a match to a mixture of 74 lbs. of pottassium nitrate and 14 lbs. of carbon nothing of interest would happen. But, had you mixed in a *third* ingredient – 12 lbs. of sulfur – you would have made 100 lbs. of gunpowder and a great deal would happen!

Look again at Figures 1a, 1b and also Figure 2, you will see that

1) Simple problem solving tools, such as are shown in Figure 2, together with elementary experimental design, were available.

2) There existed a large pool of untapped human resources that could be trained to use them to improve processes. But the breakthrough could not happen without the addition of a third vital ingredient.

3) This was when the CEO's of such companies as Motorola, GE, Honeywell and Texas Instruments realized the enormous potential of those two facts and made it clear that everyone in the organization would be trained and expected to use these simple tools for improvement on projects likely to save money.

The explosion that resulted when these three ingredients were combined was called Six Sigma. It produced spectacular results not only in manufacturing and business applications but in such organizations as banks, schools, and hospitals.

It is not my intention to catalog the profound organizational changes needed to bring about such a revolution. These are detailed in such texts as Eckes (2001); Harry and Schroeder (2000); Pande, Neuman, and Cavanagh (2000); and Snee and Hoerl (2003).

An important goal of this book however is to make available in clear and simple language ideas and methods that you may find useful in such efforts.

CHAPTER A.2

When Murphy Speaks—Listen

Murphy's Law says that "anything that *can* go wrong *will* go wrong" and this is usually regarded as bad news. If you think about it though, it's not really bad news, it's good news. Nothing is perfect and Murphy's Law says that the day-to-day operation of the system itself can help to tell us what's wrong with it. The catch is that it will only tell us *if we listen*. If we don't listen then the bug that's in the system will cause the same glitch to happen again and again. The inevitable result is frustration. But if we listen and take appropriate action then we may be able to get rid of the bug and be free of it forever.

Another way of saying this is that "every operating system supplies information on how it can be improved and if we use that information it can be a source for continuous improvement." Figure 2 in Chapter A.1 illustrated this diagrammatically. It showed a system radiating information much like a radio transmitter radiates electromagnetic waves.

WHAT IS A SYSTEM?

But what do we mean by a system? The important example that immediately comes to mind is an industrial manufacturing process. But many systems are not concerned with manufacture but with invoicing, dispatching, billing, etc. It is important to remember that it is just as essential to get the bugs out of *those* systems as out of the process of manufacturing itself. But a system of the kind I'm talking about might have nothing to do with manufacturing. It might be baggage handling in an airline, student registration in a college, or health care delivery in a hospital. If one listens to Murphy all these systems can be improved.

From Box, G. E. P. (1989), *Quality Progress*, 22(10), 79–84.

THE MYSTERIOUS "THEY"

So why isn't Murphy listened to more often? Or to put it the other way, why isn't the message that he is trying to tell us put to use? A common difficulty is that the people closest to the system (a) may have no expectation that the system could be better, or (b) have any understanding of how to make it better and (c) have no power to change it.

When I discussed this with some friends recently here are some of the stories that came up:

Friend A told me how he had recently had a medical check-up at the hospital. This involved, in particular, a number of lab tests. Two weeks later in the follow-up appointment with his doctor it turned out that the results of the tests had so far not been entered in his records. His doctor smiled and said rather triumphantly, "Don't worry *they* often do this so I keep a duplicate of the test results myself, although I'm not supposed to."

Friend B told how, the previous week when he had flown across the country, the airline had lost his baggage which contained the slides he needed to make an important presentation. They were not returned to him for two days. When he reported the loss at the airline's lost baggage claim the clerk said, "Oh yes, you were on that flight where you have to change planes in St. Louis. People are always losing luggage on that connection because the time between planes is so short and the gates in St. Louis are such a long way apart. I'm sorry but that's the way *they* schedule them."

Friend C, told me that when he started his job with a large pharmaceutical company his boss told him that their main problem was the very long time it took *them* (that is the FDA) to approve a new drug.

In these stories, the various organizations and indeed my three friends themselves all recognized that the systems they were dealing with were faulty but believed they were totally powerless to do anything about them. Furthermore, all of them believed that there was a mysterious and omnipotent *THEY* who alone decided the way things were done. But in each case Murphy had spoken, and in each case he was not heeded. The system was itself providing data which, with proper organization, could have been used to improve it.

Notice that the impersonal "THEY" is a give-away word which should always warn us when we use it that something is wrong. Its use implies a self-granted permission to be deaf to Murphy's message.

CHANGING THE WORK CULTURE

To listen and to respond to Murphy we need the major culture change described by W. Edwards Deming in his classic book *Out of the Crisis* (1986). In particular this requires

(a) a radically different management philosophy,

(b) appropriate organization to put this philosophy into effect,

(c) some simple tools.

MANAGEMENT PHILOSOPHY

The most difficult change that was needed was a change of management philosophy. The old-fashioned idea of a good manager was one who is supposed to know all the answers, can solve every problem himself, and can give appropriate orders to his subordinates to carry out his plans. This is a role which is seldom possible to sustain with comfort or without hypocrisy. A good modern manager is like a good coach who leads and encourages his team in never-ending quality improvement. Not surprisingly, while some managers are enthusiastic about their new role others find such a change threatening and associate it with loss of power.

IF, SHOULD AND SHOULDN'T

Confronted with the failure of a poor system for which he is responsible, the manager may protest, "But *if* everyone did his job according to the rules this *shouldn't* happen—people *should* follow instructions." Again the words IF, SHOULD and SHOULDN'T are give-away words. The word "if" presages a dependent clause, as in the sentence "if the moon were made of green cheese it would be a great place for mice." Such sentences, while undoubtedly true, are unconstructive when they are conditional on the impossible. The idea that the system "shouldn't" go wrong refers to what would happen in some non-existent, perfect world.

MANAGEMENT BY DISASTER

Management by Disaster can result when Murphy is ignored most of the time but is occasionally listened to when he shouts very loud.

Traditional organizations distinguished between *managers* on the one hand, such as administrators and supervisory engineers, and on the other *operatives* such as process workers, clerks and nurses. Managers were supposed to design and appropriately adapt the system, and operatives were supposed to run it. When managers were for any reason not in close contact with the process such an arrangement resulted in the system becoming steadily more complicated without being more effective. In particular this undesirable result occured when

(a) the system developed only by reaction to occasional disasters,

(b) action, supposedly corrective, was instituted by persons remote from the system,

(c) no check was made on whether the corrective action was effective or not.

For illustration consider Figure 1. The people who actually operated the system, like A and B, whether they be process workers in a manufacturing operation or nurses working in a hospital ward, were closest to the process and have intimate knowledge of it. Theoretically, this knowledge would be available to those like C and D, who are responsible for the system's design: the managers, supervisors, doctors and administrators. Unfortunately there was frequently a considerable distance, not necessarily measured in feet, between these two sets of people. The noise in the system which distorted the communication may well have been proportional to some power of that distance. In addition there was often a deliberately constructed barrier between the parties. How this came about was illustrated by a remark made during a call-in program on the local radio station where I was being interviewed. My caller said, "I'm a factory worker and I would never provide information about poor quality because it would not only get *me* into trouble, but it would also get my *buddy* into trouble. He sometimes sends me some stuff that isn't up to par and I have to fix it up."

In such a situation it will only be when a disaster occurs so bad that it cannot possibly be covered up that the manager will get to know about it and take action. The corrective action he takes, which will permanently change the system, will deal with a situation which most likely would never occur again if day to day information was regularly fed back. Thus we arrive at "Management by Disaster" by which the system acquires more and more irrelevant complication and less and less effectiveness. The example serves to illustrate the importance of Dr. Deming's imprecation to "drive out fear."

A further illustration concerned charges of serious irregularities in the Pentagon's letting of contracts. If these allegations were true the appropriate remedial action would be to simplify the system so that those ultimately responsible for it were in closer contact with what was going on. But it is much more likely that a congressional inquiry would produce draconian measures resulting not in simplification, but in further elaboration. The only result might then be to make it even more difficult for industry to deal with the defense department.

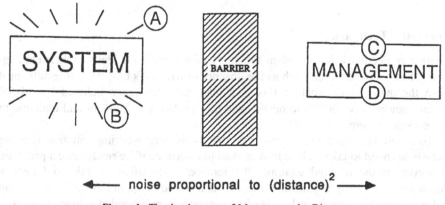

Figure 1. The development of Management by Disaster.

ORGANIZING FOR QUALITY

Improvement results when a routine is in place whereby data coming from the system are automatically used to improve it.

This happens in an organization where

(a) quality improvement is each individual person's responsibility,

(b) each individual is suitably empowered to undertake that responsibility,

(c) appropriate data are collected and discussed (not as a means of apportioning blame but to provide material for team problem-solving meetings),

(d) input from the persons closest to the system is encouraged.

In such an organization the team approach to problem solving, is an essential ingredient in the Six Sigma initiative. A team that included the doctor, the nurse, someone from the hospital laboratory and someone from the records office might solve the hospital records problem. A team that coordinated data from baggage agents, airline schedules, distances between gates, and late-arrivals might solve the baggage problem, and, so on.

To make such teamwork effective it is necessary that adequate channels for communication exist from the bottom to the top of the organization as well as from the top to the bottom. In particular, this ensures that appropriate measures can be taken even when corrective action is not within the power of the team members. It also makes it possible to call for help from an appropriate specialist when this is needed.

DEFEATING MURPHY

Three simple strategies that can be used to defeat Murphy are *Corrective Feedback*, *Preemptive Feedforward* and *Simplification*. (See Figure 2).

Corrective Feedback

Corrective Feedback occurs when the study of system faults leads to their eradication. Thus, an observation such as the absence of friend A's hospital test results must raise the question of whether this is a one time occurrence induced by special circumstances or whether the occurrence of such faults is frequent and endemic to the present system.

To avoid tinkering with a process that is already working satisfactorily, we sometimes need to take a close look at data to determine if we really have a problem. However, in the hospital example the remark, "they often do this so I keep a duplicate..." already implies that the situation would be intolerable if the physician did not keep duplicate records. Clearly an investigation is needed whose object is to

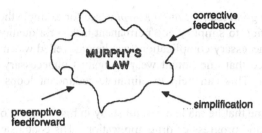

Figure 2. A three-pronged attack on Murphy's Law.

reduce the proportion defective as measured by the proportion of hospital consultations in which patient's up-to-date records are not available. Collection and study of suitable data might then show how to improve the system. For example, a Pareto chart could show the most frequent causes of failure of the present system and a run chart could point to significant patterns in the data. An attribute quality control chart kept on the improved system could ensure that the new quality level was maintained.

Preemptive Feedforward

Preemptive Feedforward is a process whereby careful forethought prevents faults from occurring. If friend B had taken the precaution of carrying his slides (and possibly some pajamas and a toothbrush) in his hand luggage the trauma of losing his luggage would have been much reduced. Preemptive Feedforward may be used as a temporary palliative for use while a fundamentally better system is being devised. However, care must be taken to see that Preemptive Feedforward is not used as a *substitute* for fixing the System. Friend A's doctor did exactly that by keeping a duplicate record of his patients' test results. Exclusive and permanent use of Preemptive Feedforward can only be justified as a policy for someone who is genuinely powerless. For most of us it should be regarded as a temporary expedient to be used only until sufficient improvement in the system makes it unnecessary. Thus, friend B should make the effort to send a written complaint to the appropriate airline official about the loss of his baggage for two days. Experience shows that public pressure of this sort can force airlines to change, but until they get a lot better he will be wise to carry a small survival kit in his hand baggage.

Simplification

As time goes by all systems if left to themselves tend to get steadily more complicated without necessarily becoming more effective. We have seen one way in which this can happen through "management by disaster." Because complication may sometimes provide work and power for particular bureaucrats, the process of simplification must not be in their hands nor subject to their obstruction.

Complication is a great temptation to Murphy (see for example the book *Parkinson's Law* (1957)). Action to simplify is a permanent task for a quality team. A valuable study on how unnecessary complications can be unravelled was made by Tim Fuller (1986). He showed that one potent weapon against unnecessary complication was the flow diagram. This can help to eliminate redundant loops and twists in the system.

Friend C told me that he made a careful study in his company of the time taken of the development and progress of drug applications. His conclusion was that at least half the delay was due *not* to the dilatory behavior of the FDA, but to extremely slow progress in collecting the required information and test results within his *own company.* In cooperation with the people involved, using flow charts and other elementary he worked out a new system whereby, without anybody working harder, the delays were cut by a very large factor.

To make use of a radio signal we need apparatus that detects, receives and decodes the messages and presents them in a recognizable form. It is much the same with the radiation from an operating system. To use the information it is sending out we need tools to catch and interpret its various aspects.

In addition to careful record-keeping and flow diagrams, useful receptors for simple problem solving are Ishikawa's Seven Tools (check sheets, Pareto charts, cause–effect charts, histograms, stratification, scatter plots and graphs) discussed in his very helpful little book, *Guide to Quality Control* (1982); see also Figure 2 of the previous chapter.

Of special value are graphical techniques including quality control charts and run charts. One important property of run charts is that, unlike the human mind, they remember the past without bias and so make it easy to compare what is happening now with what happened before. By contrast, the human memory gives undue weight to recent happenings and quickly discounts the past. As an example, another friend of mine recently constructed a run chart on his own health. He had for some time been troubled with asthma, and medical advice had not been of much help. So he decided to try to find out for himself what might alleviate the symptoms using a series of run charts. On the top chart he kept a careful daily record of how bad his asthma had been on a particular day using a scale from 1 through 10. Below it on a similar scale he kept daily records of how much he had slept, how much exercise he had taken, and he made notations of what he had to eat and to drink and when he had it. He told me that study of these charts showed him how he could alleviate his symptoms—specifically to drink less beer! For more difficult cases, records of this kind kept by patients can be of tremendous help to attending physicians.

LET'S NOT THROW AWAY OUR CREATIVE POTENTIAL

In the past a license to use creativity was awarded to only a small proportion of the workforce. Only people with MBA's, Ph.D.'s, and the like were supposed to be creative. Yet creativity is the unique quality that separates mankind from the rest of the animal kingdom and everyone possesses it. Just as "fish gotta swim and birds gotta fly," an opportunity to use creativity is a vital need for all humanity. Most people can easily understand the necessary ideas and elementary tools and can use them in helping to solve the myriad simple but expensive problems that beset any organization.

The benefits provided by worker participation are twofold:

(a) quality is greatly improved because of the finding and fixing of a very large number of problems,

(b) morale is greatly increased because it is frustrating to be treated only as a pair of hands and enormously satisfactory to be allowed to be creative.

EXPECTATION

An important determinant of quality improvement is expectation. I remember a particular car I used to own. It was a European car from a distinguished maker and I loved it. Unfortunately, it needed fixing about every two months or so. So I got to know all the mechanics in the garage, I knew their wives and their children and I became used to paying a bill of a couple of hundred dollars or so for each visit. Some years ago I bought a Japanese car with which nothing ever went wrong and I hardly met those mechanics anymore apart from an occasional social call.

Expectation of the performance I should require from my car had been totally transformed. Such a rise in expectation shared by millions of others is clearly an important reason why the quality of *non-Japanese* built cars improved.

It is very clear that in different countries expectations are very different. For example, in some undeveloped countries consumer goods are very scarce and people will tolerate low quality. But the United States is not an undeveloped country and yet expectations are inappropriately low. There is a curiously ambivalent attitude toward complaints. On the one hand it is the fashion to invite criticism but on the other hand it is regarded as bad manners to supply it. Thus when the waitress asks "Is everything alright at this table" one is *not* supposed to say "No the meat is cold and the vegetables are over-cooked." This may be because the customer suspects that the management do not expect criticism, and have not provided appropriate channels to allow critical feedback to occur. The customer does not complain because he believes that his criticism will reach no one but the waitress who is an innocent victim of the system. Thus, the expectation and acceptance of bad management exacerbates the problem of getting honest and constructive feedback.

Bad quality has been tolerated because people think that they can do nothing about it. The quality movement can raise expectations of quality and encourage individuals to be assertive in ensuring that they get it. Insistence on quality by the public must, in a free enterprise competitive system, ensure that institutions that supply it will prosper and those that do not will not survive.

ONE CITY'S EXPERIENCE

One example of the many facets of the quality improvement process was provided by what occurred in the City of Madison, Wisconsin. My late friend and colleague Bill Hunter had a vision about a model quality city. At his instigation Mayor Sensenbrenner led a quality improvement program which included the police force, the city maintenance garage, and the department of transportation. Further quality improvement programs were put in place not only in many local industries but in departments of state government including the Department of Revenue.

THE IMPORTANCE OF QUALITY IMPROVEMENT

The quality movement can produce the three desired outcomes shown in Figure 3. (See Joiner, 1994.)

We can *revitalize our economy* via the chain reaction initiated by quality improvement and discussed by Dr. Deming (1986, p. 3).

We can *improve the quality of life*, the world will be a much more satisfactory and less frustrating place if airlines do not lose our bags, if hospitals, government departments and other institutions do not waste our time, and if we could more often feel that we have received value for money.

We can all *experience the joy of creativity*; an enjoyment of what we do is vital to good performance. In the past we have acted in the fallacy that the only incentives to work were money and the fear of possible dismissal.

PUBLICITY AND POLITICS

In view of the tremendous importance for human good of the modern quality movement and of its basic simplicity, it is surprising how little the general public knows about it.

It is clearly a political time bomb. We hear much of scarce resources available to deal with health care, education, pollution, defense and so on. The wide use of quality improvement techniques could make those scarce resources go much further

Figure 3. Outcomes of quality improvement.

and to everyone's benefit, yet except for vague talk of industrial competitiveness, politicians have shown little awareness of these possibilities.

It is surely time that the media—daily papers, radio, television, and magazines—stimulated more awareness and discussion of the quality issues so important to the general public.

CONCLUSION

I leave you with the thought that Murphy is the only quality guru who is never wrong.

ACKNOWLEDGMENT

I am grateful to W. Edwards Deming, William A. Golomski, J. Stuart Hunter, Tim Kramer, Jan O'Neill and Mary Zimmerman for valuable suggestions for the revision of this paper.

CHAPTER A.3

Changing Management Policy to Improve Quality and Productivity

The extent to which methods of process improvement can be used effectively depends critically on the management culture in which they attempt to operate. Thus, Deming said

> We are living under the tyranny of the prevailing style of management. Most people imagine that this style of management has always existed, and is a fixture. Actually, it is a modern invention, a trap that has led us into decline. The workers are handicapped by the system, and the system belongs to management. . . .. We need transformation to a new kind of economics where everyone comes out better.

Although process improvement has become the "in" thing for many years now and in some industries and organizations has made considerable progress, in others it is little more than a curtain behind which business as usual can be conducted.

A GEOMETRIC REPRESENTATION OF VIEWPOINTS ON POLICY CHANGE

The management of any business or system depends on a large number of individual policies. These can involve questions of hiring and firing, of remuneration, of promotion, of who can sign off on a given amount of money, of who can give instructions to alter a process, of the length and frequency of rest periods, and so forth. (The policies I am talking about are the policies that are applied in practice,

From Box, G. E. P. (1994), *Quality Engineering*, 6(4), 719–724. Copyright © 1994 by Marcel Dekker, Inc.

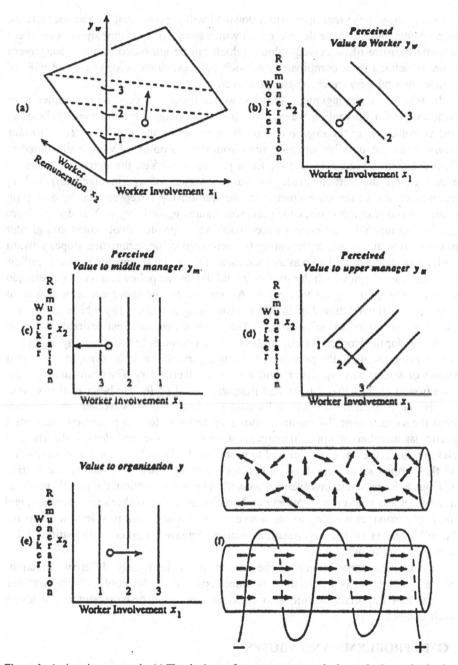

Figure 1. An imaginary example. (a) The sloping surface represents a particular worker's perceived value y_w for various combinations of worker involvement x_1 and worker remuneration x_2; (b) an equivalent contour diagram; (c) and (d) similar diagrams for a middle manager and an upper manager; (e) a diagram showing value to the organization and the direction of optimal changes; and (f) magnetization of a steel rod.

not necessarily those that are written down.) I will suppose that we can measure on some continuous scale the degree x with which some policy is employed. Also, there is some measure of perceived value y which can be attributed to any management *strategy* defined by a combination of such policies. Such perceived value will, of course, depend very much on the perceiver.

In practice, a management strategy would involve a very large number n of policies, but for illustration we consider a strategy defined by just $n = 2$ policies x_1 and x_2 with, say, x_1 the *degree of worker involvement* and x_2 the *degree of worker remuneration*. A possible situation seen from various points of view is illustrated in Figure 1. Thus, Figure 1a shows, for a particular worker, the perceived value y_w measured on the vertical scale for various values of x_1 (involvement) and x_2 (remuneration). In the case illustrated, the relationship is represented by a sloping plane on which contour lines for perceived values $y_w = 1$, $y_w = 2$, and $y_w = 3$ are drawn. Because this particular worker would value greater involvement and greater remuneration, the plane, representing the perceived value to him/her, slopes upward both as x_1 increases and also as x_2 increases. If the point 0 represents present policy, then the direction in which the worker would like to see policy *change* is represented by the arrow pointing up this plane. An equivalent and more convenient way of representing the relationship is the contour diagram (Fig. 1b) which shows the "worker's perceived value" *contours*. The arrow shows the most desirable direction of change for the worker at right angles to these contours. In a similar way, Figure 1c shows contours of y_m, the perceived value to a particular middle manager, of various values of worker participation x_1 and worker involvement x_2. The arrow indicates the direction of change that this middle manager would prefer. He believes that worker involvement threatens his job and would like less of it with worker remuneration kept the same. Figure 1d similarly shows contours y_u for the perceived value for a particular member of upper management with the associated desired direction of change. In Figure 1e, y is supposed to represent the "real" value *to the organization* of the various strategies. Thus, taken together, my illustration shows the worker anxious to increase his involvement as well as his remuneration, the middle manager in favor of less worker involvement and no change in the worker's remuneration, and the upper manager wishing to see workers more involved but paid less, whereas the "true" interests of the organization require increased worker participation at the same level of remuneration.

In this imaginary example, the arrows point in widely different directions and a situation exists where there is ample opportunity for conflict both overt and covert. It is important therefore to consider how good management might lessen such conflict.

SOME PROBLEMS AND ABUSES

Management systems have often been demonstrably inefficient. Their ailments have been the subject of much study and a variety of nostrums have been proposed. Although each of these was, in its turn, believed to provide an infallible cure, doubts remain. A few problems are discussed below.

Unnecessary Multiplication of Bureaucracy. Parkinson (1957) provided data that showed that the bureaucracy responsible for overseeing the British Empire continued to increase at the same time that the Empire itself was in decline. Present-day bureaucracy, even with the advantage of modern management systems, seems similarly afflicted. Thus, in the United States, whenever the tax system is simplified, it needs more people to run it. Whether under a Republican or a Democratic administration, bureaucracy steadily grows. Concerning this process, the *flow diagram* (see Figure 2 in Chapter A.1.) often provides food for thought. Such diagrams frequently reveal useless loops and delays which have become institutionalized over the years. But if the only way that a supervisor in charge of a staff of four can be more highly paid and have more power is to be in charge of a staff of 20, then there is a clear disincentive to replace an unnecessarily complex system by a simpler one. A popular belief is that, although such situations occur in the public sector, the private sector is not similarly afflicted; such a belief is not sustained by observation.

Mobility and Responsibility. The upwardly mobile executive who knows that she/he will almost certainly not be in the present job in 18 months time can afford to take a very short-term view. The best interests of such a person are served by making changes which, whatever their long-term effect, will produce highly visible results in the short run. As Dobzhansky (1958) says, "Extinction occurs because selection promotes what is immediately useful even if the change may be fatal in the long run" (see also Potter, 1990).

Power Tends to Corrupt and Absolute Power Corrupts Absolutely (Lord Acton, 1887). One danger in a poorly run organization is that a scramble to acquire power and to hold on to it can dominate management. Managers of this kind are certainly competitive, but their competition tends to be with one another. Once such a system is established, it may be very difficult to change.

Am I Working My Way Out of My Job? If the private belief of process workers is, for example, that the new data they have been asked to collect may threaten their employment, they may be less than enthusiastic in ensuring that a good job of data collection is done.

MOTIVATING CHANGE

Improvement of a management system requires two things:

(a) Careful study and consideration of what would be the *desirable* direction of change for the health of the organization.

(b) To act so that, at least to an approximation, everyone in the organization moves in that direction.

Suppose that the desirable direction of change for the organization can be agreed on. What we then need is some galvanizing force which can get everyone to move in that direction in very much the same way that an electric solenoid can magnetize a steel bar by reorienting its molecular magnets as in Figure 1f.

Fear. One such galvanizing force is fear. In a hierarchical organization, a worker at a lower level reports to a supervisor one step up and so on. This was a system originally developed to run an army. It is a system designed to induce people to do what they would often prefer not to do. Fear as a motivation is inefficient and expensive. For maximum effectiveness, it would require that everyone watches everyone else all the time. It is inefficient because, even with a very large and expensive force of supervisory personnel, continuous surveillance is impossible and also because such a system breeds resentment. This latent resentment produces poor morale and can be expressed in many counterproductive ways—carelessness and petty thievery, for example—which are covert and hard to deal with. We are sometimes told that what is needed is an organization which is "lean and mean." But lean and mean is a contradiction in terms—mean means fear, fear means more supervision, and more supervision causes fat not lean.

Human Motivation. It is coming to be realized that organizations can perform only as well as the people they employ and that human beings do not perform at their best unless they are most of the time enjoying their jobs, are respected for at least some of the things that they are doing, and believe that by doing the job well they are serving their own interests. In particular, most employees will welcome process improvement jobs that exercise their creativity. However, the management system which makes only a pretense of valuing employee involvement and encouraging employee empowerment merely breeds cynicism. As we were reminded by Abraham Lincoln, although it is possible in the short run to deceive people, it is not a sustainable policy.

Economic studies precisely confirm the expectations implied above (see, for example Graves (1993, Levine and Tyson, 1990, and Thurow, 1992). In particular, they show that "employee involvement/participative management efforts have tended not to last very long unless improvement was clearly in the best interest of employees."

Motivators will not necessarily take the same form at all levels of management, but a motivator aimed at one level should not demotivate people at another level.

Because "water runs downhill," the central principle of good management must be to try continually to arrange "the contours of everyone's hill" so that when water runs downhill, it runs the way that benefits the organization. Well-meaning efforts to do this are sometimes unsuccessful or even counterproductive, so careful thought and some experimentation is necessary.

I hope the geometric representation given here may prove to be useful in thinking about particular policy changes that may improve your organization.

ACKNOWLEDGMENT

This research was supported by a grant from the Alfred P. Sloan Foundation.

Scientific Method: The Generation of Knowledge

At the most fundamental level, process improvement initiatives such as "Six Sigma" are about generating knowledge. As companies learn more about their processes and customers, they can produce higher-quality products, reduce development times, and lower costs. Furthermore, the process of generating knowledge improves the morale and efficiency of the entire work force. Thus, a truly efficient organization is dedicated not only to making excellent products, but to continuously making them better (Box, 1957a). The infrastructure that this requires gives the organization flexibility and initiative, enabling it to prosper in bad times and in good.

The quality movement can be seen as the analysis, institutionalization, and democratization of scientific method, a tool for efficiently generating new knowledge. The accelerating rate of knowledge acquisition is responsible for a widening gap between the fund of human knowledge and an individual's ability to understand it. This reemphasizes the need to apply quality to the education process, ensuring that technology is used to complement, not replace, the human brain.

THE LOGIC OF CONTINUOUS IMPROVEMENT

Human beings are distinguished from the rest of the animal kingdom by their extraordinary ability to learn and to innovate. There seems, however, to be a misunderstanding about how the learning process works. For example, people trained in mathematical optimization may see continuous improvement as a foolish goal. The law of diminishing returns, they say, implies that once the optimum is reached, it is wasteful to continue seeking improvement.

From Box, G. E. P. (1997), *Quality Progress*, **30**(1), 47–50. Copyright © 1994 by Marcel Dekker, Inc.

Improving Almost Anything: Ideas and Essays, Revised Edition. By George Box and Friends
Copyright © 2006 John Wiley & Sons, Inc.

But this argument is fallacious because continuous improvement is different from mathematical optimization. In continuous improvement, the factors and measured responses are not fixed but are changed as we learn more about the system. These ideas are rooted deep in the history of science. For example, the discoveries of Michael Faraday, Charles Darwin, and Gregor Johann Mendel were not truths; they were major steps in a never-ending (and diverging) process that helped predict natural phenomena. And, when James Watson and Francis Crick discovered the structure of DNA, their results appeared in a very short paper in the journal *Nature*. A problem was solved, but inquiries did not cease. On the contrary, there are now thousands of papers describing further discoveries made possible because of that paper. Each of these discoveries leads to more questions. The practitioner's never-ending improvement is, in essence, the scientific method—analyzed, democratized, and institutionalized.

INDUCTIVE–DEDUCTIVE LEARNING

A common misconception is that discovery is a one-shot affair. This idea dies hard. It is characterized by the famous but questionable story of Archimedes exclaiming "Eureka!" when he discovered a method for determining the purity of gold. There also is the example of Samuel Pierpoint Langley, who was once regarded as one of the most distinguished scientists in the United States (Crouch, 1989). In the late 1800s, he built model airplanes that flew and believed that his results could be extrapolated to design a full-sized airplane using theory alone. The resulting full sized machines never flew but made two disastrous plunges into the water, the second of which was observed by members of Congress and almost drowned the pilot.

This approach is very different from the repetitive learning practiced by two bicycle mechanics, Orville and Wilbur Wright, who designed, built, and flew the first airplane. They conducted hundreds of experiments, each of which led to the next. They began with a large kite, then a glider, and finally a powered aircraft. One of the many discoveries they made along the way was that the accepted formula for the lift of an aerofoil was wrong. They built their own wind tunnel to study this and corrected the fallacious formula.

This iterative inductive–deductive process, which is geared to the structure of the human brain, is not esoteric, but part of one's everyday experience. To illustrate the idea I have used the following example. Suppose a man parks his car every morning in a particular place. As he leaves work he might go through a series of inductive–deductive problem-solving cycles like this:

Model: Today is like every day.

Deduction: My car will be in its parking place.

Data: It isn't.

Induction: Someone must have taken it.

Model: My car has been stolen.

Deduction: My car is not in the parking lot.

Data: No. It is over there.

Induction: Someone took it and brought it back.

Model: A thief took it and brought it back.
Deduction: My car will have been broken into.
Data: No. It's unharmed and it's locked.
Induction: Someone who had a key took it.

Model: My wife used my car.
Deduction: She probably left a note.
Data: Yes. Here it is.

More generally, suppose one wishes to solve a particular problem and initial speculation produces some relevant idea, theory, or model. One then seeks data to throw further light on the idea. This could consist of a search on the web, a trip to the library, a brainstorming meeting, passive observation of a process, or active experimentation. In any case, the facts and data gathered sometimes confirm what is already believed. More frequently, however, they suggest that the initial idea is only partly right, and sometimes, that it is totally wrong. In the latter two cases, the difference between deduction and actuality causes one to keep digging. This can point to a modified idea or model or a totally different one which, if true, might explain what has been found. This new idea or model can again be compared with both the original data and the new data.

Humans have a two-sided brain specifically designed to carry out such continuing deductive–inductive conversations. While this iterative process can lead to the solution of a problem, one should not expect the nature of the solution nor the route by which it is reached to be unique.

THE CATALYSIS OF LEARNING BY SCIENTIFIC METHOD

History shows that humans have always learned, but progress once depended on the chance coming together of an informative event and a perceptive observer. Scientific method has accelerated that process in at least four ways:

- Providing a better understanding through the interactive nature of learning
- Deducing the logical consequences of a group of facts, each individually known but not previously brought together
- Passively observing and analyzing systems already in operation and data coming from the systems
- Deliberately staging artificial experiences by experimentation

Each of these ideas relates to a particular aspect of process improvement.

1. Learning by deductive–inductive iteration is the essence of the Deming-Shewhart plan–do–check–act cycle.

2. Sharing already available knowledge to deduce previously unknown consequences is the objective of such devices as cross-functional deployment this creates a learning process by bringing together people from different disciplines with different points of view, resulting in a new understanding of how to achieve and exceed customer requirements.

3. Control charts, flowcharts, and statistical process control (SPC) tools are devices for generating hypotheses from passively observing an operating system. The system might be anything from the process of student admission to a university to the manufacture of axles.

4. Experimentation in engineering and science can be made much more efficient by using statistical design with graphical analysis, that makes direct appeal to the right brain.

DEMOCRATIZATION AND COMPREHENSIVE DIFFUSION OF SCIENTIFIC METHOD

A special contribution for which the quality movement is responsible is the democratization and comprehensive diffusion of scientific method. Important information necessary to solve a problem may be known only to people who are close to the process: the nurse with the patient or the process worker with the machine. Unhappily, there is often no system permanently in place to make this information available for use. Providing the work force with simple tools (such as control charts, flow-charts, and SPC tools) and empowering their use can ensure that thousands of deductive–inductive brains are actively producing information on how products and processes can be improved and that this information is available to management. When gains from improvements are acknowledged appropriately, this can be remarkably effective, not only by producing a better and cheaper product, but creating a more motivated work force.

AN ANTIDOTE TO KNOWLEDGE FRACTIONATION

The situation with which humans are confronted is illustrated in Figure 1. Geneticists tell us that human genetic structure has not changed over thousands of years. If it were possible to take a baby born 5,000 years ago and raise it in the present, the baby would turn out much the same as any modern-day human being. But as Figure 1 illustrates, the accelerated pace of scientific learning has led to a widening gap between the technological environment and natural genetic heritage.

Humans have attempted to fill the gap by formal education. In the developed world, this lasts a quarter of a lifetime. But the present educational process is stretched to the breaking point. It is necessary, therefore, to apply the processes of quality improvement to education—to the teaching, administration, and research of schools and universities. Accelerating knowledge generation emphasizes the need

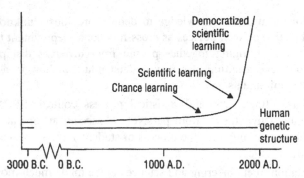

Figure 1. Three phases of steadily increased human knowledge.

for lifelong education. Teachers, scientists, and engineers can no longer regard what they learned at the university as sufficient for the rest of their lives.

It wasn't that long ago when available knowledge could be appreciated by the minds of single individuals. Today, familiarity with the tools of everyday experience, such as airplanes, computers, and fax machines, and the knowledge of how they work are becoming more and more remote. Furthermore, engineers, scientists, business executives, and physicians continue to become more knowledgeable about less and less. Quality management tools, however, can help negate the effects of this knowledge fractionation.

HUMAN POSSIBILITIES AND LIMITATIONS

One thing that computers have done is create new ways of storing knowledge, and the skills of modern knowledge retrieval are easily learned.

Figure 2 represents the past pattern for teaching. The student's mind was being used as a storage-and-retrieval system, a task for which it was not particularly well adapted.

Figure 3 shows what teaching will be like in the future. Here the teacher acts as a mentor in training the student in unstructured problem solving while the computer stores and retrieves information. This sets the mind free to do what it does best—to be inductively creative.

In management, many initiatives have come and gone—management by exception, management by objectives, and so on. There is natural concern that present efforts will experience the same rise and demise. But insofar as this movement concerns the application, diffusion, and democratization of scientific method, whether in an affinity diagram, a quality control chart, an experimental design, or a study of group dynamics, it cannot fail; we will always be better off with more knowledge than with less.

BARRIERS TO PROGRESS

In this fast-paced world, some ideas must necessarily change. Arthur William Edgar O'Shaughnessy (1939) said, "Each age is a dream that is dying or one that is coming

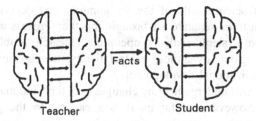

Figure 2. Traditional method of teaching.

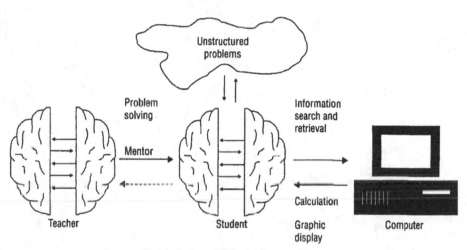

Figure 3. A model for modern teaching.

to be." Thomas Kuhn (1962) takes up this theme as it applies to science in his book *The Structure of Scientific Revolutions*. This process of change can be painful. Many people try to hold on to a treasured belief too long, while others are deflected by every wind that blows. Salvation lies somewhere in between.

Quality gurus don't necessarily help. They tend to teach a particular subset of ideas that they insist must be followed to the letter. They rarely acknowledge to what degree their particular nostrums overlap those of others or recognize that nothing is immutable except change itself.

At the other extreme, there are powerful executives who are poised to spring on or off every bandwagon that passes. Closely related to this problem is an exaggerated concern with short-term results. Since water runs downhill, an organization that encourages such behavior will lose when competing with those that are less myopic. In particular, the present craze of over-aggressive downsizing as a means of providing short-term benefits to shareholders is inviting future market losses.

The history of science tells of the invigoration that occurs when disciplines previously thought to be separate are brought together. Process improvement brings together ideas from systems analysis, operations research, problem solving, statistics, engineering, group dynamics, control theory, management science, human genetics, and organizational development.

The movement will undergo healthy changes over time and may even be called by different names; however, insofar as it is a catalyst to the generation of new knowledge, it is here to stay. In particular it can help solve many challenging problems, produce a more efficient and happier work culture, and eliminate many of the frustrations caused by poor quality services and goods.

PART B

Design of Experiments for Process Improvement

CHAPTER B.0

Introduction

It is surprising how susceptible we can be to some new buzzword heralding the very latest method for process improvement. Such methods seem to come, and then after a time, many of them go. It is as if we are looking for the Philosopher's Stone—some new technique (or some old technique disguised under a new name) that will solve all our problems. We are perhaps a little too afraid of being upstaged by someone who says "But surely you are not still using method X! Have you not heard about Y?" and, unfortunately, it sometimes serves the interests of highly paid consultants and "keynote" speakers to reinforce such thinking.

Wiser consideration contradicts such ideas. While one must always be alert for new techniques, we should be sure that uppermost in our tool kit are methods that have stood the test of time and have a long record of success. Thus Saint Paul says "Test all things, hold fast to that which is good" (*Bible*, 1 Thessalonians 5:21).

The ideas of Walter Shewhart, Kaoru Ishakawa, and Sir Ronald Fisher are not new, but we are unlikely to make much headway with process improvement if we forget them or are unaware of them. This is equally true whether we want to improve the quality of a manufactured article or of health care in a nursing home.

The understanding of what is best to do in our daily lives, as in our professional lives as engineers, scientists, quality specialists, and investigators, is bedeviled by the problem of distinguishing the signal from the noise. Much of what happens can conveniently be thought of as random variation, but sometimes hidden within that variation are important signals that could warn us of problems or alert us to opportunities. Careful analysis of existing data can sometimes help us find such signals, but frequently we need to experiment. The test runs we make inevitably determine an "experimental design." This is so even if we have never heard of the formal *science* of designing experiments. The only question is whether our approach is likely to squander, or to use effectively, available experimental effort.

Improving Almost Anything: Ideas and Essays, Revised Edition. By George Box and Friends
Copyright © 2006 John Wiley & Sons, Inc.

For a very long time I have believed that if all our engineers and scientists had some understanding of even the simplest principles of efficient experimental design a quantum leap could occur in the rate of development of our knowledge and of our ability to solve our problems. For many years my colleagues and I have taught short courses in experimental design to engineers and scientists, managers, and quality specialists from every part of this country and abroad and from a wide variety of companies. The most frequent comments are: How long has this been going on? Why haven't I heard about it before? This is valuable stuff, I can put it to use right away. Why wasn't it taught at University? When I told engineering faculty about this, they responded that their curriculum was already overloaded and that if they included a course in experimental design they would have to take something out. To which I answered: Take almost anything out. Surely an engineer who doesn't know how to run an efficient experiment is not an engineer.

When I was invited by my friend Frank Caplan to write a "column"* for the then new journal *Quality Engineering*, I quickly decided to emphasize basic ideas about experimental design. I knew that what my readers most needed to know could simply be explained with graphics and examples. It was with these ideas in mind that most of the articles that appear in this section were written.

---

* It was first called "George's Column," and after my colleague Søren Bisgaard kindly agreed to share with me responsibility for the column it was called "Quality Quandaries."

CHAPTER B.1

Do Interactions Matter?

Many engineers have been taught that when you want to study more than one factor you must experiment by changing one factor at a time. This method does not take care of interactions; to do that would need statistical methods, and many experimenters (with considerable help from academic statisticians) believe that statistical methods are impossibly complicated. Recently some engineers have been hearing from a different faction claiming that interactions were not important, or that, when they did matter, they could be transformed away. What are we to make of all this?

Many important phenomena depend, not on the operation of a single factor, but on the bringing together of two or sometimes more factors at the appropriate levels. For illustration, some experiments on rabbit breeding are shown in Figure 1(a) using the one factor at a time method. As you can see in the first experiment $(-, -)$, which was very properly run as a control, the hutch contained no rabbits. In the second experiment $(+, -)$ a doe was added, but that didn't produce anything, so the doe was taken out of the hutch and a buck $(-, +)$ was tried, unsuccessfully also. At this point things looked pretty black and it might have been concluded that you can't get rabbits that way, had not Stu Hunter come by and told the experimenter about factorial designs. When that fourth experiment was run, as in Figure 1(b) with a buck and a doe $(+, +)$, lo and behold, not only was a factorial experiment produced but eight little rabbits as well! And (I'm thankful to say) you can't get rid of this phenomenon by transformation.

Now there's always a sceptic in every crowd and I hear somebody say "Well that's alright for biology but will it work in engineering?" Figure 2(a) shows a standard deep groove bearing manufactured by SKF. This manufacturer, with plants in 14 countries, is the largest producer of rolling bearings in the world and is notable for having withstood the Japanese challenge and outdone them both in quality and price.

From Box, G. E. P. (1990), *Quality Engineering*, 2(3), 365–369. Copyright © 1990 by Marcel Dekker, Inc.

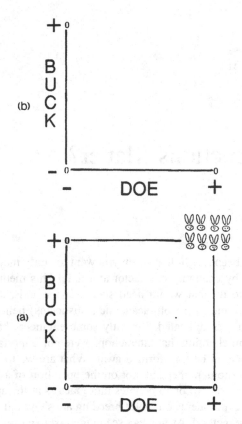

Figure 1. An experiment on rabbit breeding. A one factor at a time design (a) compared with a factorial design (b). A minus sign means absent, a plus sign means present.

Figure 2(b) shows a 2^3 factorial experiment reported by Christer Hellstrand (1989), a former graduate of the University of Wisconsin who now works for SKF. The engineers wanted to find the effect of changing to a less expensive "cage" design. The results were assessed by an accelerated life test which, however, had a rather large experimental error. The runs were expensive because they needed to be made on an actual production line and the experimenters were planning to make four runs with the standard cage and four with the modified cage. Christer asked if there were other factors they would like to test. They said there were, but that making added runs would exceed their budget. Christer showed them how they could test two additional factors "for free"—without increasing the number of runs and without reducing the accuracy of their estimate of the cage effect. In this arrangement, called a 2^3 factorial design, each of the three factors would be run at two levels and all the eight possible combinations included. The various combinations can conveniently be shown as the vertices of cube as in Figure 2(b). In each case, the standard condition is indicated by a minus sign and the modified condition by a plus sign. The factors changed were *heat treatment*, *outer ring osculation*, and *cage design*. The numbers show the relative lengths of lives of the bearings. If you look at Figure 2(b), you can

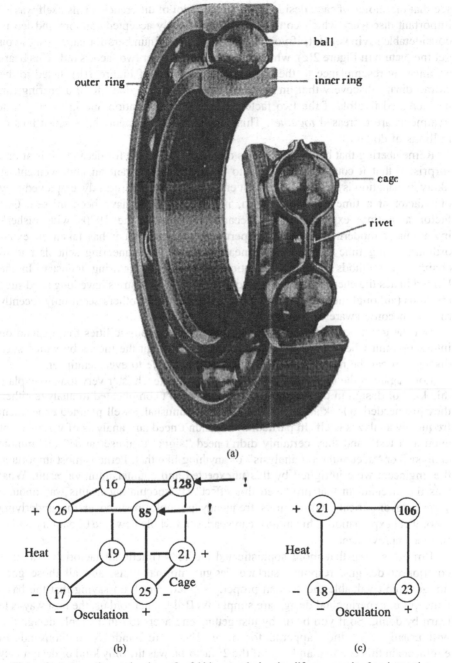

Figure 2. An experiment showing a fivefold increase in bearing life as a result of an interaction.

see that the choice of cage design did not make a lot of difference. (This itself was an important discovery which conflicted with previously accepted folklore and led to considerable savings.) But, if you average the pairs of numbers for cage design you get the picture in Figure 2(c), which shows what the other two factors did. This bears an uncanny resemblance to the experiment on rabbits of Figure 1(b). It led to the extraordinary discovery that, in this particular application, the life of a bearing can be increased fivefold if the two factor(s) outer ring osculation and inner ring heat treatments are increased *together*. This and similar experiments have saved tens of millions of dollars.

Remembering that bearings like this one have been made for decades, it is at first surprising that it could take so long to discover so important an improvement. A likely explanation is that, because most engineers have, until recently, employed only one factor at a time experimentation, interaction effects have been missed. One factor at a time experimentation became outdated in the 1920s with Fisher's invention of modern methods of experimental design, but it has taken an extraordinarily long time for this to permeate teaching in engineering schools and to change the methods of experimentation used in manufacturing industry. In the United States the chemical industry and other process industries have long used such methods (although not as extensively as they should) but others seem only recently to have become aware of them.

But the good news is (a) that many important new possibilities that depend on interaction must be waiting to be discovered and (b) that the means by which such discoveries can be made is simple and readily available to every engineer.

Look again at the Figure 2(b). I'm sure it didn't take Christer very long to explain this kind of design to the engineers at SKF. It wasn't complicated to analyze either, they just needed to look at the results. This is not unusual, a well-planned experiment frequently analyzes itself. In particular they didn't need an "analysis of variance" or even a "t test" and they certainly didn't need "signal to noise ratios" or "minute analysis" or "accumulation analysis" or anything like that. Perhaps most important, the engineers were intrigued by this unexpected and counter-intuitive result. What was the mechanism that produced this effect? By speculating in this way about a surprizing empirical finding, it is frequently possible to discover the underlying theoretical explanation. This in turn can spark off a whole new line of enquiry and of further improvement.

I'm not saying that more sophisticated methods (fractional factorials, variance component designs, response surface designs, normal plots, and all those good things) aren't valuable tools when properly applied. But I am saying that the basic concepts of experimental design are simple. As Bill Hunter told us, the best way is to learn by doing. So if you begin by just getting engineers to run a simple design this will usually whet their appetite for more. There are hundreds of thousands of engineers in this country, and even if the 2^3 factorial was the only kind of design they ever used, and even if the only method of analysis that was employed was to eyeball the data, this alone could have an enormous impact on experimental efficiency, the rate of innovation, and competitive position.

C H A P T E R   B.2

Teaching Engineers Experimental Design with a Paper Helicopter

When Søren Bisgaard, Conrad Fung, and I teach a short course about designed experiments for engineers from industry, we use a paper helicopter for illustration. We were introduced to this device some years ago by Kip Rogers then of Digital Equipment. The generic design shown in Figure 1 can be made from an $8\frac{1}{2} \times 11$ sheet of paper in a minute or so.

To dramatize some basic issues we ask the class to imagine a scenario for three experimenters, Tom, Dick, and Mary, who want to compare alternative product designs*—where in this case the product is a paper helicopter. We suppose that to make an experimental run, Tom stands on a ladder and drops a helicopter from a height of 12 feet while Dick times its fall with a stopwatch. The objective is to obtain a longer flight time.

VARIATION

To start, we imagine Tom dropping a helicopter of generic design made from blue paper. He drops it four times and the results vary somewhat. We show a typical set of data. This leads to a discussion of variation and we introduce the range and standard deviation as measures of spread, and the average as a measure of central tendency.

From Box, G. E. P. (1992), *Quality Engineering*, 4(3), 453–459. Copyright © 1992 by Marcel Dekker, Inc.
* Unfortunately the word design has to be used in two senses in this chapter. There is the *product design* of the manufactured article and the *experimental design* used to improve it. We believe that context will avoid confusion.

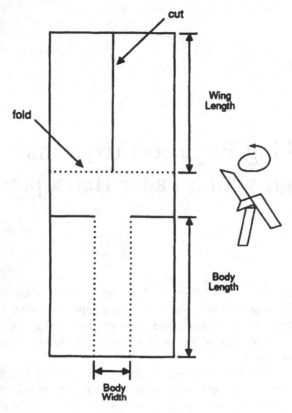

Figure 1. A paper helicopter.

COMPARING MEAN FLIGHT TIMES

At this point Dick says "I don't think much of this helicopter design; I tried a different design yesterday and made this red helicopter. I dropped it four times and got an average flight time which was considerably longer than what you just got with the blue helicopter".

So we put up the two sets of data for the four runs made with the blue helicopter and the four runs made with the red helicopter on the overhead projector. We show the two sets of averages and standard deviations and a simple test that shows that there is indeed a statistically significant difference in means in favor of the runs made with the red helicopter.

VALIDITY OF THE EXPERIMENT

At this point Mary might say "So the difference is statistically significant. So what? It doesn't necessarily mean it's because of the different helicopter design. The runs with the red helicopter were made yesterday when it was cold and wet, the runs with the blue helicopter were made today when it's warm and dry. Perhaps it's the temperature or the humidity that made the difference. What about the paper? Was the same kind of paper used to make the red helicopter as was used to make the blue one? Also, the blue helicopter was dropped by Tom and the red one by Dick. Perhaps they don't drop them the same way. And *where* did Dick drop his helicopter? I bet it was in the conference room, and I've noticed that in that particular room there is a draft which tends to make them fall toward the door. That could increase the flight time. Anyway, are you sure they dropped them from the same height?"

So we ask the class if they think these criticisms have merit and most agree that they have, and they add a few more criticisms of their own. They may even tell us about what happens in their own companies and about the many uncomfortable hours they have spent sitting around a table with a number of (possibly highly prejudiced) persons arguing about the meaning of the results from a badly designed experiment.

We tell the class how Fisher once said of data like this that "nothing much can be gained from statistical analysis; about all you can do is to carry out a postmortem and decide what such an experiment died of." And how, in the 1920s, this led him to the ideas of randomization and blocking which can provide data leading to unambiguous conclusions instead of an argument. We then discuss how these ideas can be used to compare the blue and the red helicopters by making a series of paired comparisons. Each pair (block) of experiments involves the dropping of a blue and a red helicopter by the same person at the same location; you can decide which helicopter should be dropped first by, for example, tossing a penny. The conclusions are based on the *differences* in flight time within the pairs of runs made under identical conditions. We go on to explain however that different people and different locations could be used from *pair to pair* and how, if this were done "it would widen the inductive basis" as Fisher (1935) said, for choosing one helicopter design over the other. If the red helicopter design appeared to be better, one would, for example, like to be able to say that it seemed to be consistently better no matter who dropped it or where it was dropped. As we might put it today, we would like the helicopter design to be "robust with respect to environmental factors such as the 'operator' dropping it and the location where it was dropped." This links up with later discussion of the robust design of products. (See Chapter E.5.) For a fuller discussion of experiments to compare means, see*, for example, BHHII.

A FRACTIONAL FACTORIAL USED AS A SCREENING DESIGN

Later on in the class, we use the paper helicopter to illustrate the running of a fractional factorial (orthogonal array) design. We suppose that a brainstorming session by an engineering design team on ways of improving the helicopter flight

---

* The book *Statistics for Experimenters*, Second Edition by Box, Hunter, and Hunter (2005), is such a frequent reference that in future we will simply refer to it as BHHII.

time has resulted in the selection of eight factors to be studied in a designed experiment. These selected factors are listed at the top of Figure 2 together with the two conditions (indicated by minus and plus signs) at which each will be tested. It is thought likely that only a few of these factors will have important large effects. We are thus in the familiar "Pareto" situation where, as Dr. Juran says we want to screen out "the vital few from the trivial many." The design, shown in Figure 2, is a fractional factorial design. Such designs, which were developed in England during and just after World War II, are particularly useful for this purpose of factor screening. How to generate arrangements of this kind is discussed in Chapter B.3 and B.4 that follow. This one which is a 1/16th fraction of the full 2^8 (256 run) design has two very valuable properties.

1. If there are interactions between pairs of factors, they will not bias any of the 8 main effects of the factors.

FACTORS		−	+	EFFECT
Paper Type	(P)	Regular	Bond	0.13
Wing Length	(W)	3.00"	4.75"	0.77
Body Length	(L)	3.00"	4.75"	−0.40
Body Width	(B)	1.25"	2.00"	0.02
Paper Clip	(C)	No	Yes	0.05
Fold	(F)	No	Yes	−0.10
Taped Body	(T)	No	Yes	−0.15
Taped Wing	(M)	No	Yes	0.17

Random Order	Standard Order	P	W	L	B	C	F	T	M	Flight Time
7	1	−	−	−	−	−	−	−	−	2.5
13	2	+	−	−	−	+	−	+	+	2.9
4	3	−	+	−	−	+	+	−	+	3.5
9	4	+	+	−	−	−	+	+	−	2.7
1	5	−	−	+	−	+	+	+	−	2.0
12	6	+	−	+	−	−	+	−	+	2.3
15	7	−	+	+	−	−	−	+	+	2.9
3	8	+	+	+	−	+	−	−	−	3.0
6	9	−	−	−	+	−	+	+	+	2.4
16	10	+	−	−	+	+	+	−	−	2.6
14	11	−	+	−	+	+	−	+	−	3.2
5	12	+	+	−	+	−	−	−	+	3.7
11	13	−	−	+	+	+	−	−	+	1.9
10	14	+	−	+	+	−	−	+	−	2.2
2	15	−	+	+	+	−	+	−	−	3.0
8	16	+	+	+	+	+	+	+	+	3.0

Figure 2. Results from 16 run fractional factorial experiments showing the factor levels and the calculated main effects of the eight factors.

2. If only up to 3 factors are of importance, the design will produce a complete 2^3 factorial design replicated twice in those three factors no matter which ones they are. (See Box and Hunter 1961a and b.) Such a design is said to be of *projectivity* 3.

This latter property is particularly remarkable when we consider that there are 56 different ways of choosing 3 factors from 8. You can check it for yourself by picking any 3 columns in the design of Figure 2 and verifying that whichever 3 you pick you have every combination of $(\pm, \pm, \pm)$ in these factors repeated twice over. The duplicated results can thus be plotted on the corners of a cube like Figure 2(b) in Chapter B.1.

Flight times for the 16 helicopter types obtained from an experiment run in random order are shown in Figure 2. From these flight times, 8 main effects and 7 strings of two factorial interaction effects may be calculated.* Only the eight main effects are shown in Figure 2. Each main effect is the difference between the averages of the eight runs made at the minus level of a given factor and the eight runs made at its plus level. All fifteen contrasts are plotted on probability paper (Daniel, 1959, see also BHHII) in Figure 3 suggesting that real effects are associated with W (wing length) and, less certainly, L (body length). On the basis that the remaining effects falling around the straight line are mostly due to noise, we can summarize the data simply in terms of the inset diagram in Figure 3. Going back to the original data it will be seen, for example, that there are four runs with short wing length and short body length with flight times averaging 2.6 seconds and 4 runs with long wing length and short body length averaging 3.3 seconds and so on. These averages are set out at the corners of the square in the inset diagram. A direction in which one might expect still longer flight times by using larger wings with a shorter body is indicated by the arrow. Thus the experiment immediately provides not only an improved helicopter design but also indicates the direction in which further experimentation should be carried out. This demonstrates the value of the sequential approach to experimentation—learning as you go, a theme which runs through the rest of this book.

We also discuss with the class whether they are satisfied with *flight time* as the sole criterion. Careful observation might suggest, for example, that an additional criterion that should be included in future experimentation would be flight *stability*. This teaches the lesson that more than one criterion (response) may be recorded in each of the factorial runs. It is also true that the *criteria* to be used in assessing the results may need to be modified or totally changed during an investigation as we learn more of the phenomena under study. Appropriate and feasible objectives *cannot* always be determined in advance.

---

* It is supposed in this analysis that interactions between 3 or more factors can be ignored. A fuller discussion of such analyses can be found in BHHII, p. 402.

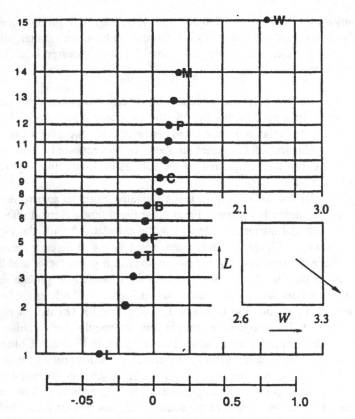

Figure 3. A normal plot of the effects from the helicopter experiment. The inset diagram summarizes the conclusions.

MANAGEMENT OF EXPERIMENTATION

In running any experiment the safest assumption is that, unless extraordinary precautions are taken, it will be run incorrectly. Therefore, we involve the class in the careful organization of the experiment; in particular, they systematically check, and recheck independently, that each of the 16 helicopter designs to be flown corresponds exactly to the specification set out in the appropriate row of Figure 2. Our course for engineers lasts only a few days so we find it necessary to prepare the paper helicopters in advance.

However, the class themselves perform the experiment. They select the random order in which the runs are to be made by drawing numbers from a bag. Usually they then select four of their members, say, A, B, C, and D, to carry out the following tasks:

A finds the appropriate helicopter and hands it up B who is at the top of the ladder.

B drops the helicopter.

C determines the flight time using a stop watch.

D records in a notebook any special characteristics of the run.

If there is an unexpected problem the class decides what to do. For example, if there is a tendency for the helicopters to hit the lower rungs of the ladder they may decide to reorient the ladder and begin again.

No elaborate analysis is needed for experiments of this kind and certainly no analysis of variance table, which at this stage serves only to waste time and confuse the class. In earlier discussions, they have already satisfied themselves, by one or two hand calculations, that the factorial main effects are the differences between the average results at the plus and minus levels of a given factor. Also the rationale of Daniel's normal plot has already been explained. So we enter the data in the computer as they become available and produce the normal plot which is immediately projected onto the overhead screen.

We find that such participatory demonstrations seize the imagination of the engineer and produce rapid learning.

Most important: notice that this experimental is telling us to *do something*, namely, to try helicopter designs with larger wings and shorter bodies. We follow up on that idea in Chapter C.1.

ACKNOWLEDGMENTS

This research was sponsored by the National Science Foundation under Grant No. DDM-8808138, and by the Vilas Trust of the University of Wisconsin, Madison.

CHAPTER B.3

What Can You Find Out from Eight Experimental Runs?

In Chapter B.1 I mentioned how Christer Hellstrand of SKF had used a simple 2^3 factorial design to lengthen the life of a particular type of bearing by a factor of five. Figure 1a shows a second example of Christer's work (Hellstrand, 1991) in which a different eight-run design was used to solve an important problem. A contract for manufacturing a specialized bearing used in a washing machine had been lost because a competitor had brought onto the market an improved design. In response, Christer's company set up a project team consisting of the technical director, the manufacturing engineer, the manufacturing statistician, the material laboratory manager, an application engineer, and an endurance testing engineer. From careful analysis of the competitor's product and after a long brainstorming session, the team decided to test four factors at standard ($-$) and modified ($+$) levels. The factors were: the manufacturing process for the balls (A), the cage design (B), the type of grease (C), and the amount of grease (D).

Now *four* factors were tested at two levels in this experiment, so a full 2^4 factorial design would have required 16 runs, but the experimental design used only half that number. It was in fact a "half replicate" of the full design. The eight chosen manufacturing conditions are shown in Figure 1a. They were run in random order and a measure of the relative average life y of bearings produced at each set of conditions is shown to the right hand side of the table. Mere eyeballing of the data suggests that the type of grease (C) and more particularly the amount of grease (D) do not appear to be doing much. If we ignore D as unimportant, the data can then be

From Box, G. E. P. (1992), *Quality Engineering*, 4(4), 619–627. Copyright © 1992 by Marcel Dekker, Inc.

	A: Balls			− Std	+ Mod	
	B: Cage			Std	Mod	
	C: Type of grease			Std	Mod	
	D: Amount of grease			Norm	Large	

Run Number	A	B	C	D	y	AB ‖ CD
1	−	−	−	−	0.31	+
2	+	−	−	+	1.38	−
3	−	+	−	+	0.73	−
4	+	+	−	−	2.17	+
5	−	−	+	+	0.95	+
6	+	−	+	−	1.37	−
7	−	+	+	−	0.92	−
8	+	+	+	+	2.57	+

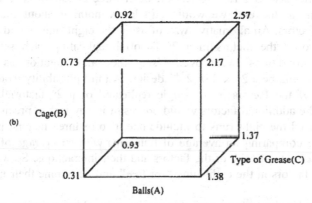

Figure 1. (a) The eight-run design to test four factors at two levels. (b) Display of the bearing life data assuming factor D (amount of grease) has little effect.

plotted in relation to the other three factors A, B, C on the cube shown in Figure 1b. The figure suggests that by appropriately modifying the manufacturing process for the balls (A) and changing the design of the cage (B), a sevenfold increase in bearing life could be obtained as compared with the standard unmodified bearing represented by run 1. These findings were doubly welcome because the modified cage design was cheaper to produce than the original design. Further tests confirmed these findings, and as a result of this simple experiment, the lost market was regained with a greatly improved bearing design which exceeded the expectations of the customer.

Now while industrial experiments such as this can often produce dramatic improvements, they are frequently expensive; particularly, as in this case, when the runs are made on actual production equipment. Statistical designs which use only a fraction of the factorial runs are, therefore, of great practical importance.

SOME DIFFERENT WAYS TO USE EIGHT RUNS

To understand why, for particular applications, *fractional* factorial designs are so efficient, let's consider other ways in which these eight process runs might have been used. Well, they could have been used to test just a single factor, say, type of grease (C). Four runs might have been made with standard grease and four runs with modified grease keeping the other three factors constant. Such an experiment would have the merit that tests for both the standard and modified types of grease would have been replicated (repeated) four times over and we might, therefore, feel some confidence in any conclusions that we drew since they would be based on a comparison between *averages* of two sets of four runs. We could call such an arrangement a 2^{1+2} design because it would test a single factor at two levels with $4 = 2^2$ replications. The trouble with this arrangement would be that we would have used up all 8 runs to study only one factor. Also, even if we could get agreement to run three further sets of 8 runs to test the other three factors in the same "one at a time" fashion, in the end we would still know nothing about the *interactions* between the factors. An alternative way of using the eight runs would have been to study just two of the factors in a 2^2 factorial replicating each set of the four experimental conditions twice over. Using the same notation as before, this arrangement would be a $2^2 \times 2$ or 2^{2+1} design. A third possibility would have been to run three of the factors in a "single replicate" of a 2^3 factorial. Notice that inclusion of the additional factors would *not* result in any loss in precision; because, whether we used the eight runs to include one, two or three factors, in the end we would still be comparing an average of four runs with an average of four runs to calculate the effects of each of the factors and their interactions. So we can include the extra two factors in the eight-runs "for free" and determine their interactions as well!

Now in the bearing experiment, this process was taken one step further and *four* factors were included in the 8-run experiment. Designs of this kind were invented by Finney (1945) and called by him *fractional replicates*. In this case the design is a "half replicate" or "half fraction" of the full 2^4 design. It is in fact $2^4 \times \frac{1}{2}$ or $2^4 \times 2^{-1}$, that is, a 2^{4-1} "fractional replicate." As we shall see, quarter replicates, eighth replicates and so forth can in certain circumstances also be useful. These designs are all examples of *orthogonal arrays*, (see also Plackett and Burman (1946) and Rao (1947)).

FRACTIONAL DESIGNS

Fractional factorial designs have been used with great success in solving industrial problems particularly when the objective is "factor screening." That is to say we are in a "Pareto situation"—where, as Dr. Juran puts it, we are "looking for the vital few [factors] among the trivial many." These designs are of value for screening purposes because of what may be called their "projective" properties. For example, if you check it out, you will find that *any* three columns of the design in the Figure 1a

produces a complete set of the eight combinations $(- - -, + - -,$
$- + -, + + -, - - +, + - +, - + +, + + +)$ for whichever three factors
you've chosen. So for this design if, as is quite likely, one of the factors is just "in
there for the ride"—that's to say it is one of the "trivial many" and has zero or little
effect—*then you will have a complete 2^3 design in the other three.*

A design, any *three* columns of which will produce a complete 2^3 factorial (or a
replicated 2^3 factorial), is called a design of *resolution* four. To keep the notation
straight we usually use a roman numeral subscript to indicate resolution. Thus, the
bearing design used by Christer is a one-half fraction of a 2^4 design of resolution 4
or a 2_{IV}^{4-1} design. In general a design of resolution R projects complete factorials
(sometimes replicated) in every set of $R - 1$ factors. It is therefore said to be of
projectivity $P = R - 1$. The concept of projectivity is more fully developed in
Chapter C.8.

ALIASES

To better understand the 2_{IV}^{4-1} design used in the bearing experiment, consider what
would have happened if all the factors had turned out to be important, and you
therefore needed to consider all four main effects A, B, C, D and all six two-factor
interactions, AB, AC, AD, BC, BD, and CD.* Now let's try to calculate the interaction AB for this experiment. The appropriate column of signs will be obtained by
multiplying the signs of columns A and B in Figure 1a to get a new column AB
shown on the right of the table. However, notice that if you do the same thing for
columns C and D, you get an identical set of signs. Thus, when you try to calculate
the AB interaction by subtracting the average of the four runs opposite minus signs
from the average of the four runs opposite plus signs, the effect you get is really the
sum of the AB interaction and a CD interaction. AB is said to have an "alias" CD. If
you multiply out the signs you'll also find that AC and BD are aliased and AD and
BC also. So if you calculated the main effects of the factors and their interactions,
the analysis would look like this:

		Average of the eight runs	1.30
Main effects	{	A (balls)	(1.14)
		B (cage)	(0.60)
		C (grease type)	(0.31)
		D (grease amount)	0.22
Interactions	{	AB + CD	(0.40)
		AC + BD	−0.11
		AD + BC	−0.01

* I will suppose, as is often realistic, that interactions between three or more factors can be ignored.

From the above table you will see that the main effects are clear of two factor interactions but the two factor interactions are aliased with each other. This is true for *any* resolution IV design. You can use this same table to check out what I said before about the "projective property" of this design. If *any* one of the factors A, B, C, or D is unimportant (and therefore any effects involving that letter can be left out) the remaining effects have no aliases. For example, cross out all the effects containing, say, the letter D, as would be appropriate if factor D was inert. Then the interaction pairs are (AB + C̶D̶), (AC + B̶D̶) and (A̶D̶ + BC) and now none of the effects are aliased because you have a complete 2^3 factorial in the other factors A, B, and C. For the bearing data, the guess is that what is going on is mostly due to balls (A) and cage (B) with the possibility of a helpful AB interaction and of a small effect due to grease type (C) indicated by brackets in the table.

In this example, no formal analysis was really necessary. As is so often the case for well-designed experiments, the data almost analyze themselves and it was realized that the problem had been solved as soon as the results shown in Figures 1a and 1b were looked at. Test runs at the new conditions confirmed this.

Even if the original plan had been to run a full 2^4 design, nothing would have been lost by running these 8 runs first. If they failed to solve the problem, you could always add some further runs later and, in particular, you could run the other half of the full 2^4 arrangement.* This is a further illustration of the sequential approach: eliminating unnecessary experimentation by ensuring that larger designs are built up only to solve more complicated problems.

GREATER FRACTIONATION

The discussion so far does not exhaust the possibilities for useful eight-run designs. The easiest way to see what is available is to write down a table of plus and minus signs that can be used to estimate all the main effects and interactions in a full 2^3 factorial design in which (dummy) factors are denoted by (say) a, b, and c as is done in Figure 2.

Now forget about the a's, b's, and c's (that were just dummy variables used to generate this table of signs) and let us rename the columns 1, 2, 3, 4, 5, 6, and 7 as indicated in the tenth row of the table. What you have now is sometimes called an L_8 *orthogonal array*. An array is just a name for a rectangular table. The word *orthogonal* means that there is complete balance in the signs of every pair of columns. Thus if you select *any one* of the seven columns, you will find that opposite the four pluses in the selected column there are two pluses and two minuses in every other column. Likewise, opposite the four minuses in the selected column there will be two pluses and two minuses in every other column. This array encapsulates every

*The two halves of the design form what are called "orthogonal blocks." Because of this, it turns out that if something slipped in the time that elapsed between performing the first set and the second set of eight runs resulting in a change of mean level, this discrepancy would be exactly balanced out and would cause no change in the values of the calculated effects (see BHHII, pp. 336–351, for details).

(a)	(b)	(c)	(ab)	(ac)	(bc)	(abc)	Bicycle data (using all 7 columns) time in secs	Process data (using columns 1,2,3,6,7) yield %
−	−	−	+	+	+	−	69	77.1
+	−	−	−	−	+	+	52	69.0
−	+	−	−	+	−	+	60	75.5
+	+	−	+	−	−	−	83	72.6
−	−	+	+	−	−	+	71	67.9
+	−	+	−	+	−	−	50	68.4
−	+	+	−	−	+	−	59	71.5
+	+	+	+	+	+	+	88	65.9
1	2	3	4	5	6	7		
•	•	•					$\rightarrow$	2^3
•	•	•				•	$\rightarrow$	2^{4-1}_{IV}
•	•	•	•	•	•	•	$\rightarrow$	2^{7-4}_{III}

Basic eight run designs
$$\begin{cases} 2^3 & \text{Full Factorial} \\ 2^{4-1}_{IV} & \text{1 rep of } 2^3 \text{ in every three factors (4 choices)} \\ 2^{7-4}_{III} & \text{2 reps of } 2^2 \text{ for every two factors (21 choices)} \end{cases}$$

Figure 2. An eight-run orthogonal array showing the basic designs 2^3, 2^{4-1}_{IV}, and 2^{7-4}_{III} with two sets of data.

possible fractional factorial design involving eight runs. Although you may see tables for such designs which look different—with ones and twos replacing minus signs and plus signs, or with different orderings of columns and rows, or in which all the signs are switched in one or more of the columns, these are not really different designs but simply correspond to a relabeling of the basic design which we generated above.

Now beneath the columns of pluses and minuses in Figure 2 there are a number of rows of bullets. The first row of three bullets indicates that if you want to use the array as a full 2^3 factorial, you could use columns 1, 2, and 3. Looking at the next row, if you want to use the eight runs for four factors to produce a 2^{4-1}_{IV} design you should add column 7 (use columns 1, 2, 3, and 7). This design was in fact employed for the bearing example discussed earlier. Although you could generate a 2^{4-1} fractional factorial design with a different column for the fourth variable, using, say, columns 1, 2, 3, and 4, this design would only have resolution three rather than resolution four. Equivalently it would have projectivity two rather than three, and certain projectives in three dimensions would not be complete 2^3 factorials. Normally you will want to use designs which, for a given degree of fractionation, produce the highest possible resolution, so you would use columns 1, 2, and 3 with column 7.

If you "saturate" the orthogonal array by associating factors with every one of the seven columns, you will have a $1/16 = 1/2^4$ replicate of a 2^7 design (because you are only running 8 of the $2^7 = 128$ runs required by the full factorial). This design has resolution III, that is to say any pair of columns yields a full 2^2 factorial (replicated twice). It is therefore called a 2_{III}^{7-4} design. Thus you could use this design to screen seven factors if, initially at least, you thought it likely that the vital few factors of importance were not more than two in number.

An example taken from BHHII concerns the times for a student cyclist to climb a particular hill on the campus at the University of Wisconsin–Madison. The seven factors—seat (up/down), dynamo (on/off), handlebars (up/down), gear (low/normal), raincoat (on/off), breakfast (yes/no), and tires (hard/soft)—were associated with columns 1, 2, 3, 4, 5, 6, and 7, respectively, and varied according to the design in Figure 2. The calculated effects, showing all the two-factor interaction aliases that could occur in the unlikely event that all seven factors were active, are also shown.

seat	$3.5 \rightarrow 1 + 24 + 35 + 67$
dynamo	$(12.0) \rightarrow 2 + 14 + 36 + 57$
handlebars	$1.0 \rightarrow 3 + 15 + 26 + 47$
gear	$(22.5) \rightarrow 4 + 12 + 56 + 37$
raincoat	$0.5 \rightarrow 5 + 13 + 46 + 27$
breakfast	$1.0 \rightarrow 6 + 23 + 45 + 17$
tires	$2.5 \rightarrow 7 + 34 + 25 + 16$

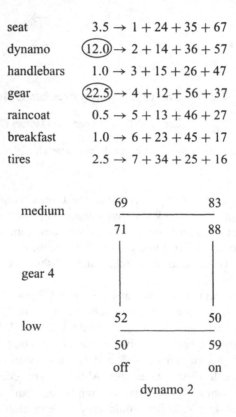

The simplest explanation of the data is obviously that the main effects of dynamo (2) and gear (4) are the vital factors that affect the time to cycle up the hill. The appropriate two-way table which shows how the data would stack up when rearranged on this basis is shown below the main table. In the original reference there is a discussion of how such tentative findings can be confirmed.

DROPPING COLUMNS: WHICH ONES?

There are, of course, intermediate cases where we wish to look at only six or five factors. Such designs are obtained by dropping one or two of the columns 4, 5, and 6 from the basic seven factor design of Figure 2. The appropriate alias pattern is found by simply omitting all the effects which contain the dropped factors.

The approach to fractional factorial designs and orthogonal arrays that I have outlined in this column was originally presented in the early 1950s in a paper entitled "On the Experimental Attainment of Optimum Conditions" (Box and Wilson, 1951). I will use an example from that paper to illustrate the idea of "dropping" factors. In the above reference, a design was employed to investigate a chemical reaction involving three chemicals, A, B, and C, and a solvent. Five factors were varied: the concentration of C, the proportion of C to A, the solvent amount, the proportion of B to A, and the reaction time. The design actually used can be obtained by dropping columns four and five from the seven factor design in Figure 2 and associating these five experimental factors with columns 1, 2, 3, 6, and 7, respectively. In this example it was argued that in the early stages of investigation when you are likely to be a long way from optimal conditions, main effects will probably dominate and interactions will be relatively small (when you are way down the side of a hill, it is the slopes of the hill that are important, curvature effects and interaction effects corresponding to ridges in the hill are not so important until you get near the summit). This quarter replicate (2^{5-2} design) was employed, therefore, on the expectation that main effects would dominate. However, a chemist thought that if any interaction did turn out to be important it would most likely be between the factors *concentration of C* and *solvent amount*. The particular allocation of factors to the basic array in Figure 2 ensures that this interaction (13) occurs in column 5 which has not been used to accommodate any expected effect and can thus be independently estimated. Once you have the basic alias pattern for seven factors, given in the earlier table, it is easy to see how by simple trial and error you can associate particular process factors with particular columns of the design so as to isolate suspected interactions in this way.

For the chemical example, the analysis and the alias pattern obtained by dropping all effects containing the numbers 4 and 5 from the full 2^7 alias pattern are as follows:

(1) concentration C	-4.0 →	$1 + 67$
(2) C/A	0.8 →	$2 + 36$
(3) Solvent	-5.1 →	$3 + 26$
	-0.2 →	$12 + 37$
(13) conc. × solvent	1.5 →	$13 + 27$
(6) B/A	-0.2 →	$6 + 23 + 17$
(7) Time	-2.8 →	$7 + 16$

It appears that factors 1, 3, and 7 all have large *negative* effects. It was concluded that higher yields might be obtained by moving in a direction such that the concentration of C, the amount of solvent and the time of reaction were all *reduced*. This checked out and a yield of 84% was eventually obtained.

In this chapter I have shown a number of ways in which 8-run fractional factorial designs can be useful and some rationales for their use.

What Can You Find Out From Sixteen Experimental Runs?

Chapter B.2 described how a paper helicopter could be used to teach engineers experimental design, and for illustration a 16-run experimental arrangement was used which tested 8 design factors that might affect its flight time. In that experiment that only two of the tested factors—wing length and body length—had detectable influence. Since a full factorial in 8 factors each at two levels would have required $2^8 = 256$ runs, the design described was a 1/16 fraction of the full 2^8 factorial design. Also, since $1/16 = 1/2^4 = 2^{-4}$, the design is called a 2^{8-4} fractional factorial design. This chapter will be about how this and other 16-run two-level designs can most easily be constructed.

Chapter B.3 showed how 8-run designs having different degrees of fractionation were readily written down using a basic and easily obtained orthogonal array. I will proceed in exactly the same way for the 16-run designs. To write down the basic 16-run orthogonal array, consider 4 (because $2^4 = 16$) dummy factors a, b, c, and d. As shown in Table 1A, you first write down the columns of signs for the four main effects using alternating signs $(-, +)$ in the first column, alternating pairs of signs $(-, -, +, +)$ in the second column, and so on. Then you multiply the signs in the four columns you have generated to get columns for the 6 two-factor interactions ab, ac, etc. In a similar way you get the 4 three-factor interactions abc, abd, etc., and the 1 four-factor interaction $abcd$. Now you forget all about the letters a, b, c, and d used to generate the table of minuses and pluses and rename the columns with capital letters $A, B, C, D, J, K, L, M, N, O, E, F, G, H, P$ as shown at the bottom of the table. (The rather strange order in which I have put these 15 capital letters is not essential but, as you will see, it makes things a little tidier later on.) The table of pluses and

From Box, G. E. P. (1993), *Quality Engineering*, 5(1), 167–178. Copyright © 1993 by Marcel Dekker, Inc.

Table 1. (A) Generation of 16-Run Orthogonal Array with Five Sets of Data

a	b	c	d	ab	ac	ad	bc	bd	cd	abc	abd	acd	bcd	abcd	1 Drill Advance	2 % Reacted	3 Flight Time	4 Shrinkage	5 Tensile
−	−	−	−	+	+	+	+	+	+	−	−	−	−	+	.23	56	2.5	.49	43.7
+	−	−	−	−	−	−	+	+	+	+	+	+	−	−	.30	53	2.9	.57	40.2
−	+	−	−	−	+	+	−	−	+	+	+	−	+	−	.52	63	3.5	.09	42.4
+	+	−	−	+	−	−	−	−	+	−	−	+	+	+	.54	65	2.7	.17	44.7
−	−	+	−	+	−	+	−	+	−	+	−	+	+	−	.70	53	2.0	.19	42.4
+	−	+	−	−	+	−	−	+	−	−	+	−	+	+	.76	55	2.3	.15	45.9
−	+	+	−	−	−	+	+	−	−	−	+	+	−	+	1.00	67	2.9	.23	42.4
+	+	+	−	+	+	−	+	−	−	+	−	−	−	−	.96	61	3.0	.17	40.6
−	−	−	+	+	+	−	+	−	−	−	+	+	+	−	.32	69	2.4	.13	42.4
+	−	−	+	−	−	+	+	−	−	+	−	−	+	+	.39	45	2.6	.12	45.5
−	+	−	+	−	+	−	−	+	−	+	−	+	−	+	.61	78	3.2	.45	43.6
+	+	−	+	+	−	+	−	+	−	−	+	−	−	−	.66	93	3.7	.53	40.6
−	−	+	+	+	−	−	−	−	+	+	+	−	−	+	.89	49	1.9	.17	44.0
+	−	+	+	−	+	+	−	−	+	−	−	+	−	−	.97	60	2.2	.27	40.2
−	+	+	+	−	−	−	+	+	+	−	−	−	+	−	1.07	95	3.0	.34	42.5
+	+	+	+	+	+	+	+	+	+	+	+	+	+	+	1.21	82	3.0	.58	46.5
A	B	C	D	J	K	L	M	N	O	E	F	G	H	P	2^4	2^{5-1}_{V}	2^{8-4}_{IV}	2^{15-11}_{III}	2^{9-5}_{III}

Table 1. (B) Examples of the Use of the Array for the "Way-Station" Designs 2^4, 2^{5-1}_V, 2^{8-4}_{IV}, 2^{15-11}_{III} Projected Properties and Calculated Effects for Examples. Example 5 is Used to Illustrate the Generation of an Improved 2^{9-5}_{III} Design for Taguchi and Wu's Experiment

Columns	a	b	c	d	ab	ac	ad	bc	bd	cd	abc	abd	acd	bcd	abcd	
1. Daniel 2^4	• A	• B	• C	• D												Full 2^4 factorial
2. BH² 2^{5-1}_V	• A	• B	• C	• D											• P	1 rep. of 2^4 in every four factors (5 choices)
3. Box 2^{8-4}_{IV}	• A	• B	• C	• D							• E	• F	• G	• H	• P	2 reps. of 2^3 in every three factors (56 choices)
4. Quinlan 2^{15-11}_{III}	• A	• B	• C	• D	• J	• K	• L	• M	• N	• O	• E	• F	• G	• H	• P	4 reps. of 2^2 in every two factors (105 choices)
5. Taguchi and Wu 2^{9-5}_{III}	• A	• B	• C	• D		• K	• L				• E	• F	• G	• H	• P	
A better design for example 5 2^{9-5}_{III}	• A	• B	• C	• D							• E	• F	• G	• H	• P	

Calculated Effects

Example	a	b	c	d	ab	ac	ad	bc	bd	cd	abc	abd	acd	bcd	abcd
1	A	B*	C*	D	AB	AC	AD	BC	BD	CD	ABC	ABD	ACD	BCD	ABCD
	0.06	0.25	0.50	0.14	−0.01	0.00	0.03	−0.02	−0.01	0.04	0.00	0.02	0.02	−0.01	0.02
2	A	B*	C	D*	AB	AC	AD	BC	BD	CD	DP*	CI	BI	AI	P*
	−2.00	20.50	0.00	12.25	1.50	0.50	−0.70	1.50	10.75	0.25	−9.50	2.25	1.25	1.25	−6.25
3	A	B*	C*	D	AB+	AC+	AD+	AE+	AF+	AG+	E	F	G	H	AH+
	0.13	0.78	−0.40	−0.03	−0.17	0.05	0.12	−0.10	0.17	−0.05	0.05	0.17	−0.15	−0.10	−0.20
4	A+	B+	C+	D+	J+	K	L+	M+	N+*	O+	E+	F+	G+	H+*	P+
	0.06	0.06	−0.06	0.07	0.03	0.00	0.04	0.08	0.24	0.09	0.04	0.03	0.07	−0.14	0.01

minuses you have produced is a 16-run orthogonal array sometimes called a L_{16} array. This one table of signs may now be used for a number of different purposes following an approach developed long ago (Box and Wilson, 1951).

You will no doubt see from time to time examples of 16-run two-level factorial, fraction, factorial, or orthogonal arrays which look different from those which I discuss in this column. However, on examination you will find that almost always* they really are the same designs, but possibly (a) with rows listed in a different order, (b) with columns listed in a different order or (c) with all the signs in one or more columns switched. None of this makes any practical difference, so long as you keep tabs on what you call things.

Remember what is meant by orthogonality. It means that any given column is balanced with respect to every other column. In Table 1B, in every column, opposite the 8 minuses you will find 4 minuses and 4 pluses in every other column. Similarly, opposite the 8 pluses, there will be 4 minuses and 4 pluses. If you have some data and, for a particular column of signs, you calculate the differences in averages between the 8 data values opposite plus signs and the 8 data values opposite minus signs, you will have what is called a data *contrast* which supplies the main effects and interactions for the various factors. Some of the possibilities will be illustrated by using data from five published experiments. These are shown on the right hand side of Table 1A.

A 2^4 FACTORIAL

The first example is of a full 2^4 factorial design due to Daniel (1976). The study concerned the performance of a small stone drill. The factors studied were: the load (A), the flow rate (B), the rotation speed (C), and the type of mud (D). The appropriate 2^4 design is obtained by associating the four factors with the first four columns of the table as indicated by the four bullets in Table 1B. The response of interest for this example was the (logged) rate of advance of the drill. The calculated effects are shown at the bottom of the table. None of the interactions is large and eyeballing the data or a normal plot, suggests that the main effects for flow rate (B), speed (C), and mud (D), indicated with asterisks, account for most of what is happening. Factor A (load), appears to be without much effect.

A HALF FRACTION

The second example, a half fraction of a 2^5 factorial taken from BHH II is a study of the performance of a chemical reactor. The five factors were Feed Rate (A), Catalyst (B), Agitation Rate (C), Temperature (D), and Concentration (I). The appropriate fractional design uses the five columns indicated by the second row of bullets. The data shown are for the "percent reacted" (compared with the theoretical maximum

---

* Different sixteen run designs can be obtained from four orthogonal arrays due to Hall, 1961. (See Box and Tyssedal, 1996.)

attainable). As I explained for eight-run designs, as soon as a design is fractionated, certain effects and interactions which could have been separately estimated from a full factorial design are entangled and must be estimated together (they are aliased). Now, as a general rule, the higher the order of an interaction the less likely is it to be real. So you would like the effects you are really interested in, usually main effects and two-factor interactions, to be aliased with effects of the highest order possible. We express this by saying we want the design to have the highest possible *resolution*.

Now a half replicate of *some* sort could be obtained from any five columns of the array; however, some choices are much better than others. A half fraction using columns A, B, C, D, and P was chosen because this gives a design of resolution five, which is the highest resolution possible for five factors tested in 16 runs. A design of resolution five has the property that main effects, "one letter" effects, are aliased only with "four letter" effects (four factor interactions) (note that $1 + 4 = 5$). Also two-factor interactions are aliased only with three-factor interactions ($2 + 3 = 5$). In general, if a design is of resolution R then effects containing m letters will be aliased with effects containing $R - m$ letters. As before we shall indicate the resolution of a design by a roman letter subscript. So the reactor design which is a half fraction of a 2^5 factorial of resolution five is a 2_V^{5-1} design. Also, from now on we will make the tentative assumption that the effects of three-factor and higher order interactions are negligible compared with main effects and two-factor interactions. On this assumption, any resolution five design can be used to estimate all main effects and two-factor interactions with no aliasing. In particular, for our 2_V^{5-1} reactor design, we can use the 15 contrasts obtained from the columns of the table to produce estimates of *all* 5 main effects and *all* 10 two-factor interactions.

To identify the various two-factor interactions, we can use Table 2. The *bracketed* items in this table (precisely one of which occurs in each line of the table) refer to the quantities which are estimated by the 2_V^{5-1} design. Thus, for example, the AB (Feed Rate × Catalyst) interaction is estimated using the contrast of column J, and the DP (Temperature × Concentration) interaction uses the contrast in column E, and so on. The effects so obtained are listed in Table 1B. If they are plotted on probability paper, it appears that only the main effects B, D, and P, and the interactions BD and DP denoted by asterisks in Table 1 are distinguishable from noise. Thus only three of the five factors—Catalyst (B), Temperature (D), and Concentration (P)—appear to be active. Feed Rate (A) and Agitation Rate (C) appear to be inactive since neither letter appears in any of the main effects or interactions that can be distinguished from noise. Although we provided for the possibility of five factors in our original experiment, only three seem to have been doing very much and what we have ended up with is essentially a *duplicated* three-level factorial in these three active factors.

This serves to illustrate another important aspect of the concept of resolution pointed out in my last chapter: any design of resolution R will project complete factorials (possibly replicated) in every set of $R - 1$ factors. For the present design of resolution five which employs column contrasts A, B, C, D, and P, this means (as you can readily check) that if any one of these columns was dropped out corresponding

Table 2. Effect Aliases on the Assumption that the Interactions Between Three Factors or More Are Negligible But that Two-Factor Interactions Could Occur

2_{III}^{15-11}	2_{IV}^{8-4}	2_V^{5-1}								
•	•	•	(A)	BJ	CK	DL	EM	FN	GO	HP
•	•	•	(B)	AJ	CM	DN	EK	FL	GP	HO
•	•	•	(C)	AK	BM	DO	EJ	FP	GL	HN
•	•	•	(D)	AL	BN	CO	EP	FJ	GK	HM
•	•		E	AM	BK	CJ	(DP)	FO	GN	HL
•	•		F	AN	BL	(CP)	DJ	EO	GM	HK
•	•		G	AO	(BP)	CL	DK	EN	FM	HJ
•	•		H	(AP)	BO	CN	DM	EL	FK	GJ
•		•	(P)	**AH**	**BG**	**CF**	**DE**	JO	KN	LM
•			J	(AB)	**CE**	**DF**	**GH**	PO	KM	LN
•			K	(AC)	**BE**	**DG**	**FH**	PN	JM	LO
•			L	(AD)	**BF**	**CG**	**EH**	PM	JN	KO
•			M	AE	(BC)	**DH**	**FG**	PL	JK	NO
•			N	AF	(BD)	**CH**	**EG**	PK	JL	MO
•			O	AG	**BH**	(CD)	**EF**	PJ	KL	MN

Bracketed ( ) letters and combinations of letters are effects which are estimated when columns A, B, C, D, and P are used to produce a 2_V^{5-1} design. Bold letters indicate effects which are estimated when columns A, B, C, D, E, F, G, and H are used to produce a 2_{IV}^{8-4} design. All combinations of letters indicate estimated effects when all 15 columns A, B, C, ..., P, are used to produce a 2_{III}^{15-11} design.

to the existence of an inactive factor, you would have a complete 2^4 in the remaining four factors (and hence a duplicated 2^3 factorial in any three factors as has happened in this example).

Sometimes from previous work or because of special expertise, we know, or think we know, which few factors are important. In such cases, a full factorial design like the 2^4 or a mildly fractionated design like the 2_V^{5-1} may be used to explore these factors. (Notice, however, that even in these examples our guesses were not entirely borne out—one of the four factors in the 2^4 design appeared to be inactive and in the 2_V^{5-1} design, two of the five factors appeared to be inactive.)

HIGHER FRACTIONS

In a common situation (occurring particularly in the study of a new process or product) a brainstorming session will reveal that there are many factors that *might* have effects, but nobody knows for sure *which* ones are important. The Pareto principle of Juran (1988) (see also Occam, 1330) says that among many factors there will usually be a *vital few* that really are important and a *trivial many* that are not. Higher order fractional designs, which are now discussed, are of great value for "screening out" these vital factors.

Consider the third example taken from Chapter B.2. This is a 2_{IV}^{8-4} design used to screen eight factors which might affect the flight time of a paper helicopter. The factors were: Paper type (A), Wing length (B), Body length (C), Body width (D), Paper clip (E), Taped wing (F), Taped body (G), and Fold (H). A statistical design of highest possible resolution (four) is obtained by using the columns indicated by the eight bullets in the third row below the table. Since the design is of resolution four, it is of projectivity $P = 3$ and if there are only three vital factors, it will produce a duplicated 2^3 factorial design in whichever three factors these turn out to be. And for those vital three factors *all* main effects and two-factor interactions can be separately estimated. You may want to check for yourself this remarkable property of the design by selecting any three of the eight factors A, B, C, D, E, F, G, H (there are 56 choices!) and satisfy yourself that there is indeed in the three selected columns, two complete sets of each of the possible factorial combinations of signs $(- - -, + - -, - + -, + + -, - - +, + - +, - + +, + + +)$. The normal plot shown in Chapter B.2 in fact suggests that there are just two important factors (indicated by asterisks in Table 1)—wing length (B) and possibly body length (C).

The *bold letters* in Table 2 show the alias pattern for this 2_{IV}^{8-4} design in factors A, B, C, D, E, F, G, and H. You will see that for this remarkable design all eight main effects are free from two-factor aliases, but two-factor interactions arrange themselves in seven sets of aliased groups of four. These are indicated in Table 1B by the first member of the set of aliases followed by a plus sign (e.g., $AB+$). Thus all the 8 main effects are fully protected from bias from all the two-factor interactions. If you want another way to check the screening properties of this design, again choose any three of the eight factors. Now suppose these three factors are active and the remaining five are inactive. In Table 2 strike out any main effect or two-factor interaction which contains *any* of the five inactive factors and you will find that all ambiguities disappear. You have the three required main effects and *all their interactions* clear of aliases.

Example 4 is from a 2_{III}^{15-11} design due to Quinlan (1985). This is a design of maximum fractionation in which 15 columns are used to accommodate 15 factors, as indicated by the bullets below the table. The experiment concerns the shrinkage of speedometer cables. The 15 factors tested were:* Liner tension (A), Liner line speed (B), Liner die (C), Liner O.D. (D), Cooling (E), Screen pack (F), Coating die (G), Wire diameter (H), Line speed ($-P$), Melt temperature ($-J$), Coating material ($-K$), Liner temperature ($-L$), Braiding tension ($-M$), Wire braid type ($-N$), and Liner material ($-O$). The plus sign following each letter in Table 1B implies that each main effect has two-factor aliases. A normal plot of the data shown, for example, in Chapter E.9 indicates that contrasts N and H in our notation are the only ones distinguishable from noise. Thus the simplest explanation of the data is that the main effects associated with—*wire braid type* and *wire diameter*—represent active

---

* The design is the same as Quinlan's but the factor letters A, B, C, etc. have been reallocated to correspond with those in Table 1 of this chapter. The minus signs associated with factors P, J, K, ... are introduced to correspond to Quinlan's design.

factors. Although the effects in this highly fractionated design have multiple aliases as shown in Table 2 (all letters), since this is a design of resolution three, if not more than two factors are important they can be identified. In this example the preferred level of wire diameter was that already employed. However, switching to the other braid type confirmed that a major improvement was possible and this turned out to be a highly successful screening experiment.

JUSTIFICATIONS FOR THE USE OF FRACTIONAL DESIGNS

In everyday life one can lose out on the one hand by being overly cautious and on the other by being foolhardy. Most of us try to run our lives wisely by using judgement in choosing some middle course. The same kind of judgement is necessary to be successful in using statistical methods, experimental design, and, in particular, fractional designs. Here is some background that may help.

There are several rationales which have been put forward for the use of highly fractioned designs and other orthogonal arrays. I'll list some of these, some good and some not so good, and then discuss them:

1. Fractional factorials are a convenient way to get representative coverage of the experimental space.
2. Most interactions can be ignored or eliminated.
3. You can predict in advance those few interactions which will be important and allow for them in the fractional design.
4. When you have a large number of factors, the concepts of design resolution and projectivity can be used to exploit the Pareto effect which you are likely to find.
5. Designs can be assembled sequentially. For example, if after the first fraction there are ambiguities of interpretation, you can perform a second fraction to resolve these.

1. Some people think of fractional designs as nothing more than a convenient way of evenly sampling the experimental space. Indeed you can imagine a policy where you run the various factor combinations of a highly fractionated design and simply "pick the winner." Such a policy would, on average, certainly result in progress; however, it is not very efficient and is a mindless strategy which usually would not lead to any understanding of the reason for what was happening and would not help the engineer to better understand the system.

2. There are circumstances where interactions between factors are likely to be unimportant. In particular this is often so at the beginning of an investigation when you are a long way away from optimum conditions. When you are "way down on the side of the hill," it is the gradients that are important and these correspond to the main effects of the factors.

Another possibility is that certain interactions can be occasionally avoided in advance by careful choice of the way the factors are defined i.e., by factor transformation. For example, in studying two ingredients A and B in a corrosion inhibitor, it may be better to define the factors as the ratio A/B and the total amount of the inhibitor $A + B$ rather than as the separate amounts of A and B. An apparent interaction in the original variables can disappear after transformation. Again suitably transforming the response variable can sometimes avoid interactions as was achieved for the drilling data (Example 1) by using the *log* of the rate of advance of the drill (see BHHII, p. 335). Having said this, however, there are very many cases where interactions (often unexpected) are extremely important but cannot be transformed away (see, e.g., Chapter B.1).

3. Occasionally it is possible to predict that certain factors are more likely than others to interact. Most predictions of this sort must be viewed with some skepticism, however. For example, it may be argued that when factors occur at different stages of the process they will not interact. This is not always the case, however, For example, the best conditions for purifying a chemical may depend very much on the conditions used for its manufacture. A reckless extension of this idea is to say that a few expected interactions can always be picked out from a much larger number of possible interactions and the remainder treated as inactive. It seems logically indefensible to say that we need an experiment to *find out* which factors have main effects (first-order effects) and at the same time claim that we *know* which factors have interactions (second-order effects). Whenever I work on planning an experiment and I draw diagrams to illustrate various possible two-factor interactions I say to the experimenter, "Could something like this happen?", whichever factors I pick for illustration I almost always get the answer, "Yes, I can see how it could."

4. When we're examining a number of factors, we can usually rely on the Pareto effect to ensure that only a few (but an *unknown* few) will be of much importance. The fact that a design of resolution R can project complete factorials in any set of $R - 1$ factors allows us to make use of such an effect. This does not necessarily prevent us from isolating certain interactions according to rationale number 3 as an additional insurance policy.

5. Not infrequently, having run a fractional factorial design, the conclusions are ambiguous because of certain aliases. It is shown (e.g., in BHHII, p. 396) how a further runs may be chosen precisely to unravel specific ambiguities. This possibility of *sequential assembly* of designs ensures that elaborate designs are only used for complicated problems and simple designs for simple problems, thus minimizing unnecessary experimentation.

INTERMEDIATE NUMBER OF FACTORS

The factorial 16-run designs so far discussed (resolution V with five factors, resolution IV with eight factors and resolution III with 15 factors) are optimal* in the

---

* More recent work of Box and Tyssedal (1995) has shown that 16-run orthogonal arrays of a different kind can accommodate more factors at the same projectivity.

sense of all fractional factorial designs they contain the highest number of factors possible for a given resolution. (See however footnote on p. 67.) When we need to accommodate intermediate numbers of factors, we can regard these special designs as optimal "way-stations" and obtain the desired arrangement by suitably dropping or adding columns. Normally our first concern should be to preserve the highest possible resolution for the design. As a secondary concern we can, if we wish, take the opportunity of isolating certain interactions regarded as potentially most important. In discussing 8-run designs I illustrated the idea of dropping columns from a critical way-station design. I'll give here an example of *adding* an extra column.

For illustration consider Example 5 which is an experiment on welding reported by Taguchi and Wu (1980) in which nine factors were studied and (in our notation) were allotted to the columns of the array as follows: *Thickness* of welded material (*A*), Welding *method* (*B*), *Current* (*C*), Kind of welding *rod* (*D*), *Preheating* (*G*), *Drying* method for rods (*H*), welded material (*−P*), *Opening* of welding device (*−K*), and *Angle* of welding device (*−L*). There are, in all, 36 possible interactions that might occur between these nine factors, but the authors assumed that only four were important and the other 32 were ignored. The four chosen interactions were those between *method, current,* and *rod* (*BC, BD,* and *CD*) and that between *rod* and *material* (*DP*). They used Taguchi's "linear graph" method to find a design that, on the assumption that these were the *only* interactions, allowed them to be separately estimated without aliases. Also on this assumption two free columns denoted by e_1 and e_2 would be left for the estimation of error. The way in which their chosen design uses the columns of the orthogonal table is shown by the fifth row of bullets in Table 1B. If instead we entertain the possibility that *any* two-factor interaction could have occurred, then the resulting alias pattern can readily be found from Table 2 and is displayed in Table 3a. Interesting analyses of the data from this design were given in Box and Meyer (1986b); see also Chapter B.11. I shall concentrate on the advisability of using this particular arrangement.

The weakness of the Taguchi design is that the main effects of the nine factors are excessively vulnerable to aliasing from unexpected two-factor interactions. Thus from Table 3a, we can see that with this arrangement two of the main effects (*A* and *G*) have three two-factor aliases while five of the main effects (*C, D, I, K, L*) have two such aliases. The alias terms from two-factor interactions have to appear somewhere, but we should prefer they stay away from main effects so far as this is possible. Now had there been only eight factors, we could have used a 2_{IV}^{8-4} design and then main effects would have had *no* two-factor aliases. This optimal eight-factor design is thus a good place to start in building up a nine-factor design. In Table 3b we have associated the letters *A, B, C, D, E, F, G,* and *H* with the first eight factors, *thickness, metal, current, . . ., opening.* All we have to decide now is how to choose one remaining column to accommodate the ninth variable—the *angle* of the welding device. For illustration we use column *P* for this purpose. From Table 2, the alias pattern will now be that of Table 3b. We see that the result of what we have done is that each of the 8 main effects *A, B, C, D, E, F, G,* and *H* receives a *single* two-factor alias from the added factor *P.* Had we made a different choice, say,

Table 3. The Alias Structure (a) of Taguchi's 2_{III}^{9-5} Welding Design Obtained from Linear Graphs and (b) of an Alternative Design Obtained Using Table 2

		(a) Taguchi's 2_{III}^{9-5} Design			(b) Alternative 2_{III}^{9-5} Design Using Table 2
Thickness	A	$(+CK + DL + HP)$		A	$(+HP)$
Method	B	$(+GP)$		B	$(+GP)$
Current	C	$(+AK + GL)$		C	$(+FP)$
Rod	D	$(+AL + GK)$		D	$(+EP)$
Preheating	G	$(+BP + CL + DK)$		E	$(+DP)$
Drying	H	$(+AP)$		F	$(+CP)$
Material	P	$(+AH + BG)$		G	$(+BP)$
Opening	K	$(+AC + DG)$		H	$(+AP)$
Angle	L	$(+AD + CG)$		P	$(+AH + BG + CF + DE)$
Method × Current (M)	BC	$(+DH + PL)$	(M)	BC	$(+AE + DH + FG)$
Method × Rod (N)	BD	$(+CH + PK)$	(N)	BD	$(+AF + CH + EG)$
Current × Rod (O)	CD	$(+AG + BH + KL)$	(O)	CD	$(+AG + BH + EF)$
Rod × Material (E)	DI	$(+BK + HL)$	(K)	DG	$(+AC + BE + FH)$
(J)	e_1	$(+AB + GH)$	(J)	e_1	$(+AB + CE + DF + GH)$
(F)	e_2	$(+BL + CP + HK)$	(L)	e_2	$(+AD + BF + CG + EH)$

column J, then each of these eight main effects would have had a single alias containing J and so on. Now by reallocating factors to letters we can make any one of the variables the ninth factor. A good strategy is therefore to choose that one factor out of the nine that we think is most likely to be inactive, and so least likely to interact with the remaining factors. In this way we stand the best chance of preserving most of the advantages of the optimal eight-factor design. Notice also that Table 3b shows us that, for this example, we can do all of this and still preserve Taguchi's requirement that the two-factor interactions between *method, current,* and *rod* and the interaction between *rod* and *material*—the interactions BC, BD, CD, and DG in the new notation—are free of aliases (given his assumption that these are the only two-factor interactions that we need to worry about).

Taguchi's "linear graphs" for 16-run designs are complicated and this and other examples suggest that even in his hands they can produce designs which are significantly inferior to those obtained by looking at a simple alias table like Table 2 or, equivalently, those in BHH II or in Bisgaard (1988).

CHAPTER B.5

What Can You Find Out from Twelve Experimental Runs?

Previous chapters have shown how 8 and 16 run two-level factorial designs could be used to study a number of factors. In particular they could be used to generate fractional factorial designs whose projective properties made them excellent screening designs for finding a few vital factors having major effects on a system. These fractional designs are particular examples of orthogonal arrays developed by Plackett and Burman which are available when the number of runs N is a multiple of 4. In particular these authors derived such a design to study 11 factors in 12 runs. It turns out that this design has the remarkable property that it yields a full 2^3 factorial plus an additional optimal half replicate for any of the 165 choices of 3 factors out of the 11 factors tested.

Chapters B.3 and B.4 have illustrated the value of two-level 8 run and 16 run fractional factorial designs. In particular it was shown that because of their "projective" properties they could be used for screening a number of candidate variables to find the vital few which had major effects on the system. It was also pointed out that the plusses and minuses in the columns of these designs were in perfect balance and that consequently they were examples of "orthogonal arrays." Now while all such fractional factorial designs are orthogonal arrays, not all orthogonal arrays are fractional factorials. In particular, Table 1 shows a useful 12 run design which, although not a simple fraction of a factorial design, has perfect orthogonal balance. To see this you can check that opposite the six plus signs in any given column are 3 plusses and 3 minuses in every other column and similarly for

From Box, G. E. P. and Bisgaard, S. (1993), *Quality Engineering*, 5(4), 663–668. Copyright © 1993 by Marcel Dekker, Inc.

Table 1. A Plackett and Burman Design for Study of 11 Factors in 12 Runs with Data Due to Adam

Run	A 1	B 2	C 3	J 4	I 5	H 6	E 7	e 8	F 9	D 10	G 11	Response y
1	+	−	+	−	−	−	+	+	+	−	+	26
2	+	+	−	+	−	−	−	+	+	+	−	43
3	−	+	+	−	+	−	−	−	+	+	+	20
4	+	−	+	+	−	+	−	−	−	+	+	19
5	+	+	−	+	+	−	+	−	−	−	+	5
6	+	+	+	−	+	+	−	+	−	−	−	13
7	−	+	+	+	−	+	+	−	+	−	−	38
8	−	−	+	+	+	−	+	+	−	+	−	13
9	−	−	−	+	+	+	−	+	+	−	+	27
10	+	−	−	−	+	+	+	−	+	+	−	27
11	−	+	−	−	−	+	+	+	−	+	+	16
12	−	−	−	−	−	−	−	−	−	−	−	26
Effect	−1.2	−.5	−2.5	2.8	−10.5	1.2	−3.8	.5	14.8	.5	−7.8	

the six minus signs. This design has sometimes been referred to as an L_{12} orthogonal array. Orthogonal arrays of this kind were first used as statistical designs by two English statisticians, Robin Plackett and Peter Burman (1946) during World War II.* They were developed in response to the need to study many variables in the creation of an anti-aircraft proximity fuse. These designs could be obtained when the number of runs N is a multiple of 4, whereas orthogonal arrays can only be derived from fractional factorials when N is a *power* of 2 (e.g., $N = 4, 8, 16, \ldots$).

The design for $N = 12$ has what is called a cyclic structure. For such designs it is easy to write down a whole design matrix once we have the first row. You can see from Table 1 that the second row is the same as the first moved over one step to the right (the sign left over at the end of the row is the first sign of the new row). This "cyclic" shifting produces the first 11 rows, and the last row is just a row of minus signs. The designs with $N = 20$ (L_{20}) and $N = 24$ (L_{24}) are also cyclic designs with first rows:

$$N = 20: \quad + + - - + + + + - + - + - - - - + + -$$
$$N = 24: \quad + + + + + - + - + + - - + + - - + - + - - - - -$$

The interested reader will find larger designs in the original reference. Plackett and Burman (1946), C. R. Rao (1947) and Sloane (2005).

For illustration we re-analyze some data due to Adam (1987). He used the L_{12} design to study the effect of ten factors denoted by A, B, C, D, $\ldots$, J, which might affect surface defects on automobile instrument panels. Since the data are

*They are examples of Hadamard matrices.

based on *counts* of defects, the square root of the counts are most likely to be appropriate for analysis (for an explanation, see, e.g., p(HHII, p. ?)). These square roots multiplied by ten are shown as the response y in Table 1. Note that Adam's design has the rows and columns of the L_{12} arranged in different orders and we have appropriately taken account of this.

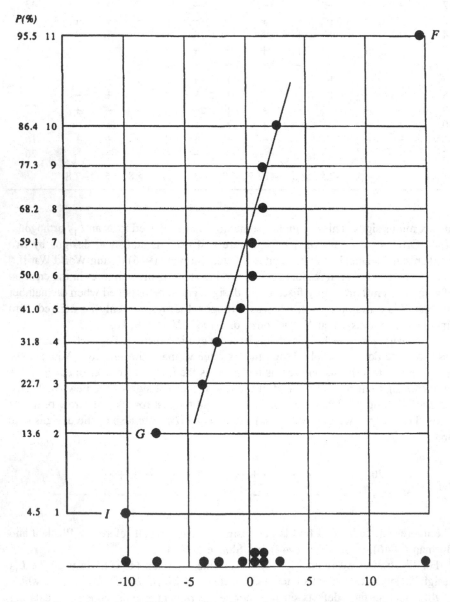

Figure 1. Daniel (normal probability) Plot of the effects calculated from Adam's data showing the main effects of G, I, and F distinguishable from the noise.

The rationale for using this design was that only main effects would be of importance. This assumption can be risky (Box, 1990), but for the moment we will also proceed on this basis. A more sophisticated method for analysis of designs of this kind that does not make this assumption due to Box & Meyer 1993 is discussed in Chapter C.6. The calculated effects for the factors and for the error column* are shown beneath the table and plotted on probability paper as a Daniel Plot (1976) in Figure 1. We see that factors F, I, and G stand out clearly from the noise and this is in agreement with Adam's conclusion.

PROJECTIVE PROPERTIES OF THE DESIGN

If we now plot the 12 data points on a cube whose axes are the three variables F, I, and G, we obtain Figure 2. From the figure we notice a remarkable fact. The design "projected" into the space of F, I, and G forms a complete cube (a complete 2^3 factorial in F, I, and G with half the points repeated. Furthermore, the repeated points provide a fractional factorial capable of estimating the main effects of the factors F, I, and G on its own, so reinforcing the main effects estimates.

Now there are 165 different ways of choosing three factors out of 11 and it turns out that *every one* of these "projections" produces a complete factorial plus an additional half replicate of the kind we see in Figure 2. So if only three factors turn out to be important, we will have a complete factorial in these three factors with four points replicated which optimally reinforce the estimates of the main effect. (See also Lin and Draper, 1992). This is an example of the surprising projective properties of Plackett–Burman designs which have great simplicity even though the alias structure (Margolin, 1968) for these designs is complicated. Notice a clue to which are the important factors is provided by determining: at which points the data are approximately replicated.

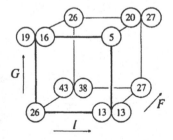

Figure 2. Adam's L_{12} experiment projected into the space of factors I, F, and G producing a complete cube plus an optimal half fraction in those variables.

---

* No factor is associated with column 8 so that if there were no interactions it would merely estimate the error of an effect.

CONCLUSIONS

In the past, there has sometimes been a reluctance to use Plackett and Burman designs for industrial experimentation because of their extremely complicated alias structure; in particular every main effect is aliased with portions of many two-factor interactions. However, the interesting projective properties of these designs provide a compelling rationale for their use. It is shown in Chapters C.6 and C.7 how Bayesian analysis can provide appropriate analyses.

ACKNOWLEDGMENT

This research was supported by a grant from the Alfred P. Sloan Foundation.

CHAPTER B.6

Sequential Experimentation and Sequential Assembly of Designs

Sir Ronald Fisher once said that "the best time to design an experiment is after you've done it." One manifestation of this seeming paradox is that after a preliminary experimental design has been run, questions are often raised about the results with an acuity of hindsight which is quite extraordinary. Questions like:

"That factor doesn't seem to be doing anything. Wouldn't it have been better if you had included this other variable?"

"You don't seem to have varied that factor over a wide enough range."

"The experiments with high pressure and high temperature seem to give the best results; it's a pity you didn't experiment with these factors at even higher levels."

And so on and so on.

Such questions come up because the results from any experiment depend critically on decisions requiring *judgment*: Which factors should be studied? How should the response be measured? At which levels of a given factor should experiments be run? Which experimental design should be used? Human judgment is fallible, and these matters need to be reconsidered as you proceed and learn more about what is going on. If it can be avoided, therefore, it is best not to plan a large "all-encompassing" experiment at the outset. This is the time when you know *least* about the system.

Also before the first set of experimental runs is planned, you need to be sure that you are properly prepared. Some of the preliminary issues are whether the operation

From Box, G. E. P. (1993), *Quality Engineering*, 5(2), 321–330. Copyright © 1993 by Marcel Dekker, Inc.

of the apparatus, pilot plant, or process on which you plan to make the experimental runs is reasonably *stable*; whether you are going to be able to *make* the changes required by the design; and whether your present methods of *measurement, analysis,* and *sampling* are satisfactory. If such considerations are attended to initially, you are less likely to be embarrassed by the need to backtrack and fix problems half-way through the experimental program.

When experimental runs can be made sequentially a rough general rule is that only a fraction, say, 25% of the experimental effort and budget should be invested in a first design. After this first design has been completed, you usually will have learned a great deal and will be able to plan a better second part and so on. Thus you follow a mode of investigation very like that of a detective solving a mystery (see Bisgaard, 1989). Figure 1 shows some of the different ways in which a second stage of investigation might go depending on what is shown by data from the first stage. The illustration is for just three variables but the strategy can be used for any number of variables. In this example, the initial design at the center of the diagram is a

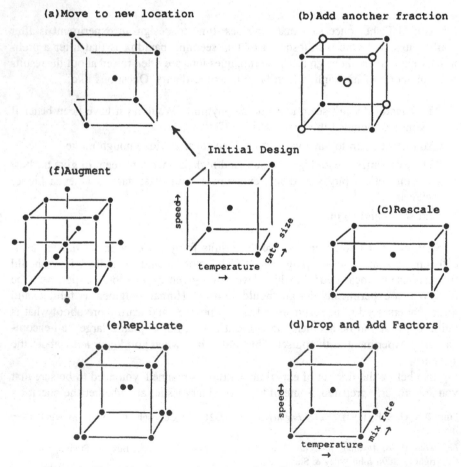

Figure 1. Alternatives for a second set of runs depending on results from the initial set.

fractional factorial. Sometimes the results from this initial design solve the problem and no further experimentation is necessary, but when this is not the case the diagram shows how, depending on the results obtained from this first set of runs, in the second set, reading clockwise, it might be appropriate:

(a) To move to a new location—because the results indicate a desirable trend in that particular direction.

(b) To stay at the present location and add further points—in particular, to resolve ambiguities found from the analysis of a first fractional design by adding a further fraction.

(c) To rescale the design—when, for example, it appears that certain variables have not been varied over wide enough ranges.

(d) To drop and add factors—in the situation illustrated, the experimenter decided that further information was needed about the effects of speed and temperature but the *gate size* should be dropped as a variable and a new variable *mix rate* substituted.

(e) To repeat some of the runs—because they were believed to be wrongly carried out, or because the experimenter did not have a good enough handle on the size of the replication error. In the latter case, depending on resources, it might be decided (i) to make a few additional runs at a single point, for example, the center point, (ii) to replicate a fraction of the design, as is illustrated in Figure 1e, or (iii) to replicate all the points.

(f) To augment the design with "star points"—getting information about curvature. To do this requires a larger number of runs. However, instead of reverting to three-level designs, the experimenter may augment an initial two-level design with a star arrangement producing what is called a *composite* design (Box and Wilson, 1951). A particular situation pointing strongly to the need to do this is when large two-factor interactions occur among the variables. Such interactions measure the change in the effect of one variable as another variable is changed and so are "second-order" effects. Large two-factor interactions point to the likely presence of curvature effects as well, for these are also second-order effects measuring the change in the effect of one variable as a change in level is made *in that same variable*. In particular, large second-order effects of both kinds occur in the neighborhood of a maximum which can be further explored, using the completed composite design (see, e.g., Chapter C.5 and Box and Draper, 1987).

SEQUENTIAL ASSEMBLY

The possibilities illustrated in Figure 1 are all examples of what may be called *sequential assembly* of designs. Table 1 illustrates in more detail the sequential assembly of fractional factorial designs using an example from BHH II. This was concerned with an experiment done on a chemical plant in which a production bottleneck occurred because of an unusually lengthy filtration cycle of up to 80

Table 1. A 2_{III}^{7-4} Industrial Experiment to Test the Effect of Seven Process Factors on Filtration Time

			−	+
(a) Factors	1 *water* source		town reservoir	well
	2 raw *material*		on site	other
	3 *temperature*		low	high
	4 *recycle*		yes	no
	5 caustic *soda*		fast	slow
	6 filter *cloth*		new	old
	7 *holdup* time		low	high

	Test	1	2	3	4	5	6	7	Filtration Time (min) *y*
(b) Design 1	1	−	−	−	+	+	+	−	68.4
	2	+	−	−	−	−	+	+	77.7
	3	−	+	−	−	+	−	+	66.4
	4	+	+	−	+	−	−	−	81.0
	5	−	−	+	+	−	−	+	78.6
	6	+	−	+	−	+	−	−	41.2
	7	−	+	+	−	−	+	−	68.7
	8	+	+	+	+	+	+	+	38.7

(c) Analysis of design 1	Water	−10.9	→ 1 + 24 + 35* + 67
	Material	−2.8	→ 2 + 14 + 36 + 57
	Temp.	−16.6	→ 3 + 15* + 26 + 47
	Recycle	3.2	→ 4 + 12 + 37 + 56
	Soda	−22.8	→ 5 + 13* + 27 + 46
	Cloth	−3.4	→ 6 + 17 + 23 + 45
	Holdup	0.5	→ 7 + 16 + 25 + 34

minutes when previous experience on similar plants had shown that a cycle of about 40 minutes should have been adequate. A list of the factors which were studied is shown in Table 1a. The first design is shown in Table 1b with corresponding filtration times. This was an eight-run 2_{III}^{7-4} arrangement. If you wish to be reminded how this kind of design is arrived at, look again at Chapter B.3. Five different rationales for using fractional designs which imply different strategies for analysis and for further experimentation were discussed in Chapter B.4. We can use the filtration example to illustrate some of these.

PICK THE WINNER

A simple "pick the winner" strategy might be used if all you wanted to do was to *fix* the problem without necessarily understanding the reason for what went wrong or *why* what you did worked. Notice, for example, that mere eyeballing of the data in Table 1b shows that two of the runs did in fact produce the desired shorter filtration times. These two tested conditions were as follows:

	1	2	3	4	5	6	7	Filtration Time
Test 6	+	−	+	−	+	−	−	41.2
Test 8	+	+	+	+	+	+	+	38.7

Now the (plus) signs in columns 1, 3, and 5 are the same for both runs whereas for columns 2, 4, 6, and 7 the signs are opposite. We might tentatively conclude, therefore, that we can get shorter filtration times simply by using the plus levels of factors 1, 3, and 5. That is, use well water, high temperature, and slow addition of caustic soda with whichever levels of the remaining variables, 2, 4, 6, and 7, are most convenient. Indeed, in this example, a switch to such manufacturing conditions would have given the low filtration rate needed. The problem would thus have been "solved." But more precisely it would have been *fixed*. If a quick fix was all that was needed, after one or two confirmatory checks, experimentation could have ended at this point.

However, there would have remained a great deal of ambiguity about the reason *why* we got these shorter times. In particular, as we see from inspection of Table 1c, although the effect that was found could have been due to the joint main effects of water supply, temperature, and rate of additional caustic soda (1, 3, and 5), an equally plausible explanation would be that only two of these factors *and their interaction* were responsible. For example, the data could equally well be explained by effects 1, 3, and 13: that is, the water source, the temperature and the water-× temperature interaction, possibly with *no contribution* from rate of caustic soda addition (factor 5). Equally, the effects could be explained by 1, 5, and 15 with no contribution from 3, or by 3, 5, and 35 with no contribution from 1. These ambiguities arise, for example, because the interaction 13 is an alias of the main effect 5, the interaction 15 is an alias of the main effect 3, and so on. (These critical aliases are indicated by asterisks in Table 1c.) Thus there are at least four plausible explanations of the results from Design 1, and if you wanted to better *understand* the process then you would need to get additional data. (However, note that, although this resolution III design has not *on its own* completely sorted out this fairly complicated situation, it was, nevertheless, very useful in greatly narrowing the possibilities.)

UPGRADING THE RESOLUTION III DESIGN TO ONE OF RESOLUTION IV

What was actually done at the second stage to clear up the ambiguities was to add a further eight runs that would convert the original eight-run resolution III design to a 16-run design of resolution IV. This was done by adding the eight additional runs shown as Design 2 in Table 2a. This second design was obtained by switching all the signs in the first design; an operation sometimes referred to as *fold over* (Box and Wilson, 1951). Notice that no factor combination (row of plus and minus signs) which occurs in the second design occurs in the first. The alias structure of this second set of eight runs and the calculated effects are shown in Table 2b. The same combinations of aliased interactions appear in each row but with minus signs instead of plus signs. This can easily be checked by multiplying the appropriate columns of signs together in Table 2a.

Table 2. A Second Fraction, with Analysis, in Which the Signs of Each Column Are Switched. Also Shown Is An Analysis of All 16 Runs

	Test	1	2	3	4	5	6	7	Filtration Time (min) y
	9	+	+	+	−	−	−	+	66.7
	10	−	+	+	+	+	−	−	65.0
	11	+	−	+	+	−	+	−	86.4
(a) Design 2	12	−	−	+	−	+	+	+	61.9
	13	+	+	−	−	+	+	−	47.8
	14	−	+	−	+	−	+	+	59.0
	15	+	−	−	+	+	−	+	42.6
	16	−	−	−	−	−	−	−	67.6

	Water	$-2.5 \rightarrow 1 - 24 - 35 - 67$
	Material	$-4.9 \rightarrow 2 - 14 - 36 - 57$
	Temp.	$-15.8 \rightarrow 3 - 15 - 26 - 47$
(b) Analysis of Design 2 Alone	Recycle	$2.3 \rightarrow 4 - 12 - 37 - 56$
	Soda	$-15.6 \rightarrow 5 - 13 - 27 - 46$
	Cloth	$-3.3 \rightarrow 6 - 17 - 23 - 45$
	Holdup	$-9.2 \rightarrow 7 - 16 - 25 - 34$

	Water	$-6.7 \rightarrow 1$
	Material	$-3.9 \rightarrow 2$
	Temp.	$-0.4 \rightarrow 3$
	Recycle	$2.8 \rightarrow 4$
	Soda	$-19.2 \rightarrow 5$
	Cloth	$-0.1 \rightarrow 6$
(c) Analysis of Design 1 & 2	Holdup	$-4.4 \rightarrow 7$
Together		$0.5 \rightarrow 12 + 37 + 56$
		$-3.6 \rightarrow 13 + 27 + 46$
		$1.1 \rightarrow 14 + 36 + 57$
		$-16.2 \rightarrow 15 + 26 + 47$
		$4.9 \rightarrow 16 + 25 + 34$
		$-3.4 \rightarrow 17 + 23 + 45$
		$-4.2 \rightarrow 24 + 35 + 67$

If we now combine the two sets of estimates we can separate main effects from the interaction strings. Thus, for example, for the two estimates associated with temperature we have

$$3 + 15 + 26 + 47 = -16.6$$
$$3 - 15 - 26 - 47 = 15.8$$

Adding, and dividing the result by two: then subtracting, and dividing the result by two; produces

$$3 = -0.4 \qquad 15 + 26 + 47 = -16.2$$

which makes it clear that the large effect almost certainly comes from the string of aliases and not from the main effect of temperature (3) itself. Proceeding in this way, we get the analysis for the complete set of 16 runs, for Designs 1 and 2 taken together, shown in Table 2c. (Precisely the same result can be obtained by direct analysis of the complete set of 16 runs. This is, in fact, a resolution IV design in seven factors that could be obtained (see Chapter B.4) by writing down a 2_{IV}^{8-4} design and dropping out one of the columns.)

From the analysis of the complete set of 16 runs in Table 2c you can see that the most likely explanation for the data is that the water source (1) and the rate of caustic soda addition (5) are the important factors and they produce an interaction corresponding to the large effect (-16.2) contained in the alias string ($15 + 26 + 47$).

EMPIRICAL OR SCIENTIFIC FEEDBACK?

We have considered two approaches: (i) "picking the winner" and adding one or more extra runs to confirm that the problem really has been fixed and (ii) the addition of eight extra runs to show us what is really happening. These approaches are designed to induce what I have called, respectively, *empirical* and *scientific* feedback. The first is useful to get a "quick fix," the latter is needed to understand what is going on. The former, if it works, usually needs fewer extra runs; but fixing the problem without knowing precisely *how* it was fixed has its disadvantages. In particular, our lack of understanding may result in having to experiment all over again if the problem recurs. Furthermore, it may lead to making changes that are not strictly necessary. In the filtration example, for instance, the empirical approach leads to an unnecessary increase in manufacturing temperature and hence an unnecessary and continuing cost in maintaining that higher temperature. Finally, and perhaps most importantly, better understanding of the system can lead to new ideas not thought of initially.

A CHEAPER COMPROMISE

The experimental error in this example appears to be quite high. Having available a second set of eight runs is attractive therefore because, as well as resolving alias ambiguities by upgrading the design resolution, it provides a worthwhile increase in the accuracy of the estimates of the effects. To better understand why the additional eight runs reduced the uncertainties look at Figure 2a where for factors 1, 3, and 5 the data from Design 1 are plotted at the corners of a cube. You will see that the ambiguities arise from the fact that this first set of runs produces only a duplicated

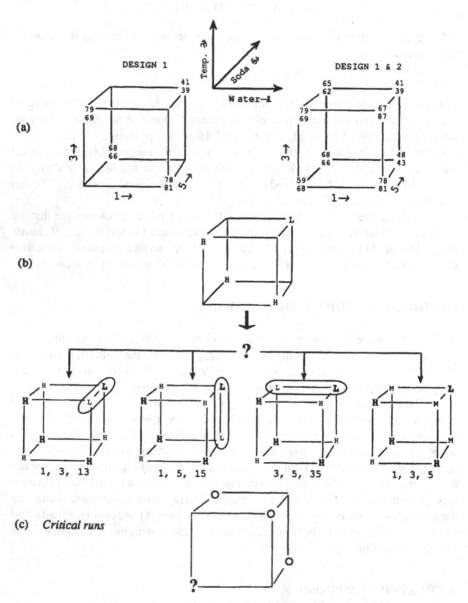

Figure 2. (a) Results from design 1 and designs 1 + 2 shown as cube plots for factors 1, 3, 5; (b) a "what if" analysis for plausible explanations of the data from design 1 only; (c) critical runs determined from (b).

half fraction of a 2^3 design in these factors. The right hand diagram shows how the second set of eight runs resolves the problem by upgrading the design to provide a duplicated and *complete* 2^3 arrangement. The resulting 16-run design is of resolution IV and would, therefore, give a complete 2^3 design in *any* set of three factors.

Now if experimentation was very expensive, a compromise using fewer additional runs might be considered. These could consist of four additional runs providing a half fraction that fills the vacant spaces in Figure 2a. To obtain even greater economy it is often helpful in solving problems in experimental design to do a "what if" analysis, setting out the kind of data we should expect from *runs not yet made* if various alternative conjectures were true. To see how this applies in the present case, look at Figure 2b. Roughly speaking, what the data from the first design is telling us is that three of the factor combinations give high (H) filtration times and one combination gives a low (L) filtration time. Each of the four conjectures mentioned above would be expected to produce a different data pattern for the remaining runs. Two low data values in the top right hand corner of the cube would be consistent with the existence of factors 1, 3, and interaction 13 and so on. The situation on the far right with three runs marked M corresponds to the fact that medium filtration times might be expected if there were just main effects for each of the three variables 1, 3, and 5 producing a general trend across the cube. It is now easily seen how the situation could be resolved, at least in theory, by running just three extra points instead of eight. These would be the "critical runs" at the positions indicated by circles in Figure 2c. An additional run at the point denoted by the question mark would provide a "3/2 fraction" in the three important factors and is somewhat preferable on grounds of simplicity and statistical efficiency. For a more fundamental discussion of the resolution of ambiguities by augmentation of fractional designs see Chapter C.7.

CHAPTER B.7

Must We Randomize Our Experiment?

"My guru says you must always randomize."

"My guru says you don't need to."

"You should never run an experiment unless you are sure the system (the process, the pilot plant, the lab operation) is in a state of control."

"Randomization would make my experiment impossible to run."

"I took a course in statistics and all the theory worked out perfectly without any randomization."

We have all heard statements like these. What are we to conclude?

THE PROBLEM OF RUNNING EXPERIMENTS IN THE REAL WORLD

The statistical design of experiments was invented almost seventy years ago by Sir Ronald Fisher. His objective was to make it possible to run informative and unambiguous experiments not just in the laboratory, but in the real world outside the laboratory. Initially his experimental material was agricultural land and his process the production of crops. Not only did this land vary a great deal as you went from one experimental plot to the next, but the yields from successive plots were related to each other. Suppose that if *no treatments* were applied so that the changes you saw in yield were merely "error," then the yields from successive plots would *not* look like the random stationary disturbance (S) in Figure 1; this is a picture of an ideal process in a state of control with results varying randomly about a fixed mean. Instead they might look like the nonstationary disturbance (N) where a high yielding plot was

From Box, G. E. P. (1990), *Quality Engineering*, 2(4), 497–502. Copyright © 1990 by Marcel Dekker, Inc.

Improving Almost Anything: Ideas and Essays, Revised Edition. By George Box and Friends
Copyright © 2006 John Wiley & Sons, Inc.

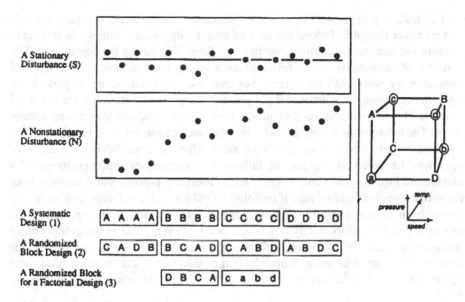

Figure 1. A stationary disturbance (S) and a nonstationary disturbance (N) together with a systematic design (1), a randomized block design (2), and a randomized block design for a 2^3 factorial arrangement (3).

likely to be followed by another high yielding plot and vice versa. So the background disturbance against which the effect of different treatments had to be judged varied in highly nonrandom—unstable—nonstationary ways. Fisher's problem was how to run valid comparative experiments in this situation.

IS YOUR EXPERIMENTAL SYSTEM IN A STATE OF STATISTICAL CONTROL?

The problem that Fisher considered is not confined to agricultural experimentation. To see if you believe in stationarity in your particular kind of work, think of the chance differences you expect in measurements taken m steps apart (in time or in space). Random variation about a fixed mean implies that the variation in such differences remains the same whether the measurements are taken one step apart or one hundred steps apart. If you believe instead that the variation would, on the average, increase as the distance apart became greater, then you don't believe in stationarity.*

---

* Standard courses in mathematical statistics have not helped, since they often *assume* a stationary disturbance. Specifically, they usually suppose that the disturbance consists of errors that vary *independently* about a *fixed* mean. Studies of the effects of discrepancies from such a model have usually concerned such questions as what would happen if the data followed a non-normal (but still stationary) process. However, discrepancies due to non-normality are usually trivial compared with those arising from nonstationarity and lack of independence.

The truth is that we all live in a nonstationary world; a world in which external factors never stay still. Indeed the idea of stationarity—of a stable world in which, without our intervention, things stay put over time—is a purely conceptual one. The concept of stationarity is useful only as a background against which the real nonstationary world can be judged. For example, the manufacture of parts is an operation involving machines and people. But the parts of a machine are not fixed entities. They are wearing out, changing their dimensions, and losing their adjustment. The behavior of the people who run the machines is not fixed either. A single operator forgets things over time and alters what he does. When a number of operators are involved, because of failures to communicate the opportunities for change are further multiplied. Thus, if left to itself any process will drift away from its initial state. It would be nice if uniformity could be achieved, once and for all, by carefully adjusting a machine, giving appropriate instructions to operators, and letting it run, but unfortunately this would rarely, if ever, result in the production of uniform product. Extraordinary efforts are needed in practice to ensure that the system does not drift away from the target value and that it behaves at least approximately like the stationary process (S). These efforts are called Quality Control!

So the first thing we must understand is that stationarity, and hence the uniformity of everything depending on it, is an *unnatural* state that requires a great deal of effort to achieve. That is why good quality control takes so much *sustained* effort. All of this is true, not only for manufacturing processes, but for any operation that we would like to be done consistently, such as the taking of blood pressures in a hospital or the performing of chemical analyses in a laboratory. Having found the best way to do it, we would like it to be done that way consistently, but experience shows that very careful planning, checking, recalibration, and appropriate intervention are needed to ensure that this happens.

PERFORMING EXPERIMENTS IN A NONSTATIONARY ENVIRONMENT

So what did Fisher suggest we do about running experiments in this nonstationary world?

For illustration suppose we had four different methods/treatments/procedures (A, B, C, and D) to compare, and we wanted to test each method four times over. Fisher suggested that rather than run a *systematic* design like Design 1 in Figure 1 in which we first ran all the A's, then all the B's, and so on, we should run a *randomized block* arrangement like Design 2. In this latter arrangement within each "block" of four you would run each of the treatments A, B, C, and D in random order.*

---

* The idea of getting rid of external variation by blocking has other uses. The blocks might refer not to runs made at different periods of time but to runs made on different machines, or to runs made with different operators. In each case randomization within the blocks would validate the experiment, the subsequent analysis, and the conclusions.

Now frequently Design 1 would be a lot easier to run than Design 2, so it is reasonable to ask "What does Design 2 do for me that Design 1 doesn't?" Well, if we could be absolutely sure that the process disturbance was like S—that *throughout the experiment* the process was in a perfect state of control—it would make absolutely no difference which design was used. Designs 1 and 2 would be equally good.

But suppose the process disturbance was nonstationary like that marked N in Figure 1. Then it is easy to see that the systematic Design 1 would give invalid results. For example, even if the different treatments really had no effect, B, C, and D could look very good compared with A, simply because the process disturbance happened to be "up" during the time when the B, C, and D runs were being made. On the other hand, if the experiment were run with the randomized Design 2 it would provide valid comparisons of A, B, C, and D. Also, the standard methods of statistical analysis would be approximately valid even though they were based on the *usual assumptions including stationarity.* That is, the data could be analyzed *as if* the process had been in a state of control. Not only this, but differences in block averages, which could account for a lot of the variation, would be totally eliminated and the relevant experimental error would be only that which occurred between plots in the *same block.*

If the experiment was not simply a comparison of different methods like A, B, C, and D, but a factorial design like the one in the right of the figure in which speed, pressure, and temperature were being tested, you could still use the same idea. As in Design 3, you could put the factorial runs marked A, B, C, and D in the first block and those labeled a, b, c, and d in the second. In both cases you would run in random order within the blocks. It would then again turn out that you could calculate the usual contrasts for all the main effects and all the two-factor interactions free from differences between blocks and that you could carry out a standard analysis as if the process had been in a state of control throughout the experiment. If the details of these ideas are unfamiliar, you can find out more about them in any good book on experimental design.

Randomized arrangements of this kind made it possible for the first time to conduct valid comparative experiments in the real world outside the laboratory. Thus beginning in the 1920s, valid experimentation was possible, not only in the field of agriculture, but also in medicine, biology, forestry, drug testing, opinion surveys, and so forth. Fisher's ideas were quickly introduced into all these areas.

COMPARATIVE EXPERIMENTS

Why do I keep using the term *comparative* experiments? Experimentation in a nonstationary environment cannot claim to tell us what is the precise result of using, say, treatment A. This must depend to some extent on the state of the environment at the particular time of the experiment (e.g., in the agricultural experiment, it would depend on the weather and on the fertility of the particular piece of soil on which treatment A was tested). What it can do is to make valid *comparisons* between A, B,

C, etc. that are likely to hold up in *similar circumstances of experimentation.** This is usually all that is needed.

Beginning in the 1930s and 1940s, these ideas of experimental design also began to be used in industrial experiments both in Britain and in the United States. Other designs were developed in Britain and America in the 1940s and 1950s, such as fractional factorial designs and other orthogonal arrays and also variance component designs and response surface designs to meet new problems. These were used particularly in the chemical industry. In all of this, Fisher's basic ideas of random-ization and blocking retained their importance because most of the time one could not be sure that disturbances were even approximately in a state of statistical control.

More recently, experimental design has been used extensively in the "parts" industries (e.g., automobile manufacture). Some of the ideas were reimported as part of what have been called "Taguchi methods" where scant attention was paid to randomization and blocking. So the question is often raised as to whether randomization is still important in the "parts" industries.

WHAT TO DO

To understand what we should do in practice, let's look at the problem a little more closely. We have seen that we live in a nonstationary world in which nothing stays put whether we are talking about the process of manufacturing or the taking of a blood pressure reading. But all models are approximations, so the real question is whether the stationary approximation is good enough for your particular situation; according to my argument, if it is, you don't need to randomize. Such questions are not easy—for while the degree of nonstationarity in making parts, for example, is usually very different from that occurring in an agricultural field, the precision needed in making a part is correspondingly much greater and the differences we may be looking for much smaller. Also, even in the automobile industry not all problems are about making parts. Many experiments (e.g., those arising in the paint shop) are conducted in circumstances less easy to control.

The crucial factor in the effective use of any statistical method is good judgment. All I can do here is to help you use it, and in particular to make clear what it is you must use your good judgment about.

Alright then. You do not need to randomize if you believe your system is, to a sufficiently good approximation, in a state of control and can be relied on to *stay* in that state *while you make experimental changes* that presumably you have never made before. Sufficiently good approximation to a state of control means that over the period of experimentation differences due to a slight degree of nonstationarity will be small compared with differences due to the treatments. In making such a judgment, bear in mind that belief is not the same as wishful thinking.

---

* Note that the question of what *are* similar circumstances of experimentation must invariably rest on the technical *opinion* of the person using the results. It is *not* a statistical question, as has been pointed out by Deming (1950, 1975) in his discussion of enumerative and analytic studies.

My advice then would be:

1. In those cases where randomization only slightly complicates the experiment, always randomize.
2. In those cases where randomization would make the experiment impossible or extremely difficult to do but you can make an honest judgment about approximate stationarity, run the experiment anyway without randomization.
3. If you believe the process is so unstable that without randomization the results would be useless and misleading, and that randomization would make the experiment impossible or extremely difficult to do, then do not run the experiment. Work first on stabilizing the process or get the information some other way.
4. A compromise design that sometimes helps to overcome some of these difficulties is the split plot arrangement discussed in Chapter E.4.

CHAPTER B.8

Botched Experiments, Missing Observations, and Bad Values

I'm often asked by engineers and other experimenters what to do if as sometimes happens an experiment design is imperfectly run. Thus you might have intended to run a 2^2 factorial in factors A and B so that the coded factor levels were those in Fig. 1a, but because of mistakes or lack of control what was actually run were those in Fig. 1b.

$$
(a) \quad
\begin{bmatrix}
A & B \\
-1 & -1 \\
1 & -1 \\
-1 & 1 \\
1 & 1
\end{bmatrix}
\quad (b) \quad
\begin{bmatrix}
x_1 & x_2 \\
-1.3 & -0.9 \\
0.8 & -0.7 \\
-1.0 & 1.3 \\
1.7 & 1.2
\end{bmatrix}
$$

Figure 1. (a) A 2^2 factorial design with levels coded as ± 1 (b) A botched design with the same coding.

Such problems can be solved using regression analysis: this is the method of Least Squares. For example, by fitting the regression equation to the data y from either design

$$\hat{y} = b_0 + b_1 x_1 + b_2 x_2 + b_{12} x_1 x_2$$

you could obtain estimate of the coefficients b_1, b_2 and b_{12}. Then multiplying each of these coefficients by 2* will produce estimates of the main effects and the interaction A, B, and AB.

---

*You need to multiply by 2 because the *coefficient* measure the predicted change in the response if as you change a factor level by one unit. But an effect is the predicted change in response as the factor level changed from -1.

From Box, G. E. P. (1990–91), *Quality Engineering*, 3(2), 249–254. Copyright © 1990 by Marcel Dekker, Inc.

Improving Almost Anything: Ideas and Essays, Revised Edition. By George Box and Friends
Copyright © 2006 John Wiley & Sons, Inc.

Table 1. The Plus and Minus Signs Used to Calculate the Effects from a 2^4 Factorial Design

	−	+
A. catalyst charge (lb)	10	15
B. temperature (°C)	220	240
C. pressure (psi)	50	80
D. concentration (%)	10	12

Run #	A	B	C	D	AB	AC	AD	BC	BD	CD	ABC	ABD	ACD	BCD	ABCD	Conversion (%)
1	−	−	−	−	+	+	+	+	+	+	−	−	−	−	+	71
2	+	−	−	−	−	−	−	+	+	+	+	+	+	−	−	61
3	−	+	−	−	−	+	+	−	−	+	+	+	−	+	−	90
4	+	+	−	−	+	−	−	−	−	+	−	−	+	+	+	82
5	−	−	+	−	+	−	+	−	+	−	+	−	+	+	−	68
6	+	−	+	−	−	+	−	−	+	−	−	+	−	+	+	61
7	−	+	+	−	−	−	+	+	−	−	−	+	+	−	+	87 (w)
8	+	+	+	−	+	+	−	+	−	−	+	−	−	−	−	80
9	−	−	−	+	+	+	−	+	−	−	−	+	+	+	−	61
10	+	−	−	+	−	−	+	+	−	−	+	−	−	+	+	50
11	−	+	−	+	−	+	−	−	+	−	+	−	+	−	+	89
12	+	+	−	+	+	−	+	−	+	−	−	+	−	−	−	83
13	−	−	+	+	+	−	−	−	−	+	+	+	−	−	+	59 (x)
14	+	−	+	+	−	+	+	−	−	+	−	−	+	−	−	51
15	−	+	+	+	−	−	−	+	+	+	−	−	−	+	−	85
16	+	+	+	+	+	+	+	+	+	+	+	+	+	+	+	78
	8	8	8	8	8	8	8	8	8	8	8	8	8	8	8	←divisor

In particular you can use the regression method if there are one or more missing observation. Alternatively a simple method to deal with data where there are one or two missing observation is illustrated by the following example. Table 1 shows the plus and minus signs for the analysis of a 2^4 design from the BHH II book. The runs were actually made in random order but, for convenience, they are listed here in regular standard order. The columns A, B, C, D display the various factor combinations run in the experiment and the last column shows the *response* "% conversion." The other columns can be used to calculate the fifteen *effects* of the various factors and their interactions AB, AC, ..., ABCD. These are shown in Table 2 in the column marked "(a) Full Design." They could be obtained from the columns of signs in the usual way by adding the results corresponding to the plus signs and subtracting the results corresponding to the minus signs and then dividing by 8.

As we explained one technique is to fit an appropriate model to the data by least squares (regression). A simpler procedure due to Yates (1933), which gives the same estimated effects, is to plug in a "fitted value" for the missing observation and carry out the analysis as before. An easy way to do this for factorial-type designs is described by Draper and Stoneman (1964). It goes like this.

Consider a 16-run experiment like that in Table 1, but suppose there is a missing observation. Since we now do not have a complete set of data we can no longer estimate *all* 15 effects. So if we were using the least-squares method we might take some effect coefficient which we believed would be negligible and omit it from the model. This turns out to be equivalent to substituting a fitted value obtained by *setting* the supposedly negligible effect equal to zero. It sounds more complicated than it is, so let's try it.

Suppose in Table 1 that observation number (13) is missing. Now, in fact, the response actually observed for this run was 59, but let's assume we don't know this. Call the response from this run "x"—the unknown quantity. In the absence of any evidence to the contrary, we might expect that the highest order interaction ABCD is the effect most likely to be negligible. On this assumption we can estimate the missing value x by *setting* this ABCD effect equal to zero. Using the ABCD column of signs in Table 1 and writing x for the unknown value of the thirteenth observation we get

$$71 - 61 - 90 + 82 - 68 + 61 + 87 - 80 - 61 + 50$$
$$+ 89 - 83 + x - 51 - 85 + 78 = 0$$

So that adding the numbers together

$$-61 + x = 0 \quad \text{and} \quad x = 61$$

An analysis with this "fitted" value substituted for observation (13) is shown in column (b) of Table 2. For this set of data, where the ABCD interaction estimated for the complete data was small to begin with, this has little effect on the estimates.

The same idea can be applied with two observations missing. For illustration, suppose that in addition to the observation missing from run (13), which we called x, the observation from run (7) is also missing. Let's call this one w. It will be possible to obtain fitted values for both these missing observations if we can assume that two *suitably chosen* effects are negligible.

What I mean by "suitably chosen" is best explained by an example. Suppose, as before, we select ABCD as our first "null" column. If you look at the plus and minus signs for this column you will notice that in rows (7) and (13), where there are missing data, the *same sign* (+) occurs. Therefore, we must arrange that, for our second null column, the signs in rows (7) and (13) are *different*. Thus we could choose ABC or ACD but not ABD or BCD as the second null column. For illustration let's choose ABC. Then

$$\text{ABCD} = 0 \quad \text{gives} \quad w + x = 148$$
$$\text{ABC} = 0 \quad \text{gives} \quad w - x = 22$$

Table 2. Estimates of Effects for the 2^4 Factorial Designs: (a) from Full Design, (b) with Observation (13) Omitted and ABCD Assumed to Be Zero, and (c) with Observations (13) and (7) Omitted and Both ABCD and ABC Assumed to Be Zero

	(a) Full Design	(b) With (13) Omitted	(c) With (7) and (13) Omitted
Average	72.25	72.375	72.375
→A	−8.00	−8.25	−8.25
→B	24.00	23.75	23.25
C	−2.25	−2.00	−2.00
→D	−5.50	−5.25	−4.75
AB	1.00	1.25	1.75
AC	0.75	0.50	0.50
AD	0.00	−0.25	−0.75
BC	−1.25	−1.50	−2.00
→BD	4.50	4.25	4.25
CD	−0.25	0.00	0.50
ABC	−0.75	−0.50	Assumed zero
ABD	0.50	0.75	0.75
ACD	−0.25	−0.50	−1.00
BCD	−0.75	−1.00	−1.00
ABCD	−0.25	Assumed zero	Assumed zero

You can see now why we need the signs to switch for the missing rows in the two "null" columns. This is to ensure that the two equations we end up with will have a solution. In this example solving the equations by first adding them together and then subtracting one from the other we get

$$w = 85 \qquad x = 63$$

The result of plugging in these fitted values in Table 1 is shown in column (c) of Table 2. The various effects do not differ materially from those calculated originally. Again, this is because, from the full set of data, both the ABCD and the ABC interactions have small effects.

This simple computational device can also be used for the analysis of *fractional factorials* and other orthogonal arrays when observations are missing. The method will give the same estimates as a least-squares analysis in which the effects we have put equal to zero are *omitted* from the model. It has the advantage that it may be easier to do. Notice that the result of this procedure depends somewhat on which columns we choose to set equal to zero. This is because the least-squares solution depends on which effects we choose to drop out of the model.

Usually the best way to *analyze* designs of this kind, whether you have missing observations or not, is to make a Daniel plot using normal probability paper. In such a plot, points falling off the straight line are likely to represent real effects not easily explained as merely due to noise (experimental error). But remember that, for example, if you have two missing observations from a 16-run design you have really *estimated* only 13 effects (you *set* the other two equal to zero). These zero values should not be plotted therefore. The remainder may be arranged in order of size and plotted in the usual way at % probability values given by $P_i = 100(i - 1/2)/m$ with $m = 13$ (see e.g., BHH II). This will provide an adequate approximate analysis so long as only a few observations are missing. In particular if you do this for the data in Table 1 you will find that the same effects A, B, D, and BD, indicated in Table 2 by arrows, show up as distinguishable from noise with or without missing observations.

Remember, as for any other technique, that the results are only as good as the assumptions. Thus, the "null columns" ought to be chosen not only so as to make the equations solvable, but so that the effects (or with fractional designs the strings of aliased effects) correspond to quantities that you really think are likely to be small. Remember also that what we have here is simply a convenient computational device. It does not of course recover the information that has been lost. For example, in Chapter B.1 I referred to an eight-run experiment on ball bearings by Hellstrand. The discovery that the life of the bearings could be increased fivefold rested entirely on the two experimental results in which the two factors—heat treatment and outer ring osculation—were increased *together*. If these two particular results had been missing, no statistical analysis of any kind could have recovered this vital information. So don't rely on your results if you have too many missing observations. Usually, I would start to feel uncomfortable with the analysis when there was more than one missing observation in an 8-run experiment, or more than two observations missing from a 16-run experiment.

So much for the technique. Now let's talk about some, perhaps more important, philosophical issues concerning missing observations. When we have missing observations it is always best if possible *to get them repeated* (usually with some other runs repeated for comparison, in case something has slipped). But often the most important question about a missing observation is "Why is it missing?" Because questions such as this about the conduct of the experiment almost always come up, it is very important for the experiments to keep a notebook with a detailed record of what happened in each run. Perhaps the missing value has occurred *simply* because of a failure to record, or, of much more concern, because the machine or process could *not be run* at this particular set of conditions. In either case it is important to follow up on such possibilities immediately. The fact that the process cannot be run at particular conditions is, in itself, important information. Suppose, for example, that the other runs suggest that the set of conditions that "cannot be run" might be especially favorable. In that case we ought to ask whether the problem of making this run is an insuperable one, or whether by some not too difficult modification these conditions could, in fact, be tried.

Sometimes an observation is not actually missing but is, for some reason or another, *suspect*. In that case you may want to use this technique to discover what the effect of dropping that particular observation might be. But if a large discrepancy is found between the observed and the fitted value you should *not* just automatically substitute the fitted value in the analysis. The difference you find may contain important information. The first thing to do is to look at the notes made at the time of performing the suspect run to see if there were unusual circumstances that might explain the result. If so it may be worthwhile to run a repeat which replicates the recorded peculiarities of the original run to see if the result can be duplicated. For example, suppose the problem was to define conditions that gave a *high* value for the response. If the discrepancy indicated an exceptionally high value, this *could* be telling us that something we accidentally did differently was especially desirable. On the other hand, if the discrepancy indicated an unusually low value, this could be telling us that these conditions should be especially avoided. Although most of the time it is the main effects and two-factor interactions that are important in a factorial design, there is no law that says that every phenomenon fits into that pattern. Occasionally, a particular combination of factor levels may give an unusually good or bad result which is not explicable in terms of main effects and low order interactions. For example, the exceptional hardness of certain steel alloys and the necessary conditions for an atomic explosion depend on the unique coming together of certain specific levels of a large number of factors. We must therefore always be ready to learn from repeatable occurrences however odd they may look at first sight.

ACKNOWLEDGMENTS

This research was sponsored by the National Science Foundation under Grant No. DDM-8808138, and by the Vilas Trust of the University of Wisconsin–Madison.

CHAPTER B.9

Graphics for Finding Bad Values in Factorial Designs

Look at the data in Table 1. It shows a 2^4 factorial experiment from Box and Meyer (1987). Shown in Figure 1 are the effects plotted on normal probability paper. As you may know, this plot (which I prefer to call a Daniel plot in honor of its originator) provide for fractional factorial designs a simple, but most valuable, tool for discriminating between effects likely to be due to noise and those effects which are almost certainly real. The former will plot as points on a straight line, the latter will fall off the line. At first sight it looks as if you might draw a rough straight line through all the points, indicating that there were no effects detectably different from noise. But, if for the moment you ignore the extreme points B and C, you can see that a better fit might be obtained by drawing *two* straight lines. Cuthbert Daniel (1976) pointed out that such a plot provides a strong clue that you have a discrepant data value.

You can see why this would be by imagining what would happen with a *good* set of data if you miswrote one of the data values. For illustration let's say for observation number 3 you had written down 53.13 when it should have been 43.13, thus making that value ten units too high. Now remember you can calculate the effects column by column by adding together all the data values opposite plus signs and subtracting those opposite minus signs. The resulting contrasts are divided by 8 to give the effects A, B, C, D, AB, AC, ..., etc. Now if observation number 3 is too high by ten units, you can see that the effect contrast for the main effect A will be ten units too *low* because the column for A has a minus sign in row 3. Similarly the main effect contrast for B will be ten units too *high* because the column for B has a plus sign in row 3, and so on. So after you've divided by 8 the A effect will be discrepant

From Box, G. E. P. (1991), *Quality Engineering.* 3(3), 405–410. Copyright © 1991 by Marcel Dekker, Inc.

Improving Almost Anything: Ideas and Essays, Revised Edition. By George Box and Friends
Copyright © 2006 John Wiley & Sons, Inc.

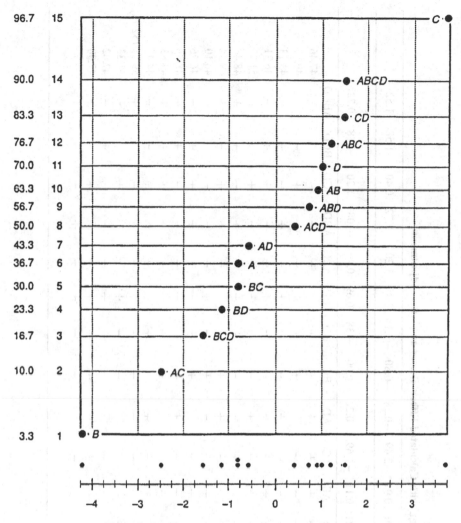

Figure 1. Daniel plot of estimated effects from 2^4 design.

by $-10/8 = -1.25$ (1.25 units too low) and the B effect by $10/8 = +1.25$ (1.25 units too high). If you look at the table of plus and minus signs opposite row 3 you can see that the effect of the discrepant value would be to make 8 of the 15 effects too low by -1.25 units and the remaining 7 too high by $+1.25$ units.

Now think about the effects which are just due to noise (experimental error). If there were no bad values, these "error" effects ought to plot as a straight line. But because of the discrepant value, some of these error effects will be biased upward and the others downward so you could expect the data in the middle of the plot to appear not as one, but as *two* straight lines.

Now think of the problem the other way around. You've made the plot and it looks like two straight lines rather than one, just as in Figure 1. What you want to

Table 1. Results from a 2⁴ Factorial Design with Calculated Effects Before and After Adjustment

	A	B	C	D	AB	AC	AD	BC	BD	CD	ABC	ABD	ACD	BCD	ABCD	
Effects	−0.80	−4.22	3.71	1.01	0.91	−2.49	−0.58	−0.80	−1.18	1.49	1.20	0.72	0.40	−1.58	1.52	
Adjusted effects	0.00	−3.42	2.91	0.21	0.11	−1.69	0.22	0.00	−.38	0.69	0.40	−.08	1.20	−.78	0.72	
1	−	−	−	−	+	+	+	+	+	+	−	−	−	−	+	47.46
2	+	−	−	−	−	−	−	+	+	+	+	+	+	−	−	49.62
3	−	+	−	−	−	+	+	−	−	+	+	+	−	+	−	43.13
4	+	+	−	−	+	−	−	−	−	+	−	−	+	+	+	46.31
5	−	−	+	−	+	−	+	−	+	−	+	−	+	+	−	51.47
6	+	−	+	−	−	+	−	−	+	−	−	+	−	+	+	48.49
7	−	+	+	−	−	−	+	+	−	−	−	+	+	−	+	49.34
8	+	+	+	−	+	+	−	+	−	−	+	−	−	−	−	46.10
9	−	−	−	+	+	+	−	+	−	−	−	+	+	+	−	46.76
10	+	−	−	+	−	−	+	+	−	−	+	−	−	+	+	48.56
11	−	+	−	+	−	+	−	−	+	−	+	−	+	−	+	44.83
12	+	+	−	+	+	−	+	−	+	−	−	+	−	−	−	44.45
13	−	−	+	+	+	−	−	−	−	+	+	+	−	−	+	59.15 (52.75)
14	+	−	+	+	−	+	+	−	−	+	−	−	+	−	−	51.33
15	−	+	+	+	−	−	−	+	+	+	−	−	−	+	−	47.02
16	+	+	+	+	+	+	+	+	+	+	+	+	+	+	+	47.90
Signs of likely "error" effects	−															

96

find out is *which* data value is responsible for the discrepancy. For the likely error effects I have indicated by plus signs at the bottom of Table 1 all the positive effects that plot as the upper straight line in Figure 1 and by minus signs all the negative effects that plot as the lower straight line. Now cast your eyes over the rows of signs corresponding to the various observations and see if you can see a row of pluses and minuses that nearly matches this (or one that matches it if all the signs are reversed, since the discrepancy could be either way). You will see that in row 13 all the signs except one match, so that the discrepant observation appears to be number 13. A more precise way to check the matching is to calculate the cross products of the signs in the rows of the table with the supposed error effects. The one that matches the best will give the largest absolute sum of cross products.

Let us now take this a step further. Assuming that there is a discrepancy in observation number 13, what is its estimated magnitude? The simplest way is to treat this observation as a missing value and estimate it using, for example, the procedure given in Chapter B.8. An alternative that illustrates the "two straight lines" idea is as follows: Look at Figure 1. The plotted effects on the two lines which are closest to zero are most likely the result of the positive and negative biases ($\pm d$, say) plus error. I have, somewhat arbitrarily, taken four of these "error" terms on each size of zero and assumed that the true effects for A, D, AB, BC, AD, BD, ABD, and ACD are probably in reality small or nonexistent.

Then we have 8 estimates for d obtained from

$$-A = 0.80, \quad +D = 1.01, \quad +AB = 0.91, \quad -BC = 0.80,$$
$$-AD = 0.58, \quad -BD = 1.18, \quad +ABD = 0.72, \quad +ACD = 0.40$$

On the assumption that the true values for these effects are zero the least squares estimate for d is obtained by averaging them to give $\bar{d} = 0.80$. Correspondingly, observation 13 is estimated to be $0.80 \times 8 = 6.40$ units too high.

We cheat a bit in getting these estimates, because we look at the data first to decide which effects to call error. A more precise method is given by Box and Meyer, but the results aren't very different. Using the adjusted value $59.15 - 6.40 = 52.75$ for observation 13 we obtain the "adjusted effects" shown above Table 1. A Daniel plot of these are shown in Figure 2. We see after making the adjustment that the main effects for factors B and C and the interaction AC are probably real.

This graphical check should be borne in mind but alternative or better still an addition is to pick a pseudo model and reestimate all the coefficients n times leaving out one observation each time.

In all of this remember what I said in the previous chapter. A discrepant value should not just be thrown away—it might be trying to tell us something. An awful warning is supplied by the "hole" in the ozone layer over the Antarctic discovered by British scientists. After the existence of the hole was confirmed the question was asked: "How come the satellite, that had been continually circling the earth for

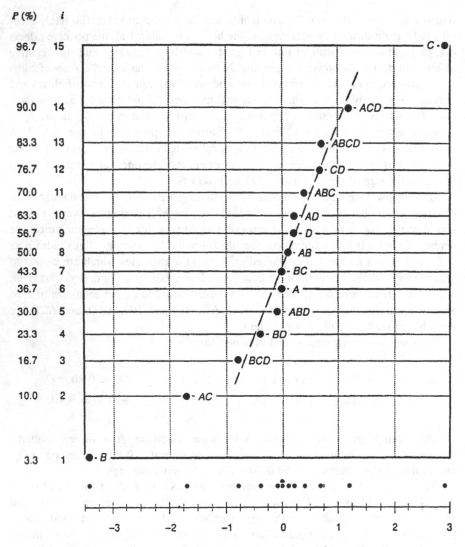

Figure 2. Daniel plot of adjusted effects from 2^4 design.

the past several years, didn't find it?" I was told that these data were automatically checked for outliers by a computer program and the hole in the ozone layer was screened out!

Let that be a lesson to us all.

ACKNOWLEDGMENTS

This research was sponsored by the National Science Foundation under Grant No. DDM-8808138, and by the Vilas Trust of the University of Wisconsin–Madison.

CHAPTER B.10

How to Get Lucky

'Tis not in mortals to command success, but we'll do more...we'll deserve it—Joseph Addison

There is a story that Napoleon once consulted some trusted colleagues concerning the advisability of appointing a particular officer to the highly responsible rank of divisional general. Napoleon's advisors were in favor of the promotion and mentioned the many qualifications of the candidate. Napoleon is said to have pondered their recommendations for some time. Finally, he asked, "Yes, but is he *lucky?*"

Now the question is a very searching one. One dictionary definition of luck is "success due to chance." In this sense, anyone can be lucky once or twice, but a consistently "lucky" person must, almost certainly, be doing a number of things differently, and doing them right. Process improvement like military success contains a significant element of chance, but, by and large, those who deserve to be successful are, and vice versa.

Nothing succeeds like success—Alexander Dumas

As is implied by the story of Napoleon, one way to judge the chance of future success is to look at the past record of a particular person or a particular procedure. But initially there *is* no past record. So this leaves open the question of how to get started. In quality improvement an *immediate* success is highly desirable; for if the first project is a failure, the whole program may be set back. It is important, therefore, to choose a project that is likely to succeed and also to use techniques that maximize this possibility. One useful idea is provided by the next quote.

From Box, G. E. P. (1993), *Quality Engineering*, 5(3), 517–524. Copyright © 1993 by Marcel Dekker, Inc.

Improving Almost Anything: Ideas and Essays, Revised Edition. By George Box and Friends
Copyright © 2006 John Wiley & Sons, Inc.

Where there are three or four machines, one will be substantially better or worse than the others—Ellis Ott

Ellis Ott (1975), a true pioneer in quality improvement, said that he made this remark "with only moderate tongue in cheek." He went on to make it clear that instead of machines, he might have talked about different *heads* on a machine, different *vendors*, different *operators*, or different *shifts*. Once you have established that one entity is different, you can work on trying to find out what makes it so; if it is worse, then you can try to make it as good as the others; if it is better, then you can try to make all the others as good as the best. Sometimes you will find that more than one are good, or more than one are bad, but the idea is essentially the same.

AN EXAMPLE

Figure 1a shows a set of coded measurements designed to seek out the source of unacceptably large variability which, it was suspected, might be due to small differences in five, supposedly identical, *heads* on a machine. To test this idea, the engineer arranged that material from each of the five heads was sampled at roughly equal intervals of time in each of six successive 8-hour *periods*. The order of sampling within each period was random and is indicated by the numerical super- scripts in the table of data. The averages for heads are shown beneath the data table and those for periods to its right. *Note:* If you already know all about the analysis of variance, or if you don't want to be bothered with its anatomy, please skip ahead to the section entitled *You can see a lot by just looking*.

A standard analysis of variance for these data is shown in Figure 1. It will be seen that the appropriate F ratio ($18.4/13.9 = 1.3$) suggested that there were no differences in head means that could not readily be attributed to chance. The investigator thus failed to find what had been expected. However, the same analysis strongly suggested that *real* differences in means occurred between the six 8-hour periods of time during which the experiment was conducted. These periods corre- sponded almost exactly to six successive shifts which I have labeled A, B, C, A', B', C' and the big discrepancies are seen to be the high values obtained in both the night shifts C and C'. In Figure 1b, a further analysis of *periods** into effects associated with *days* (D), *shifts* (S), and $D \times S$ confirms that the differences seem to be associated with *shifts*. This clue was followed up and led to the important discovery that there were substantial differences in operation during the night shift—an unexpected finding because it was believed that great care had been taken to ensure that operation did not change from shift to shift.

---

* The residual ($H \times P$) can correspondingly be split into component sums of squares corresponding to $H \times D$, $H \times S$, and $H \times D \times S$ effects. Such an analysis fails to show anything of further interest and is not presented here.

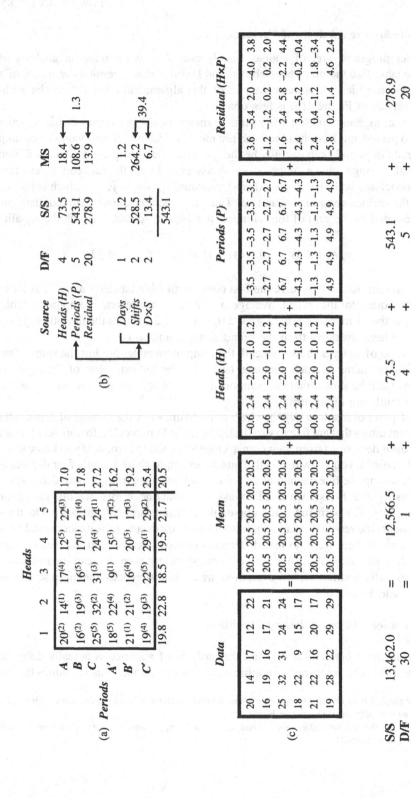

Figure 1. (a) Data from a quality improvement experiment. (b) Corresponding analysis of variance tables. (c) Additive basis for the analysis.

Rationale for the Analysis of Variance

Computer programs are, of course, readily available to calculate an analysis of variance table like the one I have shown, but I would like to remind the reader of a few underlying ideas. It is customary to do this algebraically but I think the arithmetic analysis of Figure 1(b) is less opaque.

Suppose, as seems reasonable in this example, possible changes in means from period to period and from head to head are roughly additive. A data value y_{it} coming from head i in period t can usefully be thought of as containing a contribution $\bar{y}$ from the overall average, plus a deviation $\bar{y}_i - \bar{y}$ associated with the ith head, a deviation $\bar{y}_t - \bar{y}$ associated with the tth period, and a residual $y_{it} - \bar{y}_i - \bar{y}_t + \bar{y}$ which is chosen so that the various components add up. Thus, any such rectangular table of data can be represented as the sum of four component tables in accordance with the equality

$$y_{it} = \bar{y} + (\bar{y}_i - \bar{y}) + (\bar{y}_t - \bar{y}) + (y_{it} - \bar{y}_i - \bar{y}_t + \bar{y})$$

For the present data in Figure 1c the first component table labeled "mean" has all its elements equal to the grand average $\bar{y} = 20.5$. The second component table, containing the elements $(-0.6, 2.4, -2.0, -1.0, 1.2)$, shows the deviations $\bar{y}_i - \bar{y}$ of the five head averages from the grand average, and so on.

The sums of squares shown beneath the component tables are just the sums of the 30 individual items in each separate table and, by an extension of Pythagoras's theorem, it can be shown that the component sums of squares will always add up* to give the total sum of squares.

The degrees of freedom (D/F), which also add up, show the number of items in the component tables that are free to vary independently. In particular, for *any* set of n data values, the n deviations from their own average *must* add to zero; so if you know $n - 1$ of the deviations, you know them all. Thus, for example, there are only four degrees of freedom among the five deviations $\bar{y}_i - \bar{y}$ of head averages from the grand average.

To justify the F-test, we need to assume† that the errors are distributed (a) independently, (b) with the same standard deviation, and (c) normally. On these assumptions, the residual mean square supplies an estimate of the experimental error variance σ^2. If there were no real differences between heads, then the mean square for *heads* would also estimate σ^2 and the corresponding ratio of mean squares would follow an F-distribution. If, however, there were real differences, the mean square for *heads* would be inflated.

You can see a lot by just looking—Yogi Berra

We can get a better idea of what the analysis of variance is actually doing by looking at the data graphically. Graphical analysis can also help us see things that the

---

* To save space, I have rounded these items to one decimal. Because of this rounding error, their sums of squares are not quite equal to those in Figure 1b.

† Alternatively, the test for differences in head means can be approximately justified on randomization theory (see, e.g., BHHII).

analysis of variance does *not* show us. There are a number of graphical techniques for looking at data of this kind. See in particular Ott (1975), BH^2 (1978) and Tukey (1977). In Figure 2, a scaled dot diagram representing the distribution of the head deviations is shown together with a dot diagram of the residuals. The latter then supplies a "reference distribution" against which the distribution for heads may be compared. The scale factor is chosen so that if there were no real differences between heads, the head dot diagram would have the same theoretical variance* as the residual dot diagram. The appropriate effect scale factor is given by the formula

$$\text{Effect scale factor} = \sqrt{\frac{D/F \text{ (for error)}}{D/F \text{ (for effect)}}}$$

Thus, the scale factor for heads is $\sqrt{20/4} = \sqrt{5}$. We see that when plotted in this way the scaled head deviations and the residuals look as if they might readily have come from distributions with the same variance and that the head deviations are indistinguishable from the noise. On the same basis, the scale factor for *periods* is $\sqrt{20/5} = 2$ and the scaled period deviations appear highly discrepant with the residuals.

Also bear in mind that the residual reference distribution can be thought of as "slideable" with respect to the other distributions. From this you can see, for example, that while there is no reason to suspect that the means from shifts *A, B* are different, those from shift *C* are highly discrepant.

Original data should be presented in a way that will preserve the evidence in the original data—Walter Shewhart

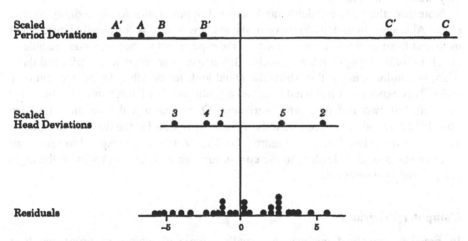

Figure 2. Data from quality improvement experiment: scaled effect deviations with residuals as reference distribution. The scale factors are such that the ratio of dot plot variances are the *F*-ratios, where each variance is obtained by dividing the sum of squares of the deviations from *zero* by the *number* of plotted dots.

*By a *theoretical variance* I mean the sum of squares of the deviation from the *known* mean of zero divided by the *number* of such deviations. The spread of the resulting dot diagrams are then comparable by eye.

Note that the graphical analysis is not a substitute for, but a valuables addition to the ANOVA table. It is capable of showing things that might be missed in the more formal analysis. It can show, for example, whether the plotted residuals look as if they came from a single stable error distribution. If there were residual outliers, this might suggest that, even when possible differences due to shifts and heads were allowed for, the process might be unstable (or perhaps that one or more numbers had been wrongly recorded, for example, 42 might have been written for 24).

Separate dot diagrams for residuals from each of the six periods and from each of the individual heads should also be plotted. For example, although no differences were found in the heads averages, individual plots might suggest that product coming from certain of the heads was more *variable* than that from the others. Also, because the heads were sampled in random order, plotting the residuals in time order could warn us of drifts taking place during a shift. The reader, if she/he wishes, may make these additional analyses and see if she/he can find anything else of interest in these data. Remember, however, that from this kind of data "mining" you should look for *clues* not final conclusions. If you find suspicious patterns that you think might be important, you should check these out in later experimentation.

Serendipity

Perhaps what Napoleon was looking for was not so much luck, as *serendipity*. This quality, named by Horace Walpole, comes from the fairy tale of *The Three Princes of Serendip* who "were always making discoveries, by accident and sagacity, of things they were not in quest of."

Note how the engineer deliberately created opportunities for serendipity to take place. Although the stated objective of the study was to look for possible differences in means from different heads, he set up the experiment so that he would be able to check for *other things* such as possible differences associated with shifts and days. Also, by randomizing within shifts, he could look for possible dependence on time order. This experiment has what is called a "split plot" structure* in which there are not one, but two different error variances. One measures the variance between "heads" or the other between periods. This is the reason for the two part analysis of variance and the two F tests in Figure 1(b). The "periods" correspond to tests done an three shifts A, B, C leading to the conclusion that it is the data taken on the night shift C that is discrepant.

Computer Programs

In practice, statistical analysis is usually done with computer programs. It is important, however, that *you* and not the computer are in charge of deciding what to look for. (In particular, an appropriate variety of *graphical* aides should be available for display and study.)

*Split plot designs and their graphical analysis are discussed in Chapters E.4 and E.6.

A Postscript

As a final illustration of Shewhart's and Berra's points, let us consider some data due to Darwin and used by Fisher to illustrate the paired *t*-test in his classic book

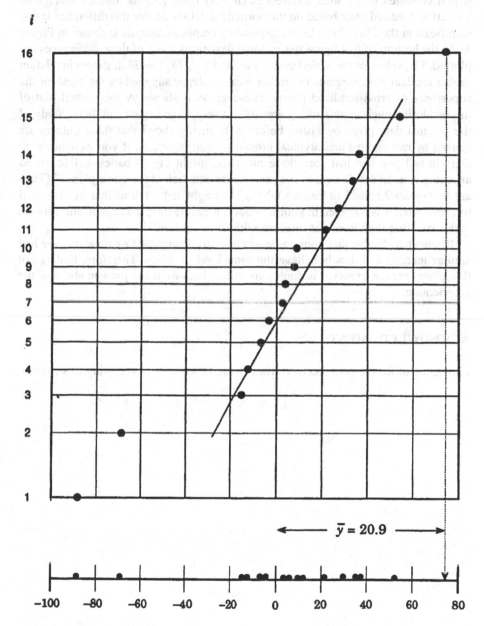

Figure 3. Fisher's paired data: scaled average $\bar{y} = 20.9$ with reference distribution of residuals $y - \bar{y}$ and Daniel plot showing two large outliers.

The Design of Experiments (1935). The object was to determine whether self-fertilized (A) and cross-fertilized (B) plants had different rates of growth. After a suitable growth period, the difference in height between an A plant and a B plant grown in the same pot was measured in units of eighths of an inch. The data (y_i), which consisted of 15 such differences ($B - A$) from 15 pots, had an average of $\bar{y} = 20.9$. A paired t-test based on the normal model shows that this difference is just significant at the 5% point. The corresponding graphical analysis is shown in Figure 3. At the bottom of this figure the residual deviations $y - \bar{y}$ of these differences are plotted. The value for the scaled average of $20.9 \times \sqrt{14/1} = 78.2$, shown in relation to this residual plot, supports the rather weak evidence supplied by the t-test for the superiority of cross-fertilized plants. However, as is shown by the normal plot of the residuals, much more striking are the two very large negative outliers. Study of the original data given by Fisher before differencing shows that these outliers are caused by two stunted and atypical cross-fertilized plants. So if you were advising Darwin and you had just seen these plots you might say, "Charles, before we do anything else, let's walk over to the greenhouse and look at these two plants." (They are from pairs 2 and 15 in Fisher's table.) You might add: "I think that we also need to check what was recorded in your notebook, measure the plants again, and perhaps make sure that they were not infected with some disease."

Remember, finally, that your engineers and your competitor's engineers have had similar training and mostly believe the same kind of things. Therefore, finding out things that are *not* expected is doubly important because it can put you ahead of the competition.

ACKNOWLEDGMENT

I am grateful for the help received from Bruce Ankenman in preparing this chapter.

Dispersion Effects from Fractional Designs

Modern process improvement techniques emphasize the designing of quality into the product and into the process that makes the product. In particular, experimental design is used to discover conditions that minimize variance and appropriately control the mean level. The direct estimation of variance by replication at each of the design points, however, can be excessively expensive in experimental runs. In this article we show how it is sometimes possible to use unreplicated fractional designs to identify factors that affect variance in addition to those that affect the mean.

INTRODUCTION

Modern process improvement techniques use statistical experimental design to study the effects of a number of factors on the variance as well as on the mean. An interesting example of such an investigation was given by Phadke, Kackar, Speeney, and Grieco (1983). If the variance, however, is to be obtained by replication at each experimental point, an excessive number of runs may be needed. In this article we show how it is sometimes possible to identify factors that affect the variance as well as those that affect the mean with *unreplicated* fractional designs. For illustration we reanalyze a highly fractionated two-level factorial design employed as a screening design in an off-line welding experiment performed by the National Railway Corporation of Japan (Taguchi and Wu, 1980); see Table 1. At the right of the table is the observed tensile strength of a weld, one of several quality characteristics

From Box, G. E. P. and Meyer, R. D. (1986), *Technometrics*, **28**(1), 19–27. Reprinted with permission from TECHNOMETRICS. Copyright © 1986 by the American Statistical Association. All rights reserved.

Table 1. A Fractional Two-Level Design Used in a Welding Experiment

Run	I 0	D 1	H 2	−e₁ 3	G 4	−F 5	GH 6	−AC 7	A 8	−E 9	AH 10	e₂ 11	AG 12	J 13	B 14	−C 15	Tensile Strength (kg/mm)
1	+	−	−	+	−	+	+	−	−	+	+	−	+	−	−	+	43.7
2	+	+	−	−	−	−	+	+	−	−	+	+	+	+	−	−	40.2
3	+	−	+	−	−	+	−	+	−	+	−	+	+	−	+	−	42.4
4	+	+	+	+	−	−	−	−	−	−	−	−	+	+	+	+	44.7
5	+	−	−	+	+	−	−	+	−	+	+	−	−	+	+	−	42.4
6	+	+	−	−	+	+	−	−	−	−	+	+	−	−	+	+	45.9
7	+	−	+	−	+	−	+	−	−	+	−	+	−	+	−	+	42.2
8	+	+	+	+	+	+	+	+	−	−	−	−	−	−	−	−	40.6
9	+	−	−	+	−	+	+	−	+	−	−	+	−	+	+	−	42.4
10	+	+	−	−	−	−	+	+	+	+	−	−	−	−	+	+	45.5
11	+	−	+	−	−	+	−	+	+	−	+	−	−	+	−	+	43.6
12	+	+	+	+	−	−	−	−	+	+	+	+	−	−	−	−	40.6
13	+	−	−	+	+	−	−	+	+	−	−	+	+	−	−	+	44.0
14	+	+	−	−	+	+	−	−	+	+	−	−	+	+	−	−	40.2
15	+	−	+	−	+	−	+	−	+	−	+	−	+	−	+	−	42.5
16	+	+	+	+	+	+	+	+	+	+	+	+	+	+	+	+	46.5
Effect	43.0	.13	−.15	.30	.15	.40	−.03	.38	.40	−.05	.43	.13	.13	−.38	2.15	3.10	

Note: A—kind of welding rods; B—period of drying; C—welded material; D—thickness; E—angle; F—opening; G—current; H—welding method; J—preheating. The estimated location effects are plotted as a dot diagram in Figure 1.

measured. For convenience in later discussion the design layout has been recast so that the columns follow standard factorial order. In addition, the preheating variable has been labeled J instead of I.*

Taguchi and Wu (1980) assumed that, in addition to main effects, only the interactions AC, AG, AH, and GH might be present. On that supposition, all nine main effects and the four selected two-factor interactions could be separately estimated by appropriate orthogonal contrasts, and the two remaining contrasts corresponding to the columns labeled e_1 and e_2 would measure only experimental error. The last row of the table shows the grand average and the 15 effect contrasts calculated in the usual manner. In this article these are referred to as "location" effects. A normal probability plot (shown in Figure 1a) has 13 effects roughly following a straight line, with main effects B and C falling markedly off the line. This suggests that, over the ranges studied, factors B and C affect tensile *location*.

On the assumption, then, that B and C are the only important location effects, the 16 runs could be regarded as four replications of a 2^2 factorial design in factors B and C only. When the results are plotted as in Figure 2 to reflect this, however, inspection suggests the existence of a dramatic effect of a different kind apparently not previously noticed. When factor C is at its minus level, corresponding to the use of an alternative material, the *spread* of the data appears much larger than when C is at its plus level. Thus in addition to detecting shifts in location due to B and C, the experiment may also have detected what we will call a *dispersion effect* associated with factor C.

[This, of course, is not the only possible explanation of the data. If, instead of adopting the assumptions of Taguchi and Wu, it had been supposed, for example, that all two-factor interactions might be appreciable, then (see, e.g. BHH II) because of the identity $-BCD = 1 \times 14 \times 15 = I$ in the defining relation of the design, the large location contrasts associated with columns 14 and 15 could have been due to B and C as postulated or alternatively to B and BD or to C and CD. Moreover, the apparent dispersion effect associated with C might be accounted for by supposing that tested factors, other than B and C, affected C only at its minus level. Analysis of the eight runs made at the minus level of C do not support any simple explanation of this kind. Screening designs like this one, however, should normally be employed in a sequential process of investigation where the alternative possibilities that they offer may be resolved in subsequent experimentation (e.g., see the discussion of Tippett's cotton experiment in Fisher, 1966, pp. 88–90). In an ongoing investigation, therefore, such possibilities ought to be considered for further study. We shall pursue the implications of the simplest explanation here while inviting the reader to bear all of the above provisos in mind.]

DISPERSION EFFECTS

Fractional arrangement and other orthogonal arrays have frequently been used in industry as screening designs when it is believed that the effects it is desired to detect

---

* It is common practice to use I for the column of ones associated with estimation of the overall average.

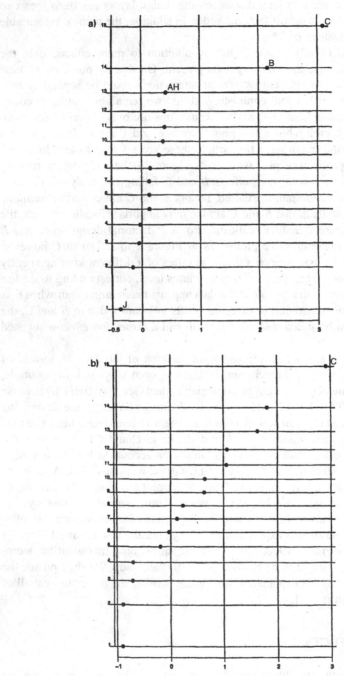

Figure 1. Normal plate for the data from the welding experiment of (a) location effects, (b) dispersion effects.

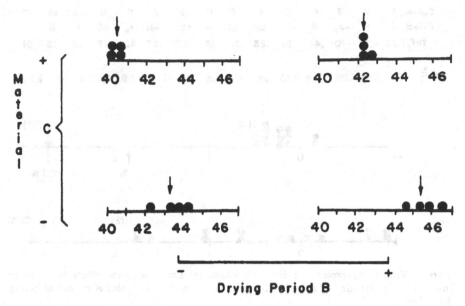

Figure 2. Welding experiment data presented as four replicates of a 2^2 factorial design in factors B and C only. Arrows indicate sample averages.

come from only a small number of the tested factors. This may be called the hypothesis of *effect sparsity* or of a *Pareto effect*. The tensile data suggest the general possibility that the use of unreplicated fractional designs might provide an economical way of detecting sparse dispersion effects as well as sparse location effects. This idea is pursued in the remainder of this chapter. The procedures we discuss are for the model *identification* stage of the problem-solving iteration (see, e.g., Box and Jenkins, 1970), suggesting tentatively which factors might have location effects and which might have dispersion effects. Efficient maximum likelihood estimation for *fitting* an identified model is briefly discussed at the end of the article.

Consider again the design of Table 1. There are 16 runs from which 16 quantities (the average and 15 effect contrasts) have been calculated. If we were interested in possible dispersion effects, we could also calculate 15 variance ratios. For example, for the ith column we could compute the sample variance $s^2(i-)$ from the eight observations associated with a minus sign and compare it with the sample variance $s^2(i+)$ from eight observations associated with a plus sign to provide the ratio $F_i = s^2(i+)/s^2(i-)$. If this is done for the 15 contrast columns of welding data, the values for $ln\ F_i^*$ given in Figure 3(a) are obtained. (Normal theory significance levels are shown to provide a rough indication of size; note, however, that needed assumptions for the usual statistical assumptions are not satisfied.)

Recall that in the earlier analysis a large dispersion effect was associated with factor C (column 15). In Figure 3(a), however, the dispersion effect for this factor is

---

*The log terms formation of the F ratios is used to produce dispersion effects which have greater variance stability.

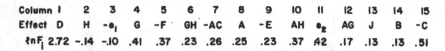

Column	1	2	3	4	5	6	7	8	9	10	11	12	13	14	15
Effect	D	H	$-e_1$	G	$-F$	GH	$-AC$	A	$-E$	AH	e_2	AG	J	B	$-C$
$\ln F_i$	2.72	$-.14$	$-.10$	.41	.37	.23	.26	.25	.23	.37	.42	.17	.13	.13	.51

$$\ln \dot{F}_i \quad -.03 \quad 1.88 \quad .26 \quad 1.09 \quad -1.10 \quad .65 \quad -.83 \quad -.86 \quad .65 \quad -1.14 \quad 1.09 \quad .12 \quad 2.12 \quad -.19 \quad 2.92$$

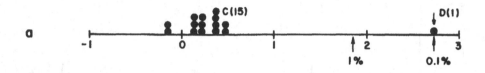

a

b

Figure 3. Welding experiment: log dispersion effects. (a) Crude dispersion effects ln F_i, before elimination of location effects B and C; (b) corrected dispersion effects ln $\dot{F}_i$ after elimination of location effects B and C.

not especially extreme; instead the effect for factor D (column 1) stands out from all the rest. We will see how this may be accounted for by the *aliasing* of location and dispersion effects, which we now consider in a preliminary way.

Since the 16 location effects are obtained by linear transformation of the original 16 data values the calculated dispersion effects must be functions of the location effects. The general nature of the location–dispersion aliasing is explained in the Section that follows. This shows that each dispersion effect is a ratio of sums of squares of sums and differences of the location effects shown at the bottom of Table 1. For immediate illustration Equation (1) shows the identity that exists for the F ratio associated with factor D, and hence for column 1, of the design. In this expression $\hat{i}$ is used to indicate the location contrast associated with the ith column.

$$F_D = F_1 = \frac{\begin{array}{c}(\hat{2}+\hat{3})^2 + (\hat{4}+\hat{5})^2 + (\hat{6}+\hat{7})^2 + (\hat{8}+\hat{9})^2 + (\hat{10}+\hat{11})^2 \\ + (\hat{12}+\hat{13})^2 + (\hat{14}+\hat{15})^2]\end{array}}{\begin{array}{c}[(\hat{2}-\hat{3})^2 + (\hat{4}-\hat{5})^2 + (\hat{6}-\hat{7})^2 + (\hat{8}-\hat{9})^2 + (\hat{10}-\hat{11})^2 \\ + (\hat{12}-\hat{13})^2 + (\hat{14}-\hat{15})^2]\end{array}}$$

$$= \frac{\begin{array}{c}[(-.15-.30)^2 + (-.15+.40)^2 + (-.03+.38)^2 \\ + (.40-.05)^2 + (.43+.13)^2]\end{array}}{\begin{array}{c}[(-.15+.30)^2 + (-.15-.40)^2 + (-.03-.38)^2 \\ + (.40+.05)^2 + (.43-.13)^2]\end{array}} \tag{1}$$

$$+ \frac{(.13-.38)^2 + (2.15+3.10)^2}{(.13+.38)^2 + (2.15-3.10)^2} = \frac{28.45}{1.95} = 14.6$$

This equation shows in particular how the extreme value for F_D can be explained by the location effects $\hat{B} = \hat{14} = 2.15$ and $-\hat{C} = \hat{15} = 3.10$, whose squared sum and squared difference appear, respectively, in its numerator and denominator.

A natural way to try to eliminate such aliasing is to compute variances from the residuals obtained after least squares modeling of large location effects. In what follows we show that after such elimination alias relations such as Equation (1) remain of the same form but with location effects from eliminated variables removed. Dispersion effects $\dot{F}_i$ calculated from residuals after eliminating the location effects of B and C are shown in Figure 3(b). The corresponding Normal plot is given in Figure 1(b). It is seen that an extreme dispersion effect is now associated with C, agreeing with our earlier analysis. Another way to proceed, suggested by Daniel (1976 pp. 135–286) and futher discussed in an unpublished report by Fuller and Bisgaard (1996), is to analyze the squared residuals or, perhaps better, their logarithms. A plot of the dispersion effects calculated in this way is very similar to Figure 1(b).

In the remainder of this chapter we discuss dispersion and location aliasing in greater detail.

DISPERSION AND LOCATION ALIASING

Identities Existing between Dispersion and Location Effects

In order to study the identity relations existing between location and dispersion effects, consider an $n \times n$ orthogonal array with $n = 2^q$ columns of -1's and $+1$'s labeled $x_0, x_1, \ldots, x_{n-1}$. Let $x_0 = I$ be a column of $+1$'s and the remaining columns delineate the usual contrasts for the main effects and interactions of a 2^q factorial design. In general we suppose that the array is to be used either as a 2^{k-p} fractional or as a full factorial, to test k factors, so $q = k - p$ with $p \geq 0$.

To generate the columns of a 2^q orthogonal array, it is convenient to begin by writing a full factorial for q letters employing Yates's standard order row-wise as well as columnwise. We then label columns from zero to $n - 1$ (as illustrated in Tables 1 and 2 for $q = 4$ and $q = 3$, respectively). In practice a design with $n = 8$ would usually be too small to allow variance effects to be usefully studied. We employ it here only to illustrate the argument. As is well known, an array generated in this way may be used as a full factorial or as a fractional design. For example, associating three factors with columns 1, 2, and 4 in Table 2 reproduces the 2^3 factorial; four factors associated with columns 1, 2, 4, and 7 produce a 2_{IV}^{4-1} fractional; and seven factors associated with columns 1–7 produce a 2_{III}^{7-4} fractional. The roman subscript is used to denote the design *resolution*—that is, the length of the shortest word in the defining relation (e.g., see Box and Hunter, 1961a,b).

Now because the columns $x_0, x_1, \ldots, x_{n-1}$ form a group closed under multiplication, defined such that the product column $x_{i \cdot j}$ has for its uth element $x_{i \cdot ju} = x_{iu}x_{ju}$, any such product column must be a column of the original array.

Table 2. The 8 × 8 Orthogonal Array in Yates's Standard Order

0	1	2	3	4	5	6	7
1	A	B	AB	C	AC	BC	ABC
+1	−1	−1	+1	−1	+1	+1	−1
+1	+1	−1	−1	−1	−1	+1	+1
+1	−1	+1	−1	−1	+1	−1	+1
+1	+1	+1	+1	−1	−1	−1	−1
+1	−1	−1	+1	+1	−1	−1	+1
+1	+1	−1	−1	+1	+1	−1	−1
+1	−1	+1	−1	+1	−1	+1	−1
+1	+1	+1	+1	+1	+1	+1	+1

Consider now the elements of a column $\frac{1}{2}(\mathbf{x}_0 \pm \mathbf{x}_i)$ $(i \neq 0)$; these are

$$\frac{1}{2}(x_{0u} - x_{iu}) = +1 \quad \text{if } x_{iu} = -1$$
$$= 0 \quad \text{if } x_{iu} = +1$$
$$\frac{1}{2}(x_{0u} + x_{iu}) = 0 \quad \text{if } x_{iu} = -1$$
$$= +1 \quad \text{if } x_{iu} = +1$$

In addition, the elements of a column $\frac{1}{2}(\mathbf{x}_j \pm \mathbf{x}_{i\cdot j})$ are

$$\frac{1}{2}(x_{ju} - x_{i\cdot ju}) = \frac{1}{2}x_{ju}(x_{0u} - x_{iu}) = x_{ju} \quad \text{if } x_{iu} = -1$$
$$= 0 \quad \text{if } x_{iu} = +1$$
$$\frac{1}{2}(x_{ju} + x_{i\cdot ju}) = \frac{1}{2}x_{ju}(x_{0u} + x_{iu}) = 0 \quad \text{if } x_{iu} = -1$$
$$= x_{ju} \quad \text{if } x_{iu} = +1$$

(3)

To see how this may be used to study location and dispersion aliasing, consider for illustration the 2^3 design. Suppose we wished to compare variances at the lower and upper level of factor $C = \mathbf{x}_4$. Then the columns $\frac{1}{2}(\mathbf{x}_j - \mathbf{x}_{4\cdot j})$ are as presented in Table 3. In general for every i the columns $(\mathbf{x}_j - \mathbf{x}_{i\cdot j})$ will appear in $n/2 = 2^{q-1}$ pairs, identical apart from sign. Now suppose data $\mathbf{y} = (y_1, \ldots, y_u, \ldots, y_n)$ are available, and let $\hat{j} = \mathbf{y}'\mathbf{x}_j$, from which the estimated effect of factor j may be obtained by dividing by an appropriate constant. Then for every i the quantities $\mathbf{y}'(\mathbf{x}_j - \mathbf{x}_{i\cdot j}) = \hat{j} - \widehat{i \cdot j}$ provide an exhaustive set of $n/2$ linearly independent contrasts of those $n/2$ observations y_u for which $x_{iu} = -1$. Correspondingly, the columns $\mathbf{x}_j + \mathbf{x}_{i\cdot j}$ provide a similar set of contrasts for the remaining observations for which $x_{iu} = +1$. Denote by $S(i-)$ and $S(i+)$ the sums of squares of the y_u for which $x_{iu} = -1$ and $+1$, respectively. Then

$$S(i-) = \frac{1}{n}\sum_{j=0}^{n-1}\left[\frac{1}{2}y'(\mathbf{x}_j - \mathbf{x}_{i \cdot j})\right]^2$$

$$= \frac{1}{n}\sum_{j=0}^{n-1}\left(\frac{\hat{j} - \widehat{i \cdot j}}{2}\right)^2$$

$$S(i+) = \frac{1}{n}\sum_{j=0}^{n-1}\left[\frac{1}{2}y'(\mathbf{x}_j + \mathbf{x}_{i \cdot j})\right]^2 \tag{4}$$

$$= \frac{1}{n}\sum_{j=0}^{n-1}\left(\frac{\hat{j} + \widehat{i \cdot j}}{2}\right)^2$$

For example, returning for illustration to the 8×8 array,

$$S(4-) = -y_1^2 + y_2^2 + y_3^2 + y_4^2$$

$$= \frac{1}{8}\left[\left(\frac{\hat{0} - \hat{4}}{2}\right)^2 + \left(\frac{\hat{1} - \hat{5}}{2}\right)^2 + \left(\frac{\hat{2} - \hat{6}}{2}\right)^2 + \left(\frac{\hat{3} - \hat{7}}{2}\right)^2\right.$$

$$= + \left(\frac{\hat{4} - \hat{0}}{2}\right)^2 + \left(\frac{\hat{5} - \hat{1}}{2}\right)^2 + \left(\frac{\hat{6} - \hat{2}}{2}\right)^2 + \left(\frac{\hat{7} - \hat{3}}{2}\right)^2\right]$$

$$= \frac{1}{4}\left[\left(\frac{\hat{0} - \hat{4}}{2}\right)^2 + \left(\frac{\hat{1} - \hat{5}}{2}\right)^2 + \left(\frac{\hat{2} - \hat{6}}{2}\right)^2 + \left(\frac{\hat{3} - \hat{7}}{2}\right)^2\right]$$

$$= \frac{1}{4}\left[\left(\frac{\hat{I} - \hat{C}}{2}\right)^2 + \left(\frac{\hat{A} - \widehat{AC}}{2}\right)^2 + \left(\frac{\hat{B} - \widehat{BC}}{2}\right)^2 + \left(\frac{\widehat{AB} - \widehat{ABC}}{2}\right)^2\right] \tag{5}$$

Elimination of Location Effects

The sums of squares in (4) and (5) would be appropriate to compute dispersion effects only if it could be assumed that all of the location effects, including the overall mean, were known to be zero. If this were not the case, then the sums of squares $S(i-)$ and $S(i+)$ could be inflated by location effects. To attempt to remove such effects, we can replace the y_u's in (5) by residuals $y_u - \hat{y}_u$ obtained after eliminating all suspected location effects, including the mean, by least squares. Since for this mode of estimation the vector of residuals is orthogonal to each column vector corresponding to an eliminated variable, it follows that the identity relation

Table 3. Eight Columns Obtained From the Expression $\frac{1}{2}(x_j - x_{4\cdot j})$ Over $j = 0, 1, \ldots, 7$

				j			
0	4	1	5	2	6	3	7
+1	−1	−1	+1	−1	+1	+1	−1
+1	−1	+1	−1	−1	+1	−1	+1
+1	−1	−1	+1	+1	−1	−1	+1
+1	−1	+1	−1	+1	−1	+1	−1
0	0	0	0	0	0	0	0
0	0	0	0	0	0	0	0
0	0	0	0	0	0	0	0
0	0	0	0	0	0	0	0

for a sum of squares calculated from such residuals is still expressed by Equation (4) but with all estimated contrasts corresponding to eliminated variables set equal to zero.

EXPECTED VALUES OF SUMS OF SQUARES OF RESIDUALS

Further understanding is gained by considering the *expected values* of $S(i-)$ and $S(i+)$ under various circumstances. Suppose a difference in variance might exist associated with the level of the single column x_i and the sums of squares $S(i-)$ and $S(i+)$ are computed from (4) but with y_u replaced by residuals after a number of location effects have been eliminated. Then, after setting to zero all of the elements $\hat{j}$ and $\widehat{i \cdot j}$ in (5) that correspond to eliminated variables, suppose there are l cases in which bracketed pairs $[\hat{j}, \widehat{i \cdot j}]$ have been eliminated and m cases in which only one element of a bracketed pair has been eliminated so that there remains $n/2 - l - m$ complete bracketed pairs.

For a bracketed pair, assuming all real location effects have been eliminated,

$$E\left[\frac{2}{n}\left\{\frac{1}{2}(\hat{j} - \widehat{i \cdot j})\right\}^2\right] = \sigma^2(i-) \tag{6}$$

and for a single element,

$$E\left[\frac{2}{n}\left\{\frac{1}{2}\hat{j}\right\}^2\right] = \frac{1}{4}[\sigma^2(i-) + \sigma^2(i+)]. \tag{7}$$

It follows that

$$E[S(i-)] = (\tfrac{1}{2}n - l - \tfrac{3}{4}m)\sigma^2(i-) + \tfrac{1}{4}m\sigma^2(i+) \tag{8}$$

$$E[S(i+)] = (\tfrac{1}{2}n - l - \tfrac{3}{4}m)\sigma^2(i+) + \tfrac{1}{4}m\sigma^2(i-) \tag{9}$$

If we now define

$$s^2(i-) = S(i-)/(\tfrac{1}{2}n - l - \tfrac{1}{2}m) \tag{10}$$

then

$$E[s^2(i-)] = \sigma^2(i-) + \frac{m}{2n - 4l - 2m}[\sigma^2(i+) - \sigma^2(i-)] \tag{11}$$

and similarly for $s^2(i+)$ with the roles of $\sigma^2(i-)$ and $\sigma^2(i+)$ reversed.

SOME ILLUSTRATIONS WITH THE 8 × 8 ARRAY

The general situation may be better understood by considering a few special cases, again using for illustration the 8 × 8 factorial array. Setting $i = 4 = C$, suppose we wish to obtain the dispersion effect $s^2(4-)/s^2(4+)$, which contrasts the variances of the first four and last four observations.

Elimination of Grand Mean

Elimination of the mean (which would usually be unknown) results in the removal of $\hat{0}$ in Equations (4). For the 8 × 8 array, $n = 8$, $l = 0$, $m = 1$,

$$s^2(4-) = \frac{\left\{\tfrac{1}{16}[(\hat{4})^2 + (\hat{1} - \hat{5})^2 + (\hat{2} - \hat{6})^2 + (\hat{3} - \hat{7})^2]\right\}}{(7/2)}$$

and using (11),

$$E[s^2(4-)] = \sigma^2(4-) + \tfrac{1}{14}[\sigma^2(4+) - \sigma^2(4-)]$$

It will be seen that the slight bias in the variance estimate arises because the isolated effect $\hat{4}$ is a function of all eight observations.

Elimination of the Mean and of Effect $\hat{4}$

If the location effect associated with factor 4 is eliminated as well as the overall mean, then a complete pair is removed in (5) and, in this example,

$$s^2(4-) = \tfrac{1}{16}\{(\hat{1} - \hat{5})^2 + (\hat{2} - \hat{6})^2 + (\hat{3} - \hat{7})^2\}/3$$

$$E[s^2(4-)] = \sigma^2(4-)$$

No bias now occurs because elimination of 0 *and* 4 is equivalent to eliminating means separately from the first four and the last four observations, and $s^2(4-)$ becomes a function of only the first four observations. Similar effects are found with all bracketed pairs. Thus if we eliminate factor 2 *and* the interaction $2 \cdot 4 = 6$, the bias term does not appear because allowance is being made for different location effects of factor 2 at the two levels of factor 4.

In the circumstances of effect sparsity considered here, the bias term in (11) involving $\sigma^2(i+) - \sigma^2(i-)$ would usually be rather small. For example, suppose, with a design having $n = 16$ runs, that $l = 2$ and $m = 1$; then the bias term will be $\{\sigma^2(i+) - \sigma^2(i-)\}/22$. It seems reasonable to conclude that for purposes of model identification, the elimination of location effects by simply taking residuals is unlikely to mislead.

If desired, however, appropriate linear combinations of Equations (8) and (9) will yield unbiased estimates $\dot{s}^2(i-)$ and $\dot{s}^2(i+)$ as follows:

$$\dot{s}^2(i-) = s^2(i-) + \frac{m}{2n - 4l - 4m}[s^2(i-) - s^2(i+)]$$

$$\dot{s}^2(i+) = s^2(i+) + \frac{m}{2n - 4l - 4m}[s^2(i+) - s^2(i-)]$$

Illustrating with the welding example, after elimination of the grand mean only,

$$S(15-) = \frac{2}{16}\left[2\left(\frac{-\widehat{15}}{2}\right)^2 + \left(\frac{\hat{1} - \widehat{14}}{2}\right)^2 + \left(\frac{\hat{2} - \widehat{13}}{2}\right)^2 + \left(\frac{\hat{3} - \widehat{12}}{2}\right)^2\right.$$

$$\left. + \left(\frac{\hat{4} - \widehat{11}}{2}\right)^2 + \left(\frac{\hat{5} - \widehat{10}}{2}\right)^2 + \left(\frac{\hat{6} - \hat{9}}{2}\right)^2 + \left(\frac{\hat{7} - \hat{8}}{2}\right)^2\right]$$

$$= \frac{1}{8}\left[\left(\frac{-3.1}{2}\right)^2 + \left(\frac{.13 - 2.15}{2}\right)^2 + \left(\frac{-.15 + .38}{2}\right)^2\right.$$

$$+ \left(\frac{-.30 - .13}{2}\right)^2 + \left(\frac{-.15 - .13}{2}\right)^2 + \left(\frac{.40 - .43}{2}\right)^2$$

$$\left. + \left(\frac{.03 + .05}{2}\right)^2 + \left(\frac{.38 - .40}{2}\right)^2\right]$$

$$= \frac{3.5}{8} = .438$$

$$s^2(15-) = \frac{1}{7.5}S(15-) = .0584$$

and

$$S(15+) = .513, \qquad s^2(15+) = .0684$$

After elimination of location effects due to B and C (columns 14 and 15),

$$S(15-) = \frac{2}{16}\left[\left(\frac{\hat{1}}{2}\right)^2 + \left(\frac{\hat{2}-\widehat{13}}{2}\right)^2 + \left(\frac{\hat{3}-\widehat{12}}{2}\right)^2 + \left(\frac{\hat{4}-\widehat{11}}{2}\right)^2\right.$$

$$\left. + \left(\frac{\hat{5}-\widehat{10}}{2}\right)^2 + \left(\frac{\hat{6}-\hat{9}}{2}\right)^2 + \left(\frac{\hat{7}-\hat{8}}{2}\right)^2\right]$$

$$= \frac{1}{8}\left[\left(\frac{.13}{2}\right)^2 + \left(\frac{-.15+.38}{2}\right)^2 + \left(\frac{-.30-.13}{2}\right)^2\right.$$

$$\left. + \left(\frac{-.15-.13}{2}\right)^2 + \left(\frac{.40-.43}{2}\right)^2 + \left(\frac{-.03+.05}{2}\right)^2 + \left(\frac{.38-.40}{2}\right)^2\right]$$

$$= \frac{.0837}{8} = .0105$$

$$s^2(15-) = \frac{1}{6.5}S(15-) = .00161$$

and

$$S(15+) = .0510, \qquad s^2(15+) = .00784$$

Finally, forming unbiased estimates of $\sigma^2(15-)$ and $\sigma^2(15+)$, assuming B and C are the only real location effects,

$$\hat{s}^2(15-) = .00161 + \tfrac{1}{24}(.00161 - .00784) = .00135$$

$$\hat{s}^2(15+) = .00784 + \tfrac{1}{24}(.00784 - .00161) = .00810$$

DISPERSION INTERACTIONS

Since more than one dispersion effect might be present, we need to consider the possibility of interaction. If the effect of changing from the minus level to the plus level of a factor i is to multiply the variance by ϕ_i irrespective of whether the plus or minus level of factor j is employed, we shall say that there is no dispersion inter-

+	$\sigma^2(i-, j+)=\phi_j\sigma^2$		$\sigma^2(i+, j+)=\phi_i\phi_j\sigma^2$
−	$\sigma^2(i-, j-)=\sigma^2$		$\sigma^2(i+, j-)=\phi_i\phi_j\sigma^2$

j is at left. Bottom axis: $-$, i , $+$

Figure 4. Two-way table illustrating the case of no dispersion interaction between factors i and j.

action between i and j. In such a case the variances for the various factor combinations are as presented in Figure 4. Equivalently, for the logged variances the dispersion effects will be additive, and in this metric dispersion interactions of all orders may be defined in the usual way. Note that when there is no dispersion interaction, the ratio of the average variance at the plus and minus levels for factor i is

$$\frac{\sigma^2(i+,j-) + \sigma^2(i+,j+)}{\sigma^2(i-,j-) + \sigma^2(i-,j+)} = \frac{\phi_i(1 + \phi_j)\sigma^2}{(1 + \phi_j)\sigma^2} = \phi_i$$

and similarly for factor j and ϕ_j. Thus even when there is more than one dispersion effect, the simple analysis just described could still be of value as a preliminary analytical device for indicating which factors needed further study. In particular, if two factors i and j appeared to exhibit dispersion effects, then further analysis would be appropriate to consider the general evidence for activity of these effects, also taking account of possible interaction. This could be done by considering general differences among the sums of squares associated with the four cells $S(i-,j-)$, $S(i-,j+)$, $S(i+,j-)$, $S(i+,j+)$ of the two-way table for the two factors. As before, these sums of squares would be calculated from residuals after eliminating location effects. The consequences of doing this is explored in the Appendix, which gives a matrix generalization of earlier results.

A convenient function for comparing a set of variances $s_1^2, \ldots, s_k^2$ having $v_1, \ldots, v_k$ degrees of freedom, respectively, is Bartlett's criterion,

$$M = N \ln\left(N^{-1} \sum_{t=1}^{k} v_t s_t^2\right) - \sum_{t=1}^{k} v_t \ln s_t^2$$

where $N = \sum_{t=1}^{k} v_t$. When, as would frequently be the case, the screening design is of only moderate size, one could not expect to study simultaneously a large number of factors in this way. For example, for $n = 16$, the individual cells from which $S(i-,j-)$, $S(i-,j+)$, and so forth, would be calculated will each contain only four observations. In circumstances of effect sparsity, however, when very few such effects are likely to be of appreciable magnitude, the preceding analysis could be of value.

We again illustrate with the welding data. Figure 5(a) shows the 35 distinct values of M computed for the data. (As a rough guide the normal-theory 5% and 1% significance levels of M are shown, although as before they are not formally justifiable.) There are $\binom{15}{2} = 105$ ways of choosing 2 columns from the 15 columns of the

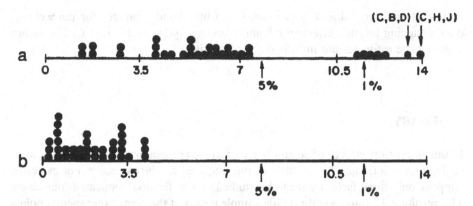

Figure 5. Values of M for distinct column triplets (a) before and (b) after elimination of the possible dispersion effect due to factor C.

design, but these are aliased in sets of three (any column is the product of two other columns). That is, if the data are divided into four groups according to the different levels defined by two columns i and j, this same partition can be achieved by grouping according to columns i and $i \cdot j$ or according to j and $i \cdot j$. Thus the largest value is associated with columns $15 = C$, $2 = H$, and $13 = J$. This effect could equally well be attributed to factors C and H with interaction $-CH = J$, or to C and J with interaction $-CJ = H$, or to H and J with interaction $-HJ = C$. It is noteworthy, however, that the seven largest values of M that stand out from the rest all include factor C in their triplets. Moreover, if the dispersion effect of C is eliminated by rescaling the residuals (residuals at the $+$ level of C are multiplied by $\sqrt{F_c} = 3.9$), the plot of Fig. 5(b) no longer shows outstanding points.

MAXIMUM LIKELIHOOD ESTIMATES OF LOCATION AND DISPERSION EFFECTS

Once a model has been identified, a more precise fitting is possible using maximum likelihood. Results of Hartley and Jayatillake (1973) show that the following method will give convergence to a stationary point of the likelihood. Conditional on the dispersion effects, location effects are obtained by weighted least squares; the dispersion effects are then recomputed from the residuals and the iteration continued until convergence is achieved. It is often convenient to assume initially that there are no dispersion effects.

Table 4. Estimates of Location and Dispersion Effects: Welding Data

Estimates	$\hat{\mu}$	$\hat{B}$	$-\hat{C}$	$\hat{\sigma}^2(C-)$	$\hat{\sigma}^2(C+)$	$\dfrac{\hat{\sigma}^2(C-)}{\hat{\sigma}^2(C+)}$
Maximum likelihood	42.96	2.04	3.10	.469	.021	22.3
Earlier approximate	43.00	2.15	3.10	.564	.031	18.2

For illustration, Table 4 shows maximum likelihood estimates for the welding data, assuming location effects for B and C and a dispersion effect for C. The earlier approximate estimates are indicated for comparison.

SUMMARY

If the dispersion effects of a number of factors are examined by straightforward replication of a design, the number of runs required can be excessive. For example, suppose only three factors were to be studied in a 2^3 factorial replicated four times. The resulting 32 runs would provide sample means at the eight experimental points and, in addition, sample variances, each based on three degrees of freedom from which dispersion effects and location effects could be calculated. At least at the preliminary screening stage of an investigation, however, it is likely that we should wish to test more than three candidate factors, and on the hypothesis of factor sparsity a fractional design might be employed. For example, the 32 runs of the replicated 2^3 factorial might instead be used to screen as many as 16 candidate factors in a 2^{16-11}_{IV} design. This resolution IV design would ensure that a complete replicated factorial design was available in any three (or fewer) of the factors identified as "location active." After eliminating these location effects, by calculating estimates and taking residuals, the design could be reexamined to detect three or fewer "dispersion active" effects. If a clear-cut analysis was not obtained, this first fraction might be used as a building block to resolve suspected ambiguities. We will not attempt details at this point, but the "column fold-over" approach (Box and Wilson, 1951; BH^2), motivated by study of the alias pattern, is one route to be followed.

It is believed that at the preliminary screening stage the strategy exemplified could provide a very economical means for identifying a few location-active and dispersion-active factors that then might be studied more carefully by more formal analysis and follow-up experimentation.

In their unpublished report Fuller and Bisgaard (1996) refer to earlier work on this problem by Daniel (1976, pp. 135–286), who analyzed the absolute value of squared residuals to test for dispersion effects and also list more recent work on the subject.

ACKNOWLEDGMENTS

This article is one in a series ("Studies in Quality Improvement") that describes research on this topic at the University of Wisconsin. It was sponsored by U.S. Army Contract DAAG29-80-C-0041, National Science Foundation Grant DMS-8420968, and the Vilas Trust of the University of Wisconsin. We were aided by access to the research computer at the University of Wisconsin's Statistics Department.

APPENDIX

The results of Section 3 can be generalized to two variables using matrix algebra. Again, $\mathbf{X}$ is a matrix of ± 1's with orthogonal columns $\mathbf{x}_0, \ldots, \mathbf{x}_{n-1}$, $\mathbf{y}$ is the vector of observations, and $\hat{j} = \mathbf{x}_j'\mathbf{y}$. Define $\mathbf{I}(*)$ to be a $n \times n$ diagonal matrix with 1's in those rows for which the condition $*$ is true, 0's elsewhere. Then, for example, $\mathbf{I}(i-)$ is diagonal with 1's in rows where $\mathbf{x}_i = -1$; $\mathbf{I}(i-,j-)$ is diagonal with 1's in rows where $\mathbf{x}_i = \mathbf{x}_j = -1$; and $\mathbf{I}$ is the identity matrix.

The identities

$$S(i\pm) = [\mathbf{I}(i\pm)\mathbf{y}]'[\mathbf{I}(i\pm)\mathbf{y}]$$

$$\mathbf{I} = (1/n)\mathbf{X}\mathbf{X}'$$

$$\mathbf{x}_i \cdot \mathbf{x} = [\mathbf{I} - 2\mathbf{I}(i-)]\mathbf{X}$$

$$\mathbf{I} = \mathbf{I}(i-) + \mathbf{I}(i+)$$

imply, after some algebra, the previously shown identity,

$$S(i-) = \frac{1}{4n}\sum_{j=0}^{n-1}(\hat{j} - \widehat{i \cdot j})^2$$

Extending to two variables, the additional identity

$$S(i\pm,j\pm) = [\mathbf{I}(i\pm,j\pm)\mathbf{y}]'[\mathbf{I}(i\pm,j\pm)\mathbf{y}]$$

implies

$$S(i-,j-) = \frac{1}{16n}\sum_{k=0}^{n-1}(\hat{k} - \widehat{i \cdot k} - \widehat{j \cdot k} + \widehat{i \cdot j \cdot k})^2$$

$$S(i-,j+) = \frac{1}{16n}\sum_{k=0}^{n-1}(\hat{k} - \widehat{i \cdot k} + \widehat{j \cdot k} - \widehat{i \cdot j \cdot k})^2$$

$$S(i+,j-) = \frac{1}{16n}\sum_{k=0}^{n-1}(\hat{k} + \widehat{i \cdot k} - \widehat{j \cdot k} - \widehat{i \cdot j \cdot k})^2$$

$$S(i+,j+) = \frac{1}{16n}\sum_{k=0}^{n-1}(\hat{k} + \widehat{i \cdot k} + \widehat{j \cdot k} + \widehat{i \cdot j \cdot k})^2$$

To compute expectations of these quantities, suppose that $\mathbf{Z}$ is the $n \times p$ matrix of columns of $\mathbf{X}$ that correspond to location effects included in the model (i.e., eliminated to obtain residuals). Then, assuming $E[\mathbf{y}] = \mathbf{Z}\tau$, and using the identity

$$E[\mathbf{y}'\mathbf{A}\mathbf{y}] = \text{tr}(\mathbf{A}E[\mathbf{y}\mathbf{y}'])$$

for A symmetric,

$$E[S(i-,j-)] = \frac{n-2p}{4}\sigma^2(i-,j-)$$

$$+ \frac{1}{n^2}\sum \text{tr}[\mathbf{Z}\mathbf{Z}'\mathbf{I}(i-,j-)\mathbf{Z}\mathbf{Z}'\mathbf{I}(i\pm,j\pm)]\sigma^2(i\pm,j\pm)$$

where summation is over the four possible combinations of $i\pm$, $j\pm$. To compute the trace in the preceding expression, we divide the columns of $\mathbf{Z}$ into four groups:

Group 4—columns $\mathbf{z}_k$ such that $\mathbf{x}_i \cdot \mathbf{z}_k$, $\mathbf{x}_j \cdot \mathbf{z}_k$ and $\mathbf{x}_i \cdot \mathbf{x}_j \cdot \mathbf{z}_k$ are also in $\mathbf{Z}$

Group 3—columns $\mathbf{z}_k$ such that exactly two of $\mathbf{x}_i \cdot \mathbf{z}_k$, $\mathbf{x}_j \cdot \mathbf{z}_k$, $\mathbf{x}_i \cdot \mathbf{x}_j \cdot \mathbf{z}_k$ are also in $\mathbf{Z}$

Group 2—columns $\mathbf{z}_k$ such that only one of $\mathbf{x}_i \cdot \mathbf{z}_k$, $\mathbf{x}_j \cdot \mathbf{z}_k$, $\mathbf{x}_i \cdot \mathbf{z}_k$ is also in $\mathbf{Z}$

Group 1—columns not in the previous three groups

Let m_k = number of columns in group k; note that m_k is a multiple of k. Further subdivide Group 2 into three subsets:

Group 2.1—pairs $\mathbf{z}_g$, $\mathbf{z}_k$ with $\mathbf{z}_g = \mathbf{x}_i \cdot \mathbf{z}_k$

Group 2.2—pairs $\mathbf{z}_g$, $\mathbf{z}_k$ with $\mathbf{z}_g = \mathbf{x}_j \cdot \mathbf{z}_k$

Group 2.12—pairs $\mathbf{z}_g$, $\mathbf{z}_k$ with $\mathbf{z}_g = \mathbf{x}_i \cdot \mathbf{x}_j \cdot \mathbf{z}_k$

Let $m_{2.k}$ be the number of columns in group $2.k$; $m_2 = m_{2.1} + m_{2.2} + m_{2.12}$. Then after some algebra,

$$E[S(i-,j-)]$$

$$= \sigma^2(i-,j-)\left[\frac{4n - 7p + 3m_4 + 2m_3 + m_2}{16}\right]$$

$$+ \sigma^2(i-,j+)\left[\frac{p - m_4 - \frac{2}{3}m_3 - m_2 + 2m_{2.1}}{16}\right]$$

$$+ \sigma^2(i+,j-)\left[\frac{p - m_4 - \frac{2}{3}m_3 - m_2 + 2m_{2.2}}{16}\right]$$

$$+ \sigma^2(i+,j+)\left[\frac{p - m_4 - \frac{2}{3}m_3 - m_2 + 2m_{2.12}}{16}\right]$$

Note that $S(i-,j-)$ will be unbiased (up to a scale factor) if all columns of $\mathbf{Z}$ are in Group 4 ($p = m_4$, $m_3 = m_2 = m_1 = 0$); that is, for each variable $\mathbf{x}_k$ eliminated, variables $\mathbf{x}_{i \cdot k}$, $\mathbf{x}_{j \cdot k}$, $\mathbf{x}_{i \cdot j \cdot k}$ are also eliminated. Similar expressions for expectations of $S(i-, j+)$, $S(i+, j-)$, and $S(i+,j+)$ can be easily obtained from the preceding formula by switching signs on $i\pm$, $j\pm$. In particular, the expression for $E[S(i-)]$ derived in Section 4 follows immediately with

$$\sigma^2(i-) = \tfrac{1}{2}[\sigma^2(i-,j-) + \sigma^2(i-,j+)]$$

$$\sigma^2(i+) = \tfrac{1}{2}[\sigma^2(i+,j-) + \sigma^2(i+,j+)]$$

Thus the calculation of estimated dispersion effects, defined as the change in the average variance, is not affected by the existence of more than one real dispersion effect.

CHAPTER B.12

The Importance of Practice in the Development of Statistics

The article shows how application and consideration of the scientific context in which statistics is used can initiate important advances such as least squares, ratio estimators, correlation, contingency tables, studentization, experimental design, the analysis of variance, randomization, fractional replication, variance component analysis, bioassay, limits for a ratio, quality control, sampling inspection, nonparametric tests, transformation theory, ARIMA time series models, sequential tests, cumulative sum charts, data analysis plotting techniques, and a resolution of the Bayes–frequentist controversy. It appears that advances of this kind are frequently made because practical context reveals a novel formulation that eliminates an unnecessarily limiting framework.

INTRODUCTION

The importance of practice in guiding the development of statistics hardly needs emphasis, and yet I think it is worth examination. For statistical methods and statistical theory, like so many other things, evolve by a process of natural selection. Least squares, invented at the beginning of the 19th century, is alive and well, but the coefficient of colligation is now seldom used. For development to occur both appropriate tools and motivation are needed. The tools are mathematics, numerical analysis, and computation. An important motivation is the practical need to solve problems once they are identified. Tools and motivation interact, of course. For example, the existence of fast computers has led to the development of new

From Box, G. E. P. (1984), *Technometrics*, **26**(1), 1–8. Reprinted with permission from TECHNO-METRICS. Copyright © 1984 by the American Statistical Association. All right reserved.

statistical methods that would be impossible without them and that presage further theoretical development. Furthermore, advance must sometimes wait for knowledge of appropriate mathematics. Thus Fisher's ability to solve the distributional problems of correlation and of the linear model rested strongly on his facility with n-dimensional geometry, which his contemporaries lacked.

It would be hard to argue, however, that deficiency in the tool-kit is disastrous. Thus least squares—although according to Gauss fully known to him in 1796—could require calculations that were dauntingly burdensome until the onset of modern computers in the 1950s. Galton, Gosset, and Wilcoxon, pioneers respectively in the concepts of correlation, studentization, and nonparametric tests, did not regard themselves as particularly competent mathematicians. In particular, Gosset's derivation of the sampling distribution of what we now call the t statistic must surely stand as the nadir of rigorous argument. But he did get the right answer: and he was first.

My theme is to illustrate how practical need often leads to theoretical development. Early examples are the development of the probability calculus, which was closely bound up with the desirability of winning at games of chance, the introduction of least squares by Gauss to reconcile astronomical and survey triangulation measurements, and the invention by Laplace of ratio estimators to determine the population of France (Cochran, 1978).

Let us consider some of the children of necessity produced in more modern times.

FURTHER EXAMPLES OF THE PRACTICE–THEORY INTERACTION

In the mid-19th century the impact of Darwin's ideas was dramatic. But Darwin, although an intellectual giant, had little mathematical ability. To Francis Galton the challenge was obvious: The rightness and further consequences of Darwin's ideas must be demonstrable using numbers. For example, given that offspring varied about some kind of parental mean, why, with each new branching of a family tree, did variation of species not continually increase? The answer to this practical question lay, he discerned, in the regression toward the mean implied by the bivariate normal surface, which ensures, for example, that, on average, sons of six-foot fathers are less than six feet tall (Galton, 1886). Again, it was the need he perceived to measure the intensity of the partial similarities between pairs of relatives that led to his introducing the concept of correlation, an idea taken up with great enthusiasm and further developed by Karl Pearson.

Pearson was a man of enormous energy and wide interests, including social reform and the general improvement of the human condition. He was, however, conscious of the fact that, in deciding what kind of reforms ought to be sought, good intentions, although necessary, were not sufficient. A course of action based on the accepted belief that alcoholism in parents produced mental deficiency in children would be ill advised if, as he demonstrated, that belief was not supported by data (Haldane, 1970). Obviously correlation might be useful in such studies, but other measures and tests of association were needed for qualitative variables. Pearson developed such tools, in particular his χ^2 test for contingency tables.

Karl Pearson's methods were developed mainly for large samples and did not meet the practical needs of W. S. Gosset when he came to study statistics for a year at University College, London, in 1906. Gosset had graduated from Oxford with a degree in chemistry and had gone to work for Guinness's, following the company's policy, begun in 1893, of recruiting scientists as brewers. He soon found himself faced with analyzing small sets of observations coming from the laboratory, field trials, and the experimental brewery of which he was placed in charge.

The general problem Gosset faced was how to deal with unknown nuisance parameters, and specifically the unknown standard deviation in the comparison of means. The method then in use was to substitute some sort of estimate for an unknown nuisance parameter,[†] and then to assume that one could treat the result as if the true value had been substituted. While this might provide an adequate approximation for large samples, it was clearly inadequate when the sample was small. Furthermore, he did not find or expect to find much interest in his problems. He later wrote to R. A. Fisher of the *t* tables, "You are probably the only man who will ever use them" (J.F. Box, 1978). It must have been clear to him in the early 1900's that if anyone was to do anything about small samples it would have to be himself.

Gosset's invention of the *t* test was a milestone in the development of statistics because it showed how, by "studentization," account might be taken of the uncertainty in an estimated nuisance parameter.[*] It thus paved the way for an enormous expansion of the usefulness of statistics, which could now begin to provide answers for agriculture, chemistry, biology, and many other subjects in which small rather than large samples were the rule.

Fisher, as he always acknowledged, owed a great debt to Gosset, both for providing the initial clue as to how the general problem of small samples might be approached, and for mooting the idea of statistically designed experiments.

When Fisher went to Rothamsted Agricultural Experimental Station in 1919 he was one of several young scientists newly recruited by Russell, the new director. He was immediately confronted with a massive set of data on rainfall recorded every day, and of harvested yields every year, for 13 Broadbalk plots that had been fertilized in the same pattern for over 60 years. As might be expected his analyses were not routine (Fisher, 1921, 1924); he introduced distributed lag models, orthogonal polynomials, an early form of the analysis of variance, and the distribution of the multiple cor-relation coefficient. Also, to check the fit of his model he considered the properties of residuals. Furthermore, he devised ingenious methods for lightening the burden-some calculations that then had to be made on a desk calculator. But the most im-portant outcome of this "raking over the muck-heap," as he called it—of analyzing field experiments that he had had no part in planning—came from the very deficiencies these data presented. The outcome was the invention of experimental design.

How, he was soon led to ask, could experiments be conducted so that they might unequivocally answer the questions posed by the investigator? One can clearly see

* Gosset published under the pseudonym Student, consequently his technique was later termed student-ization.

† By "nuisance parameters" is meant parameters necessary for formulating the problem but not of direct interest e.g. the standard deviation in testing the value of a mean.

the ideas of randomization, replication, orthogonal arrangement, blocking, factorial designs, measurement of interactions, and confounding all developing in response to the practical necessities of field experimentation (Fisher, 1926).

Design and analysis came to play complementary roles in Fisher's thinking, so that during the period 1916–1930 we see the analysis of variance first hinted at, and then developed and adapted to accompany the analysis of each new design. In 1923, the analysis of variance first appeared in the tabular form with which we are all familiar (Fisher and Mackenzie, 1923). But the object of the investigation was to solve an agricultural problem, and it is typical of Fisher that there is no reference in the title of the article either to the analysis of variance or to the other new statistical ideas it contains. However the article, called "The Manurial Response of Different Potato Varieties," introduces us not only to the analysis of variance for a replicated two-way table, but also to its partial justification by randomization theory rather than by normal theory. In addition, it presents methods of analysis (rediscovered by Wold) (1966) using models that are *nonlinear* in the parameters.

By the 1930s there existed at Rothamsted a center where careful statistical planning was going into the process of the generation and analysis of data coming from a host of important problems. Fisher left Rothamsted in 1933 and was succeeded by Yates, who had come two years earlier and was a mathematician with much practical experience in least squares calculations in geodetic survey work. Yates (1970) made many important advances. In particular he further developed factorial designs and confounding, invented new designs including balanced incomplete block arrangements, and showed how to cope when, as sometimes happened, things went wrong and there were missing data.

These ideas found wide application and inspired much new research. For example, Jack Youden (1937), then working at the Boyce Thomson Institute, was involved in an investigation of the infective power of crystalline preparations of the tobacco-mosaic virus. Some of the practical difficulties he found were that test plants varied in their tendency to become infected, leaves from the same plant varied depending on their position. Furthermore each plant could not be relied upon to provide more than five experimental leaves. In response Youden invented what came to be called the Youden Square, a design that stands in the same relationship to the latin square as the balanced incomplete block does to the randomized block design.

Another important development coming from Rothamsted was fractional replication. Fisher (1935) had pointed out that, in suitable circumstances, adequate estimates of error could be obtained in large unreplicated factorials from estimates of high-order interactions that might be assumed to be negligible. Finney (1945), responding to the frequent practical need to maximize the number of factors studied per experimental run, further exploited this possible redundancy by introducing fractional factorial designs. These designs, together with another broad class of orthogonal designs developed independently by Plackett and Burman (1946) in response to war-time problems, have since proved of great value in industrial experimentation. An isolated example of how such a design could be used for screening out a source of trouble in a spinning machine had been described as early as 1934 by L. H. C. Tippett of the British Cotton Industry Research Association (Tippett, 1935). The arrangement was a 125th fraction of a 5^5 design!

It seems that whenever a good source of problems existed in the presence of a suitably agile mind new developments were bound to occur. Thus the pressing problem of drug standardization in the hands of Gaddum (1933), Bliss (1935), and (again) Finney (1952) gave rise to modern methods of bioassay using probits, logits, and the like. And in 1940 a study of the standardization of insulin led Edgar Fieller, while working for Boots Pure Drug Company, to a resolution of the problem of finding confidence limits for a ratio and for the solution of an equation whose coefficients were subject to error (Fieller, 1940).

Earlier, Henry Daniels, then a statistician at the Wool Industries Research Association, showed how variance component models could be used to expose those parts of a production process responsible for large variations (Daniels, 1938). Variance component analysis has since proved of enormous value in the process industries and elsewhere.

Daniels's contribution was one in a series of papers on industrial statistics read in the 1930s before what was then called the Industrial and Agricultural Research Section of the Royal Statistical Society. A leading spirit in getting this section moving was Egon Pearson, whose ideas greatly influenced, and were influenced by, this body. In particular he liked data analysis and graphical illustration and used both effectively to illustrate Daniels's conclusions (Pearson, 1938).

An important influence on Pearson was the work of Walter Shewhart on quality control (Shewhart, 1931). This work and that on sampling inspection by Harold Dodge heralded more than half a century of statistical innovation coming from the Bell Telephone Laboratories (Dodge, 1969, 1970), including the rekindling of interest in data analysis in a much-needed revolution led by John Tukey (Tukey, 1977; Mosteller and Tukey, 1977).

Another innovator guided by practical matters was Frank Wilcoxon, an entomologist turned statistician at the Lederle Labs of the American Cyanamid Company. Just after the Second World War, in the age of desk calculators, he found himself confronted by the need to make thousands of tests on samples from the pharmaceutical research then in progress. He said it was simply the need for quickness that led to the famous Wilcoxon tests (Wilcoxon, 1949), the precursors of much subsequent research on nonparametric methods.

M. S. Bartlett's contributions to statistics are legion. His early contributions to the theory of transformation (Bartlett, 1936) had much to do with the fact that, when he was statistician at the Jealotts Hill agricultural research station of Imperial Chemical Industries, he was concerned with the testing of pesticides and so with data that appeared as frequencies or proportions.

Another clear example of the practice–theory interaction is seen in the development of parametric time series models. In 1927 Udny Yule was trying to understand what was wrong with William Beveridge's analysis of wheat price data. The fitting of sine waves of different frequencies by least squares had revealed significant oscillations at strange and inexplicable periods. Yule (1927) suggested that such series ought to be represented, not by a deterministic function subject to error, but by a dynamic system (represented by a linear difference equation) responding to a series of random shocks—this model was likened to a pendulum being periodically hit by peas from a pea shooter. Yule's revolutionary idea, with important further input from

Slutsky (1927), Wold (1954), and others, was the origin of autoregressive–moving average models.

Unfortunately, the practical use of these models was for some time hampered by an excessive concern with stationary processes in equilibrium about a fixed mean. Almost all of the series arising in business, economics and manufacturing do not, however, behave like realizations from such a stationary model. Consequently, for lack of anything better, operations research workers led by Holt (1957) and Winters (1960) began in the 1950s to use the exponentially weighted moving average of past data and its extensions for forecasting series of this kind. This weighted average was introduced at first on purely empirical grounds—it seemed sensible to monotonically discount the past and it seemed to work reasonably well. However, in 1960, Muth showed, rather unexpectedly, that this empirically derived statistic was an optimal forecast for a special kind of non stationary autoregressive–moving average model (Muth, 1960). Generalizations of this model later turned out to be extremely valuable for representing many kinds of practically occurring series, including seasonal series (Box and Jenkins, 1962).

The Second World War was a stimulus to all kinds of invention. Allen Wallis has described the dramatic consequence of a practical query made by a serving officer about a sampling inspection scheme (Wallis, 1980). The question was of the kind "Suppose, from a sample of twenty items, that three is the critical number of duds. If it should happen that the first three components tested are all duds, why do we need to test the remaining seventeen?" Allen Wallis and Milton Friedman were quick to see the apparent implication that "super-powerful" tests were possible! However, their suggestion that Abraham Wald be invited to work on the problem was resisted for some time. It was argued that this would clearly be a waste of Wald's time, because to do better than a most powerful test was impossible! What the objector had failed to see was that the test considered was most powerful only if it was assumed that n was fixed, and what the officer had seen was that n did *not need to be fixed*. It is well known how this led to the development of sequential tests (Wald, 1947). It is heartening that this particular happening even withstood the scientific test of repeatability, for at about the same time and with similar practical inspiration, sequential tests (of a somewhat different kind) were discovered independently in Great Britain by George Barnard (1946).

Nor was this the end of the story. Some years later Ewan Page, then a student of Frank Anscombe, while considering the problem of finding more efficient quality control charts, was led to the idea of plotting the cumulative sum of deviations from the target value (Page, 1954). The concept was further developed by Barnard (1959), who introduced the idea of a V mask to decide when action should be taken. The procedure is similar to a backwards-running two-sided sequential test. Cusum charts have since proved to be of great value in the textile and other industries. In addition, this graphical test has proved its worth in the "post mortem" examination of data where it can point to the dates on which critical events may have occurred. This sometimes leads to discovery of the reason for the events. (See Chapter D.9.)

A pioneer of graphical techniques of a different kind was Cuthbert Daniel, an industrial consultant who used his wide experience to make many contributions to statistics. An early user of unreplicated and fractionally replicated designs, he was concerned with the practical difficulty of determining which effects could not readily be explained by system noise. In particular he was quick to realize that higher order interactions sometimes do occur, and when they do it is important to be able to isolate and study them. His introduction of graphical analysis of factorials by plotting effects and residuals on probability paper (Daniel, 1976) has had major consequences. It has encouraged the development of many other graphical aids, and together with the work of John Tukey it has contributed to the growing understanding that at the hypothesis *generation* or model-modification stage of the cycle of discovery, it is the imagination that needs to be stimulated, and that this often best be done by graphical methods.

SOME INTERIM CONCLUSIONS

Obviously I could go on with other examples, but at this point I should like to draw some interim conclusions.

I think it is possible to see important ingredients leading to statistical advance. They are (a) the presence of an original mind that can perceive and formulate a new problem and move to its solution, and (b) a challenging and active environment for that mind, conducive to discovery.

Gosset at Guinness's; Fisher, Yates, and Finney at Rothamsted; Tippett at the Cotton Research Institute; Youden at the Boyce Thomson Institute (with which organization Wilcoxon and Bliss were also at one time associated); Daniels and Cox at the Wool Industries Research Association; Shewhart, Dodge, Tukey, and Mallows at Bell Labs; Wilcoxon at American Cyanamid; Daniel in his consulting practice: these are all examples of such fortunate conjunctions.

Further examples are Don Rubin's work at the Educational Testing Service, Jerry Friedman's computer intensive methods developed at the Stanford linear accelerator, George Tiao's involvement with environmental problems, Brad Efron's interaction with Stanford Medical School, the late Gwilym Jenkins's applications of time series analysis in systems applications, and John Nelder's development of statistical computing at Rothamsted.

The message seems clear: a statistician who believes himself capable of genuinely original research can find the necessary inspiration in a stimulating investigational environment.

Also, I think it possible to perceive an aspect of the specific nature of the contribution coming from applications—frequently, it is the establishment of a new frame of reference for a problem. This may involve extension, modification, or even abandonment of a previous formulation. It has to be understood that statistical problems are frequently not like, for example, chess problems that may require "white to mate in three moves," given a particular configuration of the pieces. Here a solution based on the pretense that a knight can move like a queen would be unacceptable. Yet the changes in the rules that have sometimes been adopted in reformulation of statistical problems must, at the time of their introduction, have been thought of as little short of cheating. Some examples are:

- Fisher's replacement of the method of moments by maximum likelihood;
- Yates's use of designs in which the number of treatments exceeded the block size;
- Yule's introduction of stochastic difference equations replacing deterministic models;
- Wald's and Barnard's introduction of sequential tests to replace fixed sample tests;
- Page's and Barnard's introduction of quality control charts in which the cumulative sum of the deviations rather than the deviations themselves was plotted;
- Finney's use of fractional, rather than full, factorials;
- Fisher's use of the randomization test to justify normal theory tests as approximations; and
- Daniel's and Tukey's initiation of informal graphical techniques rather than more formal procedures in data analysis.

A POSSIBLE RESOLUTION OF THE BAYES CONTROVERSY

One further matter that I think is greatly clarified by the practical context of its application is the problem of statistical inference. Here the consideration of scientific context provides, I believe, a resolution of what is sometimes called the Bayesian controversy. At its most extreme this controversy is a dispute between those who think that all statistical inferences should be made using a Bayesian posterior distribution, and others who believe that sampling theory (i.e., frequentist theory) has universal inferential applicability.

I have argued (Box, 1980, 1983a) that the Bayes–sampling theory controversy arises because of an erroneous tacit assumption that there is only one kind of scientific inference for which there are two candidates, whereas I believe that scientific investigation requires two quite distinct kinds of inference for each of which, one candidate and not the other is appropriate. One kind of inference that may be called *criticism* involves the *contrasting* of the data that actually occured with what might be expected if the assumptions A of some tentative model of interest were true. This is conveniently symbolized by subtraction: $y_d - A$. The other kind of inference, which may be called *estimation*, involves the *combination* of observed data y_d with the assumptions A of some model tentatively assumed to be true. This process is conveniently symbolized by addition: $y_d + A$.

In a statistical context, analysis of residuals, tests of fit, and diagnostic checks, both graphical and numerical, formal and informal, are all examples of techniques of model criticism intended to stimulate the scientist to model building and model modification, or to the generation of more relevant data should this prove desirable. These techniques must, I believe, appeal for formal justification to sampling theory.

By contrast, least squares estimation, likelihood estimation, shrinkage estimation, robust estimation, and ridge estimation are all solutions to estimation problems that I think would be better motivated and justified by applying Bayes's theorem with an appropriate model.

There seem to be three distinct considerations supporting this dualistic view of inference: the nature of scientific method, the physiology of the brain, and the mathematics of Bayes's theorem. I consider them in turn.

The Nature of Scientific Method

It has long been recognized that the process of learning is a motivated iteration between theory and practice. By practice I mean reality in the form of data or facts. In this iteration deduction and induction are employed in alternation. Progress is evidenced by a developing model that by appropriate exposure to reality continually evolves until some currently satisfactory level of understanding is reached. At any given stage the current model helps us to appreciate not only what we know, but what else it may be important to find out. It thus motivates the collection of new data appropriate to illuminate dark but possibly interesting corners of present knowledge.

We can find illustration of these matters in everyday experience, or in the evolution of the plot of any good mystery novel, as well as in any reasonably honest account of the events leading to scientific discovery.

Experimental science accelerates the learning process by isolating its essence. Potentially informative experiences are deliberately *staged* and made to occur in the *presence* of a trained investigator.

The instrument of all learning is the brain, an incredibly complex structure, the working of which we have only recently begun to understand. One thing that is clear is the importance to the brain of models where past experience is accumulated. At any given stage of experience some of the models M_1, M_2, ..., M_i, ..., are well established, others less so, while still others are in the early stages of creation. When some new factor or body of facts y_d comes to our attention, the mind tries to associate the new experience with an established model. When, as is usual, it succeeds in doing so, this new knowledge is incorporated in the appropriate model and can set into motion appropriate action.

Obviously, to avoid chaos the brain must be good at allocating data to an appropriate model and at initiating the construction of a new model only if this should prove to be necessary. To conduct such business the mind must be concerned with the two kinds of inferences, criticism and estimation.

The Physiology of the Brain

With two kinds of inferences to consider, it seems of great significance that research, which gathered great momentum in the past 20 years, suggested that the human brain behaved not as a single entity but as two cooperating instruments (Blackeslee, 1980; Springer and Deutsch, 1981). It appeared that in many people the left half of the cerebral cortex was concerned primarily with language and logical deduction, which plays a major role in estimation, while the right half was concerned primarily with images, patterns, and inductive processes, which play a major role in criticism. The two sides of the brain are joined by millions of connections in the corpus callosum, where information exchange takes place. It is hard to escape the conclusion that the iterative inductive–deductive process of discovery is indeed wired into us.

It was suggested, for example, that the apparently instinctive knowledge of what to do and how to do it enjoyed by, say, an experienced tennis player came from the right brain and that this skill can be temporarily lost if we invite the tennis player to explain how he does it, and thus call the left brain into a dominant and interfering mode.

In this context we see that data analyst's insistence on "letting the data speak to us" by plots and displays as an instinctive understanding of the need to encourage and to stimulate the pattern recognition and model generating capability of the brain. Also, it expresses the concern that we not allow the pushy deductive part of our brain to take over too quickly and perhaps forcibly produce unwarranted conclusions based on an inadequate model.

While the accomplishment of the brain in finding patterns in data and residuals is of enormous consequence to scientific discovery, some check is obviously needed on its pattern-seeking ability; common experience shows that some pattern or other can be seen in almost any set of data or facts. A check that we certainly apply in our everyday life is to consider whether what has occurred is really exceptional in the context of some relevant reference set of circumstances. Similarly, in statistics, diagnostic checks and tests of fit require, at a formal level, frequentist theory significance tests for their justification.

The Mathematics of Bayes's Theorem

It seems reasonable to require that by a statistical model M we mean a complete probability statement of what is currently supposed to be known a priori (i.e., tentatively entertained) about the mode of generation of data $\mathbf{y}$ and of the uncertainty about the parameters $\boldsymbol{\theta}$ given the assumptions A of the model. At some stage i of an investigation the current model M_i would therefore be defined as

$$p(\mathbf{y}, \ \boldsymbol{\theta}|A_i) = p(\mathbf{y}|\boldsymbol{\theta}, \ A_i)p(\boldsymbol{\theta}|A_i)$$

which can alternatively be factorized as

$$p(\mathbf{y}, \ \boldsymbol{\theta}|A_i) = p(\boldsymbol{\theta}|\mathbf{y}, \ A_i)p(\mathbf{y}|A_i)$$

The last factor in the second expression is the predictive distribution. This is the distribution of all possible samples $\mathbf{y}$ that could occur if the model M_i were true.

After the actual data y_d become available,

$$p(y_d, \theta|A_i) = p(\theta|y_d, A_i)p(y_d|A_i)$$

The first factor on the right is now the posterior distribution of θ, conditional on the proposition that the actually occurring data y_d are a realization from the predictive distribution that results from the assumptions of the theoretical model M_i. *If we accept this proposition,* all that can be said about θ must come from this posterior distribution, and the predictive density is without informational content. However, if, as is always in practice the case, the proposition may be seriously wrong, then, correspondingly, residual information may be contained in the predictive density, and this can not only indicate inadequacy but even point to its nature. In particular the relevance of the model may be called into question by an unusually small value of the predictive density for the observed sample y_d as measured, for example, by

$$\Pr[\, p(y|A) < p(y_d|A)]$$

or by an unusually small value of the predictive density $p\{g(y_d)|A\}$ of some suitable checking function $g(y_d)$, as measured by

$$\Pr[\, p\{g(y)|A\} < p\{g(y_d)|A\}]$$

Figure 1 illustrates the idea for a single parameter θ and a single observation y_d. For the situation illustrated, *if the model were true* the posterior distribution would provide an excellent estimate of the parameter θ; but the predictive distribution indicates that the data is almost certainly not generated by this model, so the Bayes estimate is invalid and irrelevant. After the data have become available, it is first

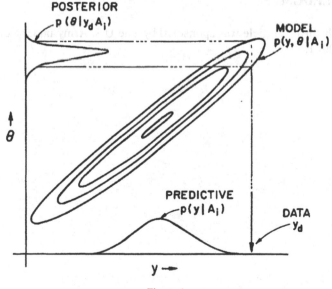

Figure 1

necessary therefore to investigate the adequacy of the model before proceeding with the estimation of θ from its posterior distribution.

There are many conclusions that flow from this approach, which are discussed and illustrated elsewhere. The most important in the present context is that the investigational background against which statistics is applied seems to require that when Bayes's procedure is employed, the proposition on which it is conditioned ought to be considered in the light of the data. This can be done by appropriate consideration of the predictive density associated with the data y_d. Such an approach can, for example, justify and suggest appropriate analyses of residuals, and at a more formal level produce sampling theory significance tests.

CONCLUSION

In summary, then, I have tried to show how application and consideration of the scientific context in which statistics is used can initiate important advances in theory such as least squares, ratio estimators, correlation, contingency tables, studentization, experimental design, the analysis of variance, randomization, fractional replication, variance component analysis, bioassay, limits for a ratio, quality control, sampling inspection, nonparametric tests, transformation theory, ARIMA time series models, sequential tests, cumulative sum charts, data analysis plotting techniques, and a resolution of the Bayes–frequentist controversy.

It appears that advances of this kind are frequently made because practical context reveals a novel formulation that eliminates an unnecessarily limiting framework.

ACKNOWLEDGMENT

The research for this article was sponsored by The U.S. Army under Contract DAA C29-80-C-0041.

CHAPTER B.13

Comparisons, Absolute Values, and How I Got to Go to the Folies Bergères[†]

If you look at the classic book *The Design of Experiments* by Fisher or at the multitude of derivative books that followed it, you will find that almost every example is concerned with making comparisons. We are shown how to answer questions such as "Is treatment A better than B and by how much?" "How big are the differences, if any, between treatments P, Q, R, and S?"

Such comparative experiments are of great value. However, sometimes we need to determine not a difference but an absolute value: What *is* the rate of flow of gas through this pipe? What *is* the cholesterol level for this patient? What *is* the diameter of this rod? We must have means not only for measuring these absolute values but also for maintaining them. This is hard to do. For example, in a study of blood cholesterol measurement (where a difference of 20 units is regarded as clinically significant), it was found that tests on identical samples of blood at a number of different labs in the state of Wisconsin could differ by as much as 60 units (1). These differences were largely caused by biases from one lab to another. Typically, repeatability at a given lab was good. Thus, for example, a comparison of the cholesterol levels for patient A with that for patient B when both measurements were made by the *same lab* was comparatively reliable. As a further illustration, in the manufacture of ammonia, extraordinary efforts are made to maintain gas flow rates as close as possible at their target values. However, careful analysis of flow rate data over an 8-month period detected four distinct changes[‡] in level (2).

From Box, G. E. P. *Quality Engineering*, **14**(1), 167–169 (2001–02). Copyright © by Marcel Dekker Inc.
[†]Parts of this column are from the book by George Box and Alberto Luceño, *Control by Monitoring and Feedback Adjustment* (1997), published by John Wiley & Sons, New York.
[‡]These changes were traced down to inaccurate recalibration of the flow meters.

The fact is that, if left to themselves, the true values for flow rates, for the dimensions of parts, and for chemically determined concentrations will all drift away from their target values. In addition, so will the *measurements* of these true values. The problem is that without intervention nothing stays put. if you own a house, you know that you must work hard to keep it habitable – the tiles on the roof, the paint on the walls, the washing machine, the refrigerator, the television, and so forth all need attention from time to time. A car, a friendship, and our own bodies must similarly be nurtured continually or they will not remain in shape for very long. The same is true of industrial processes. As Deming said, "No process, except in artificial demonstrations by use of random numbers, is steady and unwavering." A fancy way of saying this is that, in accordance with the second law of thermodynamics, the entropy (disorganization) of any uncontrolled system always increases. As Sir Arthur Eddington said, "This law holds the supreme position among the laws of nature. If your theory is against the second law of thermodynamics I can give you no hope."

Thus, a stable stationary state, such as a perfect state of statistical control, is an unnatural one. It must be regarded as a purely theoretical concept whose approximate achievement requires a hard and continuous fight. This battle is called *quality control*. Some of the weapons used in the battle are the quality control charts originally due to Shewhart, sometimes reinforced with Cusum, Cuscores, exponential charts, and so forth. These charts are all designed to detect deviations from target values not likely to be due to noise. They can trigger a search for, and possible removal of, assignable causes of trouble. Such monitoring and debugging are of enormous value because they can reduce variation produce process improvements and simplify operation. However, because not all potentially removable variations can be eliminated in this way (see, for example, Refs. 3 and 4), adjustment methods are needed in addition. These can be anticipatory or reactive and are called respectively feedforward adjustment and feedback adjustment. Procedures of this kind especially tailored to the SPC environment are discussed, for example, by Box and Luceño (5). Experimental design also has important applications in the control of quality. In particular, fractional factorial and Plackett–Burman screening designs are often valuable for tracking down causes of trouble. Furthermore, prerequisite for compensatory adjustment is knowing what to adjust; experimental design can show which adjustment "knobs" are at our disposal.

Occasionally, it is possible to avoid some of the difficulties associated with absolute measurement by substituting a comparative measurement for an absolute one. One example occurred during the time I worked for ICI before I came to the United States. Among the many products we made were mothproofing agents for treating wool carpets. Other chemical companies throughput Europe had similar products and then, as now, the Europeans were greatly concerned with standardization. They wanted a standard procedure that, on some agreed basis, could decide whether a treated carpet wool was mothproof or not. The various proofing agents all worked by poisoning the moth larvae and the proposed test was as follows. A number of moth larvae were placed on a treated wool sample that had been carefully dried and weighed. After a fixed period of time, the dead larvae were washed off and the wool sample dried and reweighed. A small but measurable loss in weight resulted from the wool eaten by the larvae before being poisoned. If this was less than some agreed threshold, the wool would be regarded as mothproof.

Although environmental conditions such as temperature and humidity had been standardized and carefully controlled, it turned out to be unexpectedly difficult to get reproducible results. In fact, it had been found that tests run on the same sample, by labs in different parts of Europe could differ by a factor as large as ten! When I became involved, representatives from the various companies had been meeting for some time at a series of rather exotic places throughout Europe. They had tried to standardize the procedure by making sure that all the labs used the same strain of moth larvae. However, this had not materially improved things. A geneticist had been consulted and had produced what he said was a more uniform type of moth, but, again, little improvement was obtained. As a last resort, it was decided that statistics might help and I suddenly found myself of my way to Paris. I had been puzzling over the problem for some time, but I think it was when I was sitting in the front row of the Folies Bergères that I saw everything clearly, namely that we should make the test a comparative one. This idea became the basis for the final successful test method. A supply of treated wool was prepared that the panel agreed was acceptably mothproof. Samples were circulated to all the labels were used as controls. The wool sample to be tested was put next to the control sample and moth larvae were equally and randomly alloted to the two samples. If the loss in weight of the test wool sample was less than that of the control, then it was designated as mothproof.*

Dr. Taguchi once told me that the same concept has also been used in the parts industry. A mechanical arm takes a sample part off the line and measures its various dimensions. Then, it immediately does the same thing for an adjacent perfect part that is kept for reference. Compliance is judged not from the absolute measurements directly, but from the observed differences between the sample part and the standard part. This, of course, eliminates the effects of bias in the measuring instrument.

REFERENCES

1. Ilassemer, D.J.; Wiebe, D.A.; Kramer, T. The Wisconsin Cholesterol Study; An Assessment of Laboratory Performance. Clin. Chem. Abst. **1989**, *35* (6), 1068.

2. Box, G.E.P.; Luceño, A. *Statistical Control by Monitoring and Feedback Adjustment*, John Wiley & Sons, New York, 1997.

3. Alwan, L.C.; Roberts, H.V. Time-Series Modeling for Statistical Process Control. J. Bus. Econ. Statist. **1988**, *6*, 87–95.

4. Box, G.E.P.; Luceño, A. Six Sigma, Process Drift, Capability Indices, and Feedback Adjustment. Qual. Eng. **2000**, *12* (3), 297–302.

5. Box, G.E.P.; Luceño, A. Feedforward as a Supplement to Feedback Adjustment to Allow for Feedstock Changes. Submitted for publication, 2001.

*I became rather unpopular with my mothproofing colleagues because that was the end of those exotic trips.

PART C

Sequential Investigation and Discovery

CHAPTER C.0

Introduction

In his "Popular Lectures and Addresses" (1891–1894), the distinguished scientist Lord Kelvin, among whose many achievements was the discovery of the second law of thermodynamics, said:

> When you can measure what you are speaking about, and express it in numbers, you know something about it; but when you cannot measure it, when you cannot express it in numbers, your knowledge is of a meager and unsatisfactory kind: it may be the beginning of knowledge, but you have scarcely, in your thoughts, advanced to the stage of *science*.

Now, over one hundred years later, I think we would need to qualify these words somewhat. While it is clear that numbers are vital to scientific discovery, we also know that numbers are deceptive. One major influence that can obscure the truth and lead the experimenter astray is the existence of noise. Another is that experimental factors rarely behave independently and detection and understanding of their dependence can be crucial to success. Since statistics, and particularly the design and analysis of experiments, has specifically been developed to deal with such difficulties, one might expect that it would be recognized as an important contributor to most scientific investigation.

Its importance is certainly recognized in the social and medical sciences, where you scarcely see a published paper without some kind of statistical analysis (good or bad). But in academia in the physical sciences and engineering, the attitude of many experimenters seems to be "Statistics? Who needs it?" When it *is* taught to students it is often done very badly and the things that are taught are not the things our future scientists and engineers need to know.

Why this striking difference in attitude? I believe it is because the methods and mindset of much of present-day statistics are appropriate for testing preconceived ideas but not for discovering new ones. I should explain that by "testing" I mean not only the testing of hypotheses but the general drawing of conclusions from one specific set of data in the light of a *given* model—I will call this a *one-shot* procedure. Now it is discovery that is most important to the experimenter in the physical and engineering sciences and discovery is the end point of a process of *learning*, which is an evolutionary alternation between induction and deduction resulting in *changes* in the model. A successful investigation—finding a drug that is effective against a particular virus—designing a printer that gives better color reproduction—almost always comes about as the result of a series of ideas, tentative models, and experiments, which, in the light of appropriate data and subject matter knowledge, spark modified ideas and models. This process leads the investigator along a previously unknown path to a solution that is neither unique nor predictable: it is like that which a detective follows to solve a mystery and discover the identity of a criminal.

By contrast, once the drug has been discovered, carefully planned trials must be conducted to prove its effectiveness and safety before it can be marketed. The results of these trials and many others in the social and medical sciences usually take a considerable time to complete, and then are subjected to rigorous statistical analysis. (In a similar way, the new color printer must be market tested before it is allowed to go into full production.) This testing is like that used in the *prosecution* of a supposed criminal and is very different from the process of investigation used by the detective.

In a criminal trial all available evidence must be collected in advance and presented formally at a specific time to a jury, who must be convinced that the hypothesis of influence is untenable "beyond all reasonable doubt" before they will find the defendant guilty. A similar philosophy is necessary for the proper conduct of other one-shot investigations, such as are appropriate particularly in the social and medical sciences. However, in physical and engineering sciences, by basing statistics on the one-shot philosophy, we ensure its exclusion from the majority of experimental effort.

Are physical scientists and engineers too foolish to recognize the value of statistics? I don't think so. Rather I think they understand that the basic philosophy that motivates much of the statistics taught at universities is not appropriate for what they do. Also much teaching and research in this kind of statistics emphasizes mathematics rather than experimental science. Mathematicians are trained to think in terms of what I will call the one-shot mathematical paradigm—proposition–theorem–proof. This fits very nicely with the idea of statistical testing. For example, the Neyman–Pearson theory supposed that you have in mind specific hypotheses and accompanying assumptions from which, given the data, you can proceed to judge the plausibility of particular propositions. In a similar way, to obtain an "optimal" experimental design, it is supposed that you are *given* the appropriate set of factors to study, you are *given* the exact factor space over which they are to be varied, you *know* the exact nature of the function linking the response to the levels of the factors, and so forth. Given all this, a mathematically optimal design (of one sort or another) can be found, which, if all these assumptions were true, could illuminate the particular questions raised *before* the experiment was begun. A similar *a priori* framework is employed in mathematical optimization itself.

This one-shot paradigm is attractive to mathematical statisticians because, within it, theorems can be proved, the supposed fundamentals of statistical science can be studied, and an imposing theoretical edifice can be constructed from which the essentiality of such principles as coherence can be expounded. Learning and discovery, however, are not coherent since they are achieved by *changing* the model in a manner not predictable in advance.* I am not saying that the mathematical aspects of statistics are unimportant, but that the reason that statistics is not a mainstream subject is because it has got lost in a mathematical tributary. It would be helpful if it was categorized and recognized a uniquely important science rather than as one of the (lesser) branches of applied mathematics.

The Design of Experiments was invented by a superb *scientist* and *experimenter* who, for the first time, enunciated the principles that must govern scientific study in the presence of noise. In particular, he produced experimental designs that made it possible to understand the behavior and interaction of many factors considered together and, at the same time to reduce the level of the noise. But Fisher worked in agriculture, where the experimental environment is very specialized. The application of his ideas must be adapted somewhat to meet the different experimental environment of the physical sciences.

The essential difference is in the *immediacy* of acquisition of data that often occurs in science and engineering and provides an opportunity for *rapid* sequential learning. Experimental runs can be made in series rather than in parallel and each group of runs not only illuminates the present level of understanding but helps decide *what should be done at the next stage.* Factorials and fractional factorial designs (see Chapter B) are superb instruments for doing just that. Also as described in Chapter C, *response surface methodology* (RSM) was developed after careful study of what chemists and chemical engineers actually did. Within that framework, Wilson and I thought we could help them do it better by suitable adaptation of Fisher's ideas.

Over the years at statistical meetings, I have been glad to see a number of sessions devoted to "response surface methods" but I have often been dismayed by their content. Here is a tremendous opportunity for further research on the development of statistical methods for sequential learning and discovery; instead, some of the papers presented seemed to take one particular aspect of RSM, try to force it into the one-shot model, and strangle it. So I was glad to contribute to a discussion on this subject published in the *Journal of Quality Technology* (and reproduced here in Chapters C.1, C.2, and C.3) and to say what I thought this methodology was about and what were its more general implications.

To avoid misunderstanding, it seemed necessary first of all to provide a live example of RSM as an agent of change. What was needed was not only a

---

* Of course, in mathematics also, innovation is almost always accompanied by a change of model, for example, by the introduction of complex numbers.

demonstration but a form of exercise in which participants could easily *experience* for themselves the process of creative investigation. I could not involve my readers in designing a better color printer or discovering a more potent drug, but to involve them in improving the design of a paper helicopter was perfectly practicable. This device has long been used to demonstrate individual experimental designs; it is a "product" that is inexpensive and easy to make and its design can be changed in an essentially infinite number of ways limited only by the imagination of the experimenter. Also, the effect on flight time of changes in design can readily be assessed by experiment and the results can be used to plan the next stage of development. Such an exercise forces attention on the facts of life concerning sequential experimentation. You *don't* know which are the important variables, over what region they should be varied, what will be an adequate approximation to the relationship between the responses and the factor levels, and what course the investigation will take. These are all things initially guessed and learned more about as you go. Patrick Liu and I (Box and Liu, 1999) followed where the investigation led us; as is described in Chapter C.1.

In this volume it seemed best to present a number of separate extracts.* Thus, after the helicopter example in Chapter C.1, some history of RSM is given in Chapter C.2, some philosophical consequences in Chapter C.3, and the importance of practice as a guide to theory in Chapter C.4. This theme is further developed in Chapter C.5, which describes the invention of the composite design.

Fractional factorials are of great importance for *factor screening*. The relevance of the *projective properties* of fractional factorial designs in this process of screening was first pointed out by Box and Hunter (1961a and b). Recently, the subject has become of even greater interest because of the newly discovered projective properties of other orthogonal arrays and, in particular, Plackett and Burman designs. Screening to find the active factors is an important part of iterative learning, especially when, as is usually the case, several responses y_1, y_2, y_3, ... are measured. The active factor space for the response y_1 may be quite different from that for y_2 and y_3. Making clear to the investigator with *subject matter knowledge* the identity of, and the behavior within, these active factor spaces can greatly help him/her in gaining a *physical understanding* of the system. With this in mind, Chapter C.6 (Box and Meyer 1993) sought better ways of identifying these active factor spaces. Also, Chapter C.7 (Meyer, Steinberg and Box 1996) shows how appropriate follow-up runs can be chosen to resolve ambiguities. As we have seen, many important techniques of experimental design are simple and can safely rely on graphical analysis. However, although projective analyses of orthogonal arrays are of great practical importance they sometimes require more sophisticated analysis. Nevertheless, the screening techniques discussed in Chapters 6 and 7 have been in almost daily use in industry and can easily be put into effect using the computer programs that are described. Those practitioners that are not greatly interested in mathematical derivations may wish to proceed straight to the examples which give safe guidance in the use of these methods. It should be added, although like all analytical techniques

---

* A further extract about robustness and optimization appears in Chapter E.5.

specific assumptions need to be made in their derivations we believe that the procedures we advocate are very robust to likely departures from the assumptions.

A formal study with John Tyssedal of the projective properties of orthogonal arrays is presented in Chapter C.8.

The reasons why I am not an admirer of alphabetic optimality are described in more detail in Chapter C.9.

Chapters C.6 and C.7 employed a Bayesian approach, which I think was appropriate in that context. More generally, in Chapter C.10, it is argued that the "Bayesian *versus* frequentist" argument is pointless because the two phases of iterative learning—deduction and induction—*require* these two *different* kinds of inference.

CHAPTER C.1

A Demonstration of Response Surface Methods

Innovation in the design and manufacture of processes and products frequently comes about as a result of careful investigation—a directed process of sequential learning. Many practitioners, although familiar with "one-shot" statistical procedures, have little knowledge of the power of statistical techniques designed to catalyze investigation itself. A simple means of demonstrating and experiencing this learning process is illustrated using response surface methods to find an improved design for a paper helicopter.

INTRODUCTION

It has long been emphasized that the statistician should be involved not merely in the analysis of data, but also in the design of the experiments which generate the data. As a result, it is now common for students taking a course in experimental design to be required to plan and perform a real experiment. The concept of the statistician as one who analyzes someone else's data is flawed, but equally inappropriate is the idea of the statistician engaged only in the design and analysis of an individual experiment.

An industrial innovation of major importance, such as discovering the cause of a problem, the improvement of a process, the development of a new drug or the design of a new engineering system, often comes about as the result of an *investigation* requiring a *sequence* of experiments. Such research and development is a process of learning: dynamic, not stationary; adaptive, not one-shot. The route by which the objective can be reached is discovered only as the investigation progresses, each

From Box, G. E. P. and Liu, P. (1999), "Statistics as a Catalyst to Learning by Scientific Method Part I—An Example," *Journal of Quality Technology*, **31**(1), 1–15.

subset of experimental runs supplying a basis for deciding the next. Also, the objective itself can change as new knowledge is brought to light. To catalyze such innovation, the statistician *must be part of the investigational team.* Unfortunately, many statistics students (and their professors) have little or no training or experience to qualify them for this important role.

Consider a statistician who is analyzing some data coming from, say, a factorial arrangement that has been designed in collaboration with an experimenter. The statistician knows that, even though the data are subject to observational error, certain probability statements can be made and conclusions drawn. However, it may not be realized how such conclusions are heavily conditional on the experimental environment. For example, if the statistician had been working on the same problem with a different experimenter, then

(a) the design would almost certainly have contained some different factors,

(b) different ranges for the factors would have been chosen,

(c) different transformations for the factors might be used (such as the length, l, and width, w, of an aircraft wing as opposed to its area, $A = wl$, and length to width ratio, l/w),

(d) different initial tentative models would have been considered.

Such arbitrary differences would affect the conclusions drawn from this single experiment far more than would observational error.

However, in an investigational *sequence* of experiments, although different experimenters will begin from different starting points and take different routes, they nevertheless can arrive at similar solutions or equally satisfactory solutions. Like mathematical iteration, scientific iteration tends to be self-correcting. The concept of iterative investigation requires a mindset unfamiliar to many students of statistics. It is difficult to teach and to illustrate, and it needs to be experienced to be appreciated. The purpose of this paper is to illustrate, using a paper helicopter, how response surface methodology (RSM) may be used to practice the process of investigation itself. For detailed descriptions of RSM, see BHHII and for example, Box and Draper (1987), Khuri and Cornell (1996), Myers and Montgomery (1995), and the many references contained therein.

Exercises of the kind discussed could be used not only to illustrate response surface methods, but also compare them with any other system of experimentation (see e.g., Ledolter, 1997). Also they might use the paper helicopters employed in the investigation described below or any other convenient experimental device. A variety of other paper airplanes which may be used in your experiments are described in the *Paper Airplane Book* by Blackbeen and Lammers (1994). However in all such exercises the device used for illustration must be one that can be subjected to design changes that are *not* predetermined. A device for which only a fixed number of built-in factors can be changed is of no use for this purpose. The important thing is that the student, using factors whose nature is only limited by the extent of his or her imagination, should start from some prototype design that seems reasonable and then experience the process of experimental learning needed to improve it. It is particularly valuable for several groups of students to conduct independent investigations simultaneously. Each group should then discuss the different routes they took, their adventures and conclusions with the other groups. Remember that, in the context of continuous never-ending improvement, there is no such thing as optimality. The best helicopters described in this paper are *not* optimal. Better designs will be found when factors not previously considered are tested. Already a correspondent, Donald Olsson, has reported designs with longer flight times. These were constructed from a special drawing paper called "rough newsprint."

Students should be warned that there are at least two important respects in which this helicopter experiment does not provide a true picture of a real investigation. In practice, not one but a number of responses will be measured, recorded, and jointly considered. Furthermore, in this example progress is made almost entirely empirically. If we were really in the business of making helicopters, there would be aerodynamicists and other engineers on the team supplying subject matter knowledge; their help in interpreting the results from the designed experiments generating multiple responses would undoubtedly have produced better helicopters quicker. The following is an experiment conducted by Box and Lui used for illustration in BHHII.

DESIGN I: AN INITIAL SCREENING EXPERIMENT

The prototype design for a paper helicopter, shown in Figure 1, was kindly made available to us some years ago by Kipp Rogers then of Digital Equipment Corporation. The objective of this investigation was to find an improved helicopter design giving longer flight times. We limited our designs to those that could be constructed from readily available office supplies and, in particular, from standard paper $11 \times 8\frac{1}{2}$ inches in size. Our test flights were carried out in a room with a ceiling 102 inches ($8'\, 6''$) from the floor.* The wings of each tested helicopter were initially held against the ceiling, and the flight time was measured with a digital stop watch.

We first ran a screening experiment to get some idea of what factors might be important (see also Chapter B.2). After some discussion we decided to test the eight factors (input variables) previously used in Chapter B.2 and listed in Table 1. Each factor was tested at two levels with the plus and minus limits shown there. The response (output variable) was the flight time. The initial experimental plan defined sixteen helicopter types set out in Table 2. The experimental design is a 2_{IV}^{8-4} fractional factorial (see, Chapter B.4). Each of the sixteen types of helicopters was dropped four times, and the flight times recorded in centiseconds (units of one-hundredth of a second). The mean flight times, $\bar{y}$, and the standard deviations, s, are also shown in Table 2, together with the quantity $100 \log(s)$, which we will call the *dispersion*. Remember it is the design that is being tested, not the individual helicopter "manufactured" according to that design. In earlier work, Sandra Martin showed that indeed a small variance component associated with "manufacturing" could be detected. The dispersions given here and calculated from repeat runs of the

---

* In the paper helicopter experiment described in Chapter B.2, the flights were made from a height of 12 feet and consequently the flight times were somewhat longer.

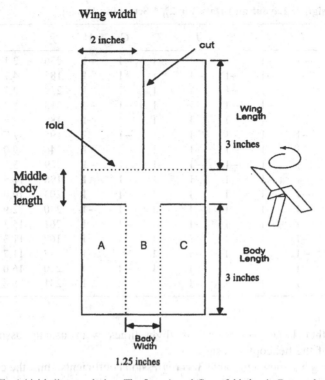

Figure 1. The initial helicopter design. The flaps A and C are folded only B to make the body.

same helicopter are therefore slight underestimates. The analyses of mean flight times with normal plots are, however, not affected since we use the averages as data. It is well known (Bartlett and Kendall, 1946) that for the analysis of variation there are considerable advantages in using the logarithm of the sample standard deviation rather than s itself. To avoid decimals, we have used $100 \log(s)$ in our analysis. The effects calculated from the mean flight time $\bar{y}$ will be called *location effects*. Effects calculated using the dispersion will be called *dispersion effects*. Visual observation

Table 1. Factor Levels Used in Design I: An Initial 2_{IV}^{8-4} Screening Experiment

Factor	Symbol	-1	$+1$
1. Paper type	P	regular	bond
2. Wing length	l	3.00 in.	4.75 in.
3. Body length	L	3.00 in.	4.75 in.
4. Body width	W	1.25 in.	2.00 in.
5. Fold	F	no	yes
6. Taped body	T	no	yes
7. Paper clip	C	no	yes
8. Taped wing	M	no	yes

Here in. stands for inches.

Table 2. Design I: Layout and Data for 2_{IV}^{8-4} Screening Design

Run	P	l	L	W	F	T	C	M	$\bar{y}$	s	$100 \log(s)$
1	−1	−1	−1	−1	−1	−1	−1	−1	236	2.1	31
2	1	−1	−1	−1	−1	1	1	1	185	4.7	67
3	−1	1	−1	−1	1	−1	1	1	259	2.7	42
4	1	1	−1	−1	1	1	−1	−1	318	5.3	72
5	−1	−1	1	−1	1	1	1	−1	180	7.7	89
6	1	−1	1	−1	1	−1	−1	1	195	7.7	89
7	−1	1	1	−1	−1	1	−1	1	246	9.0	96
8	1	1	1	−1	−1	−1	1	−1	229	3.2	50
9	−1	−1	−1	1	1	1	−1	1	196	11.5	106
10	1	−1	−1	1	1	−1	1	−1	203	10.0	100
11	−1	1	−1	1	−1	1	1	−1	230	2.9	46
12	1	1	−1	1	−1	−1	−1	1	261	15.3	118
13	−1	−1	1	1	−1	−1	1	1	168	11.3	105
14	1	−1	1	1	−1	1	−1	−1	197	11.7	107
15	−1	1	1	1	1	−1	−1	−1	220	16.0	120
16	1	1	1	1	1	1	1	1	241	6.8	83

suggested that larger *variations* of flight times were usually associated with instability of the helicopter design.

The effects are shown in Table 3 as regression coefficients; thus, the constant term is the overall average, and each of the remaining coefficients is one-half of the usual factor effect. Normal plots are shown in Figures 2(a) and 2(b). Figure 2(a) for location effects suggests that factors describing three of the dimensions of the helicopter—wing length l, body length L, and body width W—all have real effects on mean flight time, but, of the five remaining "qualitative" variables, only factor C (corresponding to the application of a paper clip to the body of the helicopter) produces an effect distinguishable from noise.

The plot for dispersion effects in Figure 2(b) suggests real effects for l, L, W, and C and for the string of two-factor interactions, $PL + lC + WT + FM$. The signs of the coefficients are such that the changes in the dimensional variables, l, L, and W, which gave increases in the mean flight time, are also associated with reductions in dispersion. However, the addition of a paper clip, while reducing the dispersion, also decreased the flight time; we made a judgment therefore that, for the moment, we would concentrate on increasing flight times and not use the paper clip. We could reconsider this later if instability became a problem. Also, we decided that we would not attempt to interpret, or to separate out by additional runs, the interaction string at this time.

On this basis, a linear model for estimating mean flight times in the immediate neighborhood of the experimental design was

$$\hat{y} = 223 + 28x_2 - 13x_3 - 8x_4 \tag{1}$$

Table 3. Design I: Estimates for a 2_{IV}^{8-4} Screening Design

	Location	Dispersion
Constant	222.8	82.7
P	5.8	3.2
l	27.7	−4.1
L	−13.2	9.7
W	−8.3	15.6
F	3.7	5.1
T	1.4	0.6
C	−10.9	−9.8
M	−4.0	5.7
$Pl + LC + WM + FT$	6.0	−0.7
$PL + lC + WT + FM$	0.2	−13.4
$PW + lM + LT + FC$	5.0	0.6
$PF + lT + LM + WC$	7.0	−4.9
$PT + lF + LW + CM$	5.2	−4.0
$PC + lL + WF + TM$	−3.3	−0.9
$PM + lW + LF + TC$	−4.2	−2.1

where the coefficients are those, suitably rounded, in Table 3 and x_2, x_3, and x_4 are the coded levels of l, L, and W (coded as in Table 2).

Equation (1) is usually called a fitted linear regression model and the coefficients are those obtained by the method of least squares. The contour diagram of Figure 3 is a convenient way of conveying visually what is implied by Equation (1); for example, those combinations of x_2, x_3, and x_4 corresponding to points on the 240 contour plane should all produce alternative helicopter designs with flight times of about 240 centiseconds.

STEEPEST ASCENT USING THE RESULTS FROM DESIGN I

Now, since increasing the wing length and reducing the body length and body width all had positive effects on mean flight time, it might be expected that helicopter designs with greater wing lengths and with reduced body lengths and body widths might give even longer flights. We can best determine such helicopter designs by exploring the direction at right angles to the contour planes indicated by the arrow in Figure 3. In the units of x_2, x_3, and x_4, this is the direction of greatest increase* at a given distance from the design center and is called the *direction of steepest ascent*.

To calculate a series of points along the direction of steepest ascent, you don't need a contour plot. You can do this by starting at the center of the design and changing the factors in proportion to the coefficients of the fitted equation. Thus, the *relative* changes in x_2, x_3, and x_4 are such that for every increase of 28 units in

*Given the relative scaling of the factors currently thought most appropriate to the experimentor.

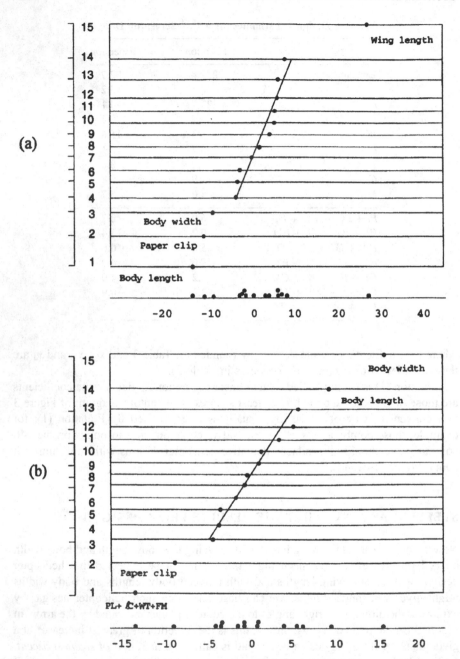

Figure 2. Design I—normal plots for (a) location effects from $\bar{y}$ and (b) dispersion effects from $100 \log(s)$.

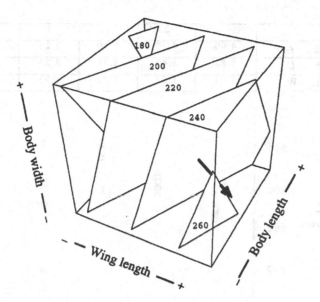

Figure 3. Design I: contours of mean flight times.

x_2, x_3 is reduced by 13 units and x_4 by 8 units. The units are the scale factors, $s_l = 0.875$, $s_L = 0.875$, and $s_W = 0.375$, which looking at Table 1 are the changes in l, L, and W corresponding to a change of one unit in x_2, x_3, and x_4, respectively.

In our investigation we chose the first design 1 to give a helicopter with a 4 inch wing length, and we then increased l by 3/4 inch increments, adjusting the other dimensions accordingly. This produced the designs 2, 3, 4, and 5 shown in Figure 4. In our investigation we ran experiments in sequence at all the five points making ten repeat drops at each point. Alternatively, experiments along such a path could have been run sequentially with the choice of the points along the path a matter of judgment guided by results as they occurred. For example, we might have decided to take a large jump initially and to try the design 5 right away. This would have given a disappointingly low result causing us to backtrack and to test designs in the neighborhood of 2 or 3. In any case, we would have ended up with more or less the same conclusion. As you see from Figure 4, design 3 gave the longest average flight time of 347 centiseconds—while designs 4 and 5 appeared to give not only lesser mean flight times, but also higher standard deviations.

Since none of the qualitative variables we tried in this and previous experiments (including heavy paper, fold at the wing tip, fold at the base, etc.) seemed to produce any positive effects, we decided that, at least at this stage, we would fix these features and explore more thoroughly the effects of the *dimensional* variables—wing length, wing width, body length, and body width using a full factorial experiment. In all subsequent runs therefore P, F, and T were set at their plus levels and C and M at their minus levels. Also at about this time, we had a discussion with an engineer whose subject matter knowledge led to the suggestion that a better way to characterize the dimensions of the wing might be in terms of wing area, $A = lw$, and length to width

Helicopter		1	2	3	4	5
Wing length	ℓ	4.00	4.75	5.50	6.25	7.00
Body length	L	3.82	3.46	3.10	2.75	2.39
Body width	W	1.61	1.52	1.42	1.33	1.24

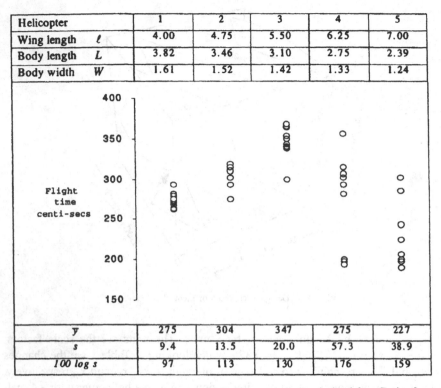

$\bar{y}$	275	304	347	275	227
s	9.4	13.5	20.0	57.3	38.9
$100 \log s$	97	113	130	176	159

Figure 4. Data for 5 helicopters on the path of steepest ascent calculated from Design I.

ratio, $Q = l/w$. Therefore, in subsequent experimentation this reparameterization was adopted.

DESIGN II: A SEQUENTIALLY ASSEMBLED
COMPOSITE DESIGN

It seemed likely that further advance to an improved experimental region might not be possible with first order steepest ascent and that a more elaborate model might be needed to represent the flight times. This was not certain, however. Therefore, a 2^4 factorial experiment in A, Q, W, and L was run with two added center points using the $(-1, 0, 1)$ levels shown in Table 4. It consisted of the first block shown in Table 5 yielding the calculated effect coefficients in Table 6. A normal plot is shown in Figure 5. The plot for dispersion effects failed to show anything of interest and is not given. We see that some two-factor interactions are quite large and approaching the size of the main effects, which suggested that we should add further runs to provide for estimation of the remaining second order (quadratic) terms. A further block, in Table 5(b), was therefore added consisting of eight axial points set at

Table 4. Factor Levels. The Levels (−1, 0, 1) Were Used in a 2^4 Factorial, Design IIa, with Center Points. Design IIb is a Second Block Adding Later Axial and Center Points with Levels (−2, 0, 2) Producing a Central Composite Design

Factor	Symbol	−2	−1	0	+1	+2
Wing area (lw [inch^2])	A	11.20	11.80	12.40	13.00	13.60
Wing length/width ratio (l/w)	Q	1.98	2.25	2.52	2.78	3.04
Body width [inch]	W	0.75	1.00	1.25	1.50	1.75
Body length [inch]	L	1.00	1.50	2.00	2.50	3.00

Table 5. Central Composite Design and Data: Block 1—Design IIa and Block 2—Design IIb

	Run	Block	A	Q	W	L	$\bar{y}$	$100\log(s)$
	1	1	−1	−1	−1	−1	367	72
	2	1	1	−1	−1	−1	369	72
	3	1	−1	1	−1	−1	374	74
	4	1	1	1	−1	−1	370	79
	5	1	−1	−1	1	−1	372	72
	6	1	1	−1	1	−1	355	81
	7	1	−1	1	1	−1	397	72
	8	1	1	1	1	−1	377	99
(a)	9	1	−1	−1	−1	1	350	90
	10	1	1	−1	−1	1	373	86
	11	1	−1	1	−1	1	358	92
	12	1	1	1	−1	1	363	112
	13	1	−1	−1	1	1	344	76
	14	1	1	−1	1	1	355	69
	15	1	−1	1	1	1	370	91
	16	1	1	1	1	1	362	71
	17	1	0	0	0	0	377	51
	18	1	0	0	0	0	375	74
	19	2	−2	0	0	0	361	111
	20	2	2	0	0	0	364	93
	21	2	0	−2	0	0	355	100
	22	2	0	2	0	0	373	80
	23	2	0	0	−2	0	361	71
(b)	24	2	0	0	2	0	360	98
	25	2	0	0	0	−2	380	69
	26	2	0	0	0	2	360	74
	27	2	0	0	0	0	370	86
	28	2	0	0	0	0	368	74
	29	2	0	0	0	0	369	89
	30	2	0	0	0	0	366	76

Table 6. Design IIa: Estimated Coefficients for Mean Flight Times

	Coefficients
Constant	367.2
A	−0.4
Q	5.4
W	0.5
L	−6.6
AQ	−3.0
AW	−3.7
AL	4.3
QW	4.7
QL	−1.5
WL	−2.0
AQW	0.0
AQL	−1.8
AWL	0.7
QWL	−0.3
AQWL	−0.2

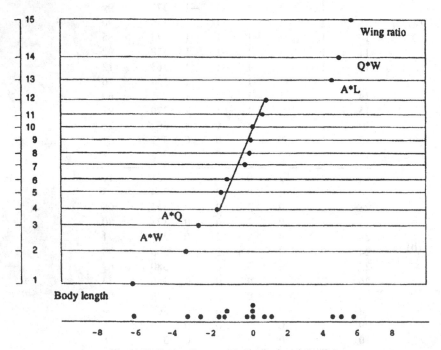

Figure 5. Design IIa: normal plot for location effects.

Table 7. Design II: Analysis of Variance for Completed Composite Design

Source	DF	SS	MS	F	P
Blocks	1	66.7	66.7	6.71	0.021
Regression	14	2907.3	207.6	20.88	<0.001
Linear terms	4	1515.2	378.7	38.09	<0.001
Interaction terms	6	1104.7	184.1	18.52	<0.001
Square terms	4	287.4	71.8	7.23	0.002
Residual error	14	139.2	9.9		
Lack-of-fit	10	126.6	12.6	4.03	0.096
Pure error	4	12.5	3.1		
Total	29	3113.2			

conditions corresponding with the levels -2 and $+2$, with four additional center points. Thus a "centered composite design" was generated from which the coefficiants for a second order model as shown in Table 8 and in equation (2)

$$\hat{y} = 372.06 - 0.08x_1 + 5.08x_2 + 0.25x_3 - 6.08x_4$$
$$- 2.04x_1^2 - 1.66x_2^2 - 2.54x_3^2 - 0.16x_4^2$$
$$- 2.88x_1x_2 - 3.75x_1x_3 + 4.38x_1x_4$$
$$+ 4.63x_2x_3 - 1.50x_2x_4 - 2.13x_3x_4 \tag{2}$$

Table 8. Central Composite Design: Estimated Coefficients for Mean Flight Times

	Coefficients	Std. Error
Constant	372.06	1.29
A	−0.08	0.64
Q	5.08	0.64
W	0.25	0.64
L	−6.08	0.64
A^2	−2.04	0.60
Q^2	−1.66	0.60
W^2	−2.54	0.60
L^2	−0.16	0.60
AQ	−2.88	0.78
AW	−3.75	0.78
AL	4.38	0.78
QW	4.63	0.78
QL	−1.50	0.78
WL	−2.13	0.78

(a) $$\hat{y} = 83.6 + 9.4x_1 + 7.1x_2 - 7.4x_1^2 - 3.7x_2^2 - 5.8x_1x_2$$

(b)

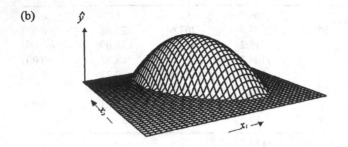

(c)

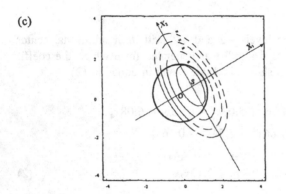

(d)

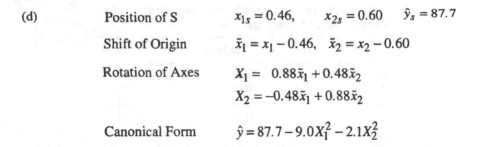

Position of S	$x_{1s} = 0.46,$ $\quad x_{2s} = 0.60$ $\quad \hat{y}_s = 87.7$	
Shift of Origin	$\tilde{x}_1 = x_1 - 0.46,$ $\quad \tilde{x}_2 = x_2 - 0.60$	
Rotation of Axes	$X_1 = 0.88\tilde{x}_1 + 0.48\tilde{x}_2$ $X_2 = -0.48\tilde{x}_1 + 0.88\tilde{x}_2$	
Canonical Form	$\hat{y} = 87.7 - 9.0X_1^2 - 2.1X_2^2$	

Figure 6. Canonical analysis of second degree equation representing a maximum.

An analysis of variance of average flight times for the completed design is given in Table 7. There is some suggestion of lack of fit. Nevertheless, for this analysis we used the overall residual mean square of 9.9 as the error variance. The overall F ratio for the fitted second degree equation is 20.88. This exceeds its five percent significance level of $F_{0.05,14,14} = 2.48$ by a factor of 8.5, thus satisfying the criterion of Box and Wetz (1973) that a factor of at least four is needed to ensure

(a)
$$\hat{y} = 84.3 + 11.1x_1 + 4.1x_2 - 6.5x_1^2 - 0.4x_2^2 - 9.4x_1x_2$$

(b)

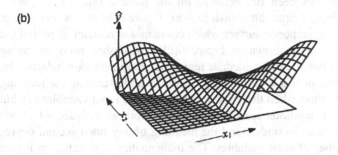

(c)

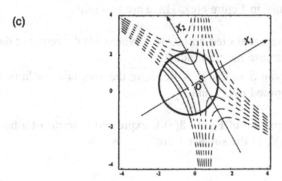

(d)   Position of S $x_{1s} = 0.38,$ $x_{2s} = 0.65$ $\hat{y}_s = 87.7$

        Shift of Origin $\tilde{x}_1 = x_1 - 0.38,$ $\tilde{x}_2 = x_2 - 0.65$

        Rotation of Axes $X_1 = 0.89\tilde{x}_1 + 0.48\tilde{x}_2$

                                $X_2 = -0.48\tilde{x}_1 + 0.89\tilde{x}_2$

        Canonical Form $\hat{y} = 87.7 - 9.0X_1^2 + 2.1X_2^2$

Figure 7. Canonical analysis of second degree equation representing a saddle.

that the fitted equation is worthy of further interpretation. See also Box and Draper (1987) and Draper and Smith (1998).

CANONICAL ANALYSIS

This second degree equation in four variables (x_1, x_2, x_3, x_4) contains 15 coefficients, and in its "raw" form is not easily understood. We briefly review methods of canonical analysis which can make its meaning clear and allow further progress. A fuller account of such analysis is given in the texts referred to earlier.

We first illustrate the analysis for a simpler constructed examples in just two variables, x_1 and x_2. Look at Figure 6. Suppose that, in the circle indicated in Figure 6(c), a suitable design has been run centered on the point O $(x_{10} = 0, \ x_{20} = 0)$, yielding the second degree equation shown in 6(a). Figure 6(b) shows a computer plot of the corresponding response surface which contains a maximum. A plot of the $\hat{y}$ contours of the surface is shown in Figure 6(c) with dashed lines indicating that they are unreliable outside the immediate region of experimentation indicated by the circle. Contour plots of this kind are very helpful in understanding the meaning of a second degree equation when there are only two or three input variables (x), but for more variables such methods are not available. Canonical analysis, which we now explain, makes it easy to understand the meaning of any fitted second degree equation for any number of such variables. The mathematics is sketched in Figure 6(d) and illustrated geometrically in Figure 6(c). There are two steps:

(i)   the origin of measurement is shifted from O to S, where S is the center of the contour system (in this case the maximum);

(ii) the axes are rotated about S so that they lie along the axes of the elliptical contours which are denoted by X_1 and X_2.

In this way the quadratic equation of Figure 8(a) is expressed in terms of a new system of coordinates X_1 and X_2 in the simpler form,

$$\hat{y} = 87.7 - 9.0X_1^2 - 2.1X_2^2 \tag{3}$$

By inspection of this canonical form you can understand the meaning of the quadratic equation without a contour plot. In this case, since the coefficients, -9.0 and -2.1, which measure the quadratic curvatures along the X_1 and X_2 axes are both negative, the point S (at which $\hat{y}_s = 87.7$) must be a maximum. Also, if you move away from S in either direction along the X_1 axis, $\hat{y}$ falls off much more rapidly than if you move similarly along the X_2 axis. This indicates that the contours are drawn out (attenuated) along the X_2 axis, which has the smaller coefficient.

Now look at Figure 7. The expressions in Figure 7(a) produces the response surface shown in Figure 7(b), which represents a "saddle" or minimax whose contours are shown in Figure 7(c). Again it is easy to understand the nature of the surface without any graphical aid using the canonical form of the equation. This turns out to be $\hat{y} = 87.7 - 9.0X_1^2 + 2.1X_2^2$. Because the coefficient of X_1^2 is negative and that of X_2^2 is positive, the center of the system S is a maximum along the X_1 axis but is a minimum along the X_2 axis. Thus, we know at once that the surface is a minimax (also called a saddle). In particular, this implies that movement away from S along the X_2 axis in either direction gives *larger* values of $\hat{y}$, perhaps suggesting the existence of more than one maximum. In response surface studies genuine saddles are rather rare, but, as we shall see, they can occur.

ANALYSIS FOR THE HELICOPTER DATA

If we apply the canonical analysis outlined above to Equation (2) for the helicopter data, then we get

$$\text{Position of } S: \quad x_{1s} = 0.86 \qquad x_{2s} = -0.33$$
$$x_{3s} = -0.84 \qquad x_{4s} = -0.12$$
$$\hat{y}_s = 371.4:$$

$$\text{Shift of Origin:} \quad \tilde{x}_1 = x_1 - 0.86 \qquad \tilde{x}_2 = x_2 + 0.33$$
$$\tilde{x}_3 = x_3 + 0.84 \qquad \tilde{x}_4 = x_4 + 0.12;$$

$$\text{Rotation of Axes:} \quad X_1 = 0.39\tilde{x}_1 - 0.45\tilde{x}_2$$
$$+ 0.80\tilde{x}_3 - 0.07\tilde{x}_4$$
$$X_2 = -0.76\tilde{x}_1 - 0.50\tilde{x}_2$$
$$+ 0.12\tilde{x}_3 + 0.39\tilde{x}_4$$
$$X_3 = 0.52\tilde{x}_1 - 0.45\tilde{x}_2$$
$$- 0.45\tilde{x}_3 + 0.57\tilde{x}_4$$
$$X_4 = -0.04\tilde{x}_1 + 0.58\tilde{x}_2$$
$$- 0.37\tilde{x}_3 - 0.72\tilde{x}_4;$$

$$\text{Canonical Form:} \quad \hat{y} = 371.4 - 4.66X_1^2 - 3.81X_2^2$$
$$+ 3.27X_3^2 - 1.20X_4^2 \tag{4}$$

Now we had thought it likely that we would find a maximum at S, in which case all four squared terms in Equation (4) would have had negative coefficients. However, the coefficient (+3.27) of X_3^2 is *positive*, and its standard error will be roughly the same as that of a quadratic coefficient in Equation (2), that is, about 0.6. This implies that the response surface almost certainly has a *minimum* in the direction represented by X_3. If this is so, we will be able to move from the point S in *either direction* along the X_3 axis and get increased flight times.

In terms of the centered $\tilde{x}$'s,

$$X_3 = 0.52\tilde{x}_1 - 0.45\tilde{x}_2 - 0.45\tilde{x}_3 + 0.57\tilde{x}_4$$

Thus, beginning at S, one direction of ascent along the X_3 axis would be such that for each increase in $\tilde{x}_1$ of 0.52 units, $\tilde{x}_2$ must be reduced by 0.45 units, $\tilde{x}_3$ reduced by 0.45 units, and $\tilde{x}_4$ increased by 0.57 units. The units are those of the design given in Table 7. To follow the opposite direction of ascent you would make precisely the opposite changes. Before we explore these possibilities further, we consider a somewhat different form of analysis.

RIDGE ANALYSIS

In the original paper by Box and Wilson (1951), the application of the method of steepest ascent to response surfaces was discussed in general and in particular for second degree equations as well as those of first degree. For two variables the general concept can be understood by considering again the two dimensional contour representation of the minimax surface in Figure 7(c). As shown in Figure 8, suppose a series of concentric circles are drawn centered at the point O with increasing radius, r. It can be shown that, as r is increased, the circles will touch the contours of any response surface at a series of points at which the rate of increase or decrease in response with respect to r will be greatest. In units of x, the path formed by such points is thus one of maximum gradient and, hence, of steepest ascent or descent. For a first degree equation, such as Equation (1), this is a straight line path at right angles to the planar contour surfaces, as in Figure 3. More generally, the path is curved. In particular, it was shown that, for a second degree equation, points along the paths of maximum gradient can be found for different values of r by solving a series of linear equations. A. E. Hoerl (1959) developed an extended technique of this kind under the general heading of "Ridge Analysis" and illustrated its use with many

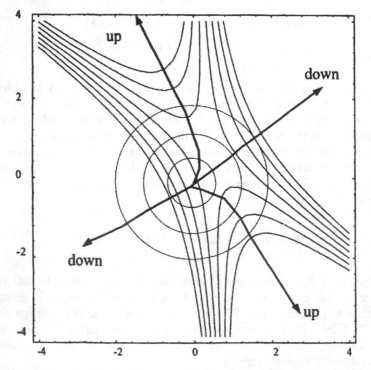

Figure 8. Second order steepest ascent and ridge analysis for the example of Figure 9.

applications (see also R. W. Hoerl, 1985). For more on the underlying theory, see Draper (1963).

Figure 8 shows, for the minimax surface of Figure 7, the paths of maximum gradient (two of steepest ascent and two of steepest descent originating from S). In this example, where O is close to S, these paths converge very rapidly onto the axes of the canonical variables X_1 and X_2. Indeed, if we start at S instead of O, these axes are themselves the paths of steepest gradient. For the helicopter example, the paths of ascent can be followed either by ridge analysis from the origin O or by following the X_3 axis from the origin S. For this example, we obtain almost identical results by either method.

Mean flight times and dispersion for a series of constructed helicopter designs along the X_3 ridge are summarized in Table 9. Each helicopter was dropped five times. To better understand these results, we also show the dimensions of the tested helicopters in terms of the original variables of wing length, wing width, body length, and body width.

These tests confirm what was implied by the earlier canonical and ridge analyses—that we can indeed get longer flight times by proceeding in either of two directions. We can increase w and L as we reduce W and l or do precisely the reverse. For sixteen helicopter designs along this path, Figure 9 shows graphically the mean flight times and standard deviations of flight times together with the dimensions of

Table 9. Experimental Data on Second Order Steepest Ascent Path[a]

Factor	A	Q	W	L	w	l	W	L	$\bar{y}$	s
Coded Factor	x_1	x_2	x_3	x_4	in.	in.	in.	in.	cent-sec.	cent-sec.
Coefficient	0.52	−0.46	−0.45	0.57						
$X_3 = 5.50$	3.73	−2.92	−3.34	2.95	2.91	5.03	0.42	3.48	332	12.7
$X_3 = 4.80$	3.37	−2.60	−3.02	2.55	2.82	5.12	0.50	3.28	373	5.8
$X_3 = 4.20$	3.05	−2.32	−2.75	2.20	2.74	5.19	0.56	3.10	395	5.9
$X_3 = 3.30$	**2.59**	**−1.91**	**−2.35**	**1.69**	**2.64**	**5.29**	**0.66**	**2.85**	**402**	**7.5**
$X_3 = 2.67$	2.25	−1.62	−2.06	1.33	2.57	5.35	0.74	2.67	395	6.6
$X_3 = 1.86$	1.83	−1.25	−1.70	0.87	2.49	5.43	0.83	2.44	385	9.0
$X_3 = 0.70$	1.23	−0.71	−1.18	0.21	2.38	5.53	0.96	2.11	374	10.2
$X_3 = 0.30$	1.03	−0.53	−1.00	−0.02	2.34	5.56	1.00	1.99	372	7.6
$X_3 = 0.00$	0.87	−0.39	−0.86	−0.19	2.31	5.59	1.04	1.91	370	6.9
$X_3 = -0.70$	0.51	−0.07	−0.55	−0.59	2.25	5.64	1.11	1.71	376	6.3
$X_3 = -1.05$	0.32	0.09	−0.39	−0.79	2.22	5.66	1.15	1.61	379	8.4
$X_3 = -1.82$	−0.17	0.53	0.04	−1.33	2.15	5.72	1.26	1.34	387	9.0
$X_3 = -2.51$	−0.43	0.76	0.27	−1.62	2.11	5.75	1.32	1.19	406	5.4
$X_3 = -3.47$	**−0.93**	**1.21**	**0.70**	**−2.17**	**2.04**	**5.81**	**1.43**	**0.92**	**416**	**6.2**
$X_3 = -3.70$	−1.05	1.31	0.81	−2.30	2.02	5.82	1.45	0.85	399	8.8
$X_3 = -4.22$	−1.32	1.55	1.04	−2.60	1.99	5.84	1.51	0.70	350	33.2

[a] The numbers in bold represent "best helicopters."

the associated helicopters. It will be seen that, in either direction, mean flight times of over 400 centiseconds can be obtained. These are almost twice the flight time of the original helicopter design. The plot shows that in both directions mean flight times go through a maximum and that the standard deviations remain reasonably constant except at the extremes where instability causes rapid increases.

Obviously, the process we have described could have been continued. However, we decided to quit at this point. In particular, we resisted the temptation to investigate further an analysis of individual degrees of freedom of the sum of squares for lack of fit. In Table 10, this lack of fit sum of squares was combined with that for "pure error" to provide that for "residual error." A more detailed analysis showed a large AQL interaction ($t = 4.3$) and a large component due to differences between curvature checks ($t = 3.8$). See, for example, Box and Draper (1987, p. 459). Both t values have 4 degrees of freedom. The reader may want to look further into these phenomena.

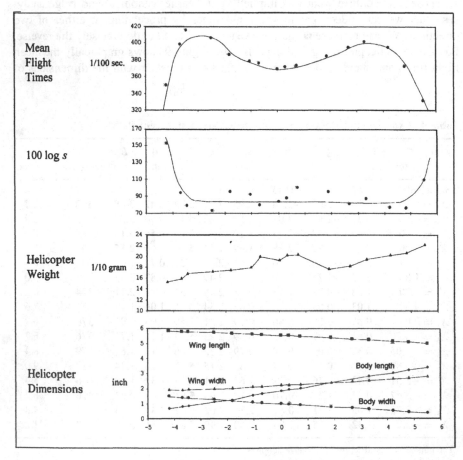

Figure 9. Characteristics of helicopters along the X_3 axis.

Of much greater importance is our hope that this example will encourage others to run *their own* iterative investigation. If paper helicopters are used, they can test their own ideas employing different starting points, varying different factors, and so forth. Also, devices other than the paper helicopter may be tried, and perhaps other methods of iterative investigation developed. Most of all we plead that the above may not be treated as providing just one more "data set" for students to further analyze and reanalyze. The art of investigation cannot be acquired by playing with someone else's data. You need to know what it feels like to make discoveries using your own. The helicopters we ended up with are not claimed to be "optimal". We know that if you use your imagination in further experimentation, you can do better and we hope you will.

CONCLUSIONS

The purpose of this paper is to provide an example of the use of statistics and in particular RSM to catalyze *investigation* and *discovery*. The iterative use of appropriate statistical methods led us along a path that *could not have been predicted* and resulted in stable helicopters with flight times almost twice those of the original prototype. Two very different helicopter designs of this kind with almost equal performance were discovered. (In practice, considerations of cost, convenience of manufacture, and so forth might decide the better choice at this point.) We believe that actual *conduct* of exercises of this kind can help students experience and understand scientific method and its catalysis using statistics.

ACKNOWLEDGMENTS

Sandra Martin's help in preliminary work on this project is gratefully acknowledged. This research was sponsored by a National Science Foundation Grant (No. DMI-9812839).

CHAPTER C.2

Response Surface Methods: Some History

The techniques demonstrated in Chapter C.1 were originally introduced in a paper read to the Royal Statistical Society many years ago (Box and Wilson, 1951) and have received considerable attention since that time. This chapter discusses some issues that arise when, as was originally intended, RSM is considered as a technique for the catalysis of iterative learning. I think I first need to explain how this paper came to be written.

SOME HISTORY

While serving in the British army during the Second World War, I was, because of my knowledge of chemistry, transferred to a research station concerned with defense against chemical warfare. Biochemical results from animal experiments were extremely variable, and, since no professional statistical help was available, I was assigned the job of designing and analyzing many statistically planned experiments. I also helped to carry them out. My efforts over the next three years were necessarily based on self-study, and most of the books and articles I was able to get were by R. A. Fisher and his followers. Later I studied theoretical statistics at University College in London and in particular became familiar with Neyman–Pearson theory.

In 1948 my first job was at a major division of ICI in England. The people there were anxious to develop methods to improve the efficiencies of their many processes, but my suggestion that statistically designed experiments might prove helpful was greeted with derision. The chemists and engineers said, "Oh, we've tried

From Box, G. E. P. (1999), "Statistics as a Catalyst to Learning by Scientific Method Part II—A Discussion," *Journal of Quality Technology*, 31(1), 16–29.

that, and it didn't work." Inquiry showed that, for them, a statistical design had meant the advance planning of an all-encompassing "one-shot" factorial experiment. This would test all combinations of the many experimental factors perceived to be important with each factor tested at a number of levels, and covering the whole of the ranges believed relevant.

A few of these large factorial arrangements had, in fact, been begun by experimenters but had quickly petered out. In the light of their knowledge of chemistry and engineering, after a few runs the statistical design was abandoned. The experimenters might say, "Now that we see these early results, we realize that we should be using much higher pressures and temperatures. Also, the data suggest that some of the factors we first thought were important are not and we should be looking at a number of others not on the original list." The failures occurred because it was presumed the use of statistics meant that experimentation had to be planned at the beginning of the investigation when the experimenters knew least about the system. The result was that statistically planned experimentation received a very bad name.

It was clear that I had much to learn, so I joined a number of teams involved in process development and improvement. I worked with them and particularly with a chemist, K. B. Wilson, who had considerable experience in that area. We watched what the experimenters did and tried to find ways to help them to do it better. It seemed that most of the principles of design originally developed for agricultural experimentation would be of great value in industry, but that most industrial experimentation differed from agricultural experimentation in two major respects. These I will call *immediacy* and *sequentiality*.

What I mean by immediacy is that for most of our investigations the results were available, if not within hours, then certainly within days and, in rare cases, even within minutes. This was true whether the investigation was conducted in a laboratory, a pilot plant, or on the full scale. Furthermore, the experimental runs were usually made in sequence,* and the information obtained from each run or small group of runs was known and could be acted upon quickly and used to plan the next set of runs. We concluded that the chief quarrel that our experimenters had with using "statistics" was that they thought it would mean giving up the enormous advantage of "learning as you go". Quite rightly, they were not prepared to make this sacrifice. The need was to find ways of using statistics to catalyze *their* process of investigation that was not static, but dynamic. RSM was introduced as a first attempt to provide a suitable adaptation of statistical methods to meet these needs. It was a great surprise to us when Professor G. A. Barnard, then ICI's statistical consultant, suggested that our work be made the subject of a paper to be read before the Royal Statistical Society.

THE KEY IDEAS OF RSM

It is necessary, I think, to reiterate the key ideas that were in the (1951) paper. The many points at which there are necessary injections of *judgement* and informed *guesswork* are indicated by italics.

*You could say these experiments were tun "in series" rather than "in parallel"

(a) Investigation is a *sequential* learning process.

(b) *When there is little or no knowledge* about the functional relationship connecting a response, y, and a group of factors, $\mathbf{x}$, *a truncated Taylor series approximation* (i.e., a polynomial in $\mathbf{x}$ of some degree d, usually 1 or 2) might produce a useful local approximation, and the *data themselves could suggest* a suitable value for d. It was shown later by Box and Cox (1964) and Box and Tidwell (1962) that the value of the simple polynomial graduation functions *often* can be considerably increased by allowing the possibility of transformation of y and/or $\mathbf{x}$.

(c) When at the beginning of an investigation *it is suspected* that considerable improvement is possible, we are *probably* "down the mountainside." In that case, most of the information concerns "which way is up" and first order terms *are likely* to dominate. Factor screening and estimation can then be achieved by using two-level Plackett–Burman and fractional factorial designs followed by first order steepest ascent.

(d) When at a later stage *first order terms appear no longer dominant*, a higher degree polynomial, in particular one of second degree, *may be useful*.

(e) When d is 2 or greater, factorial designs at $d + 1$ levels and their standard fractions obtained from group theory (Finney, 1945) are inappropriate and uneconomical for estimating the approximating polynomials. Instead, what were later called response surface designs were used. These were classified not by the number of levels used but by the degree of the approximating polynomial they estimated. Three-level designs of a different kind for fitting second degree equations were later developed by Box and Behnken (1960) that were specifically chosen to estimate the coefficients in a second degree polynomial when only three levels of the factor could be used.

(f) For comparing bias properties of possible designs, the general alias matrix was derived. In particular, if a polynomial of degree d_1 was fitted when a polynomial of degree d_2 was needed, the alias matrix determined how the estimated coefficients would be biased.

(g) By a process later called *sequential assembly*, a suitable design of higher order may be built up by *adding a further block* of runs to already existing runs from a design of lower order. For example, when the *data indicate* that this is necessary, a second order composite design may be obtained by adding axial and center points to a first order factorial or fractional. The general principle of "fold-over" is another example of sequential assembly which can be used when *it is thought necessary* to separate second order aliases from effects of first order. A more general method for determining "follow up" runs is discussed in Chapter C.7 (see also BHH II).

(h) When earlier experimentation has exploited and greatly reduced dominant first order terms, *it is likely* that a near stationary region has been reached. Rea examples showed how a second order approximating function might then be useful and how it could be estimated and *checked for model relevance and for lack of fit* (later discussed more fully in Box and Wetz, 1973, and Box and Draper, 1987).

(i) When there are only *two or three factors of major interest*, contour plots and contour overlays can be of value in understanding the system.

(j) More generally, canonical analysis of a second degree equation *can indicate* the existence of a maximum, minimum, or, as in the helicopter example, a minimax. Also the type and direction of ridges can be determined. When, as is usual, there are costs and other responses that must be considered such ridges may be *exploited* to produce better and cheaper products and processes.

(k) To check the reality of *potentially interesting characteristics* of the fitted surface, additional runs may be made at *carefully chosen* experimental conditions and, used in re-estimating the function.

Before proceeding further, I need to make a number of disclaimers:

(1) RSM as described above represented a beginning. Later, these ideas were extensively developed with other collaborators and by other researchers.

(2) The detailed methodology of RSM is appropriate to a particular species of industrial problem. It is certainly not intended as a cure-all. However, what has been made clear by my industrial experience, then and later, was that there should be more studies of statistics from the dynamic point of view. Unfortunately, with notable exceptions such as Daniel (1962), the concept of statistics as a *catalyst to iterative scientific learning* has not received much attention by statistical researchers.

(3) The illustrative investigation in Chapter C1 involves quite a number of experimental runs. Suppose, however, that we were really in the business of designing paper helicopters to achieve longer flight times and that the original design represented the previously accepted state of the art. Then we might not make the later runs initially. The increased flight time from 223 to 347 centiseconds achieved after only 21 runs using one fractional design and steepest ascent could have put us far ahead of the competition. In which case improvement efforts might, in practice, be halted temporarily. However, when competitors started to catch up, experimentation could begin again in a manner corresponding to later parts of the example. Not surprisingly, as the product got better it would take more effort to improve it.

(4) In the helicopter investigation only one response is considered (except in the early stages when dispersion is also analyzed). In most real examples there would be several responses.

(5) The helicopter investigation is conducted almost entirely empirically. In practice, the result at each stage would be considered in the light of subject matter knowledge. This would greatly accelerate the learning process. Investigation must involve *scientific feedback* as well as empirical feedback.

CHAPTER C.3

Statistics as a Catalyst to Learning

Francis Bacon (1561–1626) said: "Knowledge Itself Is Power." The application to industry of this aphorism is that by learning more about the product, the process, and the customer we can do a better job. This is true whether we are making integrated circuits, admitting patients to a hospital, or teaching a class at a university. For example, an industrial investigation to provide a new drug for a particular disease has two distinct aspects to its development: (1) the long process of *learning* by which an effective and manufacturable chemical substance is discovered and developed and (2) the process of *testing* to ensure its effectiveness and safety for human use. These two aspects have parallels in criminal procedures: (1) the tracking down of a criminal by a detective and (2) the trial of the criminal in a court of law. Solving the crime, like the discovery and development of a new, initially unidentified drug, is necessarily a sequential procedure. It emphasizes hypothesis generation based on intelligent guesswork and is inspired by clues which help decide what kind of data to seek at the next stage of the investigation.

However, the trial of the accused in a court of law is a much more formal process. It is a "one-shot" procedure in which the court must make a decision based on *already available* evidence (available data). The "null hypothesis" of innocence must be discredited "beyond all reasonable doubt" for the defendant to be found guilty. In comparison to this very formal trial process, the detective's methods for tracking down a criminal are informal and are continually concerned with such questions as: "Given what I already know and suspect, how should I proceed? What are the critical issues that need to be resolved? What new data should I try to get?" This informal investigative process cannot be put into any rigid mathematical framework.

From Box, G. E. P. (1999), "Statistics as a Catalyst to Learning by Scientific Method Part II—A Discussion," *Journal of Quality Technology*, 31(1), 16–29.

In statistics, although research on "data analysis" led by John Tukey has gone part of the way to restore respectability to methods of exploratory inquiry, it still seems to be widely believed that expertise appropriate to aid the trial judge is also appropriate to advise the detective. In particular, the concept of hypothesis testing at accepted fixed significance levels is, so far as it can be justified at all, designed for terminal testing on past data of a *believable* null hypothesis. It makes little sense in the context of exploratory inquiry. We should not be afraid of discovering something. If I know with only a small probability that there is a crock of gold behind the next tree, should I not go and look?

In the helicopter example, notice that what is plotted are the averages of repeat runs. The variation in these averages thus includes "manufacturing" variation and is appropriate for conclusions drawn about the helicopter *designs*, rather than conclusions about particular helicopters (see Chapter C.1).

The informal use of normal probability plots is quite deliberate. We use the plots simply to indicate what *might be worth trying*. The idea that before proceeding further we need to discredit the already incredible null hypothesis (that the prototype design is the best possible) is clearly ridiculous. Also when experimentation is sequential we need to think in terms of conducting a blitzkrieg rather than trench warfare. However, if desired, confidence cones about the directions of steepest ascent can be calculated (Box, 1954a; Box and Draper, 1987). For the two steepest ascent paths calculated in Chapter C.1, 95% confidence cones exclude, respectively, 97.9% and 94.7% of possible directions of advance. But it must always be remembered that (except perhaps when dining with the Borgia's) the proof of the pudding is in the eating.

The analysis of variance in Table 7 in Chapter C.1 is not intended to be used formally either. Its main purpose is to determine whether the fitted second degree equation is estimated sufficiently well to be worthy of further interpretation. The occurrence of an F multiplier of 8.61* strongly suggests that it is. The F multiplier indicates whether the overall change in response predicted by the fitted equation is reasonably large compared with the error in estimating the response (Box and Wetz, 1973; Box and Draper, 1987). For this purpose it is more appropriate than the more frequently used R^2.

PARADIGM OF SCIENTIFIC LEARNING

The paradigm for scientific learning has been known at least since the time of Robert Grosseteste (1175–1253) who attributed it to Aristotle (384–322 B.C.). The iterative inductive–deductive process between model and data is not esoteric but is part of our everyday experience.

Two equivalent representations of this process (which might be called the cycle and the sawtooth) have been given, respectively, by Deming's (1986) modification of Shewhart (1939) and by Box and Youle (1955). They are shown in Figure 1. For

---

* That is, the F value is 8.61 times as large as the value required for "5% significance."

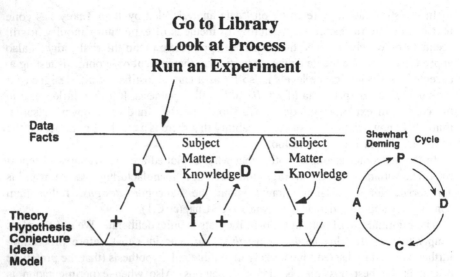

Figure 1. Diagrammatic representation of the iterative learning process.

example, the sawtooth model indicates how data which look different from what had previously been expected can lead to the conception of a new or modified idea (tentative model) by a process of induction. By contrast, consideration of what the data would imply if the model were true is achieved by a process of deduction. The first implies a contrast indicated by a minus sign; the second involves a combination of data and model indicated by a plus sign. Some of the theoretical consequences of these ideas are discussed in Box (1980, 1983a). Studies of the human brain over the last few decades have confirmed that, in fact, separate parts of the brain are engaged in a conversation with each other to perform this inductive–deductive iteration. Thus, although different people may follow different paths of reasoning, they can still arrive at the same solutions (or at different but equally useful solutions).

Notice that the acquiring of data may be achieved in many different ways, for example, by a visit to the "web" and the library, by observing an operating system, or by running a suitable experiment. But to be most fruitful, *subject matter knowledge* must be available for all such activities. For example, after the analysis of an initial experiment, a conversation between a scientific investigator and an inexperienced statistician could go something like this:

Investigator: "You know, looking at the effects of factors B and C on the response y_1, together with how they seem to affect y_4 and y_5, suggests to me that what is going on physically is thus and so. Therefore, I think that in the next design we had better introduce the new factors G and H and drop factor A."

Statistician: "But at the beginning of this investigation I asked you to list all the important variables and you didn't mention G and H."

Investigator: "Oh yes, but I had not seen these results then."

While statisticians are accepted by scientists as necessary for the testing of a new drug, their value in helping to design the series of experiments that lead to the discovery of the new drug is less likely to be recognized. For example, Lucas (1996) estimated that, of the 4,000 or so members of the American Statistical Association (ASA) who were engaged in industry at that time, about 3,000 were in the pharmaceutical industry—one suspects that a disproportionate number of these were concerned with testing rather than with the more rewarding and exciting process of discovery.

A MATHEMATICAL PARADIGM

A purely mathematical education is focused on the one-shot paradigm—"Provide me with a set of assumptions and if some proposition logically follows, then I will provide a proof." Not surprisingly this mindset can also produce a paradigm for hypothesis testing in mathematical statistics—"Provide me with the hypothesis to be tested, the alternative hypothesis, and all the other assumptions you wish to make, and I will provide an 'optimal' decision procedure." Similarly with experimental design—"Tell me what are the important variables, what is the exact experimental region of interest in the factor space, what is the functional relationship between the experimental variables and the response, and I will provide you with an alphabetically optimal design!" These are requests to which most investigators would respond, "I don't know all these things, but I hope to find them out as I run my experiments."

By historical accident, experimental design was invented in an agricultural context. For example, Fisher's earlier interest in aerodynamics could have resulted in a career in aircraft design, perhaps producing a different emphasis in the conduct of experiments. However, the circumstances of agricultural investigation in which the whole experiment must be planned one year but the results are not available until the next, are rather unusual (certain industrial life testing experiments are an exception). They should certainly not be perceived as sanctifying methods in which all assumptions are fixed *a priori* and lead to a one-shot procedure. Iterative learning, of course, goes on in agricultural trials as elsewhere. The results from each year's trials are used in planning the next. The process of iterative learning is the same but occurs more slowly.

CONTINUOUS NEVER-ENDING IMPROVEMENT

We can better understand the critical importance of sequential investigation if we consider the central principle of "continuous never-ending improvement." This might at first be confused with mathematical optimization, but mathematical optimization takes place within a fixed model; by contrast, in continuous improvement, neither the functional form of the model, nor the identity of the factors, nor even the nature of the response is fixed. They all evolve as new knowledge comes to light.

Furthermore, while optimization with a fixed model leads inevitably to the barrier posed by the law of diminishing returns, a developing experimental context provides the possibility of continuous improvement and for expanding possibilities of return.

At the end of the nineteenth century Samuel Pierpont Langley (1834–1906), a distinguished scientist and a leading expert in aerodynamics who had considerable financial support from the U.S. government, built two full scale piloted airplanes designed largely from theoretical concepts. The planes were not operated by Langley himself, and they never flew; the second fell off the end of the runway into the Potomac. By contrast, the Wright brothers, after three years of iterative learning (first flying kites, then gliders, then powered aircraft), discovered not only how to design a working airplane but also how to fly it. (In the course of their investigations they also discovered that a fundamental formula for lift was wrong; they built their own wind tunnel and corrected it.) Their airplane design may have been close to optimal in the limited factor space they considered but in a wider sense it, was not optimal any more than is that of any modern aircraft. By continuous never-ending improvement the dimensionality of the factor space in aircraft design, as in any other subject, has continually increased.

It is obviously impossible to prove mathematical theorems about the process of scientific investigation itself, for it is necessarily incoherent; there is no way of predicting the different courses that independent experimenters exploring the same problem will follow. It is understandable, therefore, that statisticians inexperienced in experimental investigation will shy away from such activities and concentrate on the development of mathematically respectable one-shot procedures. For such work it is not necessary to learn from (or cooperate with) anybody. For example to develop statistical decision theory, there was no need to consider the way in which decisions were actually made; nor to develop the (many!) mathematically "optimal" design criteria was it ever necessary to be involved in designing an actual experiment.

There has been considerable discussion of the malaise which has affected statistical application. For example, one of the sessions at the ASA annual meetings a few years ago was on the topic "The D. O. E. Dilemma: How Can We Build on Past Failure to Ensure Future Success." I believe such discussion would be more fruitful if attention was focused on the root cause of such problems—namely, the confusion between the mathematical and the scientific paradigm in determining what we do.

SUCCESS OF THE INVESTIGATION IS THE OBJECTIVE

In the context of iterative learning, optimizations of separate designs and analyses will necessarily be sub-optimizations. It is the investigation itself, involving many designs and analyses, that must be regarded as the unit, and the success of the investigation that must be regarded as the objective. Although we cannot prove any mathematical theorems about the process of iterative investigation, we can reorient some of our research to the study of the learning process itself. This should not be a matter for too much dismay. For example, after the discovery of the genetic code, geneticists realized that, in addition to their previous knowledge, they must now acquire expertise in coding theory. We, too, ought to be able to make a transformation of this kind.

THE INDUCTIVE POWER OF FACTORIAL DESIGNS

The floundering that we tend to do between the scientific and mathematical paradigms can lead to major misunderstandings. For example, I have often found myself defending the factorial design. "Surely," I am told, "with modern computers able to accomplish enormous tasks so quickly, you ought not to be content with outdated factorial designs." One can, of course, try to point out that the design of a real experiment involves judgement and the wise balancing of many different issues with the help of the investigator. But, in addition, a different and very important point is usually missed. In favor of factorial designs is their enormous *inductive* power. Even if we grant that an "optimal" design might provide a useful answer to the question posed before we did the experiment, the experimental points from such a design are often spread about in irregular patterns in factor space and are of little use as a guide to what to do next.

One of the most fundamental means by which we learn is by making comparisons. We ask, "Are these things (roughly) the *same* or are they *different*?" A preschool coloring book will show three umbrellas with the young reader invited to decide if they are the same or different. A factorial design is a superb "same or different machine." Consider, for example, a 2^3 factorial design in factors A, B, and C represented as a cube in space in which a response y is measured at experimental conditions corresponding to each corner of the cube. By contrasting the results from the two ends of any edge of the cube, the experimenter can make a comparison in which only one factor is changed. Twelve such comparisons, corresponding to the 12 edges of the cube, can be made for each response. These basic comparisons can then be combined in various additional ways. In particular, they can be used to answer such questions as: "On the average, are the results on the left hand side of the cube about the same as those on the right hand side (factor A main effect) or by about how much are they different? Are they on average the same on the front as on the back (factor B main effect) or are they different?" Also, since an interaction comparison asks whether the *differences* produced by factor A are the same or different when factor B is changed, the possibility of interaction between the factors can be assessed by similarly comparing the diagonals of the design.

As pointed out by Daniel (1966), however, there are many natural phenomena which are not best represented in terms of main effects and interactions (or by polynomial functions). In particular, a response may occur only when there is a "critical mix" of a number of experimental factors. For example, sexual reproduction requires a binary critical mix; to start an internal combustion engine requires a quaternary critical mix of gas, air, spark, and pressure, and so forth. With a 2^3 factorial design a tertiary mix is suggested as one explanation when one experimental point on the cube gives a response widely different from all the others. A binary mix is suggested when two points on an edge are different from all the rest (see, e.g., Hellstrand, 1989).* Such possibilities can be suggested by a cube plot and by a normal plot of the *original data*. For this reason it seems best to decide first the space in which there is activity, of some kind, for a group of factors and to then decide what kind of activity it is. In practice, information may become available simultaneously for not one but for many different responses measured at each experimental point. The knowledge of which factors affect which responses and graphical illustration of the kind of effects they have can provide an inspiring basis for the scientist or engineer using subject matter knowledge to figure out what might be happening and to help decide what to do next.

PROJECTIVE PROPERTIES OF FACTORIAL DESIGNS

Factorial designs were made even more attractive for purposes of screening and inductive learning by the discovery that certain two-level fractional factorials have remarkable projective properties (Box and Hunter, 1961a and b). For example, a 2_{IV}^{8-4} fractional factorial orthogonal array containing sixteen runs can be used to find up to three active factors out of eight suspects; it supplies what was later called a (16, 8, 3) screen (Box and Tyssedal, 1996). For this design, every one of the 56 possible ways of choosing three columns from the eight factor columns produces a duplicated 2^3 factorial design. The original 2_{IV}^{8-4} design can therefore be said to be of *projectivity* $P = 3$. In addition to the fractional factorial arrangements for 4, 8, 16, 32, ... runs, a different kind of two-level orthogonal array is available for any number of runs that is a multiple of four (Plackett and Burman, 1946). These arrangements will be called P.B. designs. They provide additional designs for 12, 20, 24, 28, ... runs; some of these turn out to have remarkable screening properties. Furthermore, Lin and Draper (1992) showed by computer search that some but not all of the larger P.B. designs had similar properties. The conditions necessary for such designs to produce given projectivities were categorized and proved by Box and Tyssedal (1996). Such screening designs are important because they can suggest which subset of the tested factors are, in one way or another, active. See also Lin (1993a,b, 1995).

Many designs have another interesting projective property that allows simple procedures to be used when the factor space is subject to one or more linear constraints. One such constraint that has received much attention is the additive constraint which occurs when a set of factors measures the proportions of ingredients which must sum to unity. For this and other constraints, it was shown by Box and Hau (1998) that many of the operations of RSM can be conducted by projecting standard designs and procedures onto the constrained space.

We have spent far too much time on one-shot statistical procedures designed to test rather than to learn. I have explained how I think this has come about, largely

---

*Such phenomena may mainly indicate the existence of outliers and in particular of misconducted experiments. All the possibilities must be borne in mind.

because of the idea that we can develop statistics from mathematical ideas conceived at our desks. You cannot learn to swim by blackboard instruction. In particular, for optimal experimental design, the experimenter is supposed to know in advance which factors, which region of the factor space, which function, and which constraints should be considered. The experimenter is thus credited, on one hand, with the prescience of the Oracle of Delphi, and on the other with sufficient naiveté to accept the simplistic solutions that may be offered.

When an experiment is run the most relevant question is "So now what do we do?" The structure of the experimental design should help the experimenter to decide this.

CONCLUSIONS

- Much industrial experimentation has a characteristic, here called immediacy, which means that results from an experiment are quickly known.
- In this circumstance investigations are frequently conducted sequentially with results from previous experiments interacting with subject matter knowledge to motivate the next step.
- Such investigations use what may be called the scientific learning paradigm in which data drives an alternation between induction and deduction leading to change or modification of current knowledge.
- The iterative scientific paradigm is different from the one-shot mathematical paradigm for the proofs of theorems.
- It is argued that because statistical training has unduly emphasized mathematics, confusion between the two paradigms has occurred resulting in concentration on one-shot procedures encouraged by agricultural examples in which feedback occurs very slowly.
- RSM provides one example of iterative learning using factor screening, steepest ascent, and canonical analysis of maxima.
- Factorial designs provide data which encourage inductive discovery. Projective properties of fractional factorials and other orthogonal arrays further assist this process.
- Because of these projective properties fractional factorials and other orthogonal arrays have valuable screening ability.

CHAPTER C.4

Experience as a Guide to Theoretical Development

When a new topic in statistics pops up the standard reaction is to formalize, generalize, and present alternative solutions. But one may ask: Where do such new topics (in some instances break-throughs) pop up from? I have argued in Chapter B.12 quoting some 34 examples extending from Gauss to Efron and beyond, that they have most often arisen from practical need. When a model that at first looks sensible does not lead to good practice this suggests that a change in the model may be necessary. Thus example described how in WWII an "obvious" question from a nonstatistician about an "optimal" sampling inspection scheme led to the invention of sequential tests. But nothing is less obvious than the obvious; here are some examples from my own experience, when practical experience led to appropriately modified theoretical models.

Constrained control

In 1960, when Gwilym Jenkins and I first started to think about feedback adjustment methods suitable for statistical process control, it seemed natural to develop schemes for which the deviations from the target had minimum mean square error (MMSE). We worked out schemes that did this. But not long afterwards, Dr. Park Reilly, who had been one of Gwilym's Ph.D. students and had taken an important job as an industrial statistician in Canada, told us how this "optimal" theory applied to one of his processes had led to a control algorithm which was totally nonsensical. It went something like this: observations were being taken every 30 minutes with temperature as a control variable. The algorithm which would, theoretically at least, have provided minimum mean square error (MMSE) at the output was telling him at

From Box, G. E. P. (1998), "Statistics as a Catalyst to Learning by Scientific Method. Response to the Discussion," *Journal of Quality Technology*, **31**(1), 67–72.

one point to increase the temperature by 2000°C and 30 minutes later to reduce it by 1500°C! We quickly realized that our MMSE criterion needed to be modified. Experience had shown us that the criterion should have been to produce a small mean square error at the output with minimum compensatory manipulation at the input.

Practical schemes were soon discovered (Box and Jenkins, 1968, 1970; Aström, 1970; Aström and Wittenmark, 1984) requiring remarkably little manipulation at the cost of only a slight increase in the output mean square error above the minimal level. Such schemes could be complicated, however; and in more recent work it has been possible (again, by modifying the criterion) to obtain almost equally good schemes that are very simple to design and to use (Box and Luceño, 1997).

Theoretical Models and Built-in Constraints
I have always preferred to use theoretical models when from physical, chemical, or engineering considerations approximate mechanisms could be conjectured. When available such models can lead to deeper understanding of the system being studied and to much faster progress. Frequently, however, for want of something better, we have to begin with empirical models hoping always that they might provide insight that can lead to a more satisfactory mechanistic model. In my later years at I.C.I. and after I joined the Statistical Techniques Research Group at Princeton in 1956, my colleagues and I were particularly interested in nonlinear estimation required for mechanistic models (see, e.g., Box, 1960). However, in such studies it is common to measure not one, but a number of responses simultaneously (in particular the amounts of various substituents in a sequence of different reaction times). In what can be regarded as a generalization of least squares it was shown that estimation with multiple correlated responses required the minimization of the determinant of sum of squares and products of deviations from expectation. This automatically took care of correlation between such deviations (Box and Draper, 1965).

Again practice quickly enlightened theory. A correspondent, this time from DuPont, told us that he had applied our method but the computer could find no unique minimum. As I remember it, each of three substituents y_1, y_2, and y_3, supplied information about one or more of a set of physical constants to be estimated. Because chemical balance required that $y_1 + y_2 + y_3 = 100\%$, it turned out that for the data we were given the experimenter had not actually measured y_3 but had obtained it by taking $100 - y_1 - y_2$. The determinant was then of course zero everywhere because of this linear constraint, and the computer was searching on the basis of a rounding error. By calculating the determinant only for those responses which were actually measured, satisfactory answers were obtained. My Monday Night Beer Seminar was regularly attended by a number of graduate students from chemical engineering who said that data of this kind were often published with no indication that such dependencies had been introduced. They found a publication where experimental data, on 5 responses, were again modeled in terms of kinetics and again we found that the determinant was zero everywhere as before. However, after eliminating the (1, 1, 1, 1, 1) constraint (the five responses had been forced to add to 100%), the 4×4 determinant was still hovering close to zero. The engineers searched the library and unearthed the original investigation. From this it was clear that a second dependency had been introduced because one of the substituents had been calculated by a chemical balance formula involving two of the others. Clearly, dependence of this kind was quite likely to be missed, so we devised an eigenvalue analysis that could check empirically for dependencies. A paper about this problem was later published by four of us Monday nighters (Box et al., 1973). Notice that, as commonly occurs, the likelihood of such constraints was one more thing that did not occur to us initially and had to be learned about. A somewhat similar problem arose in re-estimating parameters from data collected from a system which was undergoing feedback control (Box and MacGregor, 1974, 1976).

Empirical Ridges and Theoretical Models

When I first began to look at response surfaces, I think I assumed that maxima would be reasonably symmetric. For example, in three dimensions contours would surround the maximum like the skins of an onion. But the systems I studied almost always had ridgey maxima or consisted of stationary or rising ridges. (Ridges, of course, relate directly to interactions and to the fact that changes in variables like temperature, concentration, and pressure can often compensate each other.) Such ridges could be identified by eigenvalue analysis. In one example where only three factors were being studied, I was somewhat surprised to find two of the eigenvalues close to zero. The contour picture in three dimensions was not like an onion or like a jelly roll, but like a sandwich with the ham being the maximum and the slices of bread being contour planes on which lower yields were obtained. (I'm not sure that industry is sufficiently aware of the economic possibilities that such ridge systems offer. In particular, they raise the possibility of maximizing more than one response and so of using alternative, cheaper processes.)

My physical-chemist friend, Dr. Phillip Youle, was fascinated by this ham sandwich maximum and conjectured it could occur as the result of a particular type of consecutive reaction. You can see the idea if you think of the molecules in the reaction initially as black and white billiard balls, initially with a large preponderance of black balls, all in rapid motion on the billiard table. Now, suppose that whenever a black ball collides with a white ball, it will, with a certain probability, produce a red ball (the desired product), but, similarly, whenever one of these newly created red balls collides with a black ball, a green ball (an undesired product) will be produced. If we set the system going, the number of the desired red balls on the table will first increase then decrease until finally only black and green balls remain. To get the best yield you must stop the system at some time—the optimal reaction time—when there is a maximum of red balls.

Now running the experiment again at a higher temperature can be simulated by imagining the speed of all the balls on the table to be increased. If the probabilities of all collisions is increased *proportionately*, then precisely the same sequence will occur, with the same maximum yield of red balls, but in a shorter time. Thus reaction time and reaction temperature will compensate each other. A similar effect will be produced by increasing the initial number of black balls (corresponding to increasing the concentration). So it is easy to see that a stationary ridge system, in reaction time, temperature, and concentration, of exactly the kind we found will result. This will be so, however, only if the speed of all the balls is increased by the same proportion. The increase of speed produced by the temperature change is measured by a constant called the activation energy. For this example, the activation energies of the two reactions were evidently about equal. Consequently we got the same yield at many different combinations of temperature, reaction time, and concentration. If the activation energies were unequal, then higher yields could be obtained by changing temperature, and we would obtain a rising ridge (see Box and Youle, 1955). Also, if the "orders" of the two reactions were different, the maximum yield could also be changed by changing concentration. In our analogy this would occur, for example, if a red ball had to collide with two black balls simultaneously to produce a green ball while the rest of the system remained unchanged.

In this investigation then, an interesting empirical response surface study had shown us alternative ways of getting the same yield. Within this framework we could minimize cost and satisfy certain constraints revealed by analysis of other responses. But this analysis did not tell us how to increase the yield. However, the subject matter knowledge which led to physical understanding of the system allowed the chemical engineers to figure out how they might get higher yields and in particular to consider different catalysts which would speed up the first reaction more than the second. (It showed what new dimensions to investigate.)

Multiple Measurements, Constraints, and Graphics

Concerning multiple measurements and the occurrence of constraints, each problem, I think, has to be considered scientifically on its merits. In particular, it is the responses themselves that frequently set the constraints. A priori we don't know where they will be, but must learn about them. For over 20 years I worked with a major manufacturer of packaged foods. In experiments aimed at improving products and developing new ones we always had multiple responses, and we found the superimposition of contour maps very helpful. In particular, they helped us to learn what these constraints were and where they were. For instance, we might discover a region in which the "crumbliness" of the cake was too high. (Initially, we might not even know that "crumbliness" could be a problem.)

I don't think we should dismiss too quickly the simple idea of superimposition of contour maps but rather to use computer graphics, now available to us to extend it. We are blessed with three-dimensional eyesight, and nowadays, the computer allows us to use this greatly increased capability. So the same kind of thing can be done with any choice of three factors and any number of responses indicated, for example, by contour surfaces of different colors. For more factors several 3-D graphics of this kind can be shown side by side on the computer screen at the same time. It is impossible to over-emphasize the catalytic value of appropriate graphics of this kind which can blend with subject matter knowledge to produce lateral thinking* and new ideas. An interesting example concerned the formulation and manufacture of a paint used for automobiles. In one particular study, after over one hundred runs had been "ad hoc" made and many responses looked at, I was consulted (I think as a last

*See De Bono 1967.

resort). Going through their notebooks, the paint chemists said things like, "We can get some of the other responses right at these conditions but then these other response aren't right". The appearance of a newly painted automobile is extremely important to the salesman. It is almost equally important that the paint job should not be spoiled by minor abrasions. Some designed experiments and the careful analysis (see BHHII for more details) of what was going on pointed to two important factors that affected the glossiness and the abrasion resistance. Unfortunately, they showed that there was a diagonal band in this two-dimensional factor space where adequate abrasion resistance could be obtained, but that this band was almost parallel to the limit of the region which gave adequate glossiness. Since the regions did not overlap there was no way of using these factors to obtain both properties simultaneously— the root cause for the fruitless ad hoc runs. I showed my two-dimensional diagram to the paint technologists, and I said what you need to do, using your specialist knowledge, is to try to think of a *new* factor and consider it as constituting a third dimension. This variable must be such that when you travel through this additional dimension either the abrasion resistance band will move down or the glossiness region will move up; or, the new variable will make these constraint limits go at different angles so the two regions will overlap. They thought about it for some time and tried a few things and finally produced a variable which did exactly what was wanted. These examples all illustrate the ability of statistics, experimental design, RSM, and graphical representation to inspire induction and lateral thinking.

Model Discrimination

So far we have talked about *the* theoretical/mechanistic model, but frequently there are a number of sensible models that might explain the data. A problem important to chemical engineers with whom I've worked is how to discriminate between rival models (see, e.g., Box and Hill, 1967). Discussion of the problem of comparing models that contain a different number of parameters is given in Box and Henson (1969, 1970). More recent work with chemical engineers at Madison has led to further developments both for univariate and multivariate data (Stewart, Henson, and Box, 1996; Stewart, Shon, and Box, 1998). In these articles a number of examples are analyzed. In the second paper we discuss multivariate data arising in this particular context.

Simplification by Transformation

Concerning dispersion effects I think that, just as is the case for interactions, it is important to consider *transformable* and *non-transformable* dispersion effects. There seems to be some feeling that we should be reluctant to transform data, but it surely must be rare that the quantity most easily measured is also that in which it is appropriate and simplest to model. Often the importance of transformation tends to be obscured because, for standard transformations and if the data cover only a narrow range, the transformed data can be (almost) a linear recoding of the original. When the data cover wider ranges, however, large location effects can, of course, produce larger interaction and dispersion effects simply because the choice of metric is inappropriate (see, e.g., examples in Box and Cox (1964) and re-analysis in Box and Fung (1995)).

Correlation of What?

Correlation in data is of enormous importance and can occur in various ways. In particular, it is good to see that these days serial dependence is being taken more seriously. But even for contemporaneous correlation we need to distinguish carefully between, on the one hand, relationships between errors and, on the other relationships between mean values. Again, an example is enlightening.

I remember a designed experiment on a large rotary baker; the main object was to increase yield. The baker contained "canon balls" to break up the product, and the operation could be very noisy. The process workers told me that they already knew when to expect a high yielding batch "because it banged more." So, I arranged a one to ten scale of "banging" with them—if you had to shout 12 inches away from your co-worker's ear in order to be heard that was a "ten" and so forth. So, one of several responses reported from the experiment was the banging intensity.

The analysis showed indeed that certain factors produced large main effects and interactions for yield, but effects very similar in sign and magnitude were also found for banging. So, I temporarily forgot about the factorial analysis and plotted banging against yield and got a very good straight line. It seemed possible, therefore, that the factors we had tested were just changing the "gooey-ness" of the product and it was "gooey-ness" that determined yield.

We consulted a rheologist who thought about it and then said that for this particular process this idea might well be correct. He also suggested other simpler and cheaper ways of changing "gooey-ness". These were later tested and produced the desired result.

Notice that, unlike my previous examples, what was important was not the correlation between errors, but the correlation between mean values for yield and banging. Also notice that the solution to the problem came about as the result of a quite unexpected phenomenon which was suggested by a simple but unorthodox analysis.

SOME CONCLUSIONS

Questions concerning correlations, multiple responses, and multiple constraints have been around for many years, but there have been very few answers. It is as if we hoped that some purely mathematical solution could be applied to solve all these problems. I don't believe this is possible, and my examples above are intended to illustrate this. I think that statisticians must grit their teeth and also become practitioners. Only then can they become the respected colleagues of investigators, and only then will they discover where the truly novel problems are. If you don't know whether you are a good statistician or not, here is a simple test: If your name is Joe Blow and there is a very tough investigation coming up, and the engineers and scientists say, "We've got to have Joe on the team," then you will have arrived.

From all of this I conclude that whether statistics is taught by statistics departments, industrial engineering departments, or quality engineering departments, we need to reconsider a number of things. Mathematical capability is important, particularly if necessary training is directed toward use in engineering and science, but scientific thinking is paramount. But the prerequisites for a student studying scientific statistics should also include experience in an experimental science, and where this is lacking, remedial courses should be required and supplied. Furthermore, Ph.D. theses should not be judged on the amount of mathematics they contain (this might be small or large depending on what is needed to solve the problem), but they must demonstrate the student's ability to catalyze learning and to develop new methods which do this. Promotions, including promotions to tenure, should have similar requirements. The curriculum should be re-examined in the light of these considerations and should place much more emphasis on the usefulness to scientific inquiry of the subjects that are taught.

CHAPTER C.5

The Invention of the Composite Design

The central composite design shown for three factors in Figure 1 has come to be widely used. It was the first introduced in the original paper on response surface methods (RSM) read to the Royal Statistical Society in 1950 and published in the following year. I believe we can learn something from the history of how that design came to be. In particular, how it was that practical need motivated discovery and later theoretical development.

As described earlier in Chapter C.2, my first job in 1948 was at a large division of Imperial Chemical Industries (ICI) in England where there was a great deal of effort to improve processes using experimentation on the laboratory scale, the pilot-plant scale or, less often, the full scale. I was told it would be my job to try to develop better methods for doing this. Some chemist colleagues and I decided to make a study of how our experimenters presently conducted such investigations; we would then try to find out if and how appropriate statistical methods might help them to do it better.

Results were usually quickly available in days, hours, or, in some cases, even minutes, and our industrial experimenters naturally took advantage of this immediacy by running their process improvement investigations sequentially.* After each small group of runs, they could consider their results and decide what to do next.

It was clear that any statistical methods we developed must adopt this same approach. With this in mind, we began to run small groups of runs in sequence using two-level fractional factorials and Plackett–Burman designs. We used these designs for factor screening—to find out which of a larger number of candidate factors were

From Box, G. E. P. (1999–2000), *Quality Engineering*, **12**(1), 119–122. Copyright © 1999 by Marcel Dekker, Inc.

* Agricultural experiments are also sequential in the longer run: each year's results supplies information about the questions asked, but also raises further questions to be resolved in later years.

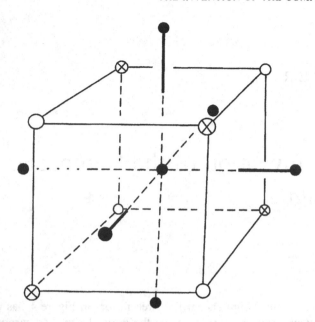

Figure 1. A composite design for three factors showing the original eight points of the 2^3 factorial indicated by open circles and the seven added axial points shown by filled circles. Also shown (by open circles containing crosses) is a half-fraction of the cube that could have been run initially.

active—but also to determine how these active factors should be changed to follow a direction of *steepest ascent*. Initially, the ideas were developed from experience with *lab scale* experiments and we had a number of striking successes. Typically, the experimenter would be a chemist working on a new route to produce a particular chemical, say, X. Using knowledge of chemistry, the experimenter might discover a route by which X might be synthesized, but often initially in very low yield, perhaps only 20% of that theoretically possible. By suitably changing levels of the factors, such as the reaction temperature, the reaction time, the reaction pressure, and the various concentrations of ingredients, it had usually been possible using "cut and try" and "one factor at a time" methods eventually to get much higher yields. Our experimenters soon found, however, that this could be done much more quickly using screening designs and steepest ascent in the manner described above. Often yields of 70% or 80% could be obtained with only one or two cycles of these procedures. Typically, the process was then further developed in a pilot plant before going to full scale.

One of my physical-chemist friends was Dr. Philip Youle. He had a remarkable mind and quickly understood the potential of statistics for process development. He was also a good politician. Philip said to me, "George, these methods certainly increase yields more quickly in the lab, but you will never get people to realize how valuable they are unless you save a lot of money by improving full-scale processes."

Philip searched round and finally came up with an important "intermediate" chemical, which I will call "product Z." This had been manufactured for many years, but at an exceptionally low yield. After earlier failures to improve the process, it was said in the research department, "Men may come and men may go, but the yield of product Z is never more than 40%." Philip said, "We will certainly get their attention if we can improve *this* process." There was clearly a strong need to do this, but experiments on the full scale were not encouraged. So it was only after a good deal of political maneuvering that Philip was able to get the people responsible for process development to run a modest experiment on the plant itself. Three factors, which we will call A, B, and C, were varied at each of two levels. Thus, we ran eight batches at experimental conditions whose levels are represented by the corners of the cube in Figure 1. From past experience, we had a fairly good estimate of the magnitude of the batch-to-batch variation. It was not particularly large, so we expected to be able to detect rather small effects.

After the eight runs had been made, everyone was anxious to know what we had found. I obtained the results but was surprised and disheartened by my analysis. In the light of hindsight, I can see what had happened. In the earlier lab experiments, we had usually started with low yields. So we were "way down the side of the mountain" and the most important question was, "Which way is up?" In such circumstances, main effects would be expected to be dominant and would be detected by the fractional design and show which way to go. Unfortunately, for this full-scale experiment, the three main effects *were all small*, but the three two-factor interactions *were all very large*. To get a better look at our data, we set up our results on what we called the "rat cage." In this device, for each run, the yield and other measurements were written on a card. The resulting eight cards were then appropriately placed on a wire grid so that we could see all our results simultaneously in three-dimensional space. We could walk around this and look at the results from various points of view in much the same way you can rotate images in a computer these days. The more we looked, the less happy we were. Although all the variables were clearly very active, we certainly were not going to get anywhere with the (first order) steepest ascent method that we had been using. With that method, it is the main effects that show which way to go. But our main effects were all essentially zero.

We had received the results in the afternoon and I knew that we were going to have a meeting next morning to discuss what we had found. I was puzzled at first, but eventually began to think along the following lines: If we are approaching a maximum, the linear terms will have been eliminated already by earlier "one factor at a time" experimentation. Therefore we may need a full second-degree approximating equation to represent the response surface (i.e., a *second-order* Taylor expansion of the unknown functional relationship). Such an equation has three linear terms, three cross-product terms, and three quadratic terms. The linear and the cross-product terms are estimated by the main effects and two-factor interactions already supplied by our 2^3 design, but what we do *not* have are estimates for the quadratic terms. These measure curvature in the three directions, and we obviously could not get them from the present two-level design. When I told Philip about this, he said, "Oh dear, that means three levels! I had a hard time getting them to do a 2^3 experiment with 8 runs, they will never do a 3^3 experiment with 27 runs." After

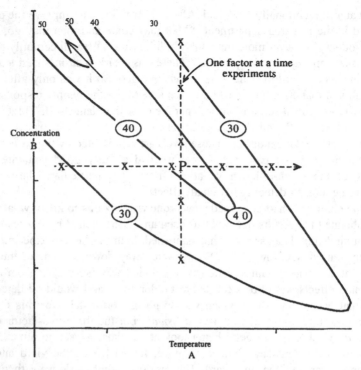

Figure 2. Contours of a rising ridge in temperature and concentration for percent conversion to the desired product. Crosses denote one factor and time experimental runs which (falsely) point to P as the maximum. Circles denote runs made with a 2^2 factorial design showing no main effects but a large two-factor interaction.

staring at the rat cage for some time, I said, "Well, maybe we don't need to do that. We might just add two extra points here, two points here, two here, and one in the middle" (just as you see in Fig. 1). Philip said, "Is that a good design?" I said, "I don't know, but I'm pretty sure it will work. So let's see if they will make these seven additional runs and then we'll see what we've got."

Well they did, and in the meantime I worked out how to fit and analyze a second-degree equation obtained from a design of this kind and also satisfied myself that the design was reasonably efficient. After we fitted a second-degree equation to our data, we found that it represented a surface that was a rising ridge. This discovery eventually showed how, although no increase in yield could be obtained by varying any one factor, progressively higher yields could be obtained by appropriately changing *all three of them together*. By doing this, we eventually saved a great deal of money.

How such a situation comes about is best explained for two factors. For simplicity Figure 2 shows the contours of an oblique rising ridge in the two factors—temperature and concentration. You will see that, as indicated by the arrow, if the concentration is increased at the same time that temperature is decreased, you will climb up the rising ridge. The circled numbers on this diagram illustrate how, for this kind of surface, a factorial experiment could show no main effects, but very large two-factor interactions.* Furthermore, the diagram shows why "one factor at a time" experimentation (which was the method that had previously been used) naturally gets stuck on such a ridge. The crosses in Figure 2 show possible runs from a "one-factor-at-a-time" experiment. The five crosses on the horizontal dotted line indicate runs made at different temperatures with the concentration fixed. These show an apparent maximum close to P. However, if temperature is now kept fixed at this "best" value and further runs are made at different concentrations on the vertical dotted line, it will be concluded that the best concentration is also close to P. Hence, it will be erroneously supposed that P is the overall maximum point.

Note that neither the one-factor-at-a-time method nor the two-level factorial experiment will tell us about the rising ridge. However, if five additional points had been added to the factorial to form a composite design, an analysis based on a second-degree equation could reveal the ridge. [In general, such an equation can describe a maximum, a minimum, or some kind of ridge.]

This building up of the composite design in two parts was an early example of what I later called *sequential assembly* of designs. With sequential assembly, designs can be built up so that the complexity of the design matches that of the problem. Thus, in Figure 1, we might not run the full 2^3 factorial design right away. We could have started with the four points marked with open circles containing a cross (a half-replicate of a 2^3 design). If interactions had been unimportant, experimental runs made at these four points would have allowed estimation of all three main effects and the fitting of a first-degree model. The direction of steepest ascent could then be calculated, and if a couple of runs along this path showed an increase, we could follow it up and appropriately shift the location of future experiments. If no increase occurred, we would suspect that a more complex situation existed, and the second block of four runs denoted by filled circles could be added, producing the full 2^3 design. If still more information was needed, the third part—the axial points and center points—could be added to form the full composite design, thus making it possible to fit a second-degree model.

By suitable positioning of the axial points and choice of the number of center points, each of the three parts of the design can be made an "orthogonal block" (see, e.g. BHHII 2005 and Box and Draper, 1987). This means that when the design is assembled sequentially, any changes in the level of the response from block to block will not alter the estimates of the effects. Such changes can occur, for example, because of small environmental differences associated with the different times that the blocks of runs are performed.

The idea of sequential assembly illustrated here by the development of the composite design is much more widely applicable. It is used, for example, in the concept of "foldover" for fractional factorial designs (see, e.g., Chapters B.6 & C.7). These and other techniques of response surface methods provide a group of statistical tools designed to meet the experimenter's needs at different stages of

---

* This throws in doubt the value of so-called "heredity" principles which are supposed to connect main effects with their corresponding interactions.

an investigation. They are sequential methods illustrated in Chapter C.1. It is also shown there how these methods can be used for teaching and *experiencing the process of investigation itself.*

These developments are, I believe, one more example showing how practical necessity has led to important new ideas in statistics and quality. (See Chapter B.12 for other examples, from Gauss onward.)

ACKNOWLEDGMENT

This work was supported by NSF Grant Number DMI-9812839.

CHAPTER C.6

Finding the Active Factors in Fractionated Screening Experiments

Highly fractionated factorial designs and other orthogonal arrays are powerful tools for identifying important, or active, factors and improving quality. We show, however, that interactions, and important factors involved in those interactions, may go unidentified when conventional methods of analysis are used with these designs. This is particularly true of Plackett–Burman designs where the number of runs is not a power of two. A Bayesian method that allows for the possibility of interactions is developed to compute the marginal posterior probability that a factor is active. The method can be applied to both orthogonal and nonorthogonal designs, as well as other troublesome situations, such as when data are missing, extra data are available, or factor settings for certain runs have deviated from those originally planned. The value of the new technique is demonstrated with three examples in which potential interactions and factors involved in those interactions are uncovered.

INTRODUCTION

Many factors affect a typical process, however it is often the case that only a few are truly important, or *active*. For example suppose we are investigating a number of process factors A, B, C, D, E etc. and we are interested in one or more responses y_1, y_2, y_3 etc. Then it may turn out that only a few factors say B, D and E, are truly important in affecting a particular response, say y_2. Then we can say that for y_2 there is a three dimensional active subspace [BDE] . By this we mean that factors B, D and E affect the level of y_2 either, through their main effects and/or through any higher

From Box, G. E. P. and Meyer, R. D. (1993), *Journal of Quality Technology,* 25(2), 94–105.

or less terms. Thus in the above example we might have detected the main effects B and E with the interactions BD and DE for the response y_2. Notice that the active subspace for some other response, say y_3, is likely to be different. It might for example be [AE]. In general it might overlap the active subspace for y_2 completely, partially or not al all. Such knowledge is invaluable for solving multiple response problems. Frequently screening experiments can identify which factors are active for the various responses. More extensive experimentation can then focus on these factors.

There are two classes of two-level orthogonal designs popularly used for screening experiments: the 2^{k-p} fractional factorials, for which the number of runs is always a power of 2 (see, e.g., Finney, 1945; Box and Hunter, 1961a,b) and the two-level orthogonal array designs of Plackett and Burman (1946), for which the number of runs is always a multiple of 4.* We call these P.B. designs. Either of these types of designs can be written as a set of columns of plus and minus signs, with an equal number of pluses and minuses in each column. Averaging the observed data opposite the plus signs for a particular column, and subtracting the average of the data opposite the minus signs, yields a *data contrast*. The expected value of a data contrast is in general a linear combination of various main effects and interactions called *aliases*.

Fractional factorial designs of the type 2^{k-p} have a relatively simple aliasing structure in which a particular alias term (main effect or interaction) appears only once in association with a single data contrast. Plackett–Burman designs for which the number of runs is also a power of 2 are the same designs as the 2^{k-p} designs. (Any 2^{k-p} design can be formed by choosing the appropriate columns from the Plackett–Burman design with 2^{k-p} runs and possibly switching minus and plus signs in some of the columns.) Plackett–Burman designs that employ a number of runs *not a power of 2* (12, 20, 24, etc.) are called nongeometric and have more complicated aliasing relationships[†] than the geometric designs. A particular alias term can be associated with a number of different data contrasts, and it can have fractional coefficients. For example, in the 12-run Plackett–Burman design, the data contrast associated with the main effect of any factor has in its alias structure all two-factor interactions not involving that factor, each with a fractional coefficient.

For fractional designs a variety of different methods can then be applied to identify contrasts and active subspaces whose effects are too large to attribute to noise alone. For example, normal probability plots (Daniel, 1959, 1976), Bayes plots (Box and Meyer, 1986a), and pseudo-standard error plots (Lenth, 1989). However unless it is assumed that only main exist these methods cannot be used for the analyses of P.B. designs.

---

*These arrangements are examples of Hademard matrices and as such have been extremely well tabled.
[†]The general way of obtaining the alias pattern for any design was given by Box and Wilson (1951), and the necessary relationship between the coefficients of the aliases for all orthogonal designs was given, by Box (1952).

A different approach, stemming from the principle of parsimony—or in Juran's words, distinguishing the "vital few" factors from the trivial many, supposes that only a small number (perhaps two or three or four) of the factors in the design are responsible for most of what is happening to a particular response but these active factors may interact with each other. Thus, say, with five factors (A, B, C, D, and E) various hypotheses might be considered. One is that a single factor A, B, C, D, or E is responsible for all that is going on, in which case one need only consider the five main effects A, B, C, D, and E. Alternatively, under the hypothesis that two factors, say, A and B, are responsible, the main effects A and B and the interaction AB are considered. Under the hypothesis that three factors, say, A, B, and C, may be active, the subset of main effects A, B, and C with interactions AB, AC, BC, and ABC are considered, and so on. We are thus attempting to find the *active factor subspace*. In the following section we develop a Bayesian method that will more completely consider all such hypotheses when analyzing the results of a screening experiment. The Bayesian model makes it possible to analyse Plackett–Burman designs where the alias structure can be quite complicated and can be used whether the design is orthogonal or nonorthogonal.

THE MODEL

Our approach is to consider all the possible explanations (including interactions) of the data from a screening experiment and to identify those that fit the data well. A Bayesian framework is used to assign an appropriate measure to each model. These posterior probabilities can then be accumulated in various ways to produce marginal posterior probabilities. It is analogous to all-subsets regression in that all possible models are evaluated.

The Bayesian approach to model identification is as follows. We consider the set of possible models labeled $M_0, \ldots, M_m$. Each model M_i has an associated vector of parameters $\boldsymbol{\theta}_i$, so that the sampling distribution of data y, given the model M_i, is described by the probability density $f(\mathbf{y}|M_i, \boldsymbol{\theta}_i)$. The prior probability of the model M_i is $p(M_i)$, and the prior probability density of $\boldsymbol{\theta}_i$ is $f(\boldsymbol{\theta}_i|M_i)$. The predictive density of y, given model M_i, is written $f(\mathbf{y}|M_i)$ and is given by the expression

$$f(\mathbf{y}|M_i) = \int_{\mathbf{R}_i} f(\mathbf{y}|M_i, \boldsymbol{\theta}_i) f(\boldsymbol{\theta}_i|M_i) \, d\boldsymbol{\theta}_i$$

where $\mathbf{R}_i$ is the set of possible values of $\boldsymbol{\theta}_i$. The posterior probability of the model M_i, given the data y, is then

$$p(M_i|\mathbf{y}) = \frac{p(M_i)f(\mathbf{y}|M_i)}{\sum\limits_{h=0}^{m} p(M_h)f(\mathbf{y}|M_h)} \tag{1}$$

The posterior probabilities $p(M_i|\mathbf{y})$ provide a basis for model identification. Tentatively plausible models are identified by their large posterior probabilities. Computationally one calculates $p(M_i)f(\mathbf{y}|M_i)$ for each model M_i (the numerator in the above expression) and then scales these quantities to sum to one.

For a screening design with k factors let M_i denote the model that a particular subset of f_i factors is active, where $0 \le f_i \le k$. There are 2^k models M_i, starting from $i = 0$ (no active factors) to $i = 2^k - 1$ (k active factors). To model the condition of factor sparsity, let π be the prior probability that any one factor is active. For a screening experiment, in which one typically expects to identify just a few (less than half) of the factors as important, appropriate values for π would be in the range from 0 to 0.5. A nominal value of $\pi = 0.25$ has given sensible results in practice, and the individual experimenter can specify a different value based on the circumstances of a particular experiment. (Sensitivity to choice of π will be discussed later.) The prior probability $p(M_i)$ of the model M_i is then $\pi^{f_i}(1 - \pi)^{k-f_i}$.

Let $\mathbf{X}_i$ be the matrix with columns for each effect under the model M_i using the convention of coded values -1 and $+1$ for two-level factors. The matrix $\mathbf{X}_i$ includes a column of 1's for the mean, and interaction columns up to any order desired. Let t_i be the number of such effects, excluding the mean. The dimensions of $\mathbf{X}_i$ are then $n \times (1 + t_i)$. Likewise, let $\boldsymbol{\beta}_i$ be the $(1 + t_i) \times 1$ vector of true (regression) effects under M_i. As before, $\mathbf{y}$ denotes the $n \times 1$ vector of responses. We now specifically assume that the probability density of $\mathbf{y}$ given M_i is that of the usual normal linear model. Thus,

$$f(\mathbf{y}|M_i, \boldsymbol{\theta}_i) = f(\mathbf{y}|M_i, \sigma, \boldsymbol{\beta}_i) \propto \sigma^{-n} \exp\left[-(\mathbf{y} - \mathbf{X}_i\boldsymbol{\beta}_i)'(\mathbf{y} - \mathbf{X}_i\boldsymbol{\beta}_i)/2\sigma^2\right]$$

The elements of $\boldsymbol{\beta}_i$ are assigned independent prior normal distributions with mean 0 and variance $\gamma^2\sigma^2$ when $z \le y \le 3$. A priori ignorance of the direction of any particular effect is represented by the zero mean, and the magnitude of the effect relative to experimental noise is captured through the parameter γ. A noninformative prior distribution is employed for the overall mean β_0 and $\log(\sigma)$, so that $f(\beta_0, \sigma) \propto 1/\sigma$. Having observed the data vector $\mathbf{y}$, the posterior probability of the model M_i can then be written

$$p(M_i|\mathbf{y}) = C\left(\frac{\pi}{1-\pi}\right)^{f_i}\gamma^{-t_i} \times \frac{|\mathbf{X}_0'\mathbf{X}_0|^{1/2}}{|\mathbf{\Gamma}_i + \mathbf{X}_i'\mathbf{X}_i|^{1/2}}\left(\frac{S(\hat{\boldsymbol{\beta}}_i) + \hat{\boldsymbol{\beta}}_i\mathbf{\Gamma}_i\hat{\boldsymbol{\beta}}_i}{S(\hat{\boldsymbol{\beta}}_0)}\right)^{-(n-1)/2} \tag{2}$$

where

$$\mathbf{\Gamma}_i = \frac{1}{\gamma^2}\begin{bmatrix} 0 & 0 \\ 0 & \mathbf{I}_i \end{bmatrix}$$

$$\hat{\boldsymbol{\beta}}_i = (\mathbf{\Gamma}_i + \mathbf{X}_i'\mathbf{X}_i)^{-1}\mathbf{X}_i'\mathbf{y}$$

$$S(\hat{\boldsymbol{\beta}}_i) = (\mathbf{y} - \mathbf{X}_i\hat{\boldsymbol{\beta}}_i)'(\mathbf{y} - \mathbf{X}_i\hat{\boldsymbol{\beta}}_i)$$

and C is the normalization constant that forces all probabilities to sum to one. (See the Appendix for further details.)

The probabilities $p(M_i|\mathbf{y})$ can be accumulated to compute the marginal posterior probability P_j that factor j is active as

$$P_j = \sum_{M_i:\ \text{factor } j \text{ active}} p(M_i|\mathbf{y}) \qquad (3)$$

The probability P_j is just the sum of the posterior probabilities of all the distinct models in which the factor j is active. The probabilities $\{P_j\}$ are thus calculated by direct enumeration over the 2^k possible models M_i. A large value for P_j would indicate that the factor j was active, and similarly, a value of P_j close to zero would indicate that the factor j was inert. After examining the $\{P_j\}$, the individual probabilities $p(M_i|\mathbf{y})$ may further identify specific combinations of factors that are most likely active. An extensive investigation of the sensitivity of the procedure to the values used for π and γ was undertaken by Barrios (2005) who found that the conclusion drawn were remarkably insensitive to these choices. Also a method is given in Box and Meyer (1992) whereby γ may be estimated from the data. A series of computed runs made for different values of γ, π and turns out that the best estimate is that which gives the smallest value for the posterior probability of no effects.

EXAMPLES

The analysis is demonstrated on three examples that will help to show its utility. The first two examples illustrate its benefits for nongeometric Plackett–Burman designs. The remaining example demonstrates its application for 2^{k-p} designs, highlighting the issue of follow-up experiments for resolving ambiguities. In each example the posterior probabilities are computed with $\pi = 0.25$ and γ chosen to minimize the probability of no active factors. Further comments on the effect of the choice of parameters are made in the Discussion.

The FORTRAN bundle mdopt by R.D. Meyer, for performing the computations described here is available at Statlib, http://lib.stat.cmu.edu. E. Barrios implemented mdopt in R language and is available as BsMD contributed package at http://cran.r-project.org/src/contrib/PACKAGES.html. Programs for Bayesian screening and model building are included in the BHH2 and BsMDR-packages available at CRAN under contributed packages. There is as well commercial software as in SAS/Jump which some readers may find easier to use. The program accepts as inputs the design matrix and the observations $\mathbf{y}$. It calculates the probabilities as shown in the following examples.

Example 1

In this example we illustrate some of the difficulties that can be encountered when analyzing the results of a nongeometric Plackett–Burman design and how the Bayesian analysis can help to overcome these difficulties. To aid further in illustrating and understanding the concepts involved, this example was constructed by extracting 12 runs from the 2^5 reactor example from Box, Hunter, and Hunter (1978,

Table 1. Factor Levels and Responses, for Example 1. Factors are A—Feed Rate, B—Catalyst, C—Agitation, D—Temperature, and E—Concentration. The Responses, Y, Measure the Percent Reacted

Run	A	B	C	D	E	Y
6	+	−	+	−	−	56
12	+	+	−	+	−	93
23	−	+	+	−	+	67
14	+	−	+	+	−	60
28	+	+	−	+	+	77
24	+	+	+	−	+	65
15	−	+	+	+	−	95
29	−	−	+	+	+	49
25	−	−	−	+	+	44
18	+	−	−	−	+	63
3	−	+	−	−	−	63
1	−	−	−	−	−	61

p. 376). Letters rather than numbers have been used to identify the experimental factors. The 12 runs chosen were those that would have been run if the first five columns of the 12-run Plackett–Burman array had been employed as a screening design for these five factors. The 12 runs are shown in Table 1. The run numbers shown are those from the original set of 32. Analysis of the full data set, in Box, Hunter, and Hunter (1978), showed main effects B, D, and E, as well as the BD and DE interactions, were important and hence the active factor space was that of B, D and E.

The 11 estimated effects, one for each column of the full Plackett–Burman design array, are shown in Table 2, and the normal plot of effects is shown in Figure 1. Only

Table 2. Estimated Coefficients and Alias Strings for Example 1

Column	Estimated Effect	Alias String
1	2.92	$A + \frac{1}{3}(-BC + \underline{BD} + BE - CD - CE - \underline{DE})$
2	10.58	$B + \frac{1}{3}(-AC + AD + AE - CD + CE - \underline{DE})$
3	−0.75	$C + \frac{1}{3}(-AB - AD - AE - \underline{BD} + BE - \underline{DE})$
4	3.58	$D + \frac{1}{3}(AB - AC - AE - BC - BE - CE)$
5	−5.25	$E + \frac{1}{3}(AB - AC - AD + BC - \underline{BD} - CD)$
6	−1.08	$\frac{1}{3}(-AB + AC - AD + AE + BC - \underline{BD} - BE + CD - CE - \underline{DE})$
7	1.08	$\frac{1}{3}(-AB - AC - AD + AE - BC + \underline{BD} - BE + CD - CE + \underline{DE})$
8	−4.42	$\frac{1}{3}(AB + AC - AD - AE - BC - \underline{BD} - BE - CD + CE + \underline{DE})$
9	3.58	$\frac{1}{3}(-AB - AC - AD - AE + BC + \underline{BD} - BE - CD - CE - \underline{DE})$
10	−0.25	$\frac{1}{3}(-AB - AC + AD - AE - BC - \underline{BD} - BE + CD + CE - \underline{DE})$
11	−4.92	$\frac{1}{3}(-AB + AC + AD - AE - BC - \underline{BD} + BE - CD - CE + \underline{DE})$

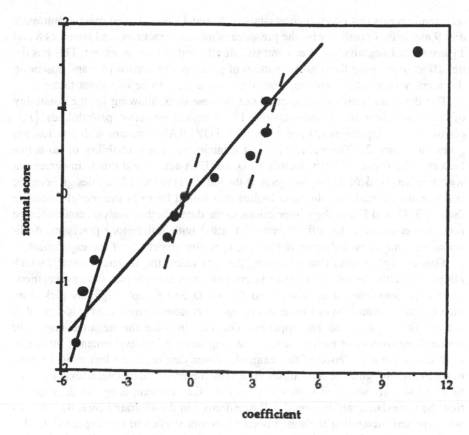

Figure 1. Normal plot of estimated coefficients for Example 1. Plackett–Burman design extracted from the 2^5 reactor example. Dashed lines indicate clusters of points along parallel lines.

the main effect of factor B, catalyst level, clearly stands out. The other notable feature of this plot is the gaps between groups of points falling along distinct parallel lines. Daniel (1976) pointed out that this indicates the possibility of outliers among the original observations (see also Chapter B.9). If a particular outlying observation is biased by a positive (negative) amount, then contrasts in which the observation enters positively (negatively) are shifted to the right, and contrasts in which the observation enters negatively (positively) are shifted to the left. Thus, the observation, if it exists, which enters positively (negatively) in the positive contrasts and negatively (positively) in the negative contrasts is identified as a potential outlier. Following this method, however, does not reveal any of the observations as clearly discordant.

The confounding structure of the Plackett–Burman design suggests another explanation for the unusual normal plot. As mentioned previously, the main effect of each factor is confounded with all two-factor interactions not involving that factor. Table 2 gives the alias relationships. The *BD* and *DE* interactions, which are known to be large from the full data analysis, are underlined. Taking into account the signs of the large interaction effects (the *BD* interaction is positive, the *DE* interaction is negative) reveals that column contrasts 1, 8, 9, and 11 are affected most. Contrasts 1 and 9 are shifted positively by the presence of these interactions, and contrasts 8 and 11 are shifted negatively. Other contrasts are affected to a lesser extent. This has the net effect of creating the unusual pattern of points in the normal plot and obscuring the identity of what are known, from the complete 2^5, to be important factors.

The Bayesian analysis was carried out on these data, allowing for the possibility of two- and three-factor interactions. The marginal posterior probabilities $\{P_j\}$, obtained from Equations (2) and (3) via the FORTRAN program with $\gamma = 1.6$, are given in Figure 2. (The value of $\gamma = 1.6$ minimized the probability of no active factors.) The results identify factors B, D, and E as active, and this is in agreement with the factors identified by analysis of the full 2^5. Had this 12-run design been the one actually carried out, the next logical step would be to fit the model involving factors B, D, and E and their interactions to the data. Further analysis could then be done to determine if the effects were estimated with satisfactory precision. Additional runs might be indicated depending upon the objectives of the experiment.

This example shows that analyzing the data using the Bayesian model, which allows explicitly for the presence of interactions, can identify plausible explanations missed by conventional analysis. Two factors, D and E, not originally picked up through the examination of the set of orthogonal column contrasts, were identified as active. The implications are apparent. One can imagine the negative impact of overlooking important factors at the screening stage of an experimental investigation. Given the experience of this example of studying only five factors in 12 runs, one can also imagine the difficulties created by the presence of interactions when up to 11 factors are screened with the 12-run Plackett–Burman array. To demonstrate that the Bayesian analysis would still be effective in the saturated case, the analysis was repeated pretending there had been 11 factors studied in this experiment. That is, for the sake of illustration, dummy factors 6 through 11 were assigned to the remaining columns of the Plackett–Burman array, and the Bayesian analysis repeated as though the design were saturated. The $\{P_j\}$ are given in Figure 3, and the same three factors B, D, and E are identified as active.

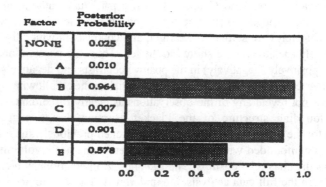

Figure 2. Marginal posterior probabilities that factors are active ($\gamma = 1.6$).

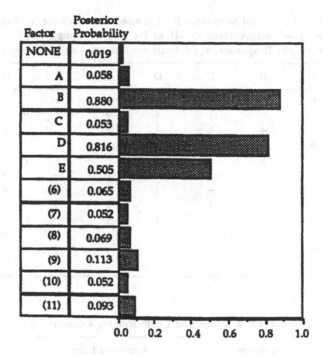

Factor	Posterior Probability	
NONE	0.019	
A	0.058	
B	0.880	
C	0.053	
D	0.816	
E	0.505	
(6)	0.065	
(7)	0.052	
(8)	0.069	
(9)	0.113	
(10)	0.052	
(11)	0.093	

Figure 3. Marginal posterior probabilities that factors are active assuming there were factors assigned to all 11 columns in the experiment ($\gamma = 1.6$).

Example 2

A 12-run Plackett–Burman design was employed to study fatigue life of weld repaired castings (Hunter, Hodi, and Eager, 1982). [The possible presence of an interaction in this example was shown by Hamada and Wu (1992).] Seven factors were varied in this experiment. The design matrix, data, estimated effects, and normal plot of effects are given in Table 3, Table 4, and Figure 4, respectively. The normal plot rather dubiously supports the original authors' conclusion that factor F and possibly factor D had significant main effects. The plot also displays to some extent the behavior noted in the previous example, gaps among the points that indicate the possibility of interaction for a Plackett–Burman design.

The Bayesian analysis was carried out on these data and the results are shown in Figure 5. (The value of $\gamma = 1.5$ minimized the probability of no active factors.) Factors F and G but not D are then identified as the active factors in this experiment with posterior probabilities of 0.979 and 0.964, respectively. In Figure 6 the data are arranged according to the levels of factors F and G. It is clear from this display that a very plausible explanation for the data is the joint effect of these two factors that F and G define the active factor subspace. Because it has a small main effect, factor G was not flagged in the normal plot of effects. As in the previous example, interactions were camouflaged.

Table 3. Factor Levels and Responses for Example 2. Factors are A—Initial Structure, B—Bead Size, C—Pressure Treat, D—Heat Treat, E—Cooling Rate, F—Polish, and G—Final Treat. The Responses, Y, Measure the Natural Log of Fatigue Life

Run	A	B	C	D	E	F	G	Y
1	+	+	−	+	+	+	−	6.058
2	+	−	+	+	+	−	−	4.733
3	−	+	+	+	−	−	−	4.625
4	+	+	+	−	−	−	+	5.899
5	+	+	−	−	−	+	−	7.000
6	+	−	−	−	+	−	+	5.752
7	−	−	−	+	−	+	+	5.682
8	−	−	+	−	+	+	−	6.607
9	−	+	−	+	+	−	+	5.818
10	+	−	+	+	−	+	+	5.917
11	−	+	+	−	+	+	+	5.863
12	−	−	−	−	−	−	−	4.809

Table 4. Estimated Coefficientsfor Example 2

Column	Estimated Effect
A	0.326
B	0.294
C	−0.246
D	−0.516
E	0.150
F	0.915
G	0.183
8	0.446
9	0.453
10	0.081
11	−0.242

Example 3

The third example is the 2^{8-4} injection molding example is discussed in the first edition of the book *Statistics for Experiments* by Box Hunter and Hunter (1978). The design matrix, observed data, estimated effects, and their alias strings (up to two-factor interactions) are given in Tables 5 and 6. The normal plot of effects showed that contrasts corresponding to effects C, E, and $AE + BF + CH + DG$ were too large to be explained by noise. There was also weak evidence to suggest that the effect H might also be important. It was orginally suggested as likely that either the AE or CH interaction would explain the large contrast associated with

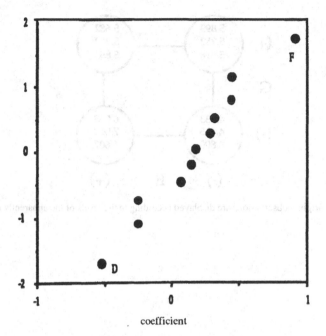

Figure 4. Normal plot of coefficients for Example 2.

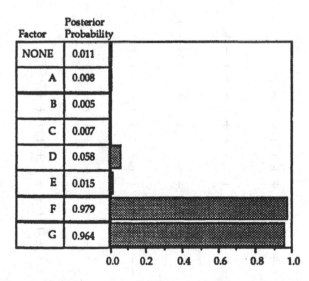

Figure 5. Marginal posterior probabilities that factors are active ($\gamma = 1.5$).

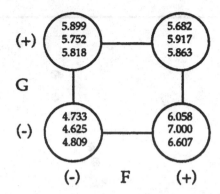

Figure 6. The original observations are displayed according to the levels of the apparently active factors F and G.

Table 5. Factor Levels and Responses for Example 3. Factors are A—Mold Temperature, B—Moisture Content, C—Holding Pressure, D—Cavity Thickness, E—Booster Pressure, F—Cycle Time, G—Gate Size, and H—Screw Speed. The Responses, Y, Measure the Shrinkage

Run	A	B	C	D	E	F	G	H	Y
1	−	−	−	+	+	+	−	+	14.0
2	+	−	−	−	−	+	+	+	16.8
3	−	+	−	−	+	−	+	+	15.0
4	+	+	−	+	−	−	−	+	15.4
5	−	−	+	+	−	−	+	+	27.6
6	+	−	+	−	+	−	−	+	24.0
7	−	+	+	−	−	+	−	+	27.4
8	+	+	+	+	+	+	+	+	22.6
9	+	+	+	−	−	−	+	−	22.3
10	−	+	+	+	+	−	−	−	17.1
11	+	−	+	+	−	+	−	−	21.5
12	−	−	+	−	+	+	+	−	17.5
13	+	+	−	−	+	+	−	−	15.9
14	−	+	−	+	−	+	+	−	21.9
15	+	−	−	+	+	−	+	−	16.7
16	−	−	−	−	−	−	−	−	20.3
17	−	+	+	+	−	−	−	+	29.4
18	−	+	−	−	−	+	+	+	19.7
19	+	+	−	−	+	−	−	+	13.6
20	+	+	+	+	+	+	+	+	24.7

Table 6. Estimated Coefficients and Alias Strings for Example 2

Alias String	Estimated Effect
A	−0.35
B	−0.05
C	2.75
D	−0.15
E	−1.9
F	−0.05
G	0.3
H	0.6
$AB + CG + DII + EF$	−0.3
$AC + BG + DF + EH$	0.45
$AD + BH + CF + EG$	−0.2
$AE + BF + CH + DG$	2.3
$AF + BE + CD + GH$	−0.15
$AG + BC + FH + DE$	−0.1
$AH + BD + CE + FG$	−0.3

$AE + BF + CH + DG$ because these involved variables with large main effects. Factors A, C, E, and H were thus identified in the original analysis as the potentially active factors in this experiment.

The marginal posterior probabilities $\{P_j\}$ are displayed in Figure 7, again allowing for two- and three-factor interactions. (The value of $\gamma = 2.0$ minimized the probability of no active factors.) Factors A, C, E, and H have marginal posterior probabilities of approximately 0.76, and thus clearly stand out as potentially active. These are the same four factors identified by the previous analysis. The remaining factors have essentially zero probability.

The 2^{8-4} design collapses to a replicated 2^{4-1} design (Resolution IV) in these four factors, so it is not possible to obtain separate estimates of their main effects and interactions from the original 16 runs. Additional runs are required. BH^2. The original account described a four-run follow-up design: runs 17–20 of Table 5. The additional four runs were selected in order to obtain separate estimates of the AE, BF, CH, and DG interactions as well as the block effect between runs 1–16 and runs 17–20. The assumption was made that no three-factor interactions were present. Solving a system of five equations in five unknowns, the $AE + BF + CH + DG$ alias string was apparently resolved and CH identified as the large interaction. Factors C, E, and H were thus identified as the active factors for this experiment.

We reconsider the problem of setting up the follow-up design based on three concerns.

(i) The Bayesian analysis indicated that factors B, D, F, and G were very likely inert; was it necessary to obtain estimates of the BF and DG interactions from the follow-up design?

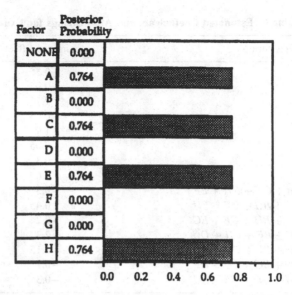

Figure 7. Marginal posterior probabilities that factors are active for initial 16 runs ($\gamma = 2.0$).

(ii) Should there be more concern about the possibility of three-factor interactions among the potentially active factors A, C, E, and H?

(iii) Should there be more concern for obtaining separate estimates of the other confounded two-factor interactions among the potentially active factors A, C, E, and H, even though the contrasts associated with them were small?

The posterior probabilities were now recomputed (Figure 8) using the combined 20 runs, defining a ninth factor to be the block effect between the first 16 runs and the last four. All four factors A, C, E, and H still stand out as potentially active. This is quite different from the conclusions drawn earlier when factor A was judged to be inert after the follow-up experiment. The difference in the two analyses is the implicit assumption made earlier that three-factor interactions were not involved in the explanation of these data. The possibilities that the large contrast associated with the C main effect could have been the AEH interaction that the contrast associated with the H main effect could have been the ACE interaction, and so forth were not considered (If the posterior probabilities are computed allowing for only two-factor interactions, factors C, E, and H have probabilities close to 1.0 and factor A's probability is closer to 0.0.)

An alternative approach for setting up the follow-up design would be to assume that factors A, C, E, and H were the potentially active factors upon completion of the 16-run initial experiment. The design when collapses to a half-fraction of the full factorial in these factors, with defining relation $I = ACEH$. Running the other half-fraction $I = (-ACEH)$, an 8-run design, would allow estimation of all main effects,

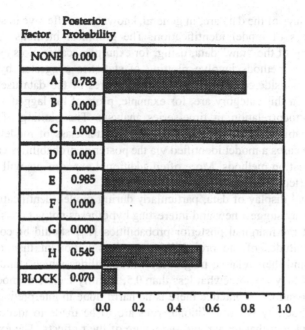

Figure 8. Marginal posterior probabilities that factors are active for combined set of 20 runs ($\gamma = 2.0$).

two- and three-factor interactions involving A, C, E, and H, and confound the four-factor interaction with the block effect. The remaining four factors B, D, F, and G, assumed inert, would be set at some convenient levels within their ranges in the original design.

In general, it seems reasonable, and we tentatively recommend, that when a subset of the original set of experimental factors is identified as potentially active via the Bayesian analysis, the design of follow-up experiments to resolve ambiguities should focus on these factors.

DISCUSSION

At the initial stage of a scientific investigation, emphasis is on model identification. Screening designs are employed in these situations to identify various hypotheses (models) that may be explored in subsequent experiments. Alternative possibilities are tentatively entertained until confirmatory data allow further refinement. It is important to attempt to identify all plausible explanations of the data obtained in a screening experiment and not overlook potentially important factors. The Bayesian analysis proposed here considers a broad, though not exhaustive, class of models and evaluates each member of the class in light of the observed data. Other models that bear consideration are, for example, those concerning dispersion effects (Box and Meyer, 1986b) and outliers (Box and Meyer, 1987).

Visual displays of the data are, in general, known to be effective in stimulating the creative process of model identification. The graphical methods employed may involve plotting of the "raw" data, using, for example, scatter plots and scatter-plot matrices. Other methods involve plotting of statistics suggested by the class of models being considered, which summarize features of the data useful for model identification. In this category are, for example, plots of the lagged autocorrelation and partial autocorrelation in time-series analysis. The plotting of the marginal posterior probabilities described here is in this latter class of model-identification tools. In some cases a model identified via the posterior probabilities can be fit using accepted estimation methods. More often additional experiments will be run before one model is (tentatively) identified and estimated.

In any visual display of data, particularly during model identification, one seeks patterns that may suggest new and interesting hypotheses not yet considered. It is in this spirit that the marginal posterior probabilities $\{P_j\}$ should be considered. The absolute magnitudes of the probabilities are less important than the pattern of probabilities and their relative magnitudes. Probabilities which stand out from the others, even if they are somewhat less than 0.5, identify factors or models worthy of further consideration. In practice there is no harm done in interpreting the posterior probabilities liberally when additional runs are to be made to identify with more certainty the factors that matter and the nature of their effects. The examples of the previous section illustrate how this might work in practice.

The Bayesian approach requires specification of prior distributions and parameters of those distributions. The choice of noninformative prior distributions for the parameters β_0 and σ is standard and sensible, and the interested reader is referred to Box and Tiao (1973) for further discussion. The choice of the normal prior distribution $N(0, \gamma^2\sigma^2)$ for the remaining elements of β_i is a natural one for the following reasons.

1. It is an informative and conjugate prior distribution; which leads to mathematical tractability.
2. There is typically prior ignorance of the direction of any particular effect, and this supports the choice of a zero mean.
3. Important main effects and interactions tend, in our experience, to be of roughly the same order of magnitude, which justifies the parsimonious choice of one common scale parameter γ. Uncertainty in γ can be handled as described in Box and Meyer (1993).

The prior probability π must also be specified. Increasing π tends to increase all of the $\{P_j\}$ but usually has little effect on which probabilities stand out from the rest. The nominal choice of $\pi = 0.25$ has served well in our experience, but individual experimenters should choose a value of π that represents their prior expectation of the proportion of active factors. The analysis can also be repeated for a few different values of π, say, 0.1, 0.25, and 0.5, and the results compared. Usually the same factors will be identified as potentially active, but if not, it is important to remember the ultimate purpose, which is to identify different plausible models that can be explored further through model-fitting and additional experiments.

The approach we have developed is completely general and can be useful for any experiment where confounding and fractionation or missing data make results difficult to interpret or when settings for certain runs have deviated from those originally planned in. We have found the Bayesian model identification scheme to be very helpful for sorting out potentially active factors in these situations. Determining follow-up designs for further model discrimination can also be facilitated by the Bayesian framework (see Example 3 and also Chapter C.7).

CONCLUSIONS

Testing several factors in only a handful of runs in a fractionated screening experiment often results in more than one plausible explanation for the data. This is where the economy of fractional factorial designs is valuable: the number of plausible hypotheses can be reduced at each round of experimentation with far fewer total runs than are needed for one, supposedly comprehensive, experiment. By including more variables in the screening experiment, arbitrarily eliminating variables at the outset of an investigation is also avoided.

Interpreting the results from these screening experiments can be a difficult task. This is particularly true of nongeometric Plackett–Burman designs since interactions are hidden in the column contrasts and important possibilities can be completely overlooked. The Bayesian analysis proposed here can greatly facilitate the identification of active factors. Those that explain the data well are identified by their large posterior probabilities. We have shown in the examples that very plausible models, undiscovered through conventional analysis, were identified through the Bayesian method.

ACKNOWLEDGMENTS

This research was sponsored by the Sloan Foundation, by the National Science Foundation Grant No. DMS-8420968, by The Lubrizol Corporation, and aided by access to the research computer of the University of Wisconsin–Madison Statistics Department. The authors would like to thank Elizabeth Schiferl, Sara Pocinki, Robert Wilkinson, and the two referees for their helpful discussion and comments, and numerous colleagues for trying the method on real examples.

APPENDIX

The posterior probability of the model M_i was given in Equation (1) to be, in general,

$$p(M_i|\mathbf{y}) = \frac{p(M_i)f(\mathbf{y}|M_i)}{\sum\limits_{h=0}^{m} p(M_h)f(\mathbf{y}|M_h)}$$

This is just Bayes's theorem applied to model selection. The denominator is simply the numerator summed over all models forcing all of the probabilities to sum to one. Thus only the numerator needs to be computed for every model M_i and then these need to be scaled to sum to one. The two items needed to compute the numerator are $p(M_i)$ and $f(\mathbf{y}|M_i)$.

To model the condition of factor sparsity, we had let π be the prior probability that any one factor is active. Since the model M_i has f_i active factors, the prior probability $p(M_i)$ of the model M_i is just

$$p(M_i) = \pi^{f_i}(1 - \pi)^{k-f_i} \tag{A1}$$

The expression for $f(\mathbf{y}|M_i)$ is more complicated. By standard Bayesian methodology, $f(\mathbf{y}|M_i)$ is given, for a general parameter vector $\boldsymbol{\theta}_i$, by the integral expression

$$f(\mathbf{y}|M_i) = \int_{\mathbf{R}_i} f(\mathbf{y}|M_i, \boldsymbol{\theta}_i) f(\boldsymbol{\theta}_i|M_i)\, d\boldsymbol{\theta}_i$$

or, in particular, where the parameters are σ plus the regression coefficient vector $\boldsymbol{\beta}_i$

$$f(\mathbf{y}|M_i) = \int_0^\infty \int_{-\infty}^\infty \cdots \int_{-\infty}^\infty f(\mathbf{y}|M_i, \boldsymbol{\beta}_i, \sigma) f(\boldsymbol{\beta}_i|M_i, \sigma) f(\sigma|M_i)\, d\boldsymbol{\beta}_i\, d\sigma$$

The steps of carrying out this integration are shown in more detail in Box and Meyer (1993). The result of the integration is

$$f(\mathbf{y}|M_i) \propto \gamma^{-f_i}|\boldsymbol{\Gamma}_i + \mathbf{X}_i'\mathbf{X}_i|^{-1/2}(S(\hat{\boldsymbol{\beta}}_i) + \hat{\boldsymbol{\beta}}_i'\boldsymbol{\Gamma}_i\hat{\boldsymbol{\beta}}_i)^{-(n-1)/2} \tag{A2}$$

where

$$\boldsymbol{\Gamma}_i = \frac{1}{\gamma^2}\begin{bmatrix} \mathbf{0} & \mathbf{0} \\ \mathbf{0} & \mathbf{I}_i \end{bmatrix}$$

$$\hat{\boldsymbol{\beta}}_i = (\boldsymbol{\Gamma}_i + \mathbf{X}_i'\mathbf{X}_i)^{-1}\mathbf{X}_i'\mathbf{y}$$

$$S(\hat{\boldsymbol{\beta}}_i) = (\mathbf{y} - \mathbf{X}_i\hat{\boldsymbol{\beta}}_i)'(\mathbf{y} - \mathbf{X}_i\hat{\boldsymbol{\beta}}_i)$$

Finally, to avoid floating-point overflow, one actually computes

$$p(M_i|\mathbf{y}) = C\frac{p(M_i) f(\mathbf{y}|M_i)}{p(M_0) f(\mathbf{y}|M_0)} \tag{A3}$$

for each model M_i, where C is the normalization constant that forces the probabilities to sum to one. M_0 is the label for the null model with no active factors. Dividing each probability by the constant $p(M_0) f(\mathbf{y}|M_0)$ does not change the final result, since everything is eventually scaled to sum to one, but it prevents getting very large numbers as intermediate results. From (A1), since $f_i = 0$

$$p(M_0) = (1 - \pi)^k \tag{A4}$$

From Equation (A2), with $t_i = 0$ and $\Gamma_i = 0$ for M_0

$$f(\mathbf{y}|M_0) \propto |\mathbf{X}_0'\mathbf{X}_0|^{-1/2}(S(\hat{\boldsymbol{\beta}}_0))^{-(n-1)/2} \tag{A5}$$

Substituting Equations (A1), (A2), (A4), and (A5) into Equation (A3), one obtains Equation (2)

$$p(M_i|\mathbf{y}) = C\left(\frac{\pi}{1 - \pi}\right)^{f_i} \gamma^{-t_i} \frac{|\mathbf{X}_0'\mathbf{X}_0|^{1/2}}{|\Gamma_i + \mathbf{X}_i'\mathbf{X}_i|^{1/2}} \left(\frac{S(\hat{\boldsymbol{\beta}}_i) + \hat{\boldsymbol{\beta}}_i'\Gamma_i\hat{\boldsymbol{\beta}}_i}{S(\hat{\boldsymbol{\beta}}_0)}\right)^{-(n-1)/2}$$

Although the expression seems somewhat complicated, it can be viewed simply as the product of two factors. The first factor

$$\left(\frac{\pi}{1 - \pi}\right)^{f_i} \gamma^{-t_i} \frac{|\mathbf{X}_0'\mathbf{X}_0|^{1/2}}{|\Gamma_i + \mathbf{X}_i'\mathbf{X}_i|^{1/2}}$$

is a penalty for increasing the number of variables in the model. The greater the number of variables, the smaller this factor is. Note that it does not depend on $\mathbf{y}$. The second factor

$$\left(\frac{S(\hat{\boldsymbol{\beta}}_i) + \hat{\boldsymbol{\beta}}_i'\Gamma_i\hat{\boldsymbol{\beta}}_i}{S(\hat{\boldsymbol{\beta}}_0)}\right)^{-(n-1)/2}$$

measures how well the model fits. The smaller the residual sum of squares $S(\hat{\boldsymbol{\beta}}_i)$, the greater this factor is. Overall, the posterior probability $p(M_i|\mathbf{y})$, as the product of these two factors, tends to be largest for those models that fit the data well with fewer active factors.

CHAPTER C.7

Follow-up Designs to Resolve Confounding in Multifactor Experiments

Sometimes, the results of a fractional experiment are ambiguous due to confounding among the possible effects, and more than one model may be consistent with the data. We have previously developed a Bayesian method based that uncovers possibly the active factors. Within the Bayesian construct, we here develop a method for designing a follow-up experiment to resolve ambiguities. The idea is to choose runs that allow maximum discrimination among the plausible models. This method is more general than methods that algebraically decouple aliased interactions and more appropriate than optimal design methods that require specification of a single model. The method is illustrated through examples of fractional experiments.

Screening designs such as two-level fractional factorials (Box and Hunter, 1961a; Finney, 1945) and Plackett–Burman (1946) designs are often applied in industry to study large numbers of factors in small numbers of runs. These designs are efficient for estimating the main effects of factors but multifactor systems rarely can be described by main effects alone. The main reason these designs are useful is the likelihood of *factor sparsity* (Juran, 1988), which can be exploited by the *projective properties* (Box and Hunter, 1961) of these arrangements.

From Meyer, R. D., Steinberg, D. M., and Box, G. E. P. (1996), *Technometrics*, **38**(4), 303–313. Reprinted with permission from TECHNOMETRICS. Copyright © 1996 by the American Statistical Association. All rights reserved.

The designs do not, however, always lead to unequivocal conclusions. The results may be inconclusive as to which factors are active and, when that happens, extra runs are needed to resolve the ambiguity. In this article we propose a model-discrimination (MD) criterion for adding extra runs and illustrate its use on several examples. Our procedure combined a design criterion proposed by Box and Hill (1967) with a method proposed for discovering which factors are active in a fractionated experiment (Box and Meyer, 1993).

In Section 1, we give an overview of the design and analysis methodology. In Sections 2 and 3, we illustrate the approach to follow-up designs with two examples. Some preliminary results on starting designs are presented in Section 4. In Section 5, we mention other approaches that have been taken to design augmentation. A discussion of some of the practical aspects of applying the method follows in Section 6. Conclusions are summarized in Section 7. Mathematical details are given in the Appendix.

OVERVIEW

In this section we present the basic strategy that guides our choice of a follow-up design. We defer a detailed derivation to the Appendix.

Our design criterion is motivated by the idea of identifying which of the k factors in the experiment are active where by "active" we mean a factor that can change the level of a particular response through a main effect and/or through any higher order effects in which it is involved. Let M_i label the model in which a particular combination of f_i factors, $0 \leq f_i \leq k$, is active. There are 2^k such models. Following the notion of factor sparsity, we assume that there is a probability π, $0 < \pi < 1$, that each factor is active. We also assume that our prior assessment that any one factor is active is independent of our beliefs about the other factors so that the prior probability of model M_i' is $P(M_i) = \pi^{f_i}(1 - \pi)^{k-f_i}$. In probabilistic terms, the set of models M_i are disjoint events, and their (prior) probabilities sum to 1.

Now suppose that we have performed a multifactor experiment, and let $\mathbf{Y}$ denote the n by 1 vector of responses. For each model M_i, there is a linear model for $\mathbf{Y}$, in which the terms in the regression function are limited to the main effects and interactions of the active factors of M_i. Interactions up to any desired order are included. We also assign prior distributions to all the coefficients in the model (see Appendix). In particular, main effects and interactions in the model are assigned $N(0, \gamma^2\sigma^2)$ priors, where σ^2 is the error variance. The parameter γ is a scale factor relating the magnitude of real effects to noise.

Having observed the response vector $\mathbf{Y}$ we can compute the posterior probability $P(M_i|\mathbf{Y})$ for each model M_i. Because the models M_i are disjoint, the probabilities $P(M_i|\mathbf{Y})$ must again sum to 1. We can also compute the marginal posterior probability that factor j is active by summing these probabilities over all the models that include factor j.

$$P_j = \sum_{M_i:\ \text{factor } j \text{ active}} P(M_i|\mathbf{Y})$$

When results are clear and unambiguous, there will be one model M_i with probability $P(M_i|\mathbf{Y})$ close to 1. The probabilities P_j will be near 1 for each factor in this model and small for all other factors.

Follow-up experiments will be most important when the initial experiment does not clearly identify the active factors. When this occurs, the posterior probabilities $P(M_i|\mathbf{Y})$ will be spread out over two or more different models. Note that it is possible for an individual factor to be present in all or most of these competing models so that its marginal probability P_j is close to 1. Results are not ambiguous with respect to such a factor—it is clearly active. There will be other factors, however, for which the evidence is less clear and a suitably chosen small follow-up experiment could produce much firmer conclusions.

Intuitively, follow-up experiments will be useful for identifying the correct model when the predicted results vary widely depending on which model is used to make the prediction. This idea was exploited by Box and Hill (1967), who proposed a model-discrimination criterion based on the predictive densities for the competing models. Here we apply the Box–Hill criterion to the problem of augmenting a multifactor design.

Let p_i denote the predictive density for a new observation(s) conditional on the observed data $\mathbf{Y}$ and on M_i being the correct model. Then our design criterion is

$$\text{MD} = \sum_{0 \le i \ne j \le m} P(M_i|\mathbf{Y})P(M_j|\mathbf{Y})I(p_i, p_j)$$

where $I(p_i, p_j) = \int p_i \ln(p_i/p_j)$ is the Kullback–Leibler information and measures the mean information for discriminating in favor of M_i against M_j when M_i is true. (Box and Hill used the letter D for the criterion. We use the notation MD for model discrimination and to avoid confusion with the familiar D-optimality criterion in experimental design.) Designs with larger values of MD are better.

The ratio p_i/p_j can be interpreted as the odds in favor of M_i against M_j given by the data from the follow-up experiment (also called the Bayes factor). Some intuition may be gained by considering the case of only two competing models M_1 and M_2. MD is then proportional to the sum of the conditional expected value of $\ln(p_1/p_2)$, given M_1, and the conditional expected value of $\ln(p_2/p_1)$, given M_2 Thus, we want a design that leads to high expected odds in favor of M_1 when it is the correct model but high expected odds in favor of M_2 when it is correct.

We have developed software that performs the Bayesian analysis, evaluates the MD criterion for candidate designs, and searches for optimal designs by an exchange algorithm. (See Meyer (1996) and Barrios (2004).) We illustrate the method in the following sections.

EXAMPLE   **215**

EXAMPLE

Typical of the problems we wish to address is the following experiment, described by Box, Hunter and Hunter (1978) which studied the effects of eight factors on percent shrinkage in an injection molding process. The experimental plan was a 2^{8-4} fractional factorial with generators $I = ABDH = ACEH = BCFH = ABCG$. This design consists of the first 16 rows of Table 1, and a normal probability plot of the contrasts is shown in Figure 1.

Contrasts corresponding to the main effects of holding pressure (factor C) and booster pressure (factor E) clearly stand out. The remaining large contrast is not associated with any main effect but is aliased with 4 two-factor interactions, AE, BF, CH, and DG. An initial analysis would probably ignore interactions of higher order, but we should also keep in mind that each main effect is aliased with 7 different three-factor interactions.

There are five models that account for 99.9% of the posterior probability, as shown in Table 2, allowing for the possibility of two- and three-factor interactions. The marginal posterior probabilities P_j for each factor are displayed in Figure 2. [These were calculated with a value of $\pi = .25$; and $\gamma = 2.0$ estimated from the data; see Appendix and Box and Meyer (1993).] Factors A, C, E, and H have marginal

Table 1. Design and Results for Example 1, the 2^{8-4} Design Injection Molding Experiment

Run	A	B	C	D	E	F	G	H	Y
1	−	−	−	+	+	+	−	+	14.0
2	+	−	−	−	−	+	+	+	16.8
3	−	+	−	−	+	−	+	+	15.0
4	+	+	−	+	−	−	−	+	15.4
5	−	−	+	+	−	−	+	+	27.6
6	+	−	+	−	+	−	−	+	24.0
7	−	+	+	−	−	+	−	+	27.4
8	+	+	+	+	+	+	+	+	22.6
9	+	+	+	−	−	−	+	−	22.3
10	−	+	+	+	+	−	−	−	17.1
11	+	−	+	+	−	+	−	−	21.5
12	−	−	+	−	+	+	+	−	17.5
13	+	+	−	−	+	+	−	−	15.9
14	−	+	−	+	−	+	+	−	21.9
15	+	−	−	+	+	−	+	−	16.7
16	−	−	−	−	−	−	−	−	20.3
17	−	+	+	+	−	−	−	+	29.4
18	−	+	−	−	−	+	+	+	19.7
19	+	+	−	−	+	−	−	+	13.6
20	+	+	+	+	+	+	+	+	24.7

Note: Runs 17–20 are the follow-up design runs.

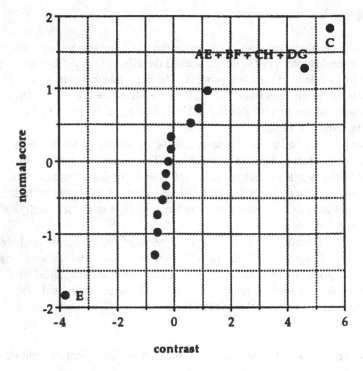

Figure 1. Normal probability plot of the contrasts for Example 1, the injection molding experiment.

posterior probabilities P_j of approximately .76 and clearly stand out as potentially active. The remaining four factors have zero probability, and we conclude that they are essentially inert.

When the 2^{8-4} design is collapsed on factors A, C, E, and H, we obtain a replicated 2^{4-1} design with defining relation $I = ACEH$. Thus these five most likely

Table 2. Five Models of Highest Posterior Probability, Example 1, Injection Molding Experiment, Based on the Initial 16 Runs Only, Accounting for 99.9% of the Total Probability

Model	Active Factors	Posterior Probability
1	A, C, E	.2356
2	A, C, H	.2356
3	A, E, H	.2356
4	C, E, H	.2356
5	A, C, E, H	.0566

EXAMPLE
217

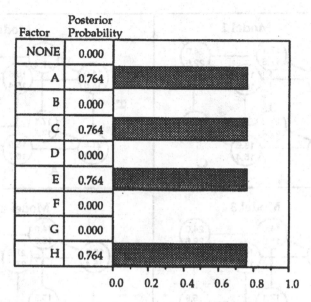

Figure 2. Posterior probabilities that the factors are active. Example 1, the injection molding experiment (initial 16 runs only).

models cannot be distinguished from one another based on the data from the initial experiment. Additional runs are needed to clarify which of the factors are active.

This is seen more clearly in Figure 3, where the results are projected into the four different subsets of factors corresponding to the four high-probability models in Table 2. In each projection, the same pairs of results are replicate experiments. The differences occur by relocating the pairs on the respective cubes.

In Table 3, we show predicted responses for all combinations of factors A, C, E, and H under each of the five competing models. The predictions are calculated using posterior means as estimates of main effects and interactions. The predictions are nearly the same for the eight factor combinations already run in the experiment. The models yield differing predictions on the eight runs not yet conducted. Note that among these latter eight runs, the five predictions are more dispersed for some runs than for others. These runs would seem to be better candidates for the follow-up experiment.

Box et al. (1978, p. 413) set up a follow-up design using a method proposed by Daniel (1962, 1976) that algebraically untangles aliased effects. They showed that adding the four runs numbered 17, 18, 19, 20 at the bottom of Table 1 makes it possible to estimate each of the 4 two-factor interactions, *assuming that no three-factor interactions exist*. The additional runs strongly indicated that the relevant interaction is between holding pressure and screw speed (CH). This effectively implies that factor A is not active.

Using the Bayesian analysis the probability P_j that each factor is active was recomputed (Fig. 4) using the combined 20 runs and including a block effect

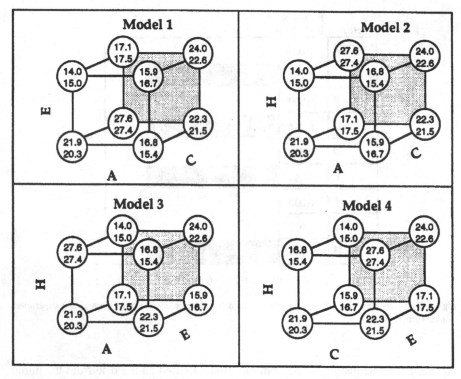

Figure 3. Four different projections of the data from Example 1, the injection molding experiment (initial 16 runs only), labeled models 1–4 as in Table 2.

between the first 16 runs and the last 4. All four factors, A, C, E, and H, still stand out as potentially active. This is quite different from the conclusions drawn previously after sorting out the aliased two-factor interactions, in which the large contrast was attributed to the CH interaction and factor A was judged to be inert. It was his apparent contradiction led us to reconsider the general problem of the design of follow-up experiments.

We observed previously that the Bayes analysis points to only factors A, C, E, and H as potentially active. Thus common sense suggested that the large interaction contrast is being driven by the mold temperature by booster pressure (AE) interaction, or the holding pressure by screw speed (CH) interaction, and that the remaining interactions, BF and DG, could be ignored. Proceeding on this assumption, it would make sense to devote the additional runs to distinguishing between only the AE and CH interactions. Finally, when considered as a design in the factors A, C, E, and H alone, the follow-up design repeats two runs from the original experiment. In general, one might expect that previously untested combinations of these factors would be more effective for identifying active factors.

We now reconsider the design of a follow-up experiment using the MD criterion. The four inert factors are ignored under this approach, and it is assumed that they are

EXAMPLE   219

Table 3. The Injection Molding Experiment, Example 1, Collapsed on the Four Factors A, C, E, and H

Design Point	Run	A	C	E	H	Y	Prediction from Model				
							1	2	3	4	5
1	14, 16	−	−	−	−	21.9, 20.3	21.08	21.08	21.08	21.08	21.09
2	1, 3	−	−	+	+	14.0, 15.0	14.58	14.58	14.58	14.58	14.54
3	5, 7	−	+	−	+	27.6, 27.4	27.38	27.38	27.38	27.38	27.44
4	10, 12	−	+	+	−	17.1, 17.5	17.34	17.34	17.34	17.34	17.32
5	2, 4	+	−	−	+	16.8, 15.4	16.16	16.16	16.16	16.16	16.13
6	13, 15	+	−	+	−	15.9, 16.7	16.35	16.35	16.35	16.35	16.33
7	9, 11	+	+	−	−	22.3, 21.5	21.87	21.87	21.87	21.87	21.88
8	6, 8	+	+	+	+	24.0, 22.6	23.25	23.25	23.25	23.25	23.27
9		−	−	−	+		21.08	14.58	27.38	16.16	19.75
10		−	−	+	−		14.58	21.08	17.34	16.35	19.75
11		−	+	−	−		27.38	17.34	21.08	21.87	19.75
12		−	+	+	+		17.34	27.38	14.58	23.25	19.75
13		+	−	−	−		16.16	16.35	21.87	21.08	19.75
14		+	−	+	+		16.35	16.16	23.25	14.58	19.75
15		+	+	−	+		21.87	23.25	16.16	27.38	19.75
16		+	+	+	−		23.25	21.87	16.35	17.34	19.75

Note: For each possible combination of these factors, the table shows the predicted responses under each of the five most probable models. The first eight rows correspond to the combinations actually used in the initial design and also show the run numbers and observed responses.

set to some convenient levels within the range of the initial experiment. A four-run follow-up design will be generated, for purposes of comparison with the follow-up design described earlier (we call this the BHH design). The MD criterion is defined earlier and the Appendix, and the models and their respective posterior probabilities are given in Table 2. A possible block effect between the initial design and the follow-up design is included in each model. We define the 16 runs of the full 2^4 factors A, C, E, and H, Table 3, to be the candidate runs for the follow-up experiment. Eight of these runs appeared in the initial experiment as well, denoted by the Run column.

There are 3,876 possible follow-up designs of four runs chosen from 16 candidates. We calculated MD for each one. The 10 best designs of four runs with respect to the MD criterion are shown in Table 4, along with the BH^2 design. The best design consists of design points 9, 9, 12, and 15 (design point 9 is replicated). The BH^2 follow-up design consisted of design points 3, 8, 9, and 14, and ranked 1,242nd among the possible follow-up designs. Note that all of the 10 best designs used only design points 9–16, the runs not performed in the initial experiment.

For general problems, there will be a large number of possible follow-up designs. To avoid the necessity of calculating MD for every one as we have done for this example, we instead employ the exchange algorithm (see Mitchell and Miller, 1970;

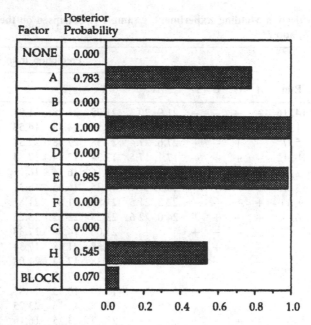

Figure 4. Posterior probabilities that the factors are active. Example 1, the injection molding experiment (including follow-up runs 17–20).

Table 4. The 10 Best Four-Run Follow-up Designs for Example 1, the Injection Molding Experiment, Ranked by the MD Criterion with the BH^2 Design, Which Ranked 1,242, Shown Also for Comparison

Rank	Design Points				MD
1	9	9	12	15	85.7
2	9	12	14	15	84.4
3	9	11	12	15	83.6
4	9	11	12	12	82.2
5	9	9	12	12	79.6
6	9	11	12	14	77.1
7	9	9	11	12	77.1
8	0	12	15	16	77.1
9	9	12	13	15	76.7
10	9	12	12	14	76.6
⋮					
1,242	3	8	9	14	27.1

Wynn, 1972) to search for the design with highest MD. The algorithm begins by generating a design at random from a set of candidate points and then sequentially modifies it until a convergence criterion is satisfied. For a more detailed description, see the Appendix. To examine the effectiveness of the exchange algorithm for this example, we ran it with 20 different random starts. Six times out of 20, convergence was achieved to what we knew from complete enumeration to be the best design (runs 9-9-12-15). The remaining 14 times, the algorithm converged at the second-best design (runs 9-12-14-15). Because it is typical practice to start the algorithm from 20 to 50 random starting designs or more, the results are satisfying.

ANOTHER EXAMPLE

In the first example, our comparison was, necessarily, limited to examining the MD-design criterion for the different follow-up designs. There was no way, however, to actually go back and conduct the runs in our recommended design and then compare the resulting analyses. Our goal in this example is to illustrate how effective the MD criterion can be in generating data that help identify active factors.

To provide data for the following example we used the 32 runs from the reactor experiment in BHHII p. 260. We extracted eight runs from the original experiment yielding a 2^{5-2} Resolution III experiment with generators $I = ABD = ACE$, We treat these runs, shown in Table 5, as an initial screening experiment. We can then simulate any follow-up runs by extracting them from the complete 2^5 experiment.

The Bayesian analysis is summarized in Figure 5, which shows the posterior probabilities for the five factors. These were computed with a value of $\pi = .25$ and a value of $\gamma = .40$ estimated from the initial data. None of the factors stands out as clearly active. At this point one might recommend another eight runs, a complementary fraction of the original design. The foldover fraction is often used and has many appealing properties. Another approach that has been taken is to find a

Table 5. The Eight Runs for Example 2, Which Constitute a 2^{5-2} Design, Extracted from the Full 2^5 Reactor Experiment Design of BH^2 (p. 376)

Run	A	B	C	D	E	Y
1	−	−	−	+	+	44
2	+	−	−	−	−	53
3	−	+	−	−	+	70
4	+	+	−	+	−	93
5	−	−	+	+	−	66
6	+	−	+	−	+	55
7	−	+	+	−	−	54
8	+	+	+	+	+	82

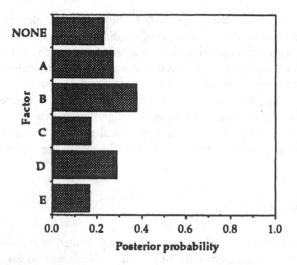

Figure 5. Posterior probabilities that the factors are active. Example 2, the 2^{5-2} reactor experiment (initial 8 runs only).

D-optimal design that consists of the eight original runs together with some follow-up runs. Still, with five factors, 10 two-factor interactions among these, an overall mean, and a block effect, there are 17 parameters. Nine additional runs at minimum would be required to estimate all these parameters.

A smaller follow-up design is possible via our approach. A four-run design was created by finding the best subset of 4 from among the full $2^5 = 32$ candidates of five factors at two levels. A block effect was included in each model to allow for a possible shift in results between the first and second experiment. The 32 possible models, and their posterior probabilities, are shown in Table 6. For this problem we employed the exchange algorithm (see Mitchell and Miller, 1970; Wynn, 1972) to find the best design, shown in Table 7. Again the responses Y are taken from the complete 2^5 experiment. For this problem there are 52,360 possible four-run designs from 32 candidates. From 50 random starts, 18 converged to the design in Table 7. The five best designs found, and their MD values, are shown in Table 8.

The posterior probabilities P_j were recomputed, using all 12 runs, and are shown in Figure 6. The probabilities now indicate the likely activity of the three factors B, D, and E, and also that factor A is clearly inert. There is perhaps weak evidence for factor C. This generally agrees with the analysis of the complete 2^5, which showed that factors B, D, and E were clearly the active factors in this experiment. (The Bayesian analysis of the complete 2^5 yields activity probabilities of 1.000 for factors B, D, and E and .000 for A and C. As would be expected the evidence is more compelling based on 32 runs than on 12 runs. However, the level of evidence shown here after 12 runs would often be sufficient for an experimenter's needs.)

Table 6. Posterior Probabilities of Each of the 32
Possible Models for Example 2, the 2^{5-2} Reactor
Experiment, After the Initial Eight Runs

Rank	Probability	Active Factors
1	.23100	None
2	.13400	B
3	.07500	D
4	.07000	A
5	.05500	AD
6	.05500	AB
7	.05500	BD
8	.05200	E
9	.05100	C
10	.03200	BC
11	.02300	BE
12	.02200	ABD
13	.01700	DE
14	.01200	CD
15	.01100	AE
16	.01100	CE
17	01100	AC
18	.00900	ACD
19	.00900	ADE
20	.00900	BCE
21	.00900	ABE
22	.00900	CDE
23	.00900	ABC
24	.00900	BCD
25	.00900	BDE
26	.00400	ABCD
27	.00300	ABDE
28	.00300	BCDE
29	.00300	ACDE
30	.00200	ACE
31	.00200	ABCE
32	.00100	ABCDE

Table 7. The MD-Optimal Four-Run Follow-up Design
for Example 2, the 2^{5-2} Reactor Experiment

Run	A	B	C	D	E	Y
9	+	+	−	−	−	61
10	+	−	−	+	+	45
11	−	+	−	+	−	94
12	+	−	−	+	−	61

Table 8. The Five Best Four-Run Follow-up Designs for Example 2, the 2^{5-2} Reactor Experiment, Ranked by the MD Criterion

Rank		Design Points			MD
1	4	10	11	26	.6153
2	4	10	11	28	.6104
3	4	10	26	27	.6079
4	4	10	12	27	.6059
5	4	11	12	26	.6033

This exercise was repeated starting from other eight-run fractions. Initial designs without *BDE* in the defining relation yield equivalent results to those shown previously. When *BDE* is in the defining relation of the initial design, results are similar to the example in Section 2: factors *B*, *D*, and *E* stand out as potentially active after eight runs, but the three possible two-factor models involving *B*, *D*, or *E* are equally consistent with the data and cannot be discriminated. However, the four additional runs chosen according to the MD criterion lead to a "correct" identification of the model with all three factors active.

Since writing this, we have experimented with adding one run at a time. Some of the results are given on page 300 of BHH II (see also Barrios, 2004a). The strategy IAA (improving almost anything) is sometimes valuable when the result from each run is available before the next is planned. Following this strategy the sequence of runs was 12, 10, 11, 15. For this particular example the same conclusions reached from the block of four runs are reached after only 3 sequential runs.

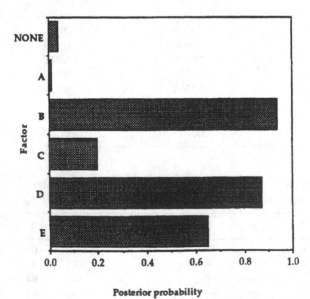

Figure 6. Posterior probabilities that the factors are active. Example 2, the 2^{5-2} reactor experiment (follow-up runs 9–12 included).

STARTING DESIGNS

The MD design criterion can also be used to generate starting designs for screening experiments. We have conducted a preliminary study of this question, limited in part by the computing time required. Our conclusions for the limited cases we have examined are that the 2^{k-p} fractional factorials of high resolution are the best designs. We are continuing to study in more depth the discriminatory properties of starting designs.

Eight-Run Designs

We conducted a search for the MD-optimal eight-run designs for studying from three to seven factors. We set $\pi = .25$ and used γ of both 1 and 2 to cover the range typically encountered in practice. Interactions involving up to three factors were included. The results are summarized in Table 9. In each case the best design was a 2^{k-p} fractional factorial. The six-factor ($\gamma = 1$) and seven-factor ($\gamma = 1$ and 2) cases

Table 9. MD-Optimal Starting Designs for Various Numbers of Runs and Factors ($\pi = .25$, $\gamma = 1$ and 2)

Random Starts	Number of Factors	Number of Runs	γ	Design	Notes
50	3	8	1	2^3	
50	3	8	2	2^3	
50	4	8	1	2^{4-1}_{IV}	
50	4	8	2	2^{4-1}_{IV}	
50	5	8	1	2^{5-2}_{III}	
50	5	8	2	2^{5-3}_{III}	
50	6	8	1	2^{5-2}_{III}	Hold one factor constant. (1)
50	6	8	2	2^{6-3}_{III}	
20	7	8	1	2^{6-3}_{III}	Hold one factor constant
	7	8	2	2^{6-3}_{III}	Hold one factor constant. (2)
20	5	16	1	2^{5-1}_{V}	
20	5	16	2	2^{5-1}_{V}	
10	6	16	1	2^{6-2}_{IV}	Also other orthogonal arrays. (3)
10	6	16	2	2^{6-2}_{IV}	Also other orthogonal arrays. (3)
10	7	16	1	2^{7-3}_{IV}	Also other orthogonal arrays. (3)
(4)	7	16	2	2^{7-3}_{IV}	Also other orthogonal arrays. (3)

Note: (1) For six factors in eight runs, the MD values for the 2^{6-3} and 2^{5-2} (hold 1 constant) designs are very close, within 1% of each other. (2) For seven factors in eight runs, $\gamma = 2$, we simply reevaluated the top designs for $\gamma = 1$. (3) The other orthogonal arrays are subsets of columns from $H_{16}(2)$ through $H_{16}(5)$ (Box and Tyssedal, 1994), which are not of the 2^{k-p} structure. (4) For seven factors in 16 runs, $\gamma = 2$, we evaluated all subsets of columns from $H_{16}(1)$ through $H_{16}(5)$ as well as the best six-factor design, holding one factor constant.

were interesting because the best design was to leave out one of the factors and run a 2^{k-p} in the remaining factors. Our interpretation of this result is that, when screening a set of six or seven factors in eight runs, with the possibility of two- and three-way interactions, a reasonable approach in the initial design may be to hold out one factor, which could be brought in later in a follow-up experiment.

16-Run Designs

Our study of 16-run designs proceeded as follows. We considered designs for five, six, and seven factors, respectively. Again, interactions involving up to three factors were included. A limited number of searches were conducted from random starting designs. In addition, we evaluated all possible subsets of five, six, and seven columns from the known orthogonal arrays of 16 runs. There are five of these (Hall, 1961), which Box and Tyssedal (1995) labeled $H_{16}(1)$ through $H_{16}(5)$. $H_{16}(1)$ is the factorial array built up by forming all interaction columns from the 2^4 factorial design. Any 16-run 2^{k-p} design is some subset of columns from $H_{16}(1)$. The remaining four arrays have interesting projective properties, as discussed by Box and Tyssedal (1994). Finally we also evaluated the best design that holds one variable constant in each case, in light of the findings for 8-run designs. The results are shown in Table 9.

In all three cases (five, six, and seven factors), the 2^{k-p} design, holding no factors constant, was the best design found. For five factors in 16 runs, the 2^{5-1} design was the unique best design. For six or seven factors, nonregular fractions formed by taking certain subsets of columns from one of $H_{16}(2)$ through $H_{16}(5)$ were equivalent to the 2^{k-p} as the best 16-run design.

32-Run Designs

We looked at one interesting case of a 32-run design for studying seven factors. Four different 2^{7-2} designs were compared, as shown in Table 10. The first three are all Resolution IV, the last is Resolution III. The three Resolution IV designs are listed in order according to the criterion of aberration (Fries and Hunter, 1980). The minimum aberration 2^{k-p} design is the design of maximum resolution R with the fewest number of words of length R in the defining relation. Intuitively, there is less

Table 10. A Comparison of Four Possible 2^{7-2} Designs for Studying Seven Factors in 32 Runs

Design	Generators	Defining Relation	Resolution	$\dfrac{MD}{\gamma = 1}$	$\dfrac{MD}{\gamma = 2}$
1	$6 = 1234, 7 = 2345$	$I = 12346 = 12357 = 4567$	IV	46.4	190.0
2	$6 = 123, 7 = 145$	$I = 1236 = 1457 = 234567$	IV	46.1	188.7
3	$6 = 123, 7 = 234$	$I = 1236 = 2347 = 1467$	IV	45.7	187.0
4	$6 = 12, 7 = 234$	$I = 126 = 23457 = 134567$	III	45.4	185.8

confounding in a design with minimum aberration. The concept of aberration, though useful, is restricted to 2^{k-p} designs.

We calculated MD for each of the designs in Table 10. The MD criterion ranks the designs in the same order as resolution/aberration considerations. The result further demonstrates how maximizing model discrimination, as measured by MD, corresponds to minimizing confounding.

OTHER APPROACHES

There has been considerable research on how to augment an initial experimental plan. Most of that research, however, has focused on choosing runs to improve estimation within the framework of a single known model (Covey-Crump and Silvey, 1970; Evans, 1979; Hebble and Mitchell, 1972; and others). Chaloner (1984) provided a review of the work on design augmentation and showed how it relates to Bayesian optimal design, in which the initial data generate the prior for the new observations.

Previous work on augmenting fractionated designs to aid in model identification has focused on de-aliasing effects by adding extra fractions. The most common application of this idea is the use of a foldover block to augment a Resolution III design (Box and Hunter, 1961; Box and Wilson, 1951; Margolin, 1969). The combined experiment has Resolution IV, so the foldover technique is an effective way to clear all main effects of two-factor interactions. It is not always economical, however, because the extra block of observations must be equal in size to the original block. Moreover, there is no obvious way to apply the foldover method to many more general problems.

Daniel (1962, 1976, Chap. 14) showed how to add an extra block of runs to resolve an ambiguous alias string in a fractional factorial experiment. We have found Daniel's method difficult to apply except in the simplest of cases. When there is more than one alias string to untangle, or when three-factor interactions are included, the method becomes unwieldy. Moreover, it cannot be applied to designs with more complex confounding such as Plackett–Burman designs.

The ideas of optimal design theory have also been applied to the model-discrimination problem. See, for example, Atkinson and Cox (1947), Atkinson and Fedorov (1975a,b), and Jones and Mitchell (1978). The optimality criterion used in these articles is related to the power to test for differences among the competing models. When applied to the factor screening setting that we consider here, these methods require designs that permit all possible models to be estimated. Thus a large follow-up design is necessary.

Alternative Bayesian approaches to the design of factorial experiments were developed by Draper and Guttman (1992), DuMouchel and Jones (1994), and Steinberg (1985). Those works assume that all the factors are active and use the Bayesian paradigm to provide flexibility in specifying the relation between the response and the factors. The response function may include many terms, with proper prior distributions assigned to all but a small subset of the coefficients. For example, in the context of a fractional factorial experiment, DuMouchel and Jones

(1994) might include all main effects and two-factor interactions in the model, with proper prior distributions assigned to the interaction coefficients. Draper and Guttman (1992) and Steinberg (1985) focused on use of the Bayesian model to scale standard designs relative to the region of experimental interest. Because these models do not include our first-stage prior, which specifies which factors are active, they necessarily use different design criteria than the one that we have proposed.

DISCUSSION

For appropriate situations the advantage of sequential assembly of designs has been emphasized; see e.g., Box and Wilson, 1951 and Box, 1999. The methodology described in this article should aid this process, by making it easier to select small but informative follow-up designs. We recognize, however, that practical constraints sometimes make sequences of smaller designs inefficient. Using a somewhat simplified view of experimenting, imagine the cost and time associated with one iteration of experimentation as including a variable component (the cost and time per run) and a fixed cost independent of the number of runs. When fixed costs and times are high, there will be benefits to using fewer iterations and larger designs. When variable costs and times dominate, more iterations of smaller designs should be used.

In lubricant development experiments, testing can be very costly, with limited test capacity. Total cost and time is mostly the variable cost and time per run. Small designs used in sequence are the rule. At the other extreme, one experimenter with whom we consulted had one night (!) of access to a specialized facility and could not afford to devote valuable time to interim analyses; he required a full experimental plan, in advance, for the entire night. Prototype testing of a mechanical or electronic apparatus offers a hybrid situation. Some factors may define the prototype itself, whereas other factors define test conditions. Units are typically expensive to build but easy to test. Once built, then, it is advantageous to test a unit at many different test conditions. Bisgaard and Steinberg (1993) discussed the design and analysis of such experiments. See also Chapters E.4, E.6 and E.7.

In our experience, these practical constraints often dictate the number of runs that can be performed at any stage of an investigation. The methodology proposed here will be useful in choosing which runs to conduct in follow-up designs.

The Bayesian model on which our criterion is based involves two levels of prior information. First, we specify a prior probability that any given set of factors is active. Second, conditional on the set of active factors, we assume a prior distribution for the coefficients in the model defined by the active factors. The prior distributions adopted at each level involve additional parameters and both the analysis and the design criterion may be sensitive to their values. At the first level, we assume that each factor is active with probability π. As Box and Meyer (1993) noted, $\pi = .25$ is a good default choice, and it may be informative to test the sensitivity of results to the choice of π. At the second level, we assume that each coefficient has a normal prior distribution whose variance is γ^2 times as large as the experimental error variance. Box and Meyer (1993) showed how γ can be estimated from the initial experiment.

We have examined the effect of γ on design augmentation in several examples. Our results suggest that the choice of γ does change the numerical value of the criterion MD substantially and only occasionally has an effect on the design selected. In selecting an initial design, the results also indicate that using smaller γ values may lead to dropping a factor from the design. In designing follow-up experiments, we recommend using the value suggested by the data from the initial design.

Some modifications of our basic model may also be useful for choosing designs. For example, one could assign main-effect coefficients a separate value of γ from the value for interaction coefficients. This choice allows the possibility that main effects will be larger than interactions. In the shrinkage experiment that we described in the Introduction, using different γ's gives higher posterior probability to the models in Table 2 that include both factors C and E, which had large main effects in the initial design, than to the models that exclude one of them. The best four-run follow-up design in this scenario would be runs 9, 11, 12, and 15 in Table 3. Runs 9, 9, 12, and 15, the best design when one value of γ was used for both main effects and inter-actions, is still very good, third best among all possible follow-up designs.

Although the examples we give are based on 2^{k-p} fractional factorials, the approach is exactly the same for any other design. For example, the designs of Plackett and Burman (1946) are often useful starting points in an investigation. We have shown previously (Box and Meyer, 1993) that the Bayesian approach to analysis is particularly effective for these designs. This is because the ambiguities resulting from confounding of Plackett and Burman designs are more complex than for 2^{k-p} designs. Likewise, the MD approach to follow-up designs will be equally effective in dealing with this complex confounding.

Going further, the methodology described here is completely general. Box and Hill (1967), of course, applied it to nonlinear models. Some experimenters may wish to consider different classes of models than the one we have used. At times it may be appropriate to consider, for example, models that include the main effects of factors without the interaction terms or models with pure quadratic terms. (We automatically include interactions when main effects are in the model for computational ease; it greatly reduces the number of candidate models while still including the most plausible ones.) In the latter case the design space would include factors at three or more levels. The general approach would work exactly the same in these other cases.

CONCLUSIONS

One of the problems in the practical application of fractional factorial designs and other screening designs has been the design of follow-up experiments to resolve confounding. The method devised by Daniel (1962) is difficult to apply except in the simplest cases. In this chapter we have shown that the idea of model discrimination can be successfully applied to the problem of augmenting an initial design. Our approach combines the analysis method of Box and Meyer (1993) with the design criterion of Box and Hill (1967). We have shown in examples that it is easily applied and leads to sensible designs and results.

Our preliminary results on starting designs in Section 5 are at the same time reassuring and intriguing. For the limited number of cases we studied, we have shown that the MD criterion leads to the commonly used fractional factorial designs. In some cases the best design involves dropping a factor. The 16-run case is interesting because some "nonfactorial" 16-run designs are shown to be as good as the 2^{k-p} regular fractional factorials. The 32-run case we examined suggests that MD follows the intuitive criteria of resolution and aberration.

We have developed FORTRAN programs that perform the Bayesian analysis, evaluate the MD criterion for candidate designs, and select optimal designs by the exchange algorithm. They are available on StatLib. (See Meyer (1996) and Barrios (2004).) This and other computations for this book can be done with the statistical language R (R Development Core Team, 2004), available at CRAN (http://cran.R-project.org). Functions for displaying anova and lambda plots, for Bayesian screening and model building are included in the BHH2 and BsMD R-packages and available at CRAN under contributed packages. There is as well commercial software SAS.Jump, which some readers will find easier to use.

ACKNOWLEDGMENTS

This work was supported in part by a grant from the Alfred P. Sloan Foundation and by The Lubrizol Corporation. The work of D. M. Steinberg was carried out in part while visiting the Center for Quality and Productivity Improvement at the University of Wisconsin. The visit to CQPI was aided by a grant from the Alfred P. Sloan Foundation. We thank Robert Mau, Jr. for programming assistance.

APPENDIX: MATHEMATICAL DETAILS

In this section we give some technical details of the Bayesian model-discrimination method and design criterion MD. For additional information on this approach to analyzing the data, see Box and Meyer (1993).

Let Y denote the n by 1 vector of responses from a fractional factorial experiment with k factors. The model for Y depends on which factors are active, and our analysis considers all the possible sets of active factors. Let M_i label the model in which a particular combination of f_i factors, $0 \le f_i \le k$, is active. Conditional on M_i being the true model, we assume the usual normal linear model $Y \sim N(X_i\beta_i, \sigma^2 I)$.

Here X_i is the regression matrix for M_i and includes main effects for each active factor and interaction terms up to any desired order. Let t_i denote the number of effects (excluding the constant term) in M_i. We will let M_0 label the model with no active factors.

We assign noninformative prior distribution to the constant term β_0 and the error standard deviation σ, which are common to all the models. Thus $p(\beta_0, \sigma) \propto 1/\sigma$. The remaining coefficients in β_i are assigned independent prior normal distributions with mean 0 and standard deviation $\gamma\sigma$.

We complete the model statement by assigning prior probabilities to each of the possible models. Following the notion of factor sparsity, we assume that there is a probability π, $0 < \pi < 1$, that any factor is active, and that our prior assessment of any one factor being active is independent of our beliefs about the other factors. Then the prior probability of model M_i is $P(M_i) = \pi^{f_i}(1 - \pi)^{k-f_i}$.

Having observed the data vector $\mathbf{Y}$, we can update the parameter distributions for each model and the probability of each model being valid. In particular, the posterior probability that M_i is the correct model will be

$$P(M_i|\mathbf{Y}) \propto \pi^{f_i}(1 - \pi)^{k-f_i}\gamma^{-t_i}|\Gamma_i + \mathbf{X}_i'\mathbf{X}_i|^{-1/2}S_i^{-(n-1)/2}, \tag{A1}$$

where

$$\Gamma_i = \frac{1}{\gamma^2}\begin{pmatrix} 0 & 0 \\ 0 & \mathbf{I}_{t_i} \end{pmatrix} \tag{A2}$$

$$\hat{\beta}_i = (\Gamma_i + \mathbf{X}_i'\mathbf{X}_i)^{-1}\mathbf{X}_i'\mathbf{Y} \tag{A3}$$

and

$$S_i = (\mathbf{Y} - \mathbf{X}_i\hat{\beta}_i)'(\mathbf{Y} - \mathbf{X}_i\hat{\beta}_i) + \hat{\beta}_i'\Gamma_i\hat{\beta}_i \tag{A4}$$

The constant of proportionality in (A1) is that value that forces the probabilities to sum to unity. The probabilities $P(M_i|\mathbf{Y})$ can be summed over all models that include factor j, say, to compute the posterior probability P_j that factor j is active,

$$P_j = \sum_{M_i:\text{ factor } j \text{ active}} P(M_i|\mathbf{Y}) \tag{A5}$$

The probabilities $\{P_j\}$ provide a convenient summary of the importance, or activity, of each of the experimental factors.

The MD design criterion is motivated by the idea of identifying a useful model and, thereby, the active factors. In terms of the Bayesian analysis, the experiment clearly suggests a particular model M_i when the posterior probability $P(M_i|\mathbf{Y})$ is close to 1. Conclusions will be ambiguous when there are several models with substantial probabilities.

The MD design criterion proposed by Box and Hill (1967) has the following form. Let $\mathbf{Y}^*$ denote the data vector that will be obtained from the additional runs and let $p(\mathbf{Y}^*|M_i, \mathbf{Y})$ denote the predictive density of $\mathbf{Y}^*$ given the initial data $\mathbf{Y}$ and the model M_i. Then

$$\text{MD} = \sum_{0 \le i \ne j \le m} P(M_i|\mathbf{Y})P(M_j|\mathbf{Y}) \int_{-\infty}^{\infty} p(\mathbf{Y}^*|M_i, \mathbf{Y}) \ln\left(\frac{p(\mathbf{Y}^*|M_i, \mathbf{Y})}{p(\mathbf{Y}^*|M_j, \mathbf{Y})}\right) d\mathbf{Y}^* \tag{A6}$$

(Box and Hill used the letter D for the criterion. We use the notation MD for model discrimination.)

The MD criterion has a natural interpretation in terms of Kullback–Leibler information. Let p_i and p_j denote the predictive densities of $\mathbf{Y}^*$ under models M_i and M_j, respectively. The Kullback–Leibler information is $I(p_i, p_j) = \int p_i \ln(p_i/p_j)$ and measures the mean information for discriminating in favor of M_i against M_j when M_i is true. Then

$$MD = \sum_{0 \leq i \neq j \leq m} P(M_i|\mathbf{Y})P(M_j|\mathbf{Y})I(p_i, p_j) \tag{A7}$$

MD is thus the Kullback–Leibler information, averaged over pairs of candidate models with respect to the posterior probabilities of the models.

We now simplify MD, proceeding conditionally on σ^2 and then integrating out σ^2 at the last step. We first derive the predictive distribution of $\mathbf{Y}^*$ for each model. Let $\mathbf{X}_i$ and $\mathbf{X}_i^*$ denote the regression matrices for the initial runs and the additional runs, respectively, when M_i is the model.

Then

$$\mathbf{Y}^*|M_i, Y, \sigma^2 \sim N(\hat{\mathbf{Y}}_i^*, \sigma^2\mathbf{V}_i^*)$$

where $\hat{\mathbf{Y}}_i^* = \mathbf{X}_i^*\hat{\beta}_i$, $\hat{\beta}_i$ defined by (A3), and

$$\mathbf{V}_i^* = \mathbf{I} + \mathbf{X}_i^*(\mathbf{\Gamma}_i + \mathbf{X}_i'\mathbf{X}_i)^{-1}\mathbf{X}_i^{*'} \tag{A8}$$

The log ratio of the predictive densities for two models, M_i and M_j, is then

$$\ln\left(\frac{p(\mathbf{Y}^*|M_i, \mathbf{Y}, \sigma^2)}{p(\mathbf{Y}^*|M_j, \mathbf{Y}, \sigma^2)}\right) = \frac{1}{2}\ln\left(\frac{|\mathbf{V}_j^*|}{|\mathbf{V}_i^*|}\right) = \frac{1}{2\sigma^2}((\mathbf{Y}^* - \hat{\mathbf{Y}}_j^*)'\mathbf{V}_j^{*-1}(\mathbf{Y}^* - \hat{\mathbf{Y}}_j^*)$$
$$- (\mathbf{Y}^* - \hat{\mathbf{Y}}_i^*)'\mathbf{V}_i^{*-1}(\mathbf{Y}^* - \hat{\mathbf{Y}}_i^*))$$

Integrating with respect to $p(\mathbf{Y}^*|M_i, \mathbf{Y}, \sigma^2)$ yields

$$\frac{1}{2}\ln\left(\frac{|\mathbf{V}_j^*|}{|\mathbf{V}_i^*|}\right) - \frac{1}{2\sigma^2}(n^*\sigma^2 - \sigma^2\,\text{tr}\{\mathbf{V}_j^{*-1}\mathbf{V}_i^*\} - (\hat{\mathbf{Y}}_i^* - \hat{\mathbf{Y}}_j^*)'\mathbf{V}_j^{*-1}(\hat{\mathbf{Y}}_i^* - \hat{\mathbf{Y}}_j^*))$$

Finally, integrating with respect to $p(\sigma^2|\mathbf{Y}, M_i)$, we have the MD criterion defined by

$$MD = \frac{1}{2}\sum_{0 \leq i \neq j \leq m} P(M_i|\mathbf{Y})P(M_j|\mathbf{Y})(-n^* + \text{tr}\{\mathbf{V}_j^{*-1}\mathbf{V}_i^*\} + (n-1)$$
$$\times ((\hat{\mathbf{Y}}_i^* - \hat{\mathbf{Y}}_j^*)'\hat{\mathbf{V}}_j^{*-1}(\hat{\mathbf{Y}}_i^* - \hat{\mathbf{Y}}_j^*))/S_i) \tag{A9}$$

where S_i was given by (A4). (The term involving determinants sums to 0 across $0 \leq i \neq j \leq m$ and so was removed.)

Note that, for the models and prior distributions we use, the quadratic form in (A9) is 0 for starting designs. Starting designs are thus evaluated by the trace term in (A9). For mathematical convenience, we place a very small value (.00001) in the $(1, 1)$ position of the matrix $\mathbf{\Gamma}_i$. This is equivalent to approximating the locally uniform prior for the intercept β_0 with a normal distribution, with mean 0 and large variance.

To compute MD-optimal augmentation designs, we use an exchange algorithm like those proposed by Mitchell and Miller (1970) and Wynn (1972). First, all candidate design points must be listed. Typically this is the full factorial list of all factor-level combinations. An initial augmentation design is chosen at random, with replacement, from among the candidate points. The initial design is then improved by adding that point, from among all the candidates, which most improves the MD-criterion, followed by removing that point from all those currently in the design, which results in the smallest reduction in the criterion. This add/remove procedure is continued until it converges, with the same point being added and then removed. The algorithm can get trapped in local optima, so we typically make from 20 to 50 or more random starts.

Author's note

In the examples used in the chapter the initial designs and candidate runs are fractional or full factorials. However the method itself, its algebraic derivation given in the appendix and the computer programs are all quite general.

To be specific it is supposed:

1. You have an experimental space defined in terms of coordinates $X_1, X_2, \ldots, X_k$, or any subset thereof.
2. You have a response of interest determined by an unknown relationship $y = M(X, e)$ where e is noise and where X may refer to all the x's or to any subset thereof which is then said to contain the active factors.
3. You speculate that one of a number of competing models $M_1, M_2, \ldots, M_m$ can represent M. The differences between the speculative models may, but need not, be concerned only with the identity of the active factors.
4. There are some existing experimental runs made at locations $x_1, x_2, \ldots, x_n$ that have yielded observed responses $y_1, y_2, \ldots, y_n$ call this prior data.
5. Suppose that a Bayes analysis of this prior data provided no strong preference among the models.
6. Suppose now that q follow-up candidate runs are proposed at locations $x_1^*, x_2^*, \ldots, x_q^*$ and of these $p(\geq 1)$ runs are chosen that maximize the *MD* criterion.
7. A new Bayes analysis is now made including the p new runs.
8. If there is still insufficient differentiation between the models, steps 4, 5 and 6 are repeated, and so on.
9. In the second edition of *Statistics for Experimenters* by Box, Hunter and Hunter, page 299, there is an example of special interest in which runs are added one at a time.

Thus: in 4, the existing runs do not need to be part of an experimental design; in 6, the list of candidate follow-up runs need not be chosen from any experimental design (it could be they were suggestions from a team of investigators based on intuition, subject matter knowledge curve fitting etc.); in a further iteration the list of candidate follow-up runs could be different from those used before as could the group of competing models.

CHAPTER C.8

Projective Properties of Certain Orthogonal Arrays

SUMMARY

A question of importance in factor screening is when a two-level orthogonal design for a multifactor experiment can be projected into lower dimension, typically $P = 2$ or 3. New results relate to the projectivity P of saturated designs in which $n - 1$ factors are tested in n runs. It is shown that: a design obtained by "doubling" an $n \times n$ orthogonal array is always of projectivity $P = 2$; a two-level cyclic design is either a factorial array, and hence has $P = 2$, or it has $P = 3$; a two-level orthogonal design with 4m runs, m odd, has $P = 3$. In particular these results allow the designs derived by Plackett and Burman (1946) to be categorized in terms of these projective properties.

INTRODUCTION

At the preliminary stage of an experimental investigation a hypothesis of factor sparsity is often appropriate. That is to say, of a larger number k of factors to be tested only a small subset, typically 2 or 3, are expected to be active. An active factor is one that individually or interactively produces change in the response. A factor which is not active is said to be inert. This means that no detectable main effect or interaction involving that factor occurs. Screening designs are then needed to identify the active subset.

From Box, G. E. P. and Tyssedal, J. (1996), *Biometrika*, **83**(4), 950–955.

An important source of screening designs are the two-level orthogonal arrays. See, for example, Finnay (1945) Plackett and Burman (1946), Rao (1947), M. J. Hall (1961), in Jet Propulsion Laboratory Research Summary 36-10 "Hadamard matrices of order 16," and in Jet Propulsion Laboratory Technical Report 32-761 "Hadamard matrices of order 20," Raghavarao (1971), and Hedayat and Wallis (1978). An orthogonal two-level array H_n is a $n \times n$ matrix with orthogonal columns, where $n = 4m$ with m a positive integer, and with a first column of $+1$'s. The remaining $n - 1$ contrast columns thus consist of an equal number of $+1$'s and -1's. Arrays obtained by renumbering rows, renumbering contrast columns, or switching all the signs in a contrast column are regarded as equivalent. An orthogonal array can be used to generate a statistical design by associating k of its contrast columns with the levels of k experimental factors. In this paper we consider in particular designs for which $k = n - 1$. The design is then sometimes called a saturated design (Box and Wilson, 1951) or a main effect plan, as given by S. Addelman and O. Kempthorne in Arlington Hall Station ASTIA reports "Orthogonal main-effect plans," pp. 220, 224, 226. When interactions between factors must be taken into account the extensive aliasing of effects that results has sometimes led to questions as to the practical usefulness of such designs. However, it was argued by Box and Hunter, 1961; that the appropriate rationale for the use of designs for factor screening was their projection properties. Since every pair of contrast columns must contain adjacent elements of the form $(- -)$, $(+ -)$, $(- +)$ or $(+ +)$ each replicated m times, the design "projects" m replicates of a 2^2 factorial design in every pair of factors and hence we shall say it is of projectivity P of at least 2. More generally we define the projectivity of a design as follows.

Definition. *A $n \times k$ design D with n observations and k factors each at 2 levels will be said to be of projectivity P if it is such that every subset of P factors out of the possible k contain a complete 2^P factorial design, possibly with some points replicated. The resulting design will then be called a (n, k, P) screen.*

For preliminary illustration see Figure 1. This shows an H_{20} orthogonal array which produces a $(20, 19, 3)$ screen.

TWO IMPORTANT CLASSES OF TWO-LEVEL ORTHOGONAL ARRAYS

When $n = 2^r$, one type of $n \times n$ orthogonal array H_n can be generated from the corresponding 2^r factorial design by writing down a column of $+1$'s, denoted by I, followed by the r columns $c_1, c_2, \ldots, c_i, \ldots, c_j, \ldots, c_r$ of ± 1's of the complete 2^r factorial design. Now represent by $c_h = c_i c_j$ the operation whereby c_h is an entry-wise product of columns $c_i c_j$, and in particular $c_i c_i = I$. A further $n - r - 1$ columns corresponding to the interaction columns of the original factorial can be obtained from all possible products of the individual columns, thus

$$c_{r+1} = c_1 c_2, \quad c_{r+2} = c_1 c_3, \ldots, \quad c_{n-1} = c_1 c_2, \ldots, c_r \qquad (1)$$

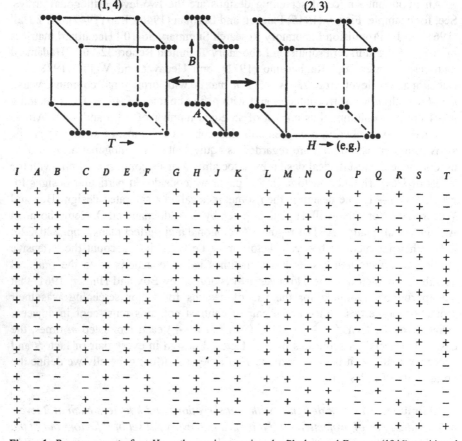

I	A	B	C	D	E	F	G	H	J	K	L	M	N	O	P	Q	R	S	T
+	+	+	+	−	−	−	+	+	−	−	+	−	−	+	+	−	+	−	−
+	+	+	−	+	−	−	−	+	+	−	−	+	−	+	−	+	−	+	−
+	+	+	−	−	+	−	+	−	−	+	−	+	+	−	+	−	−	+	−
+	+	+	−	−	−	+	−	−	+	+	+	−	+	−	−	+	+	−	−
+	+	+	+	+	+	+	−	−	−	−	−	−	−	−	−	−	−	−	+
+	+	−	+	−	−	+	−	+	+	+	−	−	+	+	+	−	−	+	+
+	+	−	−	+	+	+	−	+	−	−	+	−	−	+	+	+	−	−	+
+	+	−	+	+	−	−	+	+	−	+	−	+	+	−	−	+	+	−	+
+	+	−	−	+	+	−	+	+	+	−	+	+	−	−	−	+	+	+	+
+	+	−	+	+	+	+	−	−	−	−	+	+	+	+	+	+	+	+	−
+	−	+	−	+	−	+	+	−	+	−	−	+	+	+	+	−	+	−	+
+	−	+	−	+	+	−	+	−	−	+	+	−	+	+	−	−	+	+	+
+	−	+	+	−	−	+	+	−	−	+	+	+	−	+	−	+	−	+	+
+	−	+	+	−	+	−	−	+	+	−	+	+	+	−	+	+	−	−	+
+	−	+	+	+	+	+	+	+	+	+	−	−	−	−	+	+	+	+	−
+	−	−	−	+	−	+	+	+	−	+	+	−	+	+	+	−	−	−	−
+	−	−	+	−	+	−	−	−	+	+	−	+	−	+	−	+	−	−	−
+	−	−	+	+	−	−	+	−	+	−	+	−	+	−	−	−	+	−	−
+	−	−	+	+	−	−	+	−	+	−	+	−	+	−	−	−	+	−	−
+	−	−	−	−	−	−	−	−	−	−	−	−	−	+	+	+	+	+	+

Figure 1. Rearrangement of an H_{20} orthogonal array given by Plackett and Burman (1946) partitioned into four subarrays by the signs of arbitrarily chosen columns labelled A and B. Addition to columns A and B of any further column such as H chosen from columns C–S results in a (2, 3) projection. Addition of column T produces a (1, 4) projection.

The resulting $n \times n$ matrix will be called a factorial orthogonal array and, following Finney (1945), the saturated design produced by its $n-1$ contrast columns is a $2^{-(n-r-1)}$ fraction of a 2^{n-1} factorial with the $n-r-1$ generating relations given by the identities (1). These may be conveniently rewritten with I on the left and the generating "word" on the right as

$$I = c_1 c_2 c_{r+1}, \quad I = c_1 c_3 c_{r+2}, \dots, \quad I = c_1 c_2, \dots, c_r c_{n-1} \tag{2}$$

Multiplying these generators together in all possible ways produces the defining relation for the fractional factorial with I on the left and $2^{n-4-1} - 1$ words of the identity on the right, from which the alias relationships between the effects can be constructed. Any other fractional factorial containing $k < n - 1$ factors may be derived by omitting columns from the saturated factorial array and the defining relation for this derived design is obtained by omitting all words containing dropped factor symbols.

In a paper by Box and Hunter (1961) a criterion given for the classification of fractional factorials was their resolution R, defined as the length of the shortest word in the defining relation. These authors also stated, in effect, that a design of resolution R was of projectivity $P = R - 1$. To see this consider a parent fractional factorial design D_k of resolution R whose k columns are associated with k factors. The words in the defining relation of any fractional factorial D_P derived by dropping all but P columns from D_k are a subset of those of D_k. Therefore D_P is of resolution R or greater. But the longest word corresponding to a main effect or interaction of the factors in D_P is of length P. So if P is less than R, D_P has no defining relation and its effects have no aliases and is therefore a 2^P factorial possibly replicated. In particular, if $P = R - 1$ then any choice of P factors from the original k yield a 2^P factorial design, possibly replicated. Thus from (2) any saturated design derived from a factorial array has $R = 3$ and $P = 2$. Designs of projectivity greater than 2 can be obtained from factorial arrays but only for screening fewer factors. For example, a $(n, \frac{1}{2}n, 3)$ screen can always be obtained by dropping from the factorial array the $\frac{1}{2}n - 1$ contrast columns containing an even number of letters in the generating factorial design. But this increase in projectivity from 2 to 3 is only obtained at the cost of reducing the number of screened factors from $n - 1$ to $\frac{1}{2}n$. Thus a 16 run factorial orthogonal array can be used to screen 15 variables at projectivity 2 but only 8 variables at projectivity 3.

Whilst factorial arrays exist only when n is a power of 2, other two-level orthogonal arrays are available from Hademard matrices when $n = 4m$ and m is an integer. In particular, Plackett and Burman (1946) derived orthogonal arrays for all such cases when $n \leq 100$, with one single exception. By analogy with the saturated fractional factorials it had been conjectured that these designs would provide $(n, n - 1, P)$ screens with projectivity only 2. However, computer searches have shown that such designs could be of projectivity higher than 2 and in particular, that every one of the 165 three-dimensional projections of the $k = 11$, $n = 12$ was a full 2^3 factorial design with four replicated runs which themselves formed a half-replicate, main-effect plan. Thus for the purpose of screening the 12 run design appears to do better than the 16 run factorial array. Computer enumeration by Lin and Draper (1992) showed that similar results were possible for some, but not all, of the remaining Plackett and Burman designs. They referred to a projection in which half the vertices of the projected factorial cube were replicated t times and the remainder of the vertices $s = m - t$ times as a (t, s) projection. Thus each of the 165 three-dimensional projections of the 12-run orthogonal array design was a $(1, 2)$ projection. Also Figure 1 illustrates, for the 20 run orthogonal array given by Plackett and Burman, how both $(2, 3)$ and $(1, 4)$ projections occur. Computer studies of other aspects of the projective rationale for orthogonal arrays have been made by J. C. Wang and C. F. J. Wu, in the University of Waterloo IIQP report RR93.08 "A hidden projection property of Plackett–Burman and related designs," and by Lin (1993).

SOME GENERAL RESULTS

This paper gives three general results which provide a theoretical basis for these empirical discoveries and in particular can be used to categorize the projective properties of designs given by Plackett and Burman (1946). We first note that one way in which these authors obtained an orthogonal array for $2n$ runs was by "doubling" an orthogonal array for n runs as follows:

$$H_{2n} = \begin{pmatrix} H_n & H_n \\ H_n & -H_n \end{pmatrix}$$

Proposition 1. *A saturated design obtained from a doubled $n \times n$ orthogonal array is always of projectivity $P = 2$ and only 2.*

Proof. If the contrast columns in H_{2n} are denoted by $c_1, \ldots, c_{2n-1}$ then, noting that the first n elements of c_n are $+1$'s and the remainder are -1's, it follows for instance that $c_i c_n = c_{n+i}$, and hence that $I = c_i c_n c_{n+i}$. Thus a design formed from these columns can only have rows with sign combinations $(- - +)$, $(+ - -)$, $(- + -)$, $(+ + +)$, and consists of $n/2$ replicates of a half fraction of the 2^3 factorial. The design is therefore of projectivity $P = 2$ and only 2.   □

As a special case of these results it follows once more that every saturated fractional factorial is of projectivity $P = 2$, since, as pointed out by Plackett and Burman (1946) these designs may also be obtained by doubling. However it is not true that, with $n = 4m$ and m even, the corresponding saturated design is always of projectivity only 2. For example, while the H_{24} obtained by doubling yields a saturated design with $P = 2$, that obtained by Plackett and Burman using cyclic generation has $P = 3$.

Another way of obtaining orthogonal arrays is by cyclic generation. For a cyclic orthogonal array H_n it is possible to write down all the $n - 1$ contrast columns knowing only the sequence of signs to be applied in the first row. The cyclic 12-run orthogonal array may, for instance, be written down knowing only the 11 signs in the first row sequence $(+ + - + + + - - - + -)$. Shifting this row cyclically one place 10 times and adding a final row of minus signs and an initial column of plus signs produces a 12×12 orthogonal array.

Proposition 2. *A saturated design obtained from a cyclic orthogonal array is either a factorial orthogonal array with $P = 2$ and only 2, or else has projectivity $P = 3$.*

Proof. Recalling that any two contrast columns are regarded as identical if one can be obtained from the other by sign reversal, it follows from the cyclic generation property that, if $c_i c_{i+j} = c_l$ is in the design, the entry-wise product of any two columns at distance j must be in the design. Hence $c_i c_{i+2j} = c_i c_{i+j} c_{i+j} c_{i+2j} = c_l c_{l+j}$ is in the design and more generally $c_i c_{i+hj}$ for $h = 1, 2, \ldots$ is in the design. In particular, if c_1c_2 is in the design, the entry-wise product of any two columns is in the design.

Now suppose c_1c_2 is not in the design. Then since c_1 will be the column that follows after c_{n-1} in the cyclic generation procedure, it follows that $c_{n-1}c_1$ is not in the design either. But the distance between c_1 and c_{n-1} is an even number, $n-2$, from which it follows that no entry-wise product of columns with distance 2, or in general with distance $2h$ ($h = 1, 2, \ldots$), can be in the design. If c_1c_{1+j} is a column in the design for j odd then $c_{n-1}c_j$ must also be in the design. But since $n - (j+1)$ is an even integer, that is not possible. Hence, if c_1c_2 is not in the design no entry-wise product of any two columns is in the design.

Now assume a design constructed by cyclic generation contains c_1c_2 and thereby all of their two-factor interaction columns. If such a design contains more columns than c_1, c_2 and c_1c_2, it must contain a column c_3 orthogonal to these, and since it also contains all the possible two-factor interaction columns it must also contain c_1c_3, c_2c_3 and $c_1c_2c_3$, and none of these columns can be equal to any of the first three or to a column containing only $+1$'s. By induction the same argument now gives us that if a design generated by cyclic generation contains c_1c_2 it must have $2^r - 1$ contrast columns for some r and therefore must be a factorial orthogonal array with projectivity $P = 2$ and only 2.

Proposition 3. *Any saturated two-level design obtained from an orthogonal array containing $n = 4m$ runs, with m odd, is of projectivity $P = 3$.*

Proof. Consider a particular orthogonal array for which m is odd and let u be a column vector of m ones. Then I is a column of $n = 4m$ ones which can be written in the partitioned form $(u, u, u, u)'$. Also, the rows of the orthogonal array can be re-ordered so that the array is partitioned into four $m \times n$ subarrays in which two arbitrarily chosen contrast columns c_ic_j are partitioned in the form $(u, u, -u, -u)'$ and $(u, -u, u, -u)'$.

Now consider any other arbitrarily chosen column $c_h = (c_{h1}, c_{h2}, c_{h3}, c_{h4})'$ after this re-ordering of rows. Orthogonality implies that c_h satisfies the linear equations $c_h'I = c_h'c_i = c_h'c_j = 0$, hence

$$u'c_{h1} = -a_h, \quad u'c_{h2} = +a_h, \quad u'c_{h3} = +a_h, \quad u'c_{h4} = -a_h$$

Now let t_h represent the number of plus signs in c_{h1} and in c_{h4}, or equivalently the number of minus signs in c_{h2} and in c_{h3}. Then if $s_h = m - t_h$, we have $a_h = s_h - t_h$ and, if necessary by switching signs in the whole column c_h, a_h can be taken to be positive, so that $t_h < s_h$. Also since m is odd, a_h cannot be zero; and it cannot be m, for then c_h would correspond to the contrast c_ic_j, and no additional column could be simultaneously orthogonal to c_i, c_j and c_h. Thus, in the arbitrarily chosen space of c_i, c_j and c_h the projected experimental points will lie on the vertices of a cube, and will be distributed as shown in Table 1.

Table 1. Distribution of Projected Experimental Points

Replicated t_h Times			Replicated s_h Times		
−	−	+	−	−	−
+	−	−	+	−	+
−	+	−	−	+	+
+	+	+	+	+	−

The projected design will therefore consist of a full 2^3 factorial replicated t_h times with an additional half-replicate having defining relation $I = c_i c_j c_h$ replicated $a_h = s_h - t_h$ times. $\qquad\square$

Note that different choices of the columns c_j, c_k, c_h can result in different combinations of t and s and hence in different amounts of replication of the two design parts. This proposition is in agreement with a recently published result of Cheng (1995) who bases his proof on general properties of orthogonal arrays of a given strength.

Figure 1 shows a rearrangement of the 20×20 orthogonal array given by Plackett and Burman (1946), used earlier for preliminary illustration, with columns associated with factors A, B, C, ..., T partitioned into four subarrays obtained by setting $A = c_i$, $B = c_j$ respectively. Since the original design is obtainable by cyclic generation and also since $m = 5$ is odd, the design is a (20, 19, 3) screen with projections either of type (2, 3) or of type (1, 4). The latter are generated by a single column with four like signs and one unlike sign in each quadrant: for the case illustrated this is column T. Thus of the 969 three-dimensional projections, only $\frac{1}{17}$ are of the (1, 4) type and $\frac{16}{17}$ are of the (2, 3) type. Notice, however, that the diagram does not imply that we can omit column T to produce a (20, 18, 3) screen for which all the projections are of the (2, 3) type. This is because a different choice of two columns c_j, c_j to define the four subarrays would produce a different (1, 4) column. When r_h and s_h are each at least equal to two an additional analysis for dispersion effects is facilitated: see, for example, Box and Meyer (1986a).

The arrays tabled by Plackett and Burman are not, of course, exhaustive. Other orthogonal arrays of potential interest to the practitioner exist. For example, 5 distinct orthogonal arrays are given by M. J. Hall in the unpublished summary mentioned earlier for the case $n = 16$. All of these cases are (16, 15, 2) screens. But we have found it possible to show that one of them produces a (16, 14, 3) screen when a particular column is dropped and also that the three other arrays that are different from the 16-run factorial array can produce (16, 12, 3) screens when particular sets of three columns are dropped. We plan to discuss these and other designs in a separate paper.

To illustrate the usefulness of the three main results of this paper we show in Table 2 how they characterize the projectivity of each of the Plackett–Burman designs for $n \leq 84$. Of the 20 saturated designs considered, 14 are of projectivity 3,

Table 2. Projectivity P of Plackett and Burman (1946) Designs for $n \le 84$

									n											
	8	12	16	20	24	28	32	36	40	44	48	52	56	60	64	68	72	78	80	84
									m											
	2	3	4	5	6	7	8	9	10	11	12	13	14	15	16	17	18	19	20	21
$P = 3$:																				
m odd		*		*		*		*		*		*		*		*		*		*
cyclic	*		*	*			*		*	*			*		*	*			*	*
$P = 2$	*		*		*		*				*				*					

$P = 3$ cyclic designs are not factorial. $P = 2$ designs are factorial arrays or obtained by doubling.

four are factorial arrays of projectivity $P = 2$ and two obtained by doubling are also of projectivity $P = 2$.

Some discussion of the analysis of orthogonal array designs under the hypothesis of effect sparsity is in Box and Meyer (1986b, 1993).

ACKNOWLEDGMENT

We are grateful to a referee for helpful comments on earlier drafts of this paper. This research was supported by the Alfred P. Sloan Foundation and the Research Council of Norway.

CHAPTER C.9

Choice of Response Surface Design and Alphabetic Optimality

ABSTRACT

It is argued that the specification of problems of experimental design and, in particular, of response surface design should depend on scientific context. The specification for a widely developed theory of "alphabetic optimality" for response surface applications is analyzed and found to be unduly limiting. Ways in which designs might be chosen to satisfy a set of criteria of greater scientific relevance are suggested. Detailed consideration is given to regions of operability and interest, to the design information function, to sensitivity of criteria to size and shape of the region, and to the effect of bias. Problems are discussed of checking for lack of fit, sequential assembly, orthogonal blocking, estimation of error, estimation of transformations, robustness to bad values, using a minimum number of points, and employing simple data patterns.

INTRODUCTION

There seems to be no doubt that of all the activities in which the statistician can engage, that of designing experiments is one of the most important, since it is here that the actual mode of generation of scientific data is decided.

The importance of practice in guiding the development of the theory of experimental design (Yates, 1967) is clearly seen from the time of its invention. Fisher was engaged by Russell (J. F. Box, 1978) on a temporary basis at Rothamsted Experi-

From Box, G. E. P. (1982), *Utilitas Mathematica*, **21B**, 11–55.

mental Station in 1919 "to examine our data and elicit further information that we had missed." Records were available from the ongoing Broadbalk experiment in which particular combinations of fertilizers had been consistently applied to 13 plots for a period of almost 70 years. In his analysis, Fisher (1921, 1924) attempted to relate yield to fertilizer combination, to weather, and, in particular, to rainfall. The method he used was multiple regression with distributed lag models (!), involving an ingenious employment of orthogonal polynomials which led to important advances in the theory of regression analysis and, in particular, its distribution theory.

With only the crudest of computational aids, the work must have been burdensome, making it all the more frustrating to discover that, however ingenious the analysis, the inherent nature of the data ensured that the answers to many questions were inaccessible. A comprehension of the logical problems in drawing conclusions from such analyses led naturally to speculation on how some of the difficulties might be overcome by appropriate design. These ideas were further stimulated by the Analysis of Variance, which Fisher introduced in 1923 with W. A. Mackenzie for the elucidation of what was clearly a most unsatisfactory design which he had had no part in choosing (Fisher and Mackenzie, 1923). Thereafter, as Fisher gradually acquired more influence in the setting up of field trials, the principles of replication, and of randomization and their application to randomized blocks, Latin squares, and factorial designs quickly evolved out of the actual planning, running, and analysis of a series of experimental designs of increasing complexity and beauty. See also Yates (1937, 1970).

During my eight years as a statistician with Imperial Chemical Industries, many of the problems employed the, by now, standard designs of Fisher and Yates. However, some investigations directly concerned with the improvement of chemical processes at the laboratory, pilot plant and full scale seemed to require additional methods, which, however, still drew on the fundamental principles laid down by the originators of experimental design. This led to the development of what has come to be called response surface methodology and in particular of "response surface designs". See, for example, Box (1954a), Box and Wilson (1951), Box and Youle (1955), Herzberg and Cox (1969), Hill and Hunter (1966), and Mead and Pike (1975).

Suppose some response η of interest is believed to be locally approximated by a polynomial of low degree in k continuous experimental variables $\chi = (\chi_1, \chi_2, \ldots, \chi_k)'$. To fit such a function we need appropriate experimental designs. Let us call a design suitable for estimating a general polynomial of degree d a dth order design in k variables. Thus a design suitable for fitting the function

$$\eta = \beta_0 + \beta_1\chi_1 + \beta_2\chi_2 + \beta_{11}\chi_1^2 + \beta_{22}\chi_2^2 + \beta_{12}\chi_1\chi_2 \qquad (1)$$

would be a second-order design in two variables.

One route for choosing such designs, which has generated much mathematical research over the last twenty or so years, I shall refer to as the *alphabetic optimality* approach. For reasons I shall explain, I have reservations about the usefulness of this approach so far as response surface designs are concerned. For completeness, a brief summary of some of the main ideas are set out in below; see also Kiefer (1958, 1959, 1975), Kiefer and Studden (1976), Kiefer and Wolfowitz (1959), Whittle (1973), Pukelsheim (1980), Silvey and Titterington (1973), and Titterington (1975).

SOME ASPECTS OF OPTIMAL DESIGN THEORY FOR CONTINUOUS EXPERIMENTAL VARIABLES

Consider a response η which is supposed to be an exactly known function $\eta = x'\beta$ linear in p coefficients β, where $x = \{f_1(\chi), \ldots, f_p(\chi)\}'$ is a vector of p functions of k experimental variables, χ. Suppose a design is to be run defining n sets of k experimental conditions given by the $n \times k$ design matrix $\{\chi_u\}$ and yielding n observations y_u, so that

$$\eta_u = x_u'\beta \quad (u = 1, \ldots, n)$$

where $y_u - \eta_u = \varepsilon_u$ is distributed $N(0, \sigma^2)$ and the $n \times p$ matrix $X = \{x_u'\}$.

The elements of $\{c_{ij}\} = (X'X)^{-1}$ are proportional to the variances and covariances of the least squares estimates $\hat{\beta}$. Within this specification, the problem of experimental design is that of choosing the design $\{\chi_u\}$ so that the elements c_{ij} are to our liking. Because there are $\frac{1}{2}p(p+1)$ of these, simplification is desirable.

A motivation for simplification is provided by considering the confidence region* for β, that is,

$$(\beta - \hat{\beta})'X'X(\beta - \hat{\beta}) = \text{constant}$$

defining an ellipsoid in p parameters. The eigenvalues $\lambda_1, \ldots, \lambda_p$ of $(X'X)^{-1}$ are proportional to the squared lengths of the p principal axes of this ellipsoid. Suppose their maximum, arithmetic mean, and geometric mean are indicated by λ_{max}, $\bar{\lambda}$, and $\tilde{\lambda}$, respectively. Then it is illuminating to consider the transformation of the $\frac{1}{2}p(p+1)$ elements c_{ij} to a corresponding number of items as follows:

(i) $D = |X'X| = \tilde{\lambda}^{-p}$, so that $D^{-1/2} = \tilde{\lambda}^{p/2}$ is proportional to the volume of the confidence ellipsoid;

(ii) H, a vector of $p-1$ homogeneous functions of degree zero in the λ's, which measure the non-sphericity or state of ill-conditioning of the ellipsoid. In particular we might choose either of the quantities, $H_1 = \bar{\lambda}/\tilde{\lambda}$ and $H_2 = \lambda_{max}/\tilde{\lambda}$, both of which take the value unity for a spherical region;

(iii) $\frac{1}{2}p(p-1)$ independent direction cosines which determine the orientation of the orthogonal axes of the ellipsoid.

It is traditionally assumed that the $\frac{1}{2}p(p-1)$ elements concerned with orientation of the ellipsoid are of no interest, and attention has been concentrated on particular criteria which measure in some way or another the sizes of the eigenvalues,

---

* Obviously there are also parallel fiducial and Bayesian rationalizations.

measuring some combination of size and non-sphericity of the confidence ellipsoid. Among these criteria are

$$D = |\mathbf{X}'\mathbf{X}| = \prod \lambda_i^{-1} = \tilde{\lambda}^{-p}$$
$$A = \sum \lambda_i = \text{tr}(\mathbf{X}'\mathbf{X})^{-1} = \sigma^{-2} \sum \text{Var}(\hat{\beta}_i) = p\tilde{\lambda}H_1$$
$$E = \max\{\lambda_i\} = \tilde{\lambda}H_2$$

The desirability of a design, as measured by the D, A, and E criteria, increases as $\tilde{\lambda}$, $\tilde{\lambda}II_1$, and $\tilde{\lambda}II_2$, respectively, are decreased. But in practical situations, each of these criteria will take smaller and hence more desirable values as the ranges for the experimental variables χ are taken larger and larger. To cope with this problem it is usually assumed that the experimental variables χ_u may vary only within some *exactly known* region in the space of χ, but not outside it. I will call this permissible region **RO**.

Another characteristic of the problem which makes its study mathematically difficult is the necessary discreteness of the number of runs which can be made at any given location. In a technically brilliant paper, Kiefer and Wolfowitz (1960) dealt with this obstacle by introducing a continuous design measure ζ which determines the *proportion* of runs which should ideally be made at each of a number of points in the χ space. Realizable designs which most nearly approximated the optimal distribution could then be used in practice.

A further important result of Kiefer and Wolfowitz linked the problem of estimating $\boldsymbol{\beta}$ with that of estimating the response η via the property of *G-optimality*. G-optimal designs were defined as those which minimized the maximum value of $\text{Var}(\hat{y}_\chi)$ within **RO**. The authors were then able to show, for their measure designs, the equivalence of G-optimality and D-optimality. Furthermore, they showed that, for such a design, within the region **RO**, the maximum value of $n \, \text{Var}(\hat{y}_\chi)/\sigma^2$ was p, and that this value was actually attained at each of the design points.

For illustration we consider a second-order measure design in two variables; that is, a design appropriate for the fitting of the second-degree polynomial of Equation (1). Such a design which is both D-optimal and G-optimal for a square region **RO** with vertices (± 1, ± 1) was given by Fedorov (1972); see also Herzberg (1979). The design places 14.6% of the measure at each of the four vertices, 8.0% at each of the midpoints of the edges, and 9.6% at the origin. The design is set out in Figure 4(b).

While this approach has generated much interesting mathematics, it does not, I believe, solve the problem of choosing good response surface designs. In the hope of stimulating new initiative, I have set out what I believe is the scientific context for response surface studies and indicated some possible lines of development.

THE RESPONSE SURFACE CONTEXT

As an example suppose it is desired to study some chemical system, with the object of obtaining a higher value for a response η—such as the *yield* of the desired

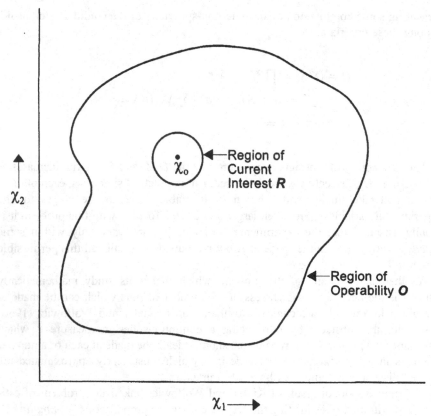

Figure 1. The region of current interest R and the unknown region of operability O in the space of two continuous experimental variables, χ_1 and χ_2.

product—which is initially believed to be some function of k continuous input variables $\chi = (\chi_1, \ldots, \chi_k)'$ such as reaction time, temperature, or concentration.

As is illustrated in Fig. 1, it is usually known initially that the system can be operated at some point χ_0 in the space of χ and that the system is expected to be capable of operating over some much more extensive region O called the *operability region*,* which, however, is usually unknown or poorly known. Response surface methods are employed when the nature of the true response function $\eta = g(\chi)$ is also unknown† or is inaccessible.

---

* One secondary object of the investigation may be to find out more about the operability region O.
† Occasionally the true functional form $\eta = g(\chi)$ may be known, or at least conjectured, from knowledge of physical mechanisms. Typically, however, $g(\chi)$ will then appear as a solution of a set of differential equations which are nonlinear in a number of parameters which may represent physical constants. Problems of nonlinear experimental design then arise which are of considerable interest although they have received comparatively little attention; see, for example, Fisher (1935), Box and Lucas (1959) and Cochran (1973).

Suppose that over some (typically much less extensive) immediate *region of current interest R* in the neighborhood of χ_0 it is guessed that a graduating function, such as a dth degree polynomial in χ

$$\eta_\chi \doteq \mathbf{x}'\boldsymbol{\beta}$$

might provide a *locally adequate approximation* to the true function $\eta(\chi) = g(\chi)$, where as before $\mathbf{x}$ is a p-dimensional vector of suitably transformed input variables $\mathbf{x}' = \{f_1(\chi), \ldots, f_p(\chi)\}$, and $\boldsymbol{\beta}$ is a vector of coefficients that may be adjusted to approximate the unknown true response function $\eta_\chi = g(\chi)$. Then progress may often be achieved by using a sequence of such approximations. For example, when a first-degree polynomial approximation could be employed it might, via the method of steepest ascent, be used to find a new region of interest $\mathbf{R}_1$ where the yield was higher. Also, a maximum in many variables is often represented by some ridge system* and a second-degree polynomial approximation when suitably analyzed might be used to elucidate, describe, and exploit such a system.

Thus we are typically involved in using a sequence of designs, each making use of information gleaned from earlier experiments, a characteristic typical of a much wider field of scientific investigation. This provides the opportunity to progressively improve not only the objective function η directly, but also the *mode* of gathering information about it. For example, at the ith stage, a design performed in a region $\mathbf{R}_i$ may suggest that a new region $\mathbf{R}_{i+1}$ is worthy of investigation, either because it can be expected to give higher values of η or because it may throw light on other important aspects of the response function. But this new region may be different (a) in its location in the space of χ, (b) in its shape (in particular, because of information fed back from previous data on transformations of χ's individually or jointly), and (c) in the identity of its component space because of feedback from the results themselves, indicating that certain variables should be dropped and/or that new variables should be added.

Thus, in any realistic view of the process of investigation, the dimensions, identity, location, and metrics of measurement of regions of interest in the experimental space, are all *iteratively evolving*. The problem of choosing suitable experimental designs in such a context is a difficult one. Some properties (Box, 1968; Box and Draper, 1975) of a response surface design, any, all, or some of which might in different circumstances be of importance in the above context, are given in Table 1.

The Design Information Function

Associated with requirements 1 and 2 of Table 1, consider the design *variance function* (Box and Hunter, 1957):

$$V_\chi = n \, \mathrm{Var}(\hat{y}_\chi)/\sigma^2 = n\mathbf{x}'(\mathbf{X}'\mathbf{X})^{-1}\mathbf{x}$$

---

* Empirical evidence suggests this. Also, integration of sets of differential equations which describe the kinetics of chemical systems almost invariably leads to ridge systems (Box, 1954; Box and Youle, 1955; Franklin, Pinchbeck and Popper, 1956; Pinchbeck, 1957). See also the discussion of Figure 8.

Table 1. Some Attributes of Designs of Potential Importance

The design should:

1. generate a satisfactory distribution of information throughout the region of interest, R;
2. ensure that the fitted value at $\hat{y}(\chi)$ be as close as possible to the true value at $\eta(\chi)$;
3. give good detectability of lack of fit;
4. allow transformations to be estimated;
5. allow experiments to be performed in blocks;
6. allow designs of increasing order to be built up sequentially;
7. provide an internal estimate of error;
8. be insensitive to wild observations and to violation of the usual normal theory assumptions;
9. require a minimum number of experimental points;
10. provide simple data patterns that allow ready visual appreciation and investigational development;
11. ensure simplicity of calculation;
12. behave well when errors occur in the settings of the χ's;
13. not require an impractically large number of predictor variable levels;
14. provide a check on the "constancy of variance" assumption.

---

or equivalently the *information function*

$$I_\chi = V_\chi^{-1}$$

It is evident that if we were to make the unrealistic assumption, made in alphabetic optimality, that the graduating function $\eta = \mathbf{x}'\boldsymbol{\beta}$ is capable of *exactly* representing the true function $g(\chi)$, then the information function would tell us all we could know about the design's ability to estimate η. For illustration, information functions and associated information contours for a 2^2 factorial used as a first-order design and for a 3^2 factorial used as a second-order design are shown in Figures 2(a) and (b) and in Figures 3(a) and (b). In these and some subsequent illustrations it is convenient to rescale χ_1 and χ_2 in a standard linear coding so that for example the levels of a 2^2 factorial design are -1 and $+1$. It will be clear from the context when coded variables are referred to.

APPLICABILITY OF ALPHABETIC OPTIMALITY

The information function for Fedorov's second-order D-optimal and G-optimal design over the permissible $\mathbf{RO}$ region (± 1, ± 1), referred to earlier, is shown in Figs. 4(a) and (b). For illustration, we suppose that the two experimental variables are temperature in °C and time in hours and that the $\mathbf{RO}$ region would permit experimentation within the limits 170–190 °C and 3–5 hours, but not outside these

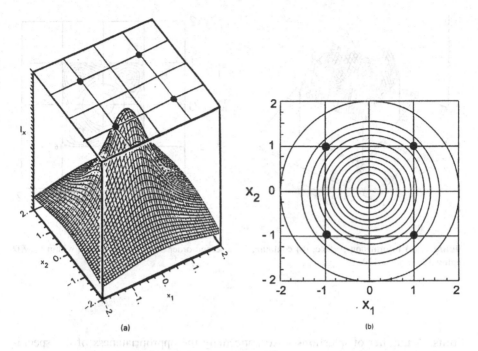

Figure 2. (a) Information surface, (b) contours, for a 2^2 factorial used as a first-order design.

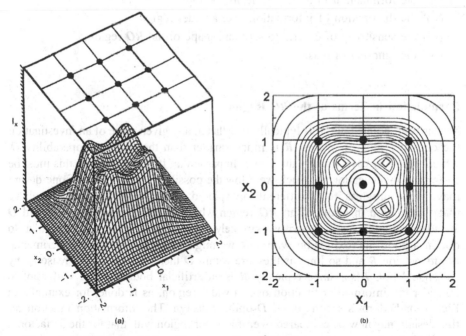

Figure 3. (a) Information surface, (b) contours, for a 3^2 factorial used as a second-order design.

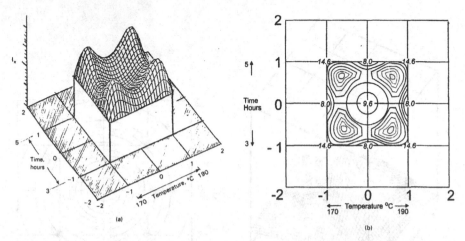

Figure 4. (a) Information surface, (b) contours, for a second-order D/G-optimal design within the RO region $170 < \chi_1 < 190$, $3 < \chi_2 < 5$.

limits. A number of questions arise concerning the appropriateness of the specification for alphabetic optimality. These concern

(i)   the formulation in terms of the RO region;
(ii) the distribution of information over a wider region;
(iii) the sensitivity of criteria to size and shape of the RO region;
(iv) the ignoring of bias.

Formulation in Terms of the RO Region

As has been pointed out, it is typically true that at any given stage of an investigation the current region of interest R is much smaller than the region of operability O, which is, in any case, usually unknown. In particular, it is obvious that this must be so for any investigation in which we allow the possibility that results of one design may allow progress to a different unexplored region. Consequently I believe that formulation in terms of an RO region which in effect implies that R and O are identical and exactly known is extremely artificial and limiting. In particular, to obtain a good approximation within R we might wish to put some experimental points outside R and so long as they are within O there is no practical reason why we should not. Also, since typically R is an artificial boundary, we will want to consider the information function over a wider region, as is done, for example, in Fig. 5 for Fedorov's second-order D-optimal design. The information function for this design may now be compared over this wider region with that for the 3^2 factorial in Fig. 3.

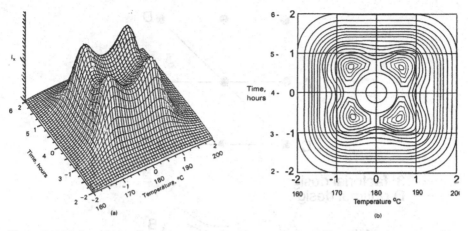

Figure 5. (a) Information surface, (b) contours, for the second-order D/G-optimal design over a wider region.

Distribution of Information Over a Wider Region

In the response surface context, the coefficients β of a graduating function $\eta_\chi \doteq x'\beta$ acting as they do merely as adjustments to a kind of mathematical French curve are not usually of individual interest except insofar as they affect η, in which case only the G-optimality criterion among those considered is of direct interest. For response surface studies, however, it is far from clear how desirable is the property of G-optimality itself.

For instance, the profiles of Fig. 6 made by taking sections of the surfaces of Figs. 3 and 5 show that neither the information function for the G/D-optimal design nor that for the 3^2 design is universally superior one to the other. In some subregions one design is slightly better, and in others the other design is slightly better. Both information functions, and particularly that of the G/D-optimal design, show a tendency to sag in the middle. This happens for the G/D-optimal design because the G-optimality characteristic guarantees that, maximized, minima for I_χ, each equal to $1/p$, occur at every design point, which includes the center point. However, this sagging information pattern of the second-order design is not of course a characteristic of the first-order design of Fig. 2 which is also D/G-optimal but contains no center point. If the idea of the desirability of designs possessing a particular kind of information profile is basic, then it seems unsatisfactory that the nature of that profile should depend very much on the order of the design. Indeed, the relevance of the *minimax criterion* which produces G-optimality is arguable. It follows from the Kiefer–Wolfowitz theorem that a second-order design for $(\pm 1, \pm 1)$ region whose information function did not sag in the middle would necessarily not be D-optimal. But as we have seen, D-optimality is only one of many single-valued criteria that might be used in attempts to describe some important characteristic of the $X'X$ matrix. Others, for example, would be A-optimality and E-optimality, and the L-optimality criterion proposed by Fedorov (1972), and these would yield different

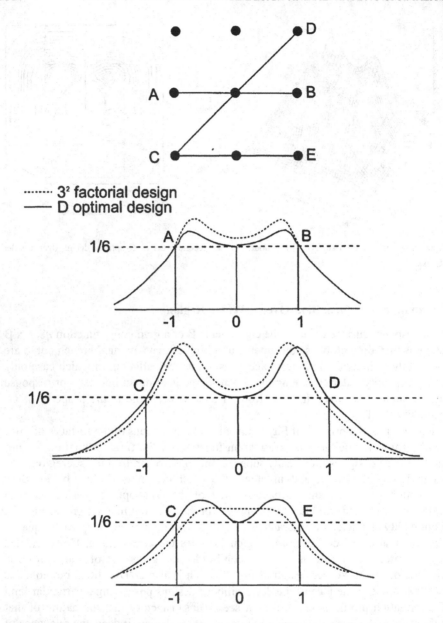

Figure 6. Profiles of I_χ for the second-order D/G-optimal design and for the 3^2 factorial design.

information profiles. But I would argue that since the information function itself is the most direct measure of desirability, our best course is to choose our design directly by picking a suitable information function, and not indirectly by finding some extremum for A, E, D, or other arbitrary criterion.

Sensitivity of Criteria to Size and Shape

In the process of scientific investigation, the investigator and the statistician must do a great deal of guesswork. In matching the region of interest R and the degree of complexity of the approximating function, they must try to take into account, for example, that the more flexible second-degree polynomial can be expected to be adequate over a larger region R than the first-degree approximation.

Obviously different experimenters would have different ideas of appropriate locations and ranges over which such approximations might be useful. In particular, ranges could easily differ from one experimenter to another by a factor of two or more.* In view of this, extreme sensitivity of design criteria to scaling is disturbing.† For example, suppose each dimension of a dth order experimental design is increased‡ by a factor c. Then the D criterion is increased by a factor of c^q, where

$$q = \frac{2k(k+d)!}{(k+1)!(d-1)!}$$

Equivalently a confidence region of the same volume as that for a D-optimal design can be achieved for a design of given D value by increasing the scale for each variable by a factor of $c = (D_{opt}/D)^{1/q}$, thus increasing the volume occupied by the design in the χ space, by a factor $c^k = (D_{opt}/D)^{k/q}$. For example, the D value for the 3^2 factorial design of Fig. 3 is 0.98×10^{-2} as compared with a D value of 1.14×10^{-2} for the D-optimal design. For $k = 2$, $d = 2$, we find $q = 16$ and $c = (1.14/0.98)^{1/16} = 1.009$. Thus, the same volume of a confidence region for the β's, as is obtained for the D-optimal design would be obtained from a 3^2 design if each side of the square region were increased *by less than 1%*. Equivalently, the area of the region would be increased by less than 2%. Using the scaling that was used in Fig. 4 for illustration, we should have to change the temperature by 20.18 °C instead of 20 °C, and the time by two hours and one minute instead of two hours, for the 3^2 factorial to give the same D value as the D/G-optimal design. Obviously no experimenter can guess with anything approaching this accuracy what are suitable

---

* Over a sequence of designs, initial bad choices of scale and location would tend to be corrected, of course.

† In particular, designs can only be fairly compared if they are first scaled to be of the *same size*. But how is size to be measured? It was suggested in Box and Wilson (1951) that designs should be judged as being of the same size when their marginal second moments $\sum(x_{iu} - \bar{x}_u)^2/n$ were identical. This convention is not entirely satisfactory, but will of course give very different results from those which assume design points to be all included in the same region RO. It is important to be aware that the apparent superiority of one design over another will often disappear if the method of scaling the design is changed. In particular, this applied to comparisons such as those made by Nalimov, Golikovo, and Mikeshina (1970) and Lucas (1976).

‡ A measure of efficiency of a design criterion (see, e.g., Atwood, 1969; M. J. Box, 1968) is motivated by considering the ratio of the number of runs necessary to achieve the optimal design to the number of runs required for the suboptimal design to obtain the same value of the criterion, supposing fractional numbers of runs to be allowed (an impossible condition). In particular, for the D criterion, this measure of D-efficiency is $(D/D_{opt})^{1/p}$. Equivalently here, to illustrate scale sensitivity, we concentrate attention on the factor c by which each scale would need to be inflated to achieve the same value of the D criterion.

ranges over which to vary these factors. In addition, of course, the 3^2 factorial design is physically realizable using a finite number (nine) of experimental runs whereas the D optimal design is not.

Thus choice of region and choice of information function are closely interlinked. For example, any set of $N = k + 1$ points in k-space which have no coplanarities is obviously a D-optimal first-order design for some* ellipsoidal region. Furthermore, it is easy to show that the information function for a design of order d is a smooth function whose harmonic average over the n experimental points (which can presumably be regarded as representative of the region of interest) is always $1/p$ *wherever* we place the points. Thus the problem of design is not so much a question of choosing the design to *increase* total information as spreading the total information around in the manner desired.

Rotatable Designs

A route for simplification different from alphabetic optimality occurs if *after transformation* of the inputs individually or jointly to produce suitably standardized variables χ, it can reasonably be asserted that nothing is known about the orientation in the χ space of the response surface we wish to study. It was argued by Box and Hunter (1957) that we should then employ designs having the property that the variance of $\hat{y}$ is a function only of $\rho = (\chi'\chi)^{1/2}$ so that

$$V_\chi = V_\rho, \qquad I_\chi = I_\rho$$

For a first-order design, rotatability implies orthogonality and vice versa, and completely decides the information function. For second- and higher-order designs, a requirement of rotatability fixes certain moment properties of the design, but V_ρ and hence I_ρ are still to some extent at our choice and can be changed by changing certain moment ratios (Box and Hunter, 1957). In particular, for a second-order design, V_ρ depends on the single moment ratio $\lambda = (n/3) \sum \chi_i^4 / (\sum \chi_i^2)^2$. For illustration, Figs. 7a and b show the information function for a second-order rotatable design with $\lambda = .75$ consisting of 8 points arranged in a regular octagon with 4 points at the center.

The truth seems to be that at any particular phase of an investigation the scientific decisions that *most* contribute to the outcome of that phase are: the identity of the factors and responses to be studied, the location, size and shapes of the current region of interest within the factor space and choice of appropriate transformations. These crucial choices that do not involve statistics. After these decisions are made, and given the assumptions that the *model fits perfectly* and that *only the variance properties of the design are of interest* any set of experiments that cover this region in some reasonably uniform way is likely to do quite well. I cannot see that the various optimality criteria are particularly relevant to this choice.

---

* Namely, for that region enclosed within the information contour $I_\chi = 1/p$ which must pass through all the $k + 1$ experimental points.

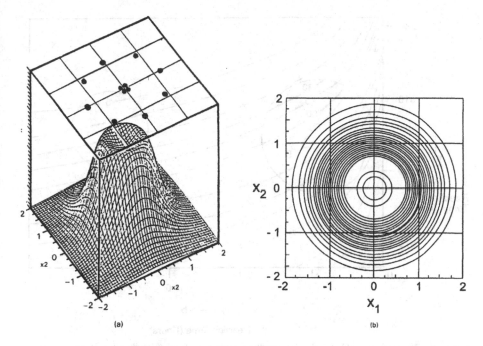

Figure 7. (a) Information surface, (b) contours, for a second-order rotatable design consisting of 8 equispaced points on a circle plus 4 center points.

Ignoring of Bias

All models are wrong; some models are useful. This aphorism is particularly true for empirical functions such as polynomials that make no claim to do more than locally graduate the true function. Some idea of what type of surface we might in reality be approximating and of the adequacy of such approximations can be gained by studying surfaces produced by chemical kinetic models. An example,* which also appears in a book by Box and Draper (1987), is shown in Fig. 8. See also Box and Youle (1955).

One conclusion I reached from many such studies was that approximations would not need to be very good for response surface methods to work. Thus within region I of Fig. 8 the locally monotonic function could be crudely approximated by a plane which could indicate a useful path of ascent. Also, valuable information might be obtained about the ridge in region II, even though the underlying surface was not exactly quadratic. Notice, however, that in the light of such examples any theory of experimental design which depended on the exactness of such approximations should be regarded with extreme skepticism.

---

* This surface was generated by considering the yield of the product B in a consecutive reaction $A \xrightarrow{k_1} B \xrightarrow{k_2} C$ following first-order kinetics with temperature sensitivity given by the Arrhenius relation $\ln k_i = \ln \alpha_i + \beta_i/T$, where temperature T is measured in kelvin units, using plausible values for the constants α_1, α_2, β_1, β_2.

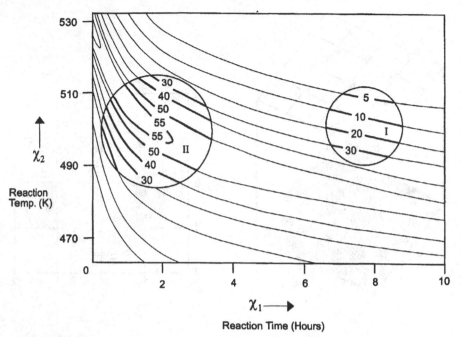

Figure 8. Contours of yield for the intermediate product B from a consecutive chemical reaction A → B → C assuming first-order kinetics and temperature sensitivity represented by the Arrhenius equation.

TAKING ACCOUNT OF BIAS

If $\hat{y} = \mathbf{x}'\hat{\boldsymbol{\beta}}$ is the fitted value using the empirical approximation, then its total error ε is

$$\hat{y} - \eta = \{\hat{y} - E(\hat{y})\} + \{E(\hat{y}) - \eta\}$$
$$\varepsilon = \varepsilon_V + \varepsilon_B$$

Thus the error ε contains a random part ε_V and a systematic, or bias, part ε_B, and we must expect that ε_B will not be negligible. Since all the theory previously discussed makes the assumption that ε_B is exactly zero, we must consider whether the resulting designs are robust to this kind of discrepancy. The optimality criteria discussed earlier which assume the response function to be exact usually produce a substantial proportion of experimental points on the boundary of **RO**. In the context of possible bias, this is not reassuring, since it is at these points that the approximating function will be most strained.

The explicit recognition that bias will certainly be present does, however, provide a more rational means for approaching the scaling problem (Box and Draper, 1959,

1963, 2005). To see this, consider again the formulation already given in terms of a region of interest R and a larger region O of operability. If we were to assume, unrealistically, that the approximation remained exact however widely the points were spread, and if some measure of variance reduction were the only consideration, then to obtain most accurate estimation within R, the size of the design would have to be increased to the boundaries of the operability region O. But, of course, the wider the points were spread, the less applicable would be the approximating function, and the bigger the bias error. This suggests that we should seek restriction of the spread of the experimental points not by artificial limitation to some region but by balancing off the competing requirements of variance on the one hand, which is reduced as the spread of the points is increased, and bias on the other hand, which is increased as the spread of the points is increased.

The mean square error associated with estimating η_χ by $\hat{y}_\chi$ standardized for the number, n, of design points and the error variance σ^2, can be written as the sum of a variance component and a squared bias component

$$nE(\hat{y}_\chi - \eta_\chi)^2/\sigma^2 = nV(\hat{y}_\chi)/\sigma^2 + n\left\{E(\hat{y}_\chi) - \eta_\chi\right\}^2/\sigma^2$$

or

$$M_\chi = V_\chi + B_\chi$$

For illustration, an example is again taken from the book of Box and Draper (1987). Figure 9 shows a situation as it might exist for a single variable when a straight line approximating function is to be used. The curved line shows an imaginary underlying function which would of course be obscured by experimental error. It is supposed that the scaling is chosen so that the region of interest R extends from -1 to $+1$ and that we consider only designs in which a single run is made at each of the levels $(-\chi_0, 0, \chi_0)$. Two such designs are shown with (a) $\chi_0 = \frac{2}{3}$ and (b) $\chi_0 = \frac{4}{3}$.

One way (Box and Draper, 1959) to obtain overall measures of variance and squared bias over any specified region of interest R is by averaging V_χ and B_χ over R to provide the quantities

$$V = \left(\int_R V_\chi \, d\chi\right)\Big/\left(\int_R d\chi\right) \quad \text{and} \quad B = \left(\int_R B_\chi \, d\chi\right)\Big/\left(\int_R d\chi\right)$$

Denoting the mean square error integrated, over R, by M, we can then write

$$M = V + B$$

For the previous example, V, B, and M are plotted against χ_0 in Fig. 10. We see how V becomes very large if the spread of the design is made very small, while if the design is made very large, V slowly approaches its minimum value of unity. The average squared bias B, on the other hand, has a minimum value when χ_0 is about 0.7, and

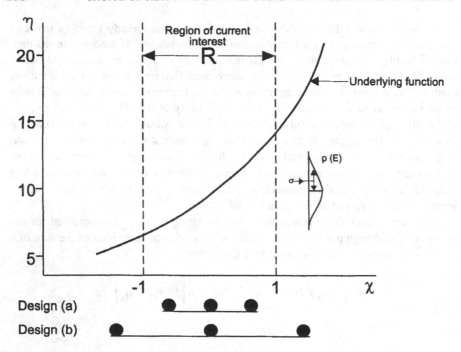

Figure 9. Two three-point designs (a) and (b) for fitting a straight line over a region of interest R, $-1 < \chi < 1$, when the underlying function is not a straight line.

increases for larger or smaller designs. A rather flat minimum for $M = V + B$ occurs near $\chi_0 = 0.79$. Thus the design which minimizes average mean squared error M is not very different from the design which minimizes average squared bias B but is extremely different from that which minimizes average variance V.

Choice of Alternative Model

A difficulty in all this is that in practice we do not know the nature of the true function η_χ. Progress may be made, however, by supposing that η_χ is to some satisfactory approximation represented by a polynomial model of higher degree d_2. Suppose then that a polynomial model of degree d_1 is fitted to n data values to give

$$\hat{\mathbf{y}}_\chi = \mathbf{X}_1 \mathbf{b}_1$$

while the true model is in fact a polynomial of degree d_2, so that

$$\eta_\chi = \mathbf{X}_1 \boldsymbol{\beta}_1 + \mathbf{X}_2 \boldsymbol{\beta}_2$$

We also need to know something about the relative magnitudes of systematic and random errors that we could expect to meet in practical cases. It was argued in Box and Draper (1959) that an investigator might typically employ a fitted approximating function such as a straight line when he believed that the average departure from the

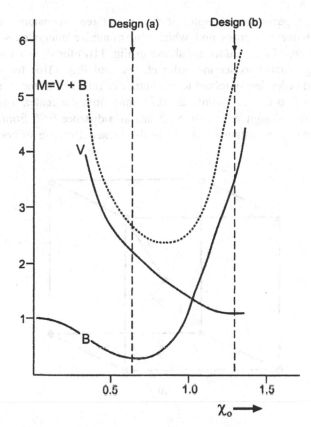

Figure 10. The behavior of the integrated variance V. The integrated squared bias B, and of $M = V + B$, for two three-point designs $(-\chi_0, 0, \chi_0)$ plotted as functions of χ_0.

truth induced by this approximating function was no worse than that induced by the process of fitting. This would suggest that the experimenter would tend to choose the size of his region R, and the degree of his approximating function in such a way that the integrated random error and the integrated systematic error were about equal. Thus we might suppose that a situation of particular interest is that where B is roughly equal to V. Examples that we studied seemed to show that designs that minimized M with the constraint $V = B$ were close to those which minimized B. Consequently, we suggested that, if a simplification were to be made in the design problem, it might be better to ignore the effects of sampling variation and concentrate on minimizing bias.

However this may be, there seems no doubt that, in making a table of useful designs, a component in our thinking should be the characteristics of the designs which minimized squared bias against feared alternatives. As a factor in our final choice, this should certainly receive more attention than the indications supplied by, say, D-optimality.

For illustration, particular examples of designs in three dimensions which have a number of attractive properties and which also minimize integrated squared bias when R is a sphere of *unit radius* are shown in Fig. 11(a) for $d_1 = 1$ and $d_2 = 2$, a first-order design robust to second-order effects, and Fig. 11(b) for $d_1 = 2$ and $d_2 = 3$, a second-order design robust to third-order effects. The former is the familiar 2^3 factorial scaled so that the points are 0.77 units from the center. The latter is a rotatable composite design with "cube" points at a distance 0.86 from the center, and "star" points at a distance 0.83 from the center. Because of the inevitable

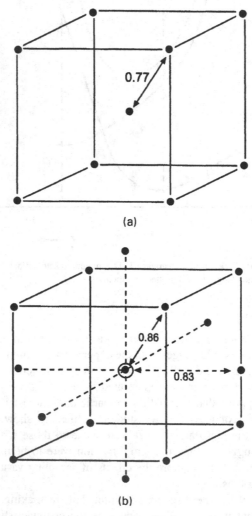

Figure 11. (a) A first-order (two-level factorial) design in three factors which minimizes squared bias from second-order terms when the region of interest is a sphere of unit radius. (b) A second-order composite rotatable design which minimizes squared bias from third-order terms when the region of interest is a sphere of unit radius.

uncertainty in choosing scales exact dimensions of the designs should not be taken too seriously, but these examples do illustrate the fact that as soon as we take account of bias, design points are *not* chosen on the boundary of *R*.

Choice of Designs Which Minimize Bias

Before considering the problem of choosing minimum bias designs it is desirable to generalize slightly the previous formulation. Although it avoids limiting the location of the design points in an artificial way, the idea of a region of interest *R* within a larger operability region *O* is still not entirely satisfactory because it implies that we have equal interest at all points within *R*. A more general formulation (Box and Draper, 1963), which subsumes what we have been discussing, employs a weight function $w(\chi)$ which extends over the operability region *O* so that $\int_O w(\chi)\,d\chi = 1$. The weighted mean square error M can now be split into a weighted variance part V and a weighted squared bias part B so that again $M = V + B$, with

$$M = \int_O w(\chi)E\{\hat{y}(\chi) - \eta(\chi)\}^2\,d\chi$$

$$V = \int_O w(\chi)E\{\hat{y}(\chi) - E(\hat{y}(\chi))\}^2\,d\chi$$

$$B = \int_O w(\chi)\{E(\hat{y}(\chi)) - \eta(\chi)\}^2\,d\chi$$

Two possible weight functions for $k = 1$ (Draper and Lawrence, 1967) are shown in Fig. 12.

Suppose as before the fitted function is a polynomial $\mathbf{X}_1\mathbf{b}_1$ of degree d_1 while the true model is a polynomial $\mathbf{X}_1\boldsymbol{\beta}_1 + \mathbf{X}_2\boldsymbol{\beta}_2$ of degree d_2 and define moment matrices for the design and for the weight function by

$$\mathbf{M}_{11} = n^{-1}\mathbf{X}_1'\mathbf{X}_1, \qquad \mathbf{M}_{12} = n^{-1}\mathbf{X}_1'\mathbf{X}_2$$

$$\mu_{11} = \int_O w(\chi)\chi_1\chi_1'\,d\chi, \qquad \mu_{12} = \int_O w(\chi)\chi_1\chi_2'\,d\chi$$

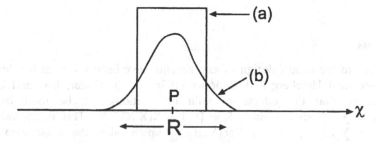

Figure 12. Two possible weight functions for $k = 1$: (a) uniform interest over *R*, no interest outside *R*; and (b) normal distribution shape, greater weight to points nearer *P*.

Then (Box and Draper, 1959) a necessary and sufficient condition for the squared bias B to be minimized is that

$$\mathbf{M}_{11}^{-1}\mathbf{M}_{12} = \mu_{11}^{-1}\mu_{12}$$

and hence a sufficient condition is that all the moments of the design, up to and including order $d_1 + d_2$, are equal to all the corresponding moments of the weight function.

SOME OTHER CONSIDERATIONS IN DESIGN CHOICE

There is insufficient space to discuss here all of the items in Table 1 that, in one circumstance or another, it might be necessary to take into account, but mention will be made of a few.

Lack of Fit (3), Sequential Assembly (6), Blocking (5), Estimation of Error (7), Transformation Estimation (4)

While the adequacy of a particular approximating function to explore a region of current interest is always to some extent a matter of guesswork, simple approximations requiring fewer runs for their elucidation will usually be preferred to more complicated ones. This leads to a strategy of building up from simpler models, rather than down from more complicated ones. A practical procedure is then: to employ the simplest approximating function which it is hoped may be adequate; (see also Atkinson, 1972, 1973; Box and Draper, 1959; DeBaun, 1956); and to switch to a more elaborate approximating function when this appears necessary. The implication for designs is (a) that they should provide for *checking* model adequacy; (b) that they should be capable of *sequential assembly*, a design of order d should be augmentable to one of order $d + 1$; (c) since conditions may change slightly from one set of runs to another, especially affecting level, ideally the pieces of the design should form *orthogonal blocks*. For illustration see Chapter B.6.

Robustness

Approaches to the robust design of experiments have been reviewed by Herzberg (1981); see also Herzberg and Andrews (1976). In particular, Box and Draper (1975) show that the effects of wild observations can be minimized by making $r = \sum r_{uu}^2$ small, where $\mathbf{R} = \{r_{tu}\} = \mathbf{X}(\mathbf{X}'\mathbf{X})^{-1}\mathbf{X}'$. This is equivalent to minimizing $\sum r_{uu}^2 - p^2/n = n \ \mathrm{Var}\{\mathrm{Var}(\hat{y})\}$ which takes the value zero when $\mathrm{Var}(\hat{y}_u) = p/n \ (u = 1, \ldots, n)$. Thus G-optimal designs are optimally robust in this sense.

Size of the Experimental Design

A good experimental design is one which focuses experimental effort on what is judged important in the current experimental context. Suppose that, in addition to estimating the p parameters of the assumed model form, it is concluded that $f \geq 0$ contrasts are needed to check adequacy of fit, $b \geq 0$ further contrasts for blocking, and that an estimate of experimental error is needed having $e \geq 0$ degrees of freedom. To obtain independent estimates of all items of interest we then require a design containing at least $p + f + b + e$ runs. However, the importance of checking fit, blocking, and obtaining an independent estimate of error will differ in different circumstances, and the minimum value of n will thus correspondingly differ. But this minimum design will in any case only be adequate if σ^2 is below some critical value. When σ^2 is larger, designs larger than the minimum design are needed to obtain estimates of sufficient precision. In this circumstance rather than merely replicate the minimum design, opportunity may be taken to employ a higher order design allowing the fitting of a more elaborate approximating function which can then cover a wider experimental region. Notice that even when σ^2 is small, designs for which n is larger than p are not necessarily wasteful. This depends on whether the additional degrees of freedom are efficiently employed to achieve the experimenter's current requirements.

Simple Data Patterns

It has sometimes been argued that we may as well choose points randomly to cover the design region or employ some algorithm that distributes them evenly *even though this does not result in a simple data pattern* such as is achieved by factorials and composite response surface designs. In favor of this idea it has been argued that the fitting of a function by least squares to a haphazard set of points is no longer a problem for modern computational devices. This is true but overlooks an important attribute of designs which form simple patterns. Two of the statistician's tasks as a member of a scientific team involve inductive model criticism and deductive model estimation. [See, for example, Box (1980).] This second task requires that he temporarily behave as if he believed the current assumptions to be true and, conditional on that belief, provide an estimate of the function (or equivalently its parameters). This can certainly be done with haphazard designs. However, to perform the first task he must devise appropriate assumptions and appropriately modify them when necessary. Thus he must decide (a) what function should be fitted in the first place, and (b) how residuals should be examined so as to detect and understand deviations from some currently entertained model—in particular, in relation to the independent variables.

Designs such as factorials and composite response surface designs employ patterns of experimental points that allow comprehensible comparisons to be made, both for the original observations and for the residuals from any fitted function and so stimulate the brain to the performance of model selection and criticism. For example, consider a 3^2 factorial design used to elucidate the effects of temperature and concentration on some response such as yield. Intelligent inductive criticism is greatly enhanced by the possibility of being able to plot the original data and the residuals against temperature for each individual level of concentration, and against concentration for each individual level of temperature. See also Chapter C.3.

CONCLUSION

(i) We must look for good design criteria which measure characteristics of the experimental arrangement in which the investigator might sensibly be interested. Because the importance of various characteristics will differ in different situations, tables of such criteria for particular designs would encourage good judgment to be used in matching the design to the scientific context. Optimum levels of these criteria can be useful as bench marks in judging the efficiencies of a particular compromise design with respect to these various criteria.

(ii) It is doubtful, however, if single criterion optimal designs are useful in locating such compromises. An optimal design is represented by a point in the multidimensional space of the coordinates of the design and a series of different criteria will give a series of such points which may be very differently located. Obviously knowledge of the location of such extrema may tell us almost nothing about the location of good compromises. For this we would need to study the joint behavior of the criterion functions at levels close to their extremal values. One limited but useful step would be to further investigate which criteria are in accord, such as G-optimality and robustness to wild observations, and which in conflict, such as variance and bias.

(iii) It is true that the problem of experimental design is full of scientific arbitrariness. No two investigators would choose the same variables, start their experiments in the same place, change variables over the same regions, and so on. But science works not by uniqueness, but by employing iterative techniques which tend to converge. Clearly we must learn to live with scientific arbitrariness, or else we are in a world of make believe. But we can make the problems worse, not better, by introducing arbitrariness for purely mathematical reasons.

This research was sponsored at the Mathematics Research Center, University of Wisconsin, Madison by the United States Army under Contract No. DAAC-29-80-C-0041.

CHAPTER C.10

An Apology for Ecumenism in Statistics

These days the statistician is often asked such questions as "Are you a Bayesian?" "Are you a frequentist?" "Are you a data analyst?" "Are you a designer of experiments?" I will argue that the appropriate answer to *all* these questions can be (and preferably should be) "yes," and that we can see why this is so if we consider the scientific context of what statisticians do.

For many years Statistics was in a rather turbulent state and the air was full of argument and controversy. The relative virtue of alternative methods of inference and, in particular, of Bayes's and Sampling (frequentist) inference was hotly debated. Data Analysis rightly received much heavier emphasis, but its more avid proponents have sometimes seemed to suggest that all else is worthless. Furthermore while biased estimators, in particular, shrinkage and ridge estimators, which have been advocated to replace the more standard varieties are clearly sensible in appropriate contexts their frequentist justification which ignores context seems unconvincing. Parallel criticism may be made of ad hoc robust procedures the proliferation of which has worried some dissidents who have argued, for example, that mechanical downweighting of peculiar observations may divert attention from important clues to new discovery.

Insofar as these debates lead us to progressive change in our ideas they are healthy and productive, but insofar as they encourage polarization they may not be. One remembers with some misgivings Saxe's poem about the six blind men of Hindustan investigating an elephant. It will be recollected that one, feeling only the elephant's trunk, thought it like a snake, another, touching its ear, thought it must be a fan, etc. The poem ends:

From Box, G. E. P. (1983), in *Scientific Inference, Data Analysis, and Robustness*, Box, G. E. P., Leonard, T., and Wu, C.-F. (eds.), Academic Press, pp. 51–84. Copyright © 1983 by Academic Press, Inc. *Improving Almost Anything: Ideas and Essays, Revised Edition.* By George Box and Friends

And so these men of Hindustan
Disputed loud and long,

Each in his opinion
Exceeding stiff and strong,

Though each was partly in the right,
And all were in the wrong.

Some of the difficulties arise from the need to simplify. But simplification included merely to produce satisfying mathematics or to reduce problems to convenient small sized pieces can produce misleading conclusions. Simplification which retains the essential *scientific* essence of the problem is most likely to lead to useful answers but this requires understanding of, and interest in, the scientific context.

SOME QUESTIONABLE SIMPLIFICATIONS

(a) It has been argued that Bayes's theorem uniquely solves *all* problems of inference. However, only part of the inferential exercises in which the statistical scientist is ordinarily engaged seem to conveniently fit the Bayesian mold. In particular, diagnostic checks of goodness of fit involving various analyses of residuals seem to require other justification. In fact I believe (Box, 1980) that the process of scientific investigation involves not one but two kinds of inference: *estimation* and *criticism*, used iteratively and in alternation. Bayes completely solves the problem of estimation and can also be helpful at the criticism stage in judging the relative plausibility of two or more models. However, because of its necessarily conditional nature, it cannot deal with inferential criticism which requires a sampling (frequentist) justification.

(b) Fisher (1956) believed that the Neyman–Pearson theory for testing statistical hypotheses, while providing a model for industrial quality control and sampling inspection, did not of itself provide an appropriate basis for the conduct of scientific research. This can be regarded as the complement to the objection raised in (a), for statistical quality control and inspection are methods of inferential *criticism* supplying a continuous check on the adequacy of fit of the model for the properly operating process. I would regard Fisher's comment as meaning that the Neyman–Pearson theory was irrelevant to problems of estimation. Certainly there is evidence in the social sciences that excessive reliance upon this theory alone, encouraged by the mistaken prejudices of referees and editors,* has led to harmful distortion of the conduct of scientific investigation in these fields.

(c) In some important contexts the scientific relevance of alphabetic optimality criteria (A, E, D, G etc.) in the choice of experimental designs has been questioned

---

*It has sometimes led to the idea that the goal is by some means or other to get a "statistically significant" result.

(see discussion of Kiefer, 1959; also Chapter C.9). Here again there is danger of deleterious feedback since users of statistical design, perhaps dazzled by impressive but poorly comprehended mathematics, may fail to realize the naive framework within which the optimality occurs.

(d) Even Data Analysis, excellent in itself, presents some dangers. It is a major step forward that in these days students of statistics are required more and more to work on real data. Indeed suitable "data sets" have been set aside for their study. But this too can produce misunderstanding. For instance, some examples have become notorious and have been analyzed by a plethora of experts; one finds three outliers, another claims that a transformation is needed and then only one outlier occurs, and so on. Too much exposure to this sort of thing can again lead to the mistaken idea that *this* represents the real context of scientific investigation. The statistician in his proper role as a *member of a scientific team* should certainly make such analyses, but realistically he would then discuss them with his scientific colleagues and present, when appropriate, not one, but alternative plausible possibilities. He need not, and usually should not, choose among them. Rather he should make sure that these possibilities were considered when he and his colleagues planned the next stage of the investigation. Together they would choose the next design so that among other things it could resolve current uncertainties judged to be important. In particular, the possible meaning and importance of discrepant values would then be discussed as well as the meaning of analyses which downweighted or excluded them.

The most dangerous and misleading of the unstated assumptions suggested to some extent by all these simplifications concerns the implied *static* nature of the process of investigation: An analysis is made; *a* hypothesis is tested; *one* model is considered; *a single* design is run; *a single* set of data is examined and reexamined.*

I believe that statisticians must be trained to have some understanding of what good scientists do and to help them do it better. It seems necessary therefore to examine at least briefly the nature of the scientific process itself.

SCIENTIFIC METHOD AND THE HUMAN BRAIN

Scientific method is a formalization of the everyday process of finding things out. For thousands of years, things were found out largely as a result of chance coincidences. Two circumstances needed to coincide: (a) a potentially informative

---

* While provision is made for adaptive feedback in data analysis, usually the possibility of acquiring further data to illuminate points at issue is not. What we do as statisticians depends heavily on expectations implied by our training. While a previous generation of graduates might have been expected to prove theorems, and to test isolated hypotheses, and perhaps to teach a new generation of students to do likewise, the present generation might be forgiven for believing that their fate is only to explore "data sets" and speculate on what might or might not explain them. We must encourage our students to accept the heritage bestowed by Fisher, who elevated the statistician from an archivist to an active designer of experiments and hence an architect and coequal investigator.

experience needed to occur, *and* (b) be known about by someone of sufficient acuity of mind to formulate, and preferably to test, a possible rule for its future occurrence.

Progress was slow because of the rarity of the two necessary individual circumstances and the still greater rarity of their coincidence. Experimental science accelerates this learning process by isolating its essence: potentially informative experiences are deliberately *staged* and made to occur in the *presence* of a trained investigator. As science has developed, we have learned how such artificial experiences may be carefully contrived to isolate questions of interest, how conjectures that are put forward may be tested, and how residual differences from what had been expected can be used to modify and improve initial ideas. So the ordinary process of learning has been sharpened and accelerated.

The instrument of all learning is the brain—an incredibly complex structure. One thing that is clear is the importance to the brain of models. To appreciate why this is so, consider how helpless we would be if, each night, all our memories were eliminated, so that we awoke to each new day with no past experiences whatever and hence no models to guide our conduct. In fact, our past experience is conveniently accumulated in models $M_1, M_2, \ldots, M_i, \ldots$. Some of these models are well established, others less so, while still others are in the very early stages of creation. When some new fact or body of facts y_d comes to our attention, the mind tries to associate this new experience with an established model. When, as is usual, it succeeds in doing so, this new knowledge is incorporated in the appropriate model and can set in train appropriate action.

Obviously, to avoid chaos the brain must be good at allocating data to an appropriate model and at initiating the construction of a new model only if this should prove necessary. To conduct such business the mind must be able to *deduce* what facts could be expected as realizations of a particular model and, what is more difficult, to *induce* what model(s) are consonant with particular facts.

Thus, it is concerned with *two* kinds of inference: (A) *Contrasting* of new facts y_d with a possible model M: an operation I will characterize by subtraction $M - y_d$. This process stimulates *induction* and will be called *criticism*. (B) *Incorporating* new facts y_d into an appropriate model: an operation I will characterize by addition $M + y_d$. This process is entirely *deductive* and will be called *estimation*.

I believe then that many of our difficulties arise because, while there is an essential need for two kinds of inference, there seems an inherent propensity among statisticians to seek for only one.

In any case, research which, following the discoveries of Roger Sperry and his associates, has gathered great momentum implies that the human brain contains two largely separate but cooperating instruments.

In many people, the left brain is more concerned with logical deduction, and the right brain with induction. The two parts of the brain are joined by millions of connections in the corpus callosum. It is known that the left brain can play a

---

*For example, the apparently instinctive knowledge of what to do and how to do it, enjoyed by, for example, an experienced tennis player comes from the right brain.

conscious and dominant role while one may be much less aware* of the working of one's right brain.

The right brain's ability to appreciate* patterns in data y_d and to find patterns in discrepancies $M_i - y_d$ between the data and what might be expected if some tentative model M_i were true is of great importance in the search for explanations of data and of discrepant events. The accomplishment of pattern recognition is of course of enormous consequence in scientific discovery.† However, some check is needed on its pattern-seeking ability, for common experience shows that some pattern or other can be seen in almost any set of data or facts.‡ This is the object of diagnostic checks and tests of fit which, I will argue, require frequentist theory significance tests for their formal justification.

THE THEORY–PRACTICE ITERATION

As I have emphasized in this book it has long been recognized that the learning process is a motivated iteration between theory and practice. By practice I mean reality in the form of data or facts. In this iteration deduction and induction are employed in alternation. Progress of an investigation is thus evidenced by a theoretical model, which is not static, but by appropriate exposure to reality continually evolves until some currently satisfactory level of understanding is reached. At any given stage in a scientific investigation the current model helps us to appreciate not only what we know, but *what else it may yet be important to find out* and so motivates the collection of new data to illuminate dark but possibly interesting corners of present knowledge. See, for example, Box and Youle (1955), Box (1976), and Box, Hunter, and Hunter (1978).

The reader can find illustration of these matters in his everyday experience, or in the evolution of the plot of any good mystery novel, as well as in any reasonably honest account of the events leading to a scientific discovery.

Different Levels of Adaptation

The adaptive iteration we have described produces change in what we believe about the system being studied, but it can also produce change in *how* we study it, and

---

* Implicit recognition of the need to stimulate the remarkable pattern-seeking ability of the right brain is evidenced by emphasis on ingenious plotting devices in the model formulation/modification phases of investigation. In particular, Chernoff's (1973) representation of multivariate data by faces and earlier Edgar Anderson's (1960) use of glyphs direct the right brain to the recognition problem at which it excels.

† Manifestations of the importance to discovery of unconscious pattern seeking by the right brain have often been noticed. For example, Beveridge (1950) remarks that happenings of the following kind are commonplace: a scientist has mulled over a set of data for many months and then, at a certain point in time, perhaps on a country walk when the problem is not being consciously thought about, he suddenly becomes aware of a solution (model) which explains these data. This point in time is presumably when the right brain sees fit to let the left brain know what it has figured out.

‡ See, for example, the King of Heart's rationalization of the poem brought as evidence in the trial of the Knave of Hearts in Lewis Carroll's *Alice in Wonderland*.

sometimes even in the objective* of the study. This multiple adaptivity explains the surprising property of convergence of a process of investigation which at first appears hopelessly arbitrary. See, for example, Box (1957). To appreciate this arbitrariness, suppose that some scientific problem were being studied independently by, say, 10 sets of investigators, all competent in the field of endeavor. It is certain that they would start from different points, conduct the investigation in different ways, have different initial ideas about which variables were important, on what scales, and in which transformation. Yet it is perfectly possible that they would all eventually reach similar conclusions. It is important to bear this context of multiple iteration in mind when we consider the scientific process and how it relates to a statistical method.

STATISTICAL ESTIMATION AND CRITICISM

In a previous paper (Box, 1980) a statistical theory was presented which, it was argued, was consonant with the view of scientific investigation outlined above. Suppose at the ith stage of such an investigation a set of assumptions A_i are tentatively entertained which postulate that, to an adequate approximation, the density function for *potential* data y is $p(y|\theta, A_i)$ and the prior distribution for θ is $p(\theta|A_i)$. Then it was argued that the *model* M_i should be defined as the *joint distribution* of y and θ

$$p(y, \theta|A_i) = p(y|\theta, A_i)p(\theta|A_i) \tag{1}$$

since it is a complete statement of prior tentative belief at stage i. In these expressions A_i is understood to indicate the assumptions in the model specification at stage i. The model of Equation (1) means to me that current belief about the outcome of contemplated data acquisition would be calibrated with adequate approximation by a *physical simulation* involving appropriate random sampling from the distributions $p(y|\theta, A_i)$ and $p(\theta|A_i)$.

This model can also be factored as

$$p(y, \theta|A) = p(\theta|y, A)p(y|A) \tag{2}$$

The second factor on the right, which can be computed *before* any data become available,

$$p(y|A) = \int p(y|\theta, A)p(\theta|A) \, d\theta \tag{3}$$

---

* If we start out to prospect for silver, we should not ignore an accidental discovery of gold. For example, one experimental attempt to find manufacturing conditions giving greater yield of a particular product failed to find any such we but did find reaction conditions giving the *same* yield with the reaction time *halved*. This meant that, by switching to the new manufacturing conditions, throughput could be doubled, and that a costly, previously planned, extension of the plant was unnecessary.

is the *predictive* distribution of the totality of all possible samples **y** that could occur if the assumptions were true.

When an *actual* data vector $\mathbf{y}_d$ becomes available

$$p(\mathbf{y}_d, \ \boldsymbol{\theta}|A) = p(\boldsymbol{\theta}|\mathbf{y}_d, \ A)p(\mathbf{y}_d|A) \tag{4}$$

The first factor on the right is the Bayes posterior distribution of $\boldsymbol{\theta}$ given $\mathbf{y}_d$

$$p(\boldsymbol{\theta}|\mathbf{y}_d, \ A) \propto p(\mathbf{y}_d|\boldsymbol{\theta}, \ A)p(\boldsymbol{\theta}|A) \tag{5}$$

while the second factor

$$p(\mathbf{y}_d|A) = \int p(\mathbf{y}_d|\boldsymbol{\theta}, \ A)p(\boldsymbol{\theta}|A) \ d\boldsymbol{\theta} \tag{6}$$

is the *predictive density* associated with the particular data $\mathbf{y}_d$ actually obtained conditional on the truth of the model and on the data $\mathbf{y}_d$ having occurred.

The posterior distribution $p(\boldsymbol{\theta}|\mathbf{y}_d, \ A)$ allows all relevant estimation inferences to be made about $\boldsymbol{\theta}$, but this posterior distribution can supply no information about the *adequacy* of the model. Information on adequacy may be provided, however, by reference of the density $p(\mathbf{y}_d|A)$ to the predictive reference distribution $p(\mathbf{y}|A)$ or of the density $p\{g_i(\mathbf{y}_d)|A\}$ of some relevant checking function $g_i(\mathbf{y}_d)$ to its predictive distribution and, in particular, by computing the probabilities

$$\text{Pr}\{ \ p(\mathbf{y}|A) \leq p(\mathbf{y}_d|A)\} \tag{7}$$

and

$$\text{Pr}[\ p\{g_i(\mathbf{y})|A\} \leq p\{g_i(\mathbf{y}_d)|A\}] \tag{8}$$

Two illustrative examples follow.

The Binomial Model

As an elementary example, suppose inferences are to be made about the proportion θ of successes in a set of binomial trials.

Suppose n trials are *about to be made* and assume a beta distribution prior with mean θ_0. Then

$$p(\theta|A) = [B\{(m\theta_0, \ m(1 - \theta_0)\}]^{-1}\theta^{m\theta_0 - 1}(1 - \theta)^{m(1-\theta_0)-1} \tag{9}$$

$$p(y|\theta, \ A) = \binom{n}{y}\theta^y(1 - \theta)^{n-y} \tag{10}$$

and the predictive distribution is

$$p(y|A) = \binom{n}{y}[B\{m\theta_0, \ m(1-\theta_0)\}]^{-1}B\{m\theta_0+y, \ m(1-\theta_0)+n-y\} \qquad (11)$$

which may be computed *before* the data are obtained.

If, now, having performed n trials, there are y_d successes, the likelihood defined up to a multiplicative constant is

$$L(\theta|y_d, \ A) = \theta^{y_d}(1-\theta)^{n-y_d} \qquad (12)$$

the predictive density is

$$p(y_d|A) = \binom{n}{y_d}[B\{m\theta_0, \ m(1-\theta_0)\}]^{-1}B\{m\theta_0+y_d, \ m(1-\theta_0)+n-y_d\} \qquad (13)$$

and the posterior distribution of θ is

$$p(\theta|y_d, \ A) = [B\{m\theta_0+y_d, \ m(1-\theta_0)+n-y_d\}]^{-1}$$
$$\times \theta^{y_d+m\theta_0-1}(1-\theta)^{n-y_d+m(1-\theta_0)-1} \qquad (14)$$

In the examples of Figures 1 and 2 full lines are used for items available *prior* to the availability of data y_d and dotted lines for items available only after the data y_d are in hand. Both Figures 1 and 2 illustrate a situation where the prior distribution $p(\theta|A)$ has mean $\theta_0 = 0.2$ and $m = 20$ and we know that $n = 10$ trials are to be performed. Knowing these facts, we can immediately calculate the predictive distribution $p(y|A)$ which is the probability distribution for all possible outcomes from such a model if we suppose the model is true.

When the experiment is actually performed suppose at first, as in Figure 1, that $y_d = 3$ of the trials are successes. The predictive probability $p(3|A)$ associated with this outcome is not unusually small. In fact, $\Pr\{p(y|A) \leq p(3|A)\} = 0.42$ and we have no reason to question the model. Thus for this sample the likelihood $L(\theta|y)$ may reasonably be combined with the prior to produce the posterior distribution shown.

In Figure 2, however, it is supposed instead that the outcome is $y_d = 8$ successes so that for this sample $\Pr\{p(y|A) \leq p(8|A)\} = 0.0013$ and the adequacy of the model, and, in particular, the adequacy of the prior distribution, is now called into question. Inspection of the figure shows how this agrees with common sense; for in the case illustrated the posterior distribution is unlike either the prior distribution or the likelihood which were combined to obtain it.

Misgivings about the use of Bayes's theorem which some have expressed in the past are certainly associated with the possibility of distorting the information coming from the data by the use of an inappropriate prior distribution. Without predictive checks, the following objections would carry great weight:

(a) that nothing in the Bayes calculation of the posterior distribution itself could warn of the incompatibility of the data and the model, and especially the prior; and

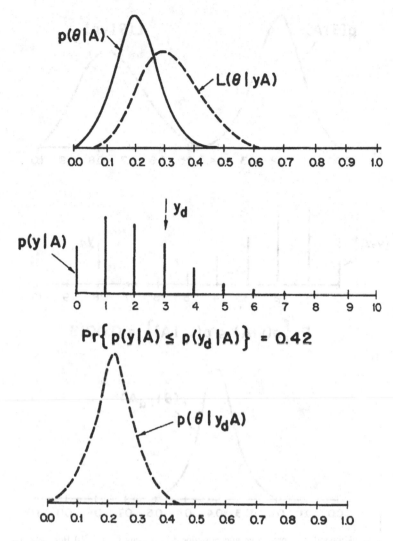

Figure 1. Prior, likelihood, and predictive and posterior distributions for $n = 10$ Bernoulli trials with $y_d = 3$ successes.

(b) that in complicated examples it would not be so obvious when this incompatibility occurred.

A case of particular interest occurs when the prior is sharply centered* at its mean value $\theta_0 = 0.2$. This happens in the above binomial setup when m is made very

---

* Such a model with a prior sharply centered at $\theta_0 = 0.2$ might be appropriate, for instance, if a trial consisted of spinning ten times what seemed to be a properly balanced pentagonal top and counting the number of times the top fell on a particular segment.

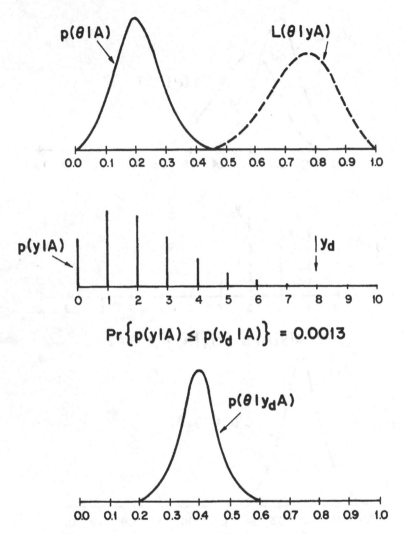

Figure 2. Prior, likelihood, and predictive and posterior distributions for $n = 10$ Bernoulli trials with $y_d = 8$ successes.

large. Then, *if* the model is unquestioned, the posterior distribution will be essentially the same as the prior leading to the conclusion that θ is close to θ_0 whatever the data. The predictive distribution in this case is $p(y|\theta_0, A)$, the ordinary binomial sampling distribution, and the predictive check is the standard binomial significance test, which can discredit the model with $\theta = \theta_0 = 0.2$ and hence discredit the application of Bayes's theorem to this case. This, to my mind, produces the most satisfactory justification for the standard significance test.

The Normal Linear Model and Ridge Estimators

Another example, discussed in Box (1980), concerns the normal linear model. In a familiar notation suppose

$$\mathbf{y} \sim N(\mathbf{1}\mu + \mathbf{X}\boldsymbol{\theta},\ \mathbf{I}_n\sigma^2) \tag{15}$$

with $\mathbf{1}$ a vector of unities and $\mathbf{X}$ of full rank k and such that $\mathbf{X}'\mathbf{1} = \mathbf{0}$ and suppose that prior densities are locally approximated by

$$\mu \sim N(\mu_0,\ c^{-1}\sigma^2), \qquad \boldsymbol{\theta} \sim N(\boldsymbol{\theta}_0,\ \boldsymbol{\Gamma}^{-1}\sigma^2), \qquad \{\sigma^2/v_0 s_0^2\} \sim \chi^{-2}(v_0) \tag{16}$$

where $\chi^{-2}(v_0)$ refers to the inverted χ^2 distribution with v_0 degrees of freedom and μ and $\boldsymbol{\theta}$ are independent conditional on σ^2.

Given a sample $\mathbf{y}_d$, special interest attaches to $\boldsymbol{\theta}$ and σ^2 which, given the assumptions, are estimated by $p(\boldsymbol{\theta},\ \sigma^2|\mathbf{y}_d,\ A)$ with marginal distributions

$$p(\boldsymbol{\theta}|\mathbf{y}_d,\ A) \propto \left(1 + \frac{(\boldsymbol{\theta} - \bar{\boldsymbol{\theta}}_d)'(\mathbf{X}'\mathbf{X} + \boldsymbol{\Gamma})(\boldsymbol{\theta} - \bar{\boldsymbol{\theta}}_d)}{(n + v_0)\hat{\sigma}_d^2}\right)^{-(n+v_0+k)/2} \tag{17}$$

$$p(\sigma^2|\mathbf{y}_d,\ A) \propto \sigma^{-(n+v_0+2)} \exp[-\tfrac{1}{2}(n + v_0)\hat{\sigma}_d^2/\sigma^2] \tag{18}$$

with

$$\bar{\boldsymbol{\theta}}_d = (\mathbf{X}'\mathbf{X} + \boldsymbol{\Gamma})^{-1}(\mathbf{X}'\mathbf{X}\hat{\boldsymbol{\theta}}_d + \boldsymbol{\Gamma}\boldsymbol{\theta}_0)$$

$$\hat{\boldsymbol{\theta}}_d = (\mathbf{X}'\mathbf{X})^{-1}\mathbf{X}'\mathbf{y}_d, \qquad v = n - k - 1 \tag{19}$$

$$(n + v_0)\hat{\sigma}_d^2 = v s_d^2 + v_0 s_0^2 + (\hat{\boldsymbol{\theta}}_d - \boldsymbol{\theta}_0)'\{(\mathbf{X}'\mathbf{X})^{-1} + \boldsymbol{\Gamma}^{-1}\}^{-1}(\hat{\boldsymbol{\theta}}_d - \boldsymbol{\theta}_0)$$
$$+ (n^{-1} + c^{-1})^{-1}(\bar{y} - \mu_0)^2$$

and

$$s^2 = \{\mathbf{I} - \mathbf{X}(\mathbf{X}'\mathbf{X})^{-1}\mathbf{X}'\}\mathbf{y}, \qquad s_d^2 = \{\mathbf{I} - \mathbf{X}(\mathbf{X}'\mathbf{X})^{-1}\mathbf{X}'\}y_d \tag{20}$$

Now let

$$s_p^2 = (v + v_0)^{-1}(v s^2 + v_0 s_0^2) \quad \text{and} \quad s_{pd}^2 = (v + v_0)^{-1}(v s_d^2 + v_0 s_0^2) \tag{21}$$

Then the joint predictive distribution can be factored into independent components for $(\hat{\boldsymbol{\theta}} - \boldsymbol{\theta}_0)/s_p$, s^2, and $v - 1$ angular elements of the standardized residuals. A predictive check based on the first of these factors

$$\Pr\{ p((\hat{\boldsymbol{\theta}} - \boldsymbol{\theta}_0)/s_p|A) < p((\hat{\boldsymbol{\theta}}_d - \boldsymbol{\theta}_0)/s_{pd}|A)\}$$

$$= \Pr\left\{ F_{k,v+v_0} > \frac{(\hat{\boldsymbol{\theta}}_d - \boldsymbol{\theta}_0)'\{(\mathbf{X}'\mathbf{X})^{-1} + \boldsymbol{\Gamma}^{-1}\}^{-1}(\hat{\boldsymbol{\theta}}_d - \boldsymbol{\theta}_0)}{k s_{pd}^2}\right\} \tag{22}$$

is the standard analysis of variance check for compatibility of two estimates $\hat{\theta}_d$ and θ_0 and was earlier proposed as a check for compatibility of prior and sample information by Theil (1963).

Now suppose the X matrix to be in correlation form and assume $\theta_0 = 0$, $\Gamma = I_k \gamma_0$, $v_0 \to 0$ so that $s_p^2 \to s^2$. Then the estimates $\bar{\theta}_d$ are the ridge estimators of Hoerl and Kennard (1970) which, given the assumptions, appropriately combine information from the prior with information from the data. The predictive check (22) now yields

$$\alpha = \Pr\left\{ F_{k,v} > \frac{\hat{\theta}_d'\{(X'X)^{-1} + I\gamma_0^{-1}\}^{-1}\hat{\theta}_d}{ks_d^2} \right\} \tag{23}$$

allowing any choice of γ_0 to be criticized.

For example, in their original analysis of the data of Gorman and Toman (1966), Hoerl and Kennard (1970) chose a value $\gamma_0 = 0.25$. But substitution of this value in Equation (23) yields $\alpha = \Pr\{F_{10,25} > 3.59\} < 0.01$ which discredits this choice.

One can see for these examples how the two functions of criticism and estimation are performed by the predictive check on the one hand and the Bayesian posterior distribution on the other.

Thus consider the ridge (Bayes mean) estimator of the second example. This estimator is a linear combination of the least squares estimate $\hat{\theta}$ and the prior mean θ_0, with weights supplied by the appropriate information matrices, and with covariance matrix obtained by inverting the *sum* of these *information matrices*. Assuming the data to be a realization of the model, this is the appropriate way of combining the two sources of information.

The predictive check, on the other hand, *contrasts* the values $\hat{\theta}$ and θ_0 with a dispersion matrix obtained by appropriately *summing* the two *dispersion matrices*.

The *combination* of information from the prior and likelihood into the posterior distribution and the *contrasting* of these two sources of information in the predictive distribution is equally clear in the binomial example and especially in its appropriate normal approximation.

SOME OBJECTIONS CONSIDERED

A recapitulation of the argument and a consideration of some objections is considered in this section.

Essential Elements of the Argument

A. Scientific investigation is an iterative process in which the model is not static but is continually evolving. At a given stage the nature of the uncertainties in a model directs the acquisition of further data, whether by choosing the design of an experiment or sample survey, or by motivating a search of a library or data bank. At, say, the ith stage of an investigation all current structural assumptions A_i, including those about the prior, must be thought of, not as being true, but rather as being subjective guesses which at this particular stage of the investigation are worth entertaining. It is consistent with this attitude that when data y_d become available checks need to be applied to assess consonance with A_i.

B. The statistical *model* at the ith stage of the investigation should be defined as the joint distribution of y and θ given the assumptions A_i

$$p(y,\ \theta|A_i) = p(y|\theta,\ A_i)p(\theta|A_i) \tag{24}$$

C. Not one but two distinct kinds of inference are involved within the iterative process: *criticism* in which the appropriateness of regarding data y_d as a realization of a particular model M is questioned; *estimation* in which the consequences of the assumption that data y_d are a realization of a model M are made manifest.

This criticism–estimation dichotomy is characterized mathematically by the factorization of the model realization $p(y_d,\ \theta|A_i)$ into the predictive density $p(y_d|A_i)$ and the posterior distribution $p(\theta|y_d,\ A_i)$. The predictive distribution $p(y|A_i)$ provides a reference distribution for $p(y_d|A_i)$. Similarly, the predictive distribution $p\{g(y)|A_i\}$ of any checking function $g(y)$ provides a reference distribution of the corresponding predictive density $p\{g(y_d)|A_i\}$. Unusually small values of this density suggest that the current model is open to question.

D. If we are satisfied with the adequacy of the assumptions A_i then the posterior distribution $p(\theta|y_d,\ A_i)$ allows for exhaustive estimation of θ and no other procedures of *estimation* are relevant. In particular, therefore, insofar as shrinkage, ridge, and robust estimators are useful, they ought to be direct consequences of an appropriate model and should not need the invocation of extraneous considerations such as minimization of mean square error.

Objections
Numbered to correspond with the various elements of the argument are responses to some objections that have been, or might be, raised.

A(i). *Iterative Investigation?* Some would protest that their own statistical experience is not with iterative investigation but with a single set of data to be analyzed, or a single design to be laid out and the results elucidated.

Some circumstances where the statistician has been involved in a "one-shot" analysis, rather than an iterative partnership, ought not to have happened. Such involvement occurs when the statistician has been drafted as a last resort, all other attempts to make sense of the data having failed. At this point data gathering will usually have been completed and there is no chance of influencing the course of the study. Statisticians whose training has not exposed them to the overriding importance of experimental design are most likely to acquiesce in this situation, or even to think of it as normal, and thus to encourage its continuance.

The statistician who has cooperated in the design of a *single* experiment which he analyzes is somewhat better off. However, one-shot designs are often inappropriate also. Underlying most investigations is a budget, stated or unstated, of time and/or money that can reasonably be expended. Sometimes this latent budget is not adequate to the goal of the investigation, but, for purposes of discussion, let us suppose that it is. Then if a sequential/iterative approach is possible it would usually be quite inappropriate to plan the whole investigation at the beginning in one large design. This is because the results from a first design will almost invariably supply new and often unexpected information about adequacy of the experiment apparatus, adequacy of measurement methods, choice of variables, metrics, transformations, regions of operability, unexpected side effects, and so forth, which will vitally influence the course of the investigation and the nature of the next experimental arrangement. A rough working rule is that not more than 25% of the time-and-money budget should be spent on the first design. Because large designs can in a limited theoretical sense be more efficient it is a common mistake not to take advantage of the iterative option when it is available. Instances have occurred of experimenters regretting that they were persuaded by an inexperienced statistician to perform a large "all inclusive" design where an adaptive strategy would have been much better. In particular, it is likely that many of the runs from such "all-embracing" designs will turn out to be noninformative because their structure was necessarily decided when *least* was known about the problem.

Scientific iteration is strikingly exemplified in response surface studies (see, e.g., Box and Wilson, 1951; Box, 1954a; Box and Youle, 1955). In particular, methods such as steepest ascent and canonical analysis can lead to exploration of new regions of the experimental space, requiring elucidation by new designs which, in turn, can lead to the use of models of higher levels of sophistication. Although in these examples the necessity for such an iterative theory is most obvious, it clearly exists much more generally, for example, in investigations employing sequences of orthodox experimental designs and to many applications of regression analysis. It has sometimes been suggested that agricultural field trials are not sequential but of course this is not so; only the time frame is longer. Obviously what is learned from one year's work is used to design the next year's experiments.

However, I agree that there are some more convincing exceptions. For example, a definitive trial which is intended to settle a controversy such as a test of the effectiveness of Laetrile as a cure for cancer. Also, the iteration can be very slow. For example, in trials on the weathering of paints, each phase can take from 5 to 10 years.

A(ii). *Subjective Probability?* The view of the process of scientific investigation as one of model evolution has consequences concerning subjective probabilities. An objection to a subjectivist position is that in presenting the *final* results of our investigation, we need to convince the outside world that we have really reached the conclusion that we say we have. It is argued that, for this purpose, subjective probabilities are useless. However, I believe that the confirmatory stage of an iterative investigation, when it is to be demonstrated that the final destination reached is where it is claimed to be, will typically occupy, perhaps, only the last 5% of the experimental effort. The other 95%—the wandering journey that has finally led to that destination—involves, as I have said, many heroic *subjective* choices (what variables? what levels? which scales? etc., etc.) at every stage. Since there is no way to avoid *these* subjective choices, which are a major determinant of success, why should we fuss over subjective probability?

Of course, the last 5% of the investigation occurs when most of the problems have been cleared up and we know most about the model. It is this rather minor part of the process of investigation that has been emphasized by hypothesis testers and decision theorists. The resultant magnification of the importance of formal hypothesis tests has inadvertently led to underestimation by scientists of the area in which statistical methods can be of value and to a wide misunderstanding of their purpose. This is often evidenced in particular by the attitudes to statistics of editors and referees of journals in the social, medical, and biological sciences.

B(i). *The Statistical Model?* The statistical model has sometimes been thought of as the density function $p(y|\theta, A)$ rather than the joint density $p(y, \theta|A)$ which reflects the influence of the prior. However, only the latter form contains all currently entertained beliefs about y and θ. It seems quite impossible to separate prior belief from assumptions about model structure. This is evidenced by the fact that assumptions are frequently interchangeable between the density $p(y|\theta)$ and the prior $p(\theta)$. As an elementary example, suppose that among the parameters $\theta = (\phi, \beta)$ of a class of distributions β is a shape parameter such that $p(y|\phi, \beta_0)$ is the normal density. Then it may be convenient, for example, in studies of robustness, to define a normal distribution by writing the more general density $p(y|\phi, \beta)$ with an associated prior for β which is concentrated at $\beta = \beta_0$. The element specifying normality which in the usual formulation is contained in the density $p(y, \theta)$ is thus transferred to the prior $p(\theta)$.

B(ii). *Do We Need a Prior?* Another objection to the proposed formulation of the model is the standard protest of non-Bayesians concerning the introduction of any prior distribution as an unnecessary and arbitrary element. However, history has shown that it is the *omission* in sampling theory, rather than the inclusion in Bayesian analysis, of an appropriate prior distribution that leads to trouble.

For instance, Stein's result (1955) concerning the inadmissibility of the vector of sample averages as an estimate of the mean of a multivariate normal distribution is well known. But consider its practical implication for, say, an experiment resulting in a one-way analysis of variance. Such an experiment could make sense when it is conducted to compare, for example, the levels of infestation of k different varieties of wheat, or the numbers of eggs laid by k different breeds of chickens, or the yields of k successive batches of chemical; in general, that is, when *a priori* we expect *similarities* of one kind or another between the entities compared. But clearly, if similarities are in mind, they ought not to be denied by the form of the model. They are so denied by the improper prior which produces as Bayesian means the sample averages, which are in turn the orthodox estimates from sampling theory. The reason that k wheat varieties, k chicken breeds, or k batch yields are being jointly considered is because they are, in one sense or another, comparable. The presence of a specific form of prior distribution allows the investigator to incorporate in the model precisely the *kind* of similarities he wishes to entertain. Thus in the comparison of varieties of wheat or of breeds of chicken it might well be appropriate to consider the variety means as randomly sampled from some prior superpopulation and, as is well known, this produces the standard shrinkage estimators as Bayesian means (Lindley, 1965; Box and Tiao, 1968b; Lindley and Smith, 1972). But notice that such a model is quite *inappropriate* for the yields of k successive batches of chemical whose mean yields might much more reasonably be regarded as a sequence from some autocorrelated time series. A prior which reflects this concept leads to functions for the Bayesian means which are quite different from the orthodox shrinkage estimators (Tiao and Ali, 1971).

In summary, then, both sampling theory and Bayes's theory can rationalize the use of shrinkage estimators, and the fact that the former does so merely on the basis of reduction of mean square error with no overt use of a prior distribution, at first, seems an advantage. However, only the explicit inclusion of a prior distribution, which sensibly describes the situation we wish to entertain, can tell us what *is* the appropriate function to consider, and avoid the manifest absurdities which seem inherent in the sampling theory approach which implies, for example, that we can improve estimates by considering together *as one group* varieties of wheat, breeds of chicken, and batches of chemical.

C(i). *Is There an Iterative Interplay Between Criticism and Estimation?* A good example of the iterative interplay between criticism and estimation is seen in parametric time series model building as described, for example, by Box and Jenkins (1970). *Critical inspection* of the plotted time series and of the corresponding plotted autocorrelation function, and other functions derivable from it, together with their rough limits of error, can suggest a model specification and, in particular, a parametric model. Temporarily behaving as if we believed this specification, we may now *estimate* the parameters of the time series model by their Bayesian posterior distribution (which, for samples of the size usually employed, is sufficiently well indicated by the likelihood). The residuals from the fitted model are now similarly *critically examined*, which can lead to respecification of the model, and so on. Systematic liquidation of serial dependence brought about by such an iteration can eventually produce a parametric time series model; that is, a linear filter which approximately transforms the time series to a white noise series. Anyone who carries through this process must be aware of the very different nature of the two inferential processes of criticism and estimation which are used in alternation in each iterative cycle.

C(ii). *Why Can't All Criticism Be Done Using Bayes's Posterior Analysis?* It is sometimes argued that model checking can always be performed as follows: let $A_1, A_2, \ldots, A_k$ be alternative assumptions; then the computation of

$$p(A_i|\mathbf{y}) = \frac{p(\mathbf{y}|A_i)p(A_i)}{\sum_{j=1}^{k} p(\mathbf{y}|A_j)p(A_j)} \quad (i = 1, 2, \ldots, k) \tag{25}$$

yields the probabilities for the various sets of assumptions.

The difficulty with this approach is that by supposing all possible sets of assumptions known *a priori* it discredits the possibility of new discovery. But new discovery is, after all, the most important object of the scientific process.

At first, it might be thought that the use of Equation (25) is not misleading, since it correctly assesses the *relative* plausibility of the models considered. But in practice this would seem of little comfort. For example, suppose that only $k = 3$ models are currently regarded as possible, and that having collected some data the posterior probabilities $p(A_i|y)$ are 0.001, 0.001, 0.998 ($i = 1$, 2, 3). Although in relation to these particular alternatives $p(A_3|y)$ is overwhelmingly large this does not necessarily imply that A_3 could be safely adopted. For, suppose unknown to the investigator, a fourth possibility A_4 exists which, given the data, is a thousand times more probable than the group of assumptions previously considered. Then, if that model had been included, the probabilities would be 0.000,001, 0.000,001, 0.000,998, and 0.999,000.

Furthermore, *in ignorance of A_4* it is highly likely that a study of the components of the predictive distribution $p(y|A_3)$ and in *particular of the residuals*, could (a) have shown that A_3 was not acceptable and (b) have provided clues as to *the identity of A_4*. The objective of good science must be to conjure into existence what has *not* been contemplated previously. A Bayesian theory which excludes this possibility subverts the principal aim of scientific investigation.

More generally, the possibility that there are more than one set of assumptions that may be considered merely extends the definition of the *model* to

$$p(y, \theta, A_j) = p(y|\theta A_j)p(\theta|A_j)p(A_j) \quad (j = 1, 2, \ldots, k)$$

which in turn will yield a predictive distribution. In a situation when this more general model is inadequate a mechanical use of Bayes's theorem could produce a misleading analysis, while suitable inspection of predictive checks could have demonstrated, on a sampling theory argument, that the global model was almost certainly wrong and could have indicated possible remedies.*

C(iii). *An Abrogation of the Likelihood Principle?* The likelihood principle holds, of course, for the estimation aspect of inference in which the model is temporarily assumed true. However, it is inapplicable to the criticism process in which the model is regarded as in doubt.

If the assumptions A are supposed true, the likelihood function contains all the information about θ coming from the particular observed data vector y_d. When combined with the prior distribution for θ it therefore tells all we can know about θ given y_d and A. In such a case the predictive density $p(y_d|A)$ can tell us nothing we have not already assumed to be true, and will fall within a given interval with precisely the frequency forecast by the predictive distribution. When the assumptions are regarded as possibly false, however, information about model inadequacy can be supplied by considering the density $p(y_d|A)$ in relation to $p(y|A)$. Thus for the

---

* I am grateful to Dr. Michael Titterington for pointing out that in discriminant analysis the atypicality indices of Aitchison and Aitken (1976) use similar ideas.

Normal linear model, the distribution of residuals contains no information if the model is true, but provides the reference against which standard residual checks, graphical and otherwise, are made on the supposition that it may be importantly false.

In the criticism phase we are considering whether, given A, the sample y_d is likely to have occurred at all. To do this we *must* consider it in relation to the *other* samples that could have occurred but did not.

For instance, in the Bernoulli trial example, had we sampled until we had r successes rather than until we had n trials, then the likelihood, and, for a fixed prior, the posterior distribution would have been unaffected, but the predictive check would (appropriately) have been somewhat different because the appropriate reference set supplied by $p(y|A)$ would be different.

C(iv). *How Do You Choose the Significance Level?* It has been argued that if significance tests are to be employed to check the model, then it is necessary to state in advance the level of significance α which is to be used and that no rational basis exists for making such a choice.

While I believe the ultimate justification of model checking is the reference of the checking function to its appropriate predictive distribution, the examples I have given to illustrate the predictive check may have given a misleading idea of the formality with which this should be done. In practice the predictive check is not intended as a formal test in the Neyman–Pearson sense but rather as a rough assessment of signal to noise ratio. It is needed to see which indications might be worth pursuing. In practice model checks are frequently graphical, appealing as they should to the pattern recognition capability of the right brain. Examples are to be found in the Normal probability plots for factorial effects and residuals advocated by Daniel (1959), Atkinson (1973), and Cook (1977). Because spurious patterns may often be seen in noisy data some rough reference of the pattern to its noise level is needed.

D. As might be expected the mistaken search for a single principle of inference has resulted in two kinds of incongruity:

(1) attempts to base estimation on sampling theory, using point estimates and confidence intervals; and

(2) attempts to base criticism and hypothesis testing entirely on Bayesian theory.

The present proposals exclude both these possibilities.

Concerning estimation, we will not here recapitulate the usual objections to confidence intervals and point estimates but will consider the latter in relation to shrinkage estimators, ridge estimators, and robust estimators. From the traditional sampling theory point of view these estimators have been justified on the ground that they have smaller mean square error than traditional estimators. But from a Bayesian viewpoint, they come about as a direct result of employing a credible rather than an incredible model. The Bayes approach provides some assurance against incredibility since it requires that all assumptions of the model be clearly visible and available for criticism.

For illustration, shown below are the assumptions that would be needed for a Bayesian justification of standard linear least squares. We must postulate not only the model

$$y_u = \boldsymbol{\theta}'\mathbf{x}_u + e_u, \quad u = 1, 2, \ldots, n \tag{26}$$

with the e_u *independently and normally* distributed with *constant variance* σ^2, but also postulate an *improper prior* for $\boldsymbol{\theta}$ and σ^2.

(a) Consider first the choice of prior. As was pointed out by Anscombe (1963), if we use a measure such as $\boldsymbol{\theta}'\boldsymbol{\theta}$ to gauge the size of the parameters, a locally flat prior for $\boldsymbol{\theta}$ implies that the larger is the size measure $\boldsymbol{\theta}'\boldsymbol{\theta}$ the more probable it becomes. The model is thus incredible. From a Bayesian view-point shrinkage and ridge estimators imply more credible choices of the model, which, even though approximate, are not incredible.

(b) For data collected serially (in particular, for much economic data) the assumption of error independence in Equation (26) is equally incredible and again its violation can lead to erroneous conclusions. See, for example, Coen, Gomme, and Kendall (1969) and Box and Newbold (1971).

(c) The assumption that the specification in Equation (26) is necessarily appropriate for *every* subscript $u = 1, 2, \ldots, n$ is surely incredible. For it implies that the experimenter's answer to the question "Could there be a small probability (such as 0.001) that any one of the experimental runs was unwittingly misconducted?" is "No; that probability is *exactly* zero."

So far as the last assumption is concerned a more credible model considered by Jeffreys (1932), Dixon (1953), Tukey (1960), and Box and Tiao (1968a) supposes that the error e is distributed as a mixture of Normal distributions

$$p(e|\boldsymbol{\theta}, \sigma) = (1 - \alpha)f(e|\boldsymbol{\theta}, \sigma) + \alpha f(e|\boldsymbol{\theta}, k\sigma) \tag{27}$$

To illustrate the operation of this model (see Bailey and Box, 1980), partial analyses are given in Table 1 for two quite separate sets of data obtained at different times by investigators who were considering different problems but were using the same "Box–Behnken" design to estimate the 15 coefficients in the fitted model

$$y = \beta_0 + \sum_{i=1}^{4}\beta_i x_i + \sum_{i=1}^{4}\sum_{j>i}^{4}\beta_{ij}x_i x_j + \sum_{i=1}^{4}\beta_{ii}x_i^2 + e \tag{28}$$

Table 1 shows some of their Bayes estimates (marginal means and standard deviations of the posterior distribution). For simplicity, only a few of the coefficients are shown; the behavior of the others is similar. Table 1a uses data from a paper by Box and Behnken (1960). These data (see Figure 3) apparently contain a single bad value (y_{10}), with a small possibility of a second bad value (y_{13}). Table 1b shows the same

Table 1. Bayesian Means with Standard Deviations (in parentheses) for Selected Coefficients Using Various Values of (ε, α) in the Contaminated Model (with k = 5)

(a) Box–Behnken Data: One or Two Suspect Values

Estimates	Least Squares	Robust				
ε	Zero	.001	.005	.010	.015	.020
α	Zero	.005	.024	.048	.070	.091
β_4	-3.7 (.5)	-3.2 (.3)	-3.1 (.2)	-3.1 (.2)	-3.1 (.2)	-3.1 (.2)
β_{44}	-2.6 (.7)	-3.0 (.4)	-3.1 (.4)	-3.1 (.4)	-3.1 (.4)	-3.1 (.4)
β_{13}	-3.8 (.8)	-3.8 (.4)	-3.8 (.4)	-3.8 (.4)	-3.8 (.3)	-3.8 (.3)
β_{14}	1.0 (.8)	-.5 (.9)	-.5 (.9)	-.5 (.9)	-.5 (.9)	-.5 (.9)

(b) Bacon Data: No Suspect Values

Estimates	Least Squares	Robust				
ε	Zero	.001	.005	.010	.015	.020
α	Zero	.005	.024	.048	.070	.091
β_4	4.7 (.3)	4.7 (.3)	4.7 (.3)	4.7 (.3)	4.7 (.3)	4.7 (.3)
β_{44}	.9 (.5)	.9 (.5)	.9 (.5)	.9 (.5)	.9 (.5)	.9 (.5)
β_{13}	.8 (.6)	.8 (.6)	.8 (.5)	.8 (.5)	.8 (.5)	.8 (.5)
β_{14}	-.4 (.6)	-.4 (.6)	-.4 (.6)	-.4 (.6)	-.4 (.7)	-.4 (.7)

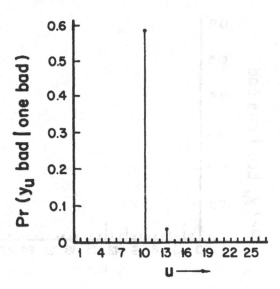

Figure 3. Posterior probability that y_u is bad given that one observation is bad (Box–Behnken data).

analysis for a second set of data arising from the same design and published by Bacon (1970), which (see Figure 4) appears to contain no bad values. It was shown by Chen and Box (1979) that for $k \geq 5$ the posterior distribution of β is mainly a function of the single parameter $\varepsilon = \alpha/(1 - \alpha)k$ and the results obtained from $k = 5$ are labelled in terms of ε as well as α. The analysis is based on locally noninformative priors on β and on $\log \sigma$ so that the estimates in the first columns of the tables ($\varepsilon = \alpha = 0$) are ordinary least squares estimates.

The important point to notice is that for the first set of data which appears to contain one or two bad values, a major change away from the least squares estimates can occur as soon as there is even a slight hint ($\varepsilon = 0.001$, $\alpha = 0.005$) of the possibility of contamination. The estimates then remain remarkably stable for widely different values* of ε. But for the second set (Bacon's data), which appears to contain no bad values, scarcely any change occurs at all as ε is changed.

It has been objected that while the Normal model is inadequate, the contaminated model (27) may be equally so, and that "therefore" we are better off using ad hoc robust procedures of downweighting such as have been recommended by Tukey and others and justified on the basis of their sampling properties. This argument loses force, however, since it can be shown by elementary examples (Chen and Box, 1979; Box, 1979, 1980; Chen and Box, 1989) that the effect of the Bayes analysis is also to produce downweighting of the observations with downweighting functions very similar to those proposed by the empiricists. However, the Bayes analysis has the advantage of being based on a visible model which is itself open to criticism and has greater adaptivity, doing nothing to samples that look normal, and reserving

---

* They are, however (see reply to the discussion of Box, 1980), considerably different from estimates obtained by omitting the suspect observation and using ordinary least squares.

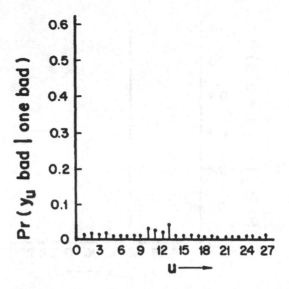

Figure 4. Posterior probability that y_u is bad given that one observation is bad (Bacon data).

robustification for samples that do not. A further advantage of the present point of view is that when an outlier occurs, while the posterior distribution will discount it, the predictive distribution will emphasize it, so that the fact that a discrepancy has occurred is not lost sight of.

CONCLUSION

In summary I believe that the scientific method employs and requires not one but two kinds of inference—criticism and estimation; once this is understood the statistical advances made in recent years in Bayesian methods, data analysis, and robust and shrinkage estimators can be seen as a cohesive whole.

PART D

Control

CHAPTER D.0

Introduction

As a student, my main interest was experimental design and I found the courses I took on time series analysis extremely boring. Somewhat unexpectedly, however, it was experimental design that eventually led me to a very different view.

It started when I was consulted about a problem arising in the operation of a catalytic converter. Somewhat simplified, the problem was as follows: On any given day the process curve representing the yield y plotted as a function of the temperature T was approximately quadratic. Over a period of weeks, however, the catalyst gradually decayed and the process curve, and hence the optimal operating temperature, slowly drifted in an unpredictable way. Suppose that at some particular time the reactor was being run at temperature T_0. My solution (which it later turned out was not original with me) was to perturb T sinusoidally about T_0. The recorded process yield was then continuously multiplied by an independently generated sine wave of the same phase and frequency, and then integrated over time. This integral would steadily increase if T_0 was below the optimum temperature and decrease if it was above, so this integrated signal could be fed back to appropriately adjust the temperature set point and follow the moving maximum.

My friend Gwilym Jenkins, whom I had met at Princeton, agreed that we might get interesting results if we could get such a system of adaptive optimization built and study its behavior. He pointed out, moreover, that to design such a reactor we would need to determine the structure of the noise and also to take account of dynamics (which could change the phase of the injected signal).

After I moved to Wisconsin, I met the wise and famous chemical engineer Olaf Hougen. He was enthusiastic about our idea; so we set up a collaborative project, supported by NSF, between the Statistics and the Chemical Engineering Departments to build and study such a reactor. Gwilym came to visit, and with the help of graduate students from both departments, the reactor was built and was persuaded to work. (The gas furnace series that later appeared in the Box and Jenkins book was data taken during some early trials and you can see a photograph of the apparatus we built in the publication Altpeter, Box and Kotnour (1966).)

Usually observations would be taken at discrete time intervals. So Gwilym and I studied the representation of dynamics by linear difference equations and of time series by stochastic difference equations. Also, we used these models to solve problems of forecasting, which are closely related to those of control. So began my reeducation in what I now found a fascinating field. In the paper that Gwilym and I presented to the Royal Statistical Society in 1962, we discussed feedback adjustment schemes for statistical process control (SPC) as well as adaptive optimization. We also provided manual feedback adjustment charts that were later simplified. One of the discussants of our paper, the engineer J. H. Westcott, said that he "welcomed this flirtation between control engineering and statistics but doubted whether they could yet be said to be going steady." I'm afraid that what happened was that for several decades they drifted apart. For this, we ourselves were partly to blame; in the "Box–Jenkins" book our discussion of control was much too general and too abstract to be of much practical use. I think also that this work came too early to attract much interest from statisticians or from the quality control community.

One thing that we correctly emphasized was the importance of *nonstationary* models. Except as a local approximation, the concept of a process in a perfect state of control, where, apart from "special (assignable) causes", *successive* measurements vary in a stationary manner about a fixed mean (the I.I.D. model), is contrary to the second law of thermodynamics. This law, to quote Sir Arthur Eddington, "holds ... the supreme position among the laws of nature, ... if your theory is against the second law of thermodynamics I can give you no hope." Thus, on theoretical grounds, drift of the process mean is to be expected.

The material in Chapter D.1 is given first place because it underlines the acknowledgment based on *experience* that, even when best efforts are made using standard quality control methods, substantial allowance must be made for process *drift*. This acknowledgment comes from the authors of the highly successful six sigma initiative—managers rather than traditional quality practitioners—who are massively engaged in quality and process improvement. It also has first place because it illustrates that available methods for controlling drift by feedback adjustment are intuitively obvious, powerful, robust, and *simple to apply*.

These methods are equivalent to continually updating an *exponentially weighted moving average* (EWMA), which is used to estimate (and then to cancel) the current deviation of the process mean from target. The EWMA fills the need for a current local average that takes less and less account of the more and more remote past. It was an empirical argument based on this concept that led Holt and others in the late 1950s to use it for forecasting and for inventory control. Chapters D.2, D.3, and D.4 provide simple explanations of EWMAs and their use in control.

In 1960, Muth showed that the EWMA provides a best estimate of the location of a drifting series represented by what was later called an *integrated moving average* (IMA) time series model.

The IMA time series model for a current observation z_t is such that

$$z_t = (\text{EWMA of } z_{t-1}, z_{t-2}, \ldots) + (\text{independent error } a_t)$$

There are a number of other equivalent representations.

Formal time series model building applied to a large number of real time series has surprisingly often led to the IMA model, at least as an adequate approximation. Thus the exponentially weighted average and the IMA time series model seem to have the same status in the analysis of nonstationary series as do the arithmetic average and the I.I.D. model in the analysis of randomized experiments. Obviously these models are not exact or universally useful, but they should clearly occupy central roles in statistical thinking. Other theoretical reasons supporting the use of the EWMA and the IMA time series model for control applications are given in Chapter D.5.

A number of years ago I was urgently summoned by a food company for which I consulted. My assignment was to try to bring peace in a war that had broken out between their SPC community and the control engineers. After I returned to Madison, bloody but still surviving, I discovered that Tim Kramer had had a similar task thrust upon him in his work for a major U.S. automobile manufacturer. Our subsequent cooperation resulted in the paper which appears here in Chapter D.5. In it we attempted to clarify the different purposes of each form of control and the importance of using them together. We also thought that the time might (at last) be ripe to reintroduce the notion that simple feedback adjustment methods could be of use in SPC.

Soon after this I met Alberto Luceño. Our papers which form Chapters D.6, D.7, and D.8, cleared the ground for our book, *Statistical Control by Monitoring and Feedback Adjustment,* which appeared in 1997. In an attempt to reach a wide audience who could put these methods to use, we took an approach that was as simple as possible. The dual strategy for control discussed in this book is to first get the bugs out using process monitoring and then remove the residual drift by feedback adjustment.

In the absence of special causes the standard I.I.D. process monitoring model assumes homogeneous random variation about a fixed mean. This cannot be justified as an exact time series model. Nevertheless it can be a useful *approximation* to represent short-term behavior. In 1959 George Barnard introduced the cumulative sum chart as a monitoring device, which was a useful adjunct to the Shewhart chart. Chapter D.9 is an elementary introduction to "Cusums" and also to their use for the "post mortem" examination of data. Properties of monitoring charts are frequently expressed in terms of average run lengths (ARLs). It is well known, however, that the ARL for false alarms can be a very deceptive measure. In particular (on the N.I.I.D. assumption), the false alarm run length distribution would be approximately exponential, implying that many of the false alarms would occur at times inconveniently early. How, for Cusums, this situation can be improved by increasing the sampling frequency is discussed in a paper with Alberto Luceño appearing in Chapter D.10.

The general problem of process monitoring is that of seeking a particular kind of signal in a particular kind of noise. In the paper in Chapter D.11 with José Ramirez, we show how this can be done using cumulative score ("Cuscore") statistics. These are based on Fisher's concept of efficient score statistics. In particular, we show under what assumptions the Shewhart, Cusum, and EWMA monitoring charts are special cases of Cuscores.

CHAPTER D.1

Six Sigma, Process Drift, Capability Indices, and Feedback Adjustment*

SUMMARY

The Six Sigma specification makes an allowance of 1.5 standard deviations for process drift on either side of the target value. Simple ways in which a major part of such drift can be removed are given. These employ feedback adjustment methods specifically designed for SPC applications.

SIX SIGMA

The Six Sigma initiative (see, e.g., Smith, 1992; Harry, 1994; Hoerl, 1998) is seen by many as the blueprint for process and quality improvement for today and for the future.

That all humankind and hence all employees at every level have the natural ability to be creative has long been understood. But it has taken some time for the enormous potential of this fact to be fully appreciated. At last, in some of the most important companies, management has come to understand that their chief responsibility is to foster such efforts. They have done this, in part, by:

(a) making it absolutely clear that quality improvement is part of each person's everyday job,

(b) providing appropriate training at all levels in the organization,

(c) making quality improvement a competitive sport using "Managerial Champions," "Black Belts," and so on.

From Box, G. E. P. and Luceño, A. (2000), *Quality Engineering*, 12(3) 297–302.
* This article was contributed to the 1999 ASA Quality and Productivity Research Conference in Schenectady, New York.

It is not surprising to learn that these principles rigorously applied in such companies as Motorola, G.E., Texas Instruments, Polaroid, and Allied Signal have produced impressive results—for example, an annual report of Allied Signal attributes a savings of about 1.5 billion dollars to their Six Sigma initiative.

PROCESS DRIFT

This article is about how such efforts might be further improved. The Six Sigma initiative uses the many tools for quality improvement to ensure that all but a tiny fraction of manufactured articles as they are shipped to the customer are within specification limits. To achieve this, it is required that the process is so well controlled that each of these limits is six standard deviations away from the target. This corresponds to the use of a capability index of 2 or equivalently a spec/sigma ratio of 12 (Box and Luceño, 1997).* The stated goal is to produce no more than 3.4 defects per million. On the assumption that the process distribution is Normal and has a *fixed mean value that is on target*, this would require a spec/sigma ratio much less than 12. However, the authors of the Six Sigma concept wisely assume that the process does *not* have a fixed mean value but undergoes drift—specifically that the local mean may drift on either side of the target by about 1.5 standard deviations. Conventional wisdom would say that such process drift must occur from special causes and that these ought to be tracked down and eliminated. But in practice such drift, although detectable, may be impossible to assign or to eliminate economically. It is refreshing to see, at last, this acknowledgment that, *even when best efforts are made* using standard quality control methods, the process mean can drift substantially. In the evolution of ideas about quality this new provision comes much closer to reality, but it is a major *departure* from what, for the last fifty years, many have believed was a fundamental concept of statistical quality control.

The ideas of Shewhart and Deming concern the continuous debugging of the system by process monitoring using control charts or other devices. These techniques are usually justified by the approximation that in the absence of "special causes" the process will be a state of control, varying stably about a fixed mean. However, the second law of thermodynamics ensures that no process could ever be in a perfect state of control about a fixed mean. Furthermore, the long-term instability of processes, after standard techniques of quality control have been applied, has been verified by extensive studies of process data (see, e.g., Alwan and Roberts, 1988). These considerations do not, of course, imply that standard quality control chart methods are valueless. On the contrary, their application has resulted in the elimination of thousands of quality problems and have produced improvement in a host of industrial processes. Rather, this is a further demonstration that models must be treated as approximations: all models are wrong, but some models are useful. Notice that this implies that no "optimal" scheme is ever *in practice* optimal. What we must aim for are schemes which are *robust* and *good* over a wide

* Later referenced as B., L.

range of circumstances. That is to say, they will operate reasonably well in circumstances which depart from the ideal. Many standard monitoring devices possess these characteristics because for the detection of assignable causes we need only to approximate fairly *short run* behavior. However, *the proportion of defective articles* depends on capability indices reflecting *long-run* process behavior for which the short-term fixed mean approximation is inadequate. When steering a boat we should certainly not allow for every wave, but, if we want to stay off the rocks, it will be necessary to compensate for the current.

Thus while continuing to use standard methods of quality monitoring for the assignment and elimination of special causes, we will frequently need to use feedback adjustment *as well*.

FEEDBACK ADJUSTMENT IN THE SPC CONTEXT

Using feedback adjustment it ought to be possible to remove a considerable part of the systematic drift which is allowed for in the Six Sigma specification. This could make possible tighter specification limits and production of an even better product. The importance of developing feedback control methods suitable for use in the SPC environment has been recognized for some considerable time (see, e.g., Box and Jenkins, 1968; MacGregor, 1987; Box and Kramer, 1992; Box, Jenkins, and Reinsel, 1994; Box and Luceño, 1997; Box, 1998).

TWO EXAMPLES

In the context of the Six Sigma model, we illustrate two simple methods for feedback adjustment with an example.

Repeated Adjustment

Figure 1(a) shows a constructed series representing some quality characteristic y for a process that drifts. This series has an overall sample standard deviation $\hat{\sigma}_y$ of 1.25. As explained in more detail in the Appendix, it has been obtained by adding a drift component d with $\sigma_d = 0.76$ to a random noise component a with $\sigma_a = 1.0$. The 6σ limits, shown by bold lines in the figures, are based on $\sigma_a = 1$. (This will be a slight underestimate of the *short-term* standard deviation of y obtained, for example, by using the rational subgroup method.) The dotted lines indicate 3σ limits. It should be noted that the standard deviation $\sigma_d = 0.76$ of the drift component we have used, is considerably *less* than $\sigma_d = 1.50$ allowed for in the 6σ specification.

A robust and good feedback adjustment procedure is to compensate repeatedly for a proportion G of each deviation from target. A value for G in the neighborhood of 0.2–0.4 is frequently effective and $G = 0.2$ is used in this illustration. Figure 1(b) shows the adjustments made on this basis and Figure 1(c) the process output after applying such adjustment. The effect of this adjustment technique is to reduce the standard deviation from 1.25 to 1.02. As shown in Figure 1(c) the sample standard deviation of the adjusted process is not much greater than that of the random component used in constructing the series. It can be shown (B.L.) that repeatedly

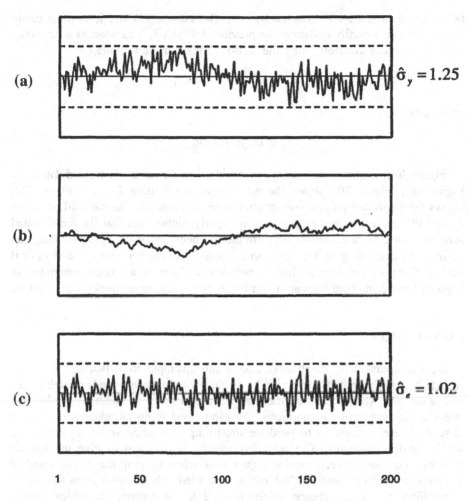

Figure 1. Performance of continuous adjustment scheme: (a) a process subject to drift, (b) adjustments by continuous control scheme, and (c) controlled process output.

applying this simple procedure is equivalent to continually applying an *overall* correction to the *uncontrolled* process which is an exponentially weighted moving average (EWMA) of past data with smoothing constant $1 - G$ (0.8 in this example). Figure 1(b) shows that this adjustment "filter" has produced a very good estimate of the drift which is automatically removed by the control scheme.

Periodic Feedback Adjustment Using a Bounded Chart

It is sometimes inconvenient to make repeated adjustments in the manner described above. In this case a *bounded adjustment* scheme may be used. It turns out that a scheme of this kind which is robust and good may be obtained as follows: action is taken only when, at time t, an EWMA $\tilde{y}_t$ of present and past deviations from target falls outside tabled limits denoted by $\pm L$. The current EWMA $\tilde{y}_t$ is most easily obtained by continually updating the previous EWMA $\tilde{y}_{t-1}$ as soon as a new data value y_t becomes available using the well-known recursive formula:

$$\tilde{y}_t = Gy_t + (1 - G)\tilde{y}_{t-1}$$

or with $G = 0.2$

$$\tilde{y}_t = 0.2y_t + 0.8\tilde{y}_{t-1}$$

Figure 2(c) illustrates this mode of control using the same constructed data as in Figure 1(a). Figure 2(b) shows the adjustments made using $L = \pm 1$. Figure 2(c) shows the controlled process output after these adjustments. The standard deviation $\hat{\sigma}_e = 1.08$ for the process output is now slightly higher than that for the repeated scheme. These calculations are very simple and need not be very precise. They can be made by eyeballing an interpolation 0.2 of the distance between $\tilde{y}_{t-1}$ and y_t, or if desired from a more formal chart, a pocket calculator, or a process computer. In Figures 1 and 2 the bold lines are the $\pm 6\sigma$ limits and the dotted lines the $\pm 3\sigma$ limits.

CONCLUSION

You can of course delve much more deeply into this topic. In particular, it is shown in B.L. how such methods are derived, why the procedures described are both robust and good, and when somewhat more sophisticated methods might be needed. On specific but reasonable assumptions, the continuous feedback adjustment scheme described here is designed to produce small output variation while requiring only small sized adjustments. The bounded scheme is designed to give the longest possible intervals between needed adjustments while keeping the output standard deviation as small as possible. Tables* are provided in B.L. which show alternative possibilities for various choices of the limits $\pm L$. For instance, the tables indicate that for the value $L = \pm 1.0\sigma_a$ used in the example the average adjustment interval (AAI)—the long-run average interval between needed adjustments—would be about 31 with an output standard deviation of about $1.09\sigma_a$. This is in reasonable agreement with what we see in Figures 2(b) and 2(c). If we had used $L = 1.5$ (i.e., if we had set limits at $\pm 1.5\sigma_a$) the tables indicate that the AAI would have increased to about 66 with an increased output standard deviation of about $1.19\sigma_a$. The theory underlying the schemes of Figures 1 and 2 is the same. Indeed, the repeated adjustment procedure is obtained if L is set equal to zero in the bounded adjustment scheme.

Of course not all feedback adjustment problems can be dealt with in this simple manner. But, schemes of this kind do seem to be widely applicable and they provide an excellent base from which to practice and study process adjustment for SPC.

---

*The model on which these tables are based is slightly different from that used in the Appendix for generating the data. However, robustness properties ensure that the tables provide a good approximation.

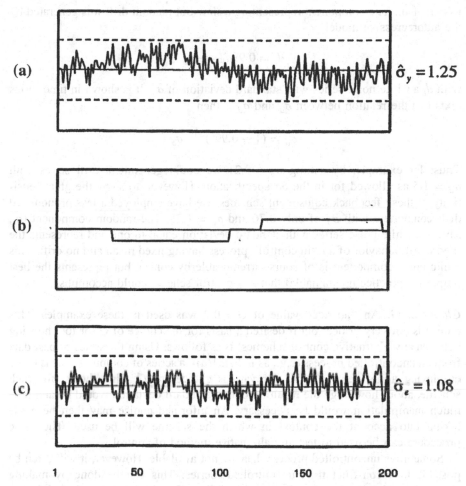

Figure 2. Performance of bounded adjustment scheme: (a) process subject to drift [same as Fig. 1(a)], (b) adjustments applied by bounded control scheme, and (c) controlled process output.

APPENDIX: SOME DETAILS OF THE CALCULATIONS

White Noise
In this chapter the term "white noise" is used to mean a series of independently distributed random variables which are (roughly) normally distributed with fixed standard deviation and mean zero.

Generation of the Constructed "Process" Series in Figures 1(a) and 2(a).
Values y_t for the series in Figures 1(a) and 2(a) simulating a drifting process were generated by adding a drift component d_t to a random component a_t so that

$y_t = d_t + a_t$. The component representing a slow but random drift was generated by the autoregressive model

$$d_t = 0.99d_{t-1} + \alpha_t$$

with d_t a white noise series with standard deviation of σ_α. It is shown in time series texts that the relation between σ_α and σ_d is then

$$\sigma_d = (1 - 0.99^2)^{-1/2}\sigma_\alpha$$

Thus, for example, by setting $\sigma_\alpha = 0.21$ we could generate a drift series with $\sigma_d = 1.5$ as allowed for in the 6σ specification. However, to show the great sensitivity of these feedback adjustment schemes, we have employed a less pronounced drift component with σ_d of only 0.76 and $\sigma_y = 1.25$. The random component (a_t, say) is a white noise series with standard deviation equal to one and represents the theoretical behavior of an "in control" process having fixed mean and no drift. This white noise component is of course irremovable by control but represents the best output (in practice unattainable) that any control scheme could accomplish.

Choice of G. An "ad hoc" value of $G = 0.2$ was used in these examples. This choice is not very critical but if desired a basis for the choice of G (or for choosing between any alternative control schemes) is as follows. Using the series of past data from an uncontrolled process (e.g., as in Fig. 1(a)), a series of computer runs is made employing adjustment schemes for various choices of G. This shows, for each scheme, about how large the resulting output standard deviation would be and how much manipulation would be necessary. An informed choice may then be made taking into account the context in which the scheme will be used. The same procedure can be used to test any alternative method of control.

Sometimes uncontrolled process data are not available. However, it will often be possible to reconstruct the "uncontrolled" series. This can be done by making appropriate allowance for the changes that are known to have been made and thus to reconstruct the "uncontrolled" series.

ACKNOWLEDGMENT

This research was partially supported by the National Science Foundation under Grant DMI-9812839 and from the Low Emissions Technologies Research and Development Partnership (LEP) of Daimler Chrysler, Ford, and General Motors. Alberto Luceño also acknowledges the support of the Spanish DGESIC under Grant PB97-0555.

CHAPTER D.2

Understanding Exponential Smoothing: A Simple Way to Forecast Sales and Inventory

Sam Sales is the sales manager and Pete Production is the production manager for Grommets Incorporated. One day Sam bursts into Pete's office and says "Guess what Pete; we sold 50,000 more boxes of grommets last month than we forecast! So let's put up next month's forecast by 50,000." Pete is somewhat cautious, however. He knows that, on the one hand, if he produces too much he will be burdened with a big inventory, but on the other if he produces too little he may not be able to meet higher sales demand should it materialize. So Pete says, "Sam we both know from experience that some of the big sales fluctuations that we see both up and down are greatly influenced by chance occurrences, so I have a rule whereby to update the forecast I always discount 60% of the difference between what we forecast last month and what we actually get. So in this case let's increase the forecast not by 50,000 boxes as you suggest, but by 20,000."

At first sight Pete's forecasting rule seems too simple-minded to be of much interest, but it's actually more sophisticated than it looks. Suppose we've reached a certain month (say, August) which we will call the tth month and that one month earlier, in July, we forecast that the sales for August would be $\hat{y}_t$ (we use a "hat" on the y to indicate an estimate). But what we found when August's figures actually became available was a value y_t, which differed somewhat from the forecast made one month earlier. Then the observed discrepancy (actual minus forecast) for August would be $y_t - \hat{y}_t$ and Pete's policy of discounting 60% of the good (or bad) news says that to update the forecast and get the estimated September sales $\hat{y}_{t+1}$ you should only take account of 40% of this discrepancy. Thus the change you make in

From Box, G. E. P. (1991), *Quality Engineering*, 3(4), 561–566. Copyright © 1991 by Marcel Dekker, Inc.

the forecast, $\hat{y}_{t+1} - \hat{y}_t$, is 0.4 of the observed discrepancy $y_t - \hat{y}_t$. Thus, Pete's forecasting formula is

$$\hat{y}_{t+1} = \hat{y}_t + 0.4(y_t - \hat{y}_t) \tag{1}$$

which after rearrangement becomes simply

$$\hat{y}_{t+1} = 0.4y_t + 0.6\hat{y}_t \tag{2}$$

Now remember that since this is Pete's consistent policy, he got last month's forecast the same way from

$$\hat{y}_t = 0.4y_{t-1} + 0.6\hat{y}_{t-1}$$

So another way of writing the forecast $\hat{y}_{t+1}$ in Eq. (2) is

$$\hat{y}_{t+1} = 0.4y_t + 0.6\{0.4y_{t-1} + 0.6\hat{y}_{t-1}\}$$

But then two months ago he got $\hat{y}_{t-1}$ from

$$\hat{y}_{t-1} = 0.4y_{t-2} + 0.6\hat{y}_{t-2}$$

So Eq. (2) could also be written as

$$\hat{y}_{t+1} = 0.4y_t + 0.6\{0.4y_{t-1} + 0.6(0.4y_{t-2} + 0.6\hat{y}_{t-2})\}$$

If you carry on this way and multiply out all the brackets, you can see that Pete's rule for getting the forecast $\hat{y}_{t+1}$ is equivalent to calculating from the previous data the quantity

$$\hat{y}_{t+1} = 0.4(y_t + 0.6y_{t-1} + 0.6^2 y_{t-2} + 0.6^3 y_{t-3} + \cdots) \tag{3}$$

That is,

$$\hat{y}_{t+1} = (0.400y_t + 0.240y_{t-1} + 0.144y_{t-2} + 0.086y_{t-3} \\ + 0.052y_{t-4} + 0.031y_{t-5} + \cdots) \tag{4}$$

This is called an *exponentially weighted moving average* (EWMA). That's rather a mouthful so let's take it bit by bit.

In general, a *weighted* average of past sales would look like this:

$$w_1 y_t + w_2 y_{t-1} + w_3 y_{t-2} + \cdots$$

where w_1, w_2, w_3, etc. are constants that determine how much weight or importance is given to previous monthly sales figures y_t, y_{t-1}, y_{t-2}, etc. For example, suppose we just took an ordinary average $\bar{y}$ of the last ten observations by adding up the last ten y's and dividing by ten. This is equivalent to calculating

$$\bar{y} = 0.1y_t + 0.1y_{t-1} + \cdots + 0.1y_{t-9} \tag{5}$$

In general this familiar kind of average of n values has all the weights equal to $1/n$. For our example with $n = 10$, $w_1 = 0.1$, $w_2 = 0.1$, ..., $w_{10} = 0.1$. This weighting is very different from that for the EWMA of Eq. (4) in which $w_1 = 0.400$, $w_2 = 0.240$, $w_3 = 0.144, \ldots$, and so on.

Why do we call the EWMA an *average*? It's because it has the property that if all the observations are increased by some fixed amount then the EWMA is also increased by that same amount. For this to happen only requires that the weights add up to one (unity). Obviously, this is true for the weights of the familiar equally weighted average $\bar{y}$ in Eq. (5) but it is equally true of the EWMA, and if you check you will find that the sum of the weights in the EWMA of Eq. (4) finally add up to one.

What do we mean by *exponentially* weighted? Well, it means simply that the weights fall off in a geometric progression—they are progressively discounted by the same factor. This factor is 0.6 in our example. As a means of forecasting the next value of a series, an EWMA clearly makes a lot of sense because it puts most weight on the last sales figure, somewhat less on the previous one, and so on. The discount factor (0.6 in our example) is often called the *smoothing constant* and is denoted by the greek letter θ (theta).

So in general Pete's forecast updating formula (2) can be written

$$\hat{y}_{t+1} = (1 - \theta)y_t + \theta\hat{y}_t \tag{6}$$

where θ is between 0 and 1. This is the most convenient equation to use to successively update the forecast. After the process gets going it is *equivalent* to using the EWMA formula (4) but it is much easier to calculate.

You can see by studying Figure 1 that the choice of θ is a question of finding the right compromise between, on the one hand, achieving fast reaction to change, and on the other, averaging out the noise. If we choose θ to be large (say, 0.8) then the averaging process reaches back a long way into the series so that the smoothing out of the noise (local variation) will be considerable. But such a forecast will be slow to react to a permanent change in level. If we got a sustained boom in sales or a sustained slump, the forecast would not react very quickly. On the other hand, if we take a smaller value for θ like 0.4, the reaction will be quicker, but if the series is noisy, the forecast will be very volatile; reacting to apparent changes that may be just temporary fluctuations.

The best compromise can be found by trial and error calculation on a reasonably good run of past data for this particular product—a hundred or more if you can get it. Just to illustrate the calculation, let's use the data plotted in Figure 1 and set out in Table 1. For simplicity, I will show a trial calculation for $\theta = 0.5$.

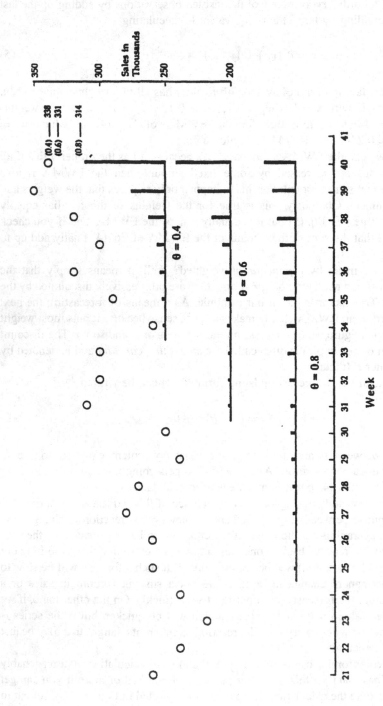

Figure 1. Twenty successive monthly sales (open circles) of grommets showing EWMA forecasts for month 41 with discount factors $\theta = 0.4$, $\theta = 0.6$, and $\theta = 0.8$.

Table 1. Sales of Grommets for Month 21 Through Month 40 in Units of 1000[a]

t	21	22	23	24	25	26	27	28	29	30	31	32	33	34	35	36	37	38	39	40	41
y_t	260	240	220	240	260	260	280	270	240	250	310	300	300	260	290	320	300	320	350	340	
$\hat{y}_t$	260	260	250	235	237	249	254	267	269	254	252	281	290	295	278	284	302	301	310	330	335
e_t	(0)	−20	−30	5	23	11	26	3	−29	−4	58	19	10	−35	12	36	−2	19	40	10	

$$S_{C.5} = 12{,}072$$

[a] Showing forecasts and forecast errors for $\theta = 0.5$ and the sum of squares of the errors $S_{0.5} = 12{,}072$.

In that case Eq. (6) gives

$$\hat{y}_{t+1} = 0.5y_t + 0.5\hat{y}_t$$

To start off this calculation I have put $\hat{y}_{21}$ equal to the actual value y_{21} occurring at time $t = 21$. Then,

$$\hat{y}_{22} = (0.5 \times 260) + (0.5 \times 260) = 260$$
$$\hat{y}_{23} = (0.5 \times 240) + (0.5 \times 260) = 250$$
$$\hat{y}_{24} = (0.5 \times 220) + (0.5 \times 250) = 235$$

and so on. In particular, you will notice that the forecast $\hat{y}_{41}$ for $\theta = 0.5$ is about 335 thousand and you may want to practice by checking out my forecasts for values $\theta = 0.8$, $\theta = 0.6$.

Now the error we make in the forecast is the difference between what we forecast one step ago and what we actually got

$$e_t = y_t - \hat{y}_t$$

We write these errors down in the fourth row of Table 1. Obviously what you would like to do is to choose as an estimate of θ a value which makes these errors as small as possible. A sensible way of doing this is to see which value of θ makes the sum of squares of the errors smallest. Ignoring the first value of e (which is necessarily zero) the sum of squares of the 19 values for our series when we put $\theta = 0.5$ is

$$S_{0.5} = (-20)^2 + (-30)^2 + (5)^2 + \cdots + (10)^2 = 12,072$$

Now this particular series is rather too short to give a reliable estimate of θ but for a longer series of, say, 50 observations or more we could calculate the sum of squares of the errors S_θ for, say, $\theta = 0.0, 0.1, 0.2, \ldots, 0.9, 1.0$, and pick out the value which gives the smallest value of S_θ. Or you may prefer to graph S_θ against θ, draw a smooth curve through the points, and read off the value of θ which gives the minimum. At first sight these calculations may seem a bit grim but, in fact, they can be done quite quickly even on a hand held calculator and are very easily programmed on a computer.

If you have a lot of series to worry about (as is often the case when, e.g., EWMAs are used to forecast inventories and hence to get reordering times), one way to go is to pick a few typical items, do the calculations just for these, and assume that similar items will have similar values of θ. You can usually get away with this because the sum of squares curve typically has a very high value when θ is one and then descends very rapidly as θ is decreased, becoming rather flat for values of θ less than about 0.8. This means that for θ less than 0.8 the EWMA method of forecasting is very "robust" to moderate changes in this smoothing constant. Finally if there just isn't time or resources to use the sum of squares method for finding θ, I would just pick a value "out of the air" between 0.8 and 0.6. This will work quite well for most situations; although you may not get the absolutely best forecast, usually you will not be very far from the best.

How would you know if some method such as the one I have outlined here is giving good forecasts? To find out you should look at the series of *forecast errors*. Plot them on a quality control chart and see if they look like a random series. If they don't and if it looks as if each error might to some extent be forecast from its predecessors, then your forecasting method is not producing a good forecast: for if there exists a way of forecasting forecast errors, then you can obviously obtain a better forecast than the one you've got! Forecast errors from a good forecasting method must produce a nonforecastable error series. If this doesn't happen, it may simply mean that you've chosen a badly wrong value for θ or you may need a more sophisticated *method* of forecasting. You can study the method I've described here in more detail and also look at more complex procedures in a time series book such as those by Box, Jenkins and Reinsel (1994) or Abraham and Ledolter (1983). Also, computer programs for EWMA forecasting and for more sophisticated methods are available from various vendors. Some of these are quite excellent—in particular, the program available from Scientific Computing Associates. It is surprising how far you can get with this very simple EWMA technique and I will show you in the next chapter how these ideas can also be used for manually adjusting a process when it is not in a state of control.

ACKNOWLEDGMENTS

This research was sponsored by the National Science Foundation under Grant No. DDM-8808138, and by the Vilas Trust of the University of Wisconsin–Madison.

CHAPTER D.3

Feedback Control by Manual Adjustment

If you ask a statistical quality control practitioner and a control engineer what they each mean when they speak of Process Control, you are likely to receive very different answers. On the one hand, the quality control practitioner will most likely talk about the uses of Shewhart charts and possibly some more recent innovations such as Cusum charts and EWMA charts for process *monitoring*. On the other hand, the control engineer will talk about such things as feedback control, proportional-integral controllers, and so forth for process *regulation*.

Both are highly important concepts and both have long and distinguished records of practical achievement. My purpose here is to underline the nature of the different contexts, objectives, and assumptions of their operation and to show how some of the ideas of the control engineer can be readily adapted by the statistical quality practitioner to deal with certain problems of *process adjustment* with which he is often faced.

PROCESS MONITORING AND PROCESS REGULATION: CONTEXT, OBJECTIVES, AND ASSUMPTIONS

Process monitoring, whether by the use of Shewhart, Cusum, or EWMA charts, is conducted in the context that it is feasible to bring the process to an approximate state of statistical control without continual adjustment. A state of statistical control implies stable random variation about the target value produced by a wide variety of

From Box, G. E. P. (1991), *Quality Engineering*, 4(1), 143–151. Copyright © 1991 by Marcel Dekker, Inc.

"common causes" whose identity is currently unknown. This stable random variation about the target value provides a "reference distribution" against which apparent deviations can be judged, in particular, by the provision of "three sigma" and "two sigma" limit lines. It is only when improbable deviations from this postulated state of control occurs that a search for an assignable or "special" cause will be undertaken. This makes sense in the implied context that the "in control" state applies most of the time and that we must be careful to avoid spending time and money on "fixing the system when it ain't broke." The objectives of process monitoring are thus (a) to continually confirm that the established common cause system remains in operation and (b) to look for deviations unlikely to be due to chance that can lead to the tracking down and elimination of assignable causes of trouble.

Now, while we should always make a dedicated endeavor to bring a process into a state of control by fixing causes of variation such as differences in raw materials, differences in operators, and so forth; situations occur, which cannot be dealt with in this way; in spite of our best efforts there remains a tendency for the process to wander off target. This may be due to naturally occurring phenomena such as variations in ambient temperature, humidity, and feedstock quality or they may occur from causes which currently are unknown. In such circumstances, some system of adjustment may be necessary.

Process adjustment by feedback control* is conducted in the context that there exists some additional variable that can be used to compensate for deviations in the quality characteristic and so to continually adjust the process to be as close as possible to target. In this article "as close as possible" will mean with smallest mean square error (or essentially with the mean on target and with smallest standard deviation about the mean). What I am talking about here is very much like adjusting the steering wheel of your car to keep it near the center of the traffic lane that you wish to follow. Thus while the statistical background for process monitoring parallels that of significance testing, that for process regulation parallels statistical estimation (estimating the level of the disturbance and making an adjustment to cancel it out). It is important not to confuse these two issues. In particular, if process regulation is your objective but you wait for a deviation to be statistically significant before making a change, you can end up with a much greater mean square deviation from target than is necessary (or if driving a car you will end up with an accident).

It is unfortunate that the word "automatic" is usually associated with the methods of process regulation used by the control engineer. As with process monitoring, it is the concepts that are important and not the manner in which they are implemented. Once the principles are understood, the quality practitioner can greatly benefit by adding some expertise on simple feedback control to his tool kit. This can be put to good use whether adjustments are *determined* graphically or by the computer; and whether they are *put into effect* manually or by transducers.

---

*Feedforward control and other techniques are also employed by the control engineer but are not discussed in this chapter.

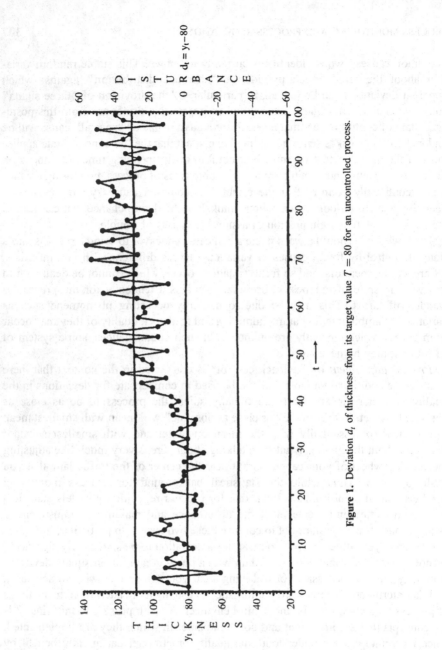

Figure 1. Deviation d_t of thickness from its target value $T = 80$ for an uncontrolled process.

A MANUAL ADJUSTMENT CHART FOR FEEDBACK CONTROL

Consider Figure 1 which shows 100 observations of the thickness of a very thin metallic deposit taken at equally spaced intervals of time. It is desired to maintain this characteristic as close as possible to the target value $T = 80$. On the right hand scale of Figure 1 is shown what I will call the "disturbance." It is equal to the deviation from the target at a particular time t when no attempt is made at adjustment, $d_t = y_t - T$. It is evident that for this process major drifts from target can occur and $\pm 3\sigma$ limits calculated from the mean range show the process to be badly out of control.

Other efforts to stabilize the process having failed, the thickness might have to be controlled by manually adjusting what we will call the deposition rate X whose level at time t will be denoted by X_t. Effective feedback adjustment could then be obtained by using the manual adjustment chart shown in Figure 2. To use it the operator records the latest value of thickness and reads off on the adjustment scale the appropriate amount by which he should now increase or decrease the deposition rate.

For example, the first recorded thickness of 80 is on target so no action is called for. The second value of 92 is 12 units above the target so $d_2 = 12$ corresponding on the left hand scale to a deposition rate adjustment $X_2 - X_1$ of -2. Thus the operator should now reduce the deposition rate by 2 units from its present level. It will be seen that the adjustment called for by the chart is equivalent to using the "adjustment equation"

$$X_t \quad X_{t-1} = \tfrac{1}{6}(y_t \quad T) \tag{1}$$

Notice that the successive recorded thickness values shown on the chart in Figure 2 are the readings which would actually occur *after adjustment*; the underlying disturbance shown in Figure 1 is, of course, not seen by the operator. The chart can be highly effective in bringing the adjusted process to a much more satisfactory state of control and would produce in this example a very large reduction in mean square error; the standard deviation $\hat{\sigma}$ of the process is now only about 11.

HOW WAS THE CHART ARRIVED AT?

Three pieces of information used in constructing the chart were:

1. A change in the deposition rate X produced all of its effect on thickness within one time interval.
2. An increase of 1 unit in the deposition rate X induced an increase of 1.2 units in thickness y. (The constant $1.2 = 6/5$ is sometimes called the *gain*, it is denoted by g and has the same function as a "regression coefficient" relating the magnitude of a change in X to that of the change in y.)
3. An uncontrolled disturbance like that in Figure 1 could be effectively forecast one step ahead using exponential smoothing (if you are unfamiliar with this concept, please look at Chapter D.2).

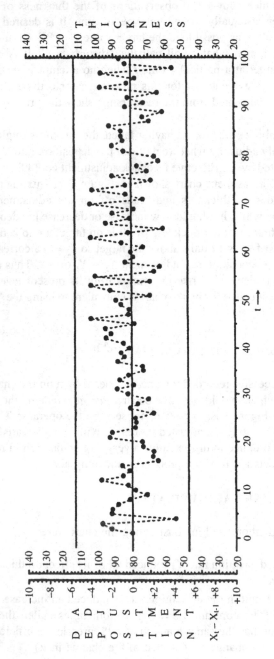

Figure 2. A manual adjustment chart for thickness which allows the operator to read off the appropriate change in deposition rate. The diagram shows the thickness whose uncontrolled state as shown in Figure 1 after control.

Notice that the action called for by the chart does not at first seem to be in accord with supposition 2 above, which might suggest that the control equation ought to be such as would cancel the whole of the deviation from target, in which case the adjustment equation would be

$$X_t - X_{t-1} = -\tfrac{5}{6}(y_t - T) \tag{2}$$

This corresponds to what Deming called "tinkering" and is never a good policy.

The manual adjustment chart of Figure 2 calls for action which is only a fifth of this [compare Eqs. (1) and (2)]. Where does the factor of $1/5 = 0.2$ come from? The short answer is that for the unadjusted disturbance, the value of the exponential smoothing constant θ, estimated from the data of Figure 1 in the manner described in my last chapter, is about 0.8. The mysterious constant is just $1 - \theta = 1 - 0.8 = 0.2$. This quantity keeps cropping up so I will give it a special symbol $(1 - \theta) = \lambda$ (greek lambda).

In Chapter D.2 about sales forecasting using exponential smoothing, I also called θ the *discount factor*. This was because when we saw an apparent increase or decrease of sales d_t above or below the previously forecast value $\hat{d}_t$ we knew that a substantial proportion of the discrepancy was almost certainly due to "noise." The discount factor θ is an estimate of that proportion. To get accurate forecasting, therefore, you must update the forecast by only a fraction $\lambda = 1 - \theta$ of the discrepancy. In symbols this says that the updating formula for exponential smoothing can be written

$$\hat{d}_{t+1} - \hat{d}_t = \lambda(d_t - \hat{d}_t) \tag{3}$$

To see precisely what is going on, we need to do a little algebra. Whatever system of adjustment we employ the model for the *adjusted* process at time $t + 1$ will be

$$y_{t+1} - T = d_{t+1} + gX_t \tag{4}$$

This equation simply says that at time $t + 1$ the deviation $y_{t+1} - T$ of thickness from its target depends on the current level of the disturbance d_{t+1} and the level X_t at which we last set the deposition rate. Obviously to achieve good control we shall want to arrange that the deviations..., d_{t-1}, d_t, d_{t+1},... are, so far as possible, continuously canceled out by the compensating actions..., gX_{t-2}, gX_{t-1}, gX_t,....

Now look again at Eq. (4). Obviously at time t we would like to have adjusted the deposition rate X_t so that the right hand side of the equation was zero and then the deviation from target at time $t + 1$ would be zero. Unfortunately we can't do this because at time t we don't know the value of d_{t+1}. However, at time t we *can* make an exponential forecast $\hat{d}_{t+1}$ of d_{t+1} and if we write $e_{t+1} = d_{t+1} - \hat{d}_{t+1}$ for the forecast error, Eq. (4) can be written

$$y_{t+1} - T = e_{t+1} + \hat{d}_{t+1} + gX_t \tag{5}$$

So what we need to do is to set X_t so that

$$gX_t = -\hat{d}_{t+1} \tag{6}$$

and since its last two terms now cancel out Eq. (5) becomes simply

$$y_{t+1} - T = e_{t+1} \tag{7}$$

This says that the deviation from target which we will see in the adjusted process of Figure 2 will be just the forecast error. Now Eq. (6) says that, at time t, the *adjustment* we need to make in the deposition rate from its previous value at time $t - 1$ is

$$X_t - X_{t-1} = -\frac{1}{g}(\hat{d}_{t+1} - \hat{d}_t) \tag{8}$$

But from Eq. (3) you will see that if the $\hat{d}$'s are exponential averages then

$$\hat{d}_{t+1} - \hat{d}_t = \lambda(d_t - \hat{d}_t) = \lambda e_t \tag{9}$$

Putting Eqs. (7), (8), and (9) together, we obtain the control equation:

$$X_t - X_{t-1} = -\frac{\lambda}{g}(y_t - T) \tag{10}$$

In the present example we set $g = 1.2$ and $\lambda = 0.2$ and we get Eq. (1), which is precisely the control that the chart of Figure 2 achieves.

In Figure 3 the horizontal dashes are the values $-\hat{d}_{t+1}$ the aggregated effect of all the adjustments up to that time. The diagram illustrates what the simple adjustment chart of Figure 2 achieves. It shows how, by the policy of repeated adjustment, the level of the deposition rate X_t will be such as will just cancel out that part $\hat{d}_{t+1}$ of the disturbance d_{t+1} that is predictable at time t. The vertical arrow indicates the situation for $t = 86$.

RECONSTRUCTING A PROCESS DISTURBANCE RECORD FROM PAST DATA

Chapter D.2 showed how to estimate an appropriate value for θ (and hence for $\lambda = 1 - \theta$) which gave you the smallest squared forecast errors. To make this

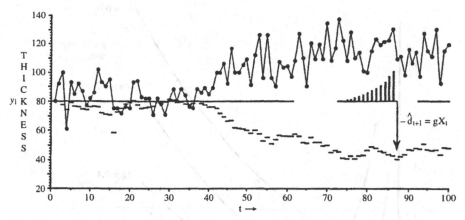

Figure 3. Compensation achieved by the manual adjustment chart of Figure 2.

calculation you need a past record of the process disturbance. In most circumstances past records will be of process operation during which adjustments have been made by one rationale or another. The disturbance series would, therefore, usually need to be back-calculated from the record of the deviations from target that actually occurred taken together with the adjustments that had been made. Usually it is possible in this way to reconstruct to a sufficient approximation what would have happened if there had been no adjustment. It is this reconstructed disturbance record that should be used to estimate the appropriate smoothing constant θ and hence λ. But in some cases you may not need to estimate λ; instead you can pick a value empirically as is shown in the next section.

EFFECT OF USING AN INCORRECT VALUE FOR THE SMOOTHING CONSTANT

A manual adjustment chart like that of Figure 2 is only mildly affected by moderate errors in the choice of $\lambda = 1 - \theta$; thus while, in the example we studied, good adjustment was obtained by plugging into Eq. (10) the estimated value $\lambda = 0.2$ it would have made little difference if we had used, say, $\lambda = 0.3$. Although, as we shall see, it would have made a great deal of difference if we had set $\lambda = 1.0$ producing a state of overcontrol (i.e., tinkering).

A good deal can be learned from careful study of the theoretical curves* in Figure 4. This chart shows how the variance about the target will be inflated if we use some value of λ to design the chart when ideally we should have used some other value λ_0. To begin with suppose the true situation is that $\lambda_0 = 0$. In this case the

---

* The curves are exact for large samples when the disturbance model is from the class of time series for which exponential smoothing is optimal.

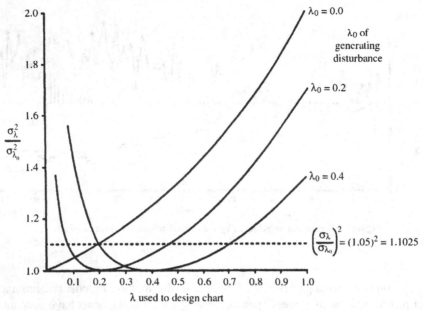

Figure 4. Inflation in the variance of the adjusted process arising from a wrong choice of λ.

process would be in a perfect state of control and no adjustment would be called for [see, e.g., Eq. (10)]. In that case, as is evident by inspection of the curve for $\lambda_0 = 0$, *any* manual adjustment would increase the variance. This is the situation discussed by Deming (1986) and illustrated with his funnel experiment. In particular, as he points out, a doubling of the variance would occur due to overcontrol, if as in Eq. (2) we set $\lambda = 1$. Notice, however, that if we design our adjustment chart using a smaller value of λ, such as the value of 0.2 appropriate for the series of Figure 1, the variance would not increase very much even if the process were (theoretically) in a perfect state of control. On the other hand, if the process were slightly out of control with the true value of λ_0 greater than zero, a *very large* increase in variance could result if *no adjustment* action was taken. Notice also that if we adopted $\lambda = 0.2$ as a compromise value then if the true value of λ_0 were *anywhere* within the range from zero to 0.4 the standard deviation for the adjusted process would be at most only 5% larger than the minimum value achievable if λ_0 were exactly known. Specifically, if the process was already in a perfect state of control and $\lambda_0 = 0$, manual adjustment would produce only a 5% increase in the value of the standard deviation. On the other hand, if λ_0 were really as large as 0.4 the standard deviation would again be only about 5% greater than it would have been had we designed our adjustment chart using the value of $\lambda = 0.4$. In practice therefore since some instability in the process is to be expected, the running of a feedback scheme with an empirically chosen value of $\lambda = 0.2$ or 0.3 might be wise. Furthermore, it turns out (see B.L. and Chapter D.6) that the use of a value of λ somewhat less than λ_0 can be advantageous, because for a small increase in the output standard deviation the magnitude of needed adjustments can often be greatly lessened.

The chart of Figure 2 uses a policy whereby some adjustment, often quite small, is made as each new observation becomes available. In Chapter D.4 I'll discuss ways of reducing the number of needed adjustments while simultaneously keeping as small as possible the resulting increase in mean square deviation about the target value.

ACKNOWLEDGMENTS

This research was sponsored by the National Science Foundation under Grant No. DDM-8808138, and by the Vilas Trust of the University of Wisconsin, Madison.

CHAPTER D.4

Bounded Adjustment Charts

One important activity of the quality practitioner is to detect and eliminate special causes of variation. Such endeavors are greatly helped by the use of process monitoring charts such as a Shewhart chart which verifies that a constant common cause system continues to produce stable variation about a fixed target value, while at the same time identifying special deviations unlikely to be due to chance. It can thus induce the gradual elimination of assignable causes of trouble and hence the improvement of the process. However, it often happens that, because of factors beyond the control of the operator, such as variation in the characteristics of feed-stocks or changes in ambient temperature, there remains a tendency for the process to wander away from target. It becomes necessary then to employ some system of adjustment also called process *regulation*. This can be done by borrowing some elementary ideas from the control engineer. In Chapters D.1 and D.3, I described a simple process adjustment chart which, in effect, fed back an estimate of the current deviation from target so that an adjustment could be made that would just cancel it out. It turned out that this estimate of the current deviation from target was an exponentially weighted average (an EWMA) of past data of the same kind as is used in sales forecasting discussed, for example, in the last two chapters.

The chart called for adjustment after each observation. Such repeated adjustment charts are of value when the cost of adjustment is essentially zero as happens, for example, when making a change consists simply of the turning of a valve by an operator who must be available anyway. In other situations, however, process

---

*One simple way to achieve this suggested by Box and Jenkins (1976) is to employ the manual adjustment chart I discussed in Chapter D.3, but with a "deadband" of about ±1 standard deviation about the target value within which no adjustment action is taken. This can reduce the needed frequency of adjustment somewhat without much increase in the standard deviation of the adjusted process. However, for deadbands of greater width, which are needed to achieve longer adjustment intervals, this approach produces unnecessarily large increases in the standard deviation and is not recommended.

From Box, G. E. P. (1991–92), *Quality Engineering*, 4(2), 331–338. Copyright © 1991 by Marcel Dekker, Inc.

adjustment may require the stopping of a machine or the changing of a tool or some other action entailing cost and it may then be necessary to employ a scheme which calls for less frequent adjustment* in exchange for a small increase in process variation.

It was shown by Box and Jenkins (1963; see also, Box, Jenkins, and MacGregor, 1974) that this objective can be achieved efficiently with what I will call a *bounded adjustment chart*, which again employs an exponentially weighted average of past data to estimate the deviation to be compensated but calls for only periodic adjustment. The chart is superficially similar to that proposed for process monitoring by Roberts (1959; see also Hunter, 1989), in that in both kinds of charts an exponentially weighted average of the data is plotted between parallel limit lines. The purpose and design of the bounded adjustment chart are, however, very different from those of the EWMA monitoring chart. The purpose is to decide when and by how much to adjust the process. The positioning of the boundary lines is not based on the discovery of statistically significant (e.g., 3σ) deviations, but on the need to balance a reduction in the frequency of adjustment against an acceptable increase in the standard deviation of the adjusted process.

Figure 1 shows such a bounded adjustment chart for the first 50 values of the thickness data used for illustration in Chapter D.3. The open circles represent the observations y_t of thickness which would be obtained after adjustments had been made by periodically changing the deposition rate X_t. The filled circles $\hat{y}_t$ are forecasts made one period previously using the appropriate exponentially weighted average. The estimated value of λ used in calculating the EWMA is 0.2 (see Chapter D.3), so the forecasts are updated by the formula $\hat{y}_{t+1} = 0.2y_t + 0.8\hat{y}_t$. The particular chart shown in Figure 1 has boundary lines at $T \pm 8$ where $T - 80$ is the target value. Bearing in mind that the standard deviation obtained from continual adjustment was about $\sigma_e = 11$, these limit lines are quite close together at $T \pm 0.72\sigma_e$. Suppose initially that the deposition rate was at some value X_0. This would remain unchanged until at time $t = 13$ when the forecasted value $\hat{y}_{14} = 88.7$ fell outside the upper limit. At observation 13 the chart signals that a change is needed in the deposition rate calculated to reduce the thickness by $-(88.7 - 80.0) = -8.7$. Thus an adjustment of $X_{13} - X_0 = -8.7/1.2$ would be made to the deposition rate, where $g = 1.2$ is the change produced in y per unit change in X. This adjustment does not upset the calculation of the next EWMA because the adjustment affects y_{14} and $\hat{y}_{14}$ equally, so that

$$\hat{y}_{15} = (0.2 \times 81.3) + (0.8 \times 80.0) = 80.3$$

where 80 is the appropriate value of $\hat{y}_{14}$ *after the adjustment has been made to bring the process on target.*

The next adjustment would occur after observation 20 when the forecasted value $\hat{y}_{21} = 71.1$ fell below the lower boundary line calling for an increase in the deposition rate of $8.9/1.2$. Proceeding in this way for the whole series of 100 values we find that for this chart with $\lambda = 0.2$ and boundaries set at 80 ± 8, the total number of needed adjustments would have been only seven, so that the average adjustment interval (AAI) $= 99/7 = 14$. Also the sum of squares of deviations from

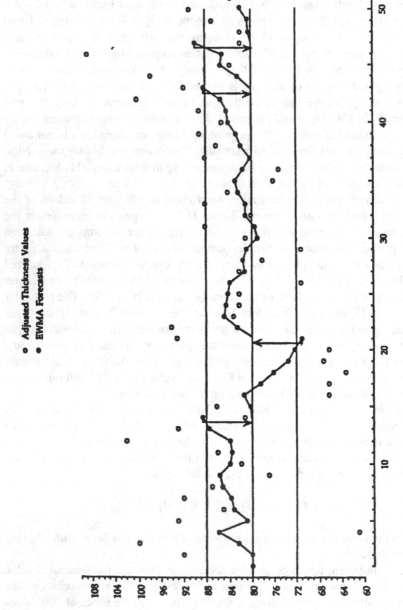

Figure 1. A bounded adjustment chart: the open circles are the thickness levels y_t after adjustment, the closed circles are their EWMA forecasts $\hat{y}_t$.

○ Adjusted Thickness Values

● EWMA Forecasts

the target value for the adjusted process (represented by *open* circles in the figure) would be increased from 12,256 for the repeated adjustment scheme to 14,433 for this bounded adjustment scheme. Thus since $\sqrt{14433/12256} = 1.085$, this bounded adjustment scheme composed to the unbounded scheme produces an increase in the standard deviation of the adjusted process of only 8.5%.

Table 1 shows the results of a few calculations of this kind made with boundary lines at different distances $\pm L$ from the target value. It will be seen that in exchange for a small increase in the standard deviation quite long average adjustment intervals are possible. Any degree of technological sophistication can be used in applying these ideas—anything from computers making the adjustment calculations with transducers taking the necessary action, to manual adjustment from "eyeball" interpolation. The chart shown in Figure 2, used a pushpin and a piece of thread to indicate the appropriate *manual* adjustments.

A good deal can be learned from the interpolation chart of Figure 2. Suppose we are currently at some time t, then the scale on the left labelled *current forecast* $\hat{y}_t$ is the value forecast at the previous time $t - 1$. The scale on the right labelled *actual value* records the value for y_t actually observed at time t. The third intermediate scale labelled *updated forecast* is arranged so as to divide the horizontal distance between the two outer scales in the ratio 0.2 : 0.8 (or in general, $\lambda : \theta$, where $\theta = 1 - \lambda$). On the far left an adjustment scale for the deposition rate is shown. This is blank between the boundary values 72 and 88 within which no action is called for. Outside these limits it is calibrated to show the change in deposition rate required. For example, if an updated forecast which had just crossed the upper boundary was equal to 92 (and consequently deviated from target by 12 units) then, since $g = 1.2$, the deposition rate would have to be reduced by 10 units, and this can be directly read off the chart. A convenient device for putting the above actions into effect uses a piece of black thread attached to a pushpin P located on the left-hand scale to make the interpolation. In the situation depicted in Figure 2 the current forecast made at time $t - 1$ is $\hat{y}_t = 86$ and the actual value observed at time t is $y_t = 66$. Just before the current time t, therefore, the location of the pushpin on the current forecast scale would be at $\hat{y}_t = 86$ with the thread hanging down from the pin. As soon as the actual value $y_t = 66$ became available the thread would be pulled tightly to join the point 66 on the right-hand scale. The updated forecast $\hat{y}_{t+1} = 82$ would then be read off on the intermediate scale. This value lies within the boundaries, so the pushpin

Table 1. Thickness Data: Effect[a] of Different Spacing of Boundary Lines $T \pm L$

L	0	4	8	12
Sum of squares	12,256	13,141	14,433	14,973
% Increase in SD	0	3.5	8.5	10.5
Number m of adjustments needed	99	20	7	3
AAI $= 99/m$	1	5	14	33

[a] These calculations are for a *particular* run of 100 observations and are of course subject to quite large sampling errors.

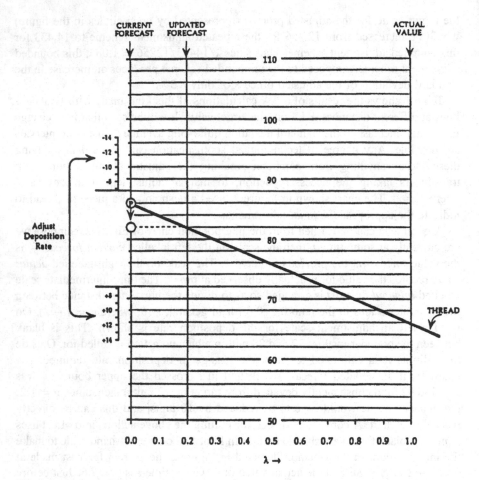

Figure 2. An interpolation chart to update the forecasted value of thickness and to indicate when and by how much the deposition rate should be adjusted.

would be moved down to this new current forecast value with the thread hanging loose again until the next actual value became available to produce a new updated forecast. If the updated forecast had fallen outside either boundary, the appropriate adjustment in deposition rate would have been made, and the pushpin would be *placed on the target value* ready for the next interpolation.

The chart makes it easy to see why when L is zero, the action called for by bounded adjustment is identical to that for repeated adjustment discussed in Chapter D.3. In the latter case since $L = 0$ there *were* no boundary lines (or rather they coincided on the target value) and the adjustment scale was continuous with no blank range in the middle. Also, if we had used this device with $L = 0$, the pushpin representing the value on the current forecast scale would at each stage have been adjusted to the target value T. Thus, at any time t, the updated forecast of the adjusted series would have been such that

$$\hat{y}_{t+1} - T = \lambda(y_t - T)$$

and consequently the required adjustment would be

$$X_t - X_{t-1} = -\frac{1}{g}(\hat{y}_{t+1} - T) = -\frac{\lambda}{g}(y_t - T)$$

which is the control equation for continuous adjustment.

The interpolation chart also makes it easy to appreciate the precise function of λ in separating the signal from the noise. It supplies a damping factor which, by smoothing out the noise, extracts an estimate of the real changes that are occurring. Changing the value of λ results in more or less smoothing and it will depend on how strong the signal really is, as to how large a value of λ we should employ. The sum of squares plot referred to in Chapters D.2 and D.3, in effect, helps to decide the optimal amount of smoothing. Also, the value of L determines how much signal we

Table 2. Theoretical Values of AAI and % Increase in SD for Various Values of L/σ_e and λ

λ	L/σ_e	AAI	% Increase in SD
0.1	0.5	32	2.4
	1.0	112	9.0
	1.5	243	18.0
	2.0	423	30.0
0.2	0.5	10	2.6
	1.0	32	9.0
	1.5	66	20.0
	2.0	112	32.0
0.3	0.5	5	2.6
	1.0	16	10.0
	1.5	32	20.0
	2.0	52	33.0
0.4	0.5	4	2.6
	1.0	10	10.0
	1.5	19	21.0
	2.0	32	34.0
0.5	0.5	3	2.5
	1.0	7	10.0
	1.5	13	21.0
	2.0	21	35.0

are prepared to allow to "leak" into the adjusted series in exchange for less frequent adjustment.

Table 2 shows theoretical Average Adjustment Intervals (AAI) and percent increases in the standard deviation of the adjusted process for various values of L/σ_e and λ. For illustration, suppose that $\lambda = 0.3$ and that we are prepared to allow, in exchange for less frequent need for adjustment, an increase of 10% in the standard deviation of the adjusted process above the value obtained when continual action is taken after each observation. From the table we see that the desired scheme requires that $L/\sigma_e = 1$ and yields an AAI of about 16.

The table is based on calculations in Box and Jenkins (1963) as extended by Box and Kramer (1992). These assume that the disturbances can be represented by the class of time series models for which exponential smoothing is optimal. However, computer simulations suggest that the tabulated values are not particularly sensitive to this assumption.

MONITORING AND ADJUSTMENT

To say that we must be clear headed about the fundamental differences between systems of monitoring and systems of adjustment is not to say that these ideas cannot or should not be combined. A process which must be adjusted because of uncontrollable drifts may also exhibit occasional large superimposed deviations arising from specific special causes whose origin it may be possible to track down. These might be detected by looking for outlying points on a control chart for the *residual error series* $e_t = y_t - \hat{y}_t$ or by means discussed in Chapter D.9.

EXPLICIT CONSIDERATION OF DOLLAR COSTS

Implicit in the balancing of less frequent adjustment against increase in mean square deviation from target is the idea that, in some sense, we are balancing the monetary *value* of less frequent adjustment against the *cost* of some increase in variation about the target. However, it is often not an easy matter to assess such values explicitly in terms of dollars. Are plant operators less or more likely to be alert with a scheme that calls for frequent adjustment? If such an effect exists what is its worth in dollars?

It is because of such questions that it may be impractical to try to work directly with dollar costs. It is often better to design a number of alternative schemes showing estimates of the frequency of adjustment and the corresponding standard deviations for the adjusted process with each scheme. The choice among these options should be made by someone (such as the supervising engineer) who is familiar with the human as well as the technical circumstances surrounding this particular facility. Occasionally, however, the costs of being off target and of making a change in the process may be arrived at, at least approximately. Ways of doing this, and of making the choice among alternative schemes so as to achieve minimum overall cost were given in the paper by Box and Jenkins (1963). A more comprehensive discussion, which also includes the possible cost of *monitoring* the process, will be found in Chapter D.5 and in B.L.

ACKNOWLEDGMENTS

This research was sponsored by the National Science Foundation under Grant No. DDM-8808138, and by the Vilas Trust of the University of Wisconsin, Madison.

CHAPTER D.5

Statistical Process Monitoring and Feedback Adjustment—A Discussion

Rationales for process monitoring using some of the techniques of statistical process control and for feedback adjustment using some techniques associated with automatic process control are explored, and issues that sometimes arise are discussed. The importance of some often unstated assumptions are illustrated. Minimum-cost feedback schemes are discussed for some simple, but practically interesting, models.

INTRODUCTION

The term *process control* is used in many ways. In particular, Shewhart charts, cumulative sum (CUSUM) charts, and exponentially weighted moving average (EWMA) charts are frequently employed for process *monitoring* as part of what is called statistical process control (SPC). By contrast, various forms of feedback and feedforward regulation are used for process *adjustment* in what is often called automatic process control (APC). The people responsible for SPC and those responsible for APC are usually from different departments and have different technical backgrounds, so it is hardly surprising that there has sometimes been controversy and misunderstanding between them.

SPC and APC originated in different industries—the *parts* industry and the *process* industry. The control objectives in these two industries were often very different. The parts industry was typically attempting to reproduce individual items as accurately as possible—for example, to manufacture large numbers of steel rods

From Box, G. E. P. and Kramer, T. (1992), *Technometrics*, 34(3), 251–267. Reprinted with permission from TECHNOMETRICS. Copyright © 1992 by the American Statistical Association. All rights reserved.

324

with diameters having *smallest possible variation* about a fixed target value T. The process industries were typically concerned with yields of product, percentage conversion of chemicals, and measures of purity and were attempting to obtain the highest possible mean values for these measures with smallest variation. Moreover, while in the parts industry properties of feed materials such as steel sheet could be reasonably well controlled, in the process industry external variables such as ambient temperature and the properties of natural feedstocks were often uncontrollable and usually had to be compensated by feedback control. Again in the parts industry, the *cost of adjustment* was frequently substantial, involving the stopping of the machine or the replacement of a tool, and *frequency of monitoring* the process might also be an appreciable cost factor. By contrast, the only noncapital cost involved in automatic process adjustment was often, but not invariably, the cost of being off target. Some of these issues were addressed earlier by Box and Jenkins (1962, 1963, 1970) and by Box, Jenkins, and MacGregor (1974), but at that time they did not excite a great deal of interest.

More recently the sharply drawn lines dividing the parts industry and the process industry have disappeared. One reason is that some processes like the manufacturing of computer chips are hybrids, having certain aspects of the parts industry and others of the chemical industry. Another reason is that conglomerate companies, in which both kinds of manufacture occur, are now much more common. A third reason is that because of the "quality revolution" a greater awareness of the importance of control has led each industry to experiment with the control technology of the other. Correspondingly, this has led to a renewed research interest, and many authors, to whom we are indebted, have been concerned with various aspects of the resulting problems. These include Taguchi (1981), Alwan and Roberts (1988), Crowder (1986), Hunter (1986), Lorenzen and Vance (1986), MacGregor (1987), Adams (1988), Harris (1988), Hahn, Faltin, Tucker, Richards, and Vander Wiel (1988), Adams and Woodall (1989), Taguchi, Elsayed, and Hsiang (1989), and Vander Wiel and Tucker (1989).

In this article, we begin by discussing simple examples of process monitoring and of process adjustment by feedback control, and to provide a framework for discussion, we consider the validity of criticisms that have sometimes been made of these techniques. In the later part of the article, we try to shed light on these questions by considering the different types of feedback-control schemes that emerge when the relative costs are taken into account (a) of being off target, (b) of making an adjustment, and (c) of taking an observation. In what follows, to concentrate attention on the issues rather than the technical details we deliberately use the simplest models that have a claim to reality.

An Example of Process Monitoring Using a Shewhart Chart

Shewhart charts, CUSUM charts, and EWMA control charts are devices for process monitoring. To fix ideas, we discuss what we feel is the simplest application of a Shewhart chart—that of monitoring a frequency of occurrence. A company has a small employee health clinic where cuts, bruises, and other minor injuries are treated.

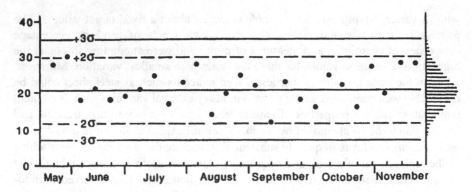

Figure 1. A control chart for weekly numbers of new patients at a clinic in relation to a Poisson reference distribution.

To monitor the situation, a count is kept of the number of new patients attending the clinic each week. Over an extended period of time, the average number is 20.6. If we hypothesize that during each small fixed increment of time the chance of injury is the same, the data would correspond to random drawings from the Poisson distribution with mean 20.6 drawn on the right of Figure 1. If the data did not look like a random sequence from the reference distribution generated by this "null" model, we would have reason to suppose that it was false and that something else was going on.

In practice, the whole reference distribution would not usually be shown. Instead upper and lower control limits and warning lines might be set and where, as in this case, a normal approximation to the reference distribution was adequate, these would be conventionally set at three and two standard deviations above and below the target. As was emphasized by Shewhart (1931) and by Deming (1950, 1986), charts of this kind are of great value in helping to distinguish between so-called *common causes* and *special* (or *assignable*) *causes*. Common causes refer to variations that are in accordance with a null model and therefore call for no special action. Special causes, on the other hand, are those producing temporary deviations from this model.

Thus a higher frequency of injury in a particular week producing a point falling above the upper control limit should trigger the search for such an assignable cause. Investigation might show that this corresponded to the temporary failing of adequate lighting in a particular part of the factory, leading to action that ensured that there would be no reoccurrence of this situation. By contrast, a change in the common-cause system could occur, for example, as a result of management's installing safer machinery. In such a case, a new common-cause system might be established approximated by Poisson variation about a lower mean value.

When the common-cause model explains the variation, the system is said to be *in an* (approximate) *state of control*. It is often supposed that this requires that successive observations behave like independent identically distributed random variables, which we refer to as a state of iid control. Although, for simplicity, we have assumed this in most of what follows, it need not be the case. For example, one can find examples of stable systems that vary about a fixed mean in which successive deviations are dependent but that can be represented by a fixed common-cause model. In statistical terms, the common-cause system refers to generation of data by an approximately stable *model*, but special causes correspond to temporary deviations from the model signaled by *outliers* or unusual patterns of points.

The purpose of process monitoring is thus first to confirm that the common-cause system remains in place and second to react to possible assignable causes. Because the search for assignable causes is usually expensive in time and in money, it is reasonable not to begin such a search unless we have real evidence of departure from the common-cause model.

Process monitoring thus resembles a system of continuous statistical *hypothesis testing*, whereas, as we see later, feedback control can be thought of as a process of statistical *estimation* of the current level of the disturbance that is then used to apply appropriate compensatory adjustment.

A Simple Example of Feedback Control

Suppose that a system is affected by a drifting change, due, for example, to changes in feedstock or in ambient temperature. If attempts to eliminate the cause of this disturbance have proved fruitless, then we may need to compensate for it by feedback or feedforward control. For simplicity in this discussion, we limit ourselves to considerations of elementary feedback-control schemes. (Feedforward control, sometimes employed in conjunction with feedback control, is used to eliminate the effect of some fluctuating measured input by making an adjustment from direct calculation of its effect on the output.) Suppose that we need to maintain the viscosity of a polymer as close as possible to some target value T by manipulating the catalyst formulation X and that the need for control arises because of disturbances whose overall effect on viscosity is represented at the output by Z_t. Thus (see Figure 2) Z_t represents what would happen to viscosity if no control action were taken. At time t the effect of present and past adjustments to the catalyst formulation is experienced at the output as a compensation of Y_t units of viscosity. If, as is convenient, we consider the disturbance Z_t and the attempted compensation Y_t as *deviations* from the target value T, then the output error—that is, the deviation from the target—of the adjusted process will be

$$e_t = Z_t + Y_t \tag{1}$$

The feedback-control equation determines how the catalyst formulation X should be adjusted and calls for action that is some function of the current and past errors. In many automatic controllers, measurements and control actions are taken continuously. In particular, for the classic proportional-integral (PI) controller, the control action X_t is a linear function of the present error e_t and of the integral over time of past errors. In what follows we will consider a *discrete system* in which the process is monitored and it is possible to take control action at equally spaced time intervals. The appropriate analog of PI control is then

$$-X_t = k_0 + k_P e_t + k_I \sum_{i=1}^{t} e_t \tag{2}$$

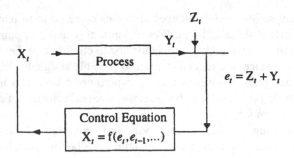

Figure 2. A feedback-control scheme.

where k_P and k_I are positive constants determining the amount of *proportional* and *integral* control, respectively.

MODELS FOR DISTURBANCES AND DYNAMICS

PI feedback-control schemes of this kind have been used for a very long time (Mayr, 1970) and were originally developed empirically. They can, however, be shown to provide minimum mean squared error (MMSE) control about the target if certain specific assumptions are true. These we now discuss.

Choice of a Disturbance Model

The need for process regulation arises when the system is afflicted with disturbances that cause it to wander off target if no action is taken. In developing a model for feedback control, therefore, we need a reasonably realistic representation of the deviation from target Z_t that would occur if no adjustment action were taken. In what follows, we will use the expression *white noise* to denote a sequence $\{a_t\}$ of iid random variables that are, to an adequate approximation, normally distributed having mean 0 and variance σ_a^2. The individual a_t's are sometimes called innovations and σ_a^2 the innovation variance.

A familiar disturbance model for a process in a state of control assumes that the deviation from target is a white-noise sequence. Thus

$$Z_t = a_t \tag{3}$$

A tendency to wander from target implies dependence between successive deviations. A more general class of models for a disturbance in which successive deviations from target are dependent can be written as a weighted (theoretically infinite) sum of such white-noise innovations $Z_t = a_t + \psi_1 a_{t-1} + \psi_2 a_{t-2} + \cdots$. An important class of such models is *stationary* models for which the mean deviation from target is 0 and the variance of $Z_t = \sigma_a^2 \sum_{i=1}^{\infty} \psi_i^2 = \sigma_Z^2$ is finite. Particular examples are stationary autoregressive models of order p of the form $Z_t = \varphi_1 Z_{t-1} + \varphi_2 Z_{t-2} + \cdots + \varphi_p Z_{t-p} + a_t$, in which the autoregressive coefficients are chosen so as to satisfy the preceding condition of stationarity. Such stationary disturbance models suffer from the often unrealistic consequence that, if the process were left to itself and no adjustments were made, it would continue to vary about the same fixed-target value. If in fact the process might permanently drift away from target, such an assumption would produce unrealistic control schemes. A class of *nonstationary* models that can represent such drifting behavior is the autoregressive integrated moving average models discussed, for example, by Box and Jenkins (1970). The simplest and most useful of such models is the integrated moving average (IMA) model, which may be defined by writing

$$Z_t = \hat{Z}_t + a_t \tag{4}$$

where $\hat{Z}_t$ is independent of a_t and is an EWMA of past data defined by

$$\hat{Z}_t = \lambda(Z_{t-1} + \theta Z_{t-2} + \theta^2 Z_{t-3} + \cdots) \tag{5}$$

and characterized by the *smoothing constant* θ or equivalently by the *measure of nonstationarity* $\lambda = 1 - \theta$, and for the purpose of this article it will be supposed that $0 \leq \theta < 1$ and hence that $0 < \lambda \leq 1$. The coefficients λ, $\lambda\theta$, $\lambda\theta^2$, ... in Equation (5) form a convergent series that sums to unity. Detailed properties of this model will be found in books on time series. Here we will only outline the characteristics we need for this discussion.

The time series defined by (4) has no mean value and infinite variance. As was shown by Muth (1960), however, given data on Z only up to time $t-1$, $\hat{Z}_t$ is the MMSE estimate of all future values of the series and is therefore an estimate of the current *location* of the series at time t. In particular, it is the MMSE *forecast* of Z_t made at time $t-1$. By algebraic manipulation of (4) and (5), it is readily shown that

$$\hat{Z}_{t+1} = \lambda Z_t + \theta \hat{Z}_t, \qquad \hat{Z}_{t+1} - \hat{Z}_t = \lambda a_t \tag{6}$$

Using either of these recursive formulas, the forecast can be conveniently updated as each new observation comes to hand. Unless θ is very close to 1, the coefficients in (5) converge fairly quickly to 0, and in practice an adequate approximation to the EWMA is obtained by suitably truncating the series. [For an EWMA truncated after $p+1$ terms, a slightly improved approximation is obtained by setting the last term equal to $\theta^p Z_{t-p-1}$ instead of $\lambda\theta^p Z_{t-p-1}$. Since the infinite sum of the residual coefficients $\lambda(\theta^p + \theta^{p+1} + \cdots)$ is θ^p, this is equivalent to taking an infinite sum with all Z's after Z_{t-p} taken equal to Z_{t-p-1}.]

Moreover, it follows from (4) and (5) that the first difference of Z_t is the first-order moving average model

$$W_t = Z_t - Z_{t-1} = a_t - \theta a_{t-1} \tag{7}$$

from which the model (4) may be referred to as an IMA model.

By summing Equation (7) and using Equation (4) with $t = 1$, we obtain

$$Z_t = \hat{Z}_1 + a_t + \lambda \sum_{i=1}^{t-1} a_t, \quad 0 < \lambda \leq 1$$

In particular, suppose that the process is set on target at time $t = 1$ by adjusting its level so that $\hat{Z}_1 = 0$. Then the subsequent course of the deviations from target is represented by

$$Z_t = a_t + \lambda \sum_{i=1}^{t-1} a_i, \quad 0 < \lambda \leq 1 \tag{8}$$

which can be thought of as an interpolation between the white-noise disturbance of Equation (3) obtained as λ approaches 0 and the highly nonstationary *random-walk* model

$$Z_t = \sum_{i=1}^{t} a_i \tag{9}$$

obtained when $\lambda = 1$. We believe the IMA model characterized by (4) and (5) or by (7) occupies a central place in representing the behavior of many nonstationary disturbances. One reason concerns its variogram.

Characterizing a Time Series Model by the Variogram. Suppose that the variance $V(Z_{t+m} - Z_t)$ of the difference between observations taken m steps apart in a time series is independent of t; then the inflation of this variance for an interval of length m relative to that for a unit interval is measured by

$$G(m) = V(Z_{t+m} - Z_t)/V(Z_{t+1} - Z_t) \tag{10}$$

The quantities $G(m)$ plotted in Figure 3 as functions of m are the (standardized) *variograms* (see e.g., Cressie, 1988; Jowett, 1952, 1955).

In an ideal world we could perhaps set the controls of a machine or give a set of instructions to an ever-alert and conscientious operator once and for all, yielding a perfectly stable process from that point on. In such a case the disturbance might be represented by the white-noise process of Equation (3), and its corresponding variogram $G(m)$ would be independent of m and equal to 1 as illustrated in Figure 3. But in reality the parts of any machine are continually losing adjustment and wearing out, and people gradually forget instructions and miscommunicate. Thus, for an uncontrolled disturbance, some kind of dependence in time must be expected. Dependence in the form of autocorrelation can be produced by linear stationary models (e.g., stationary autoregressive moving average models). For such stationary models, however, although $G(m)$ will initially increase with m, it will always

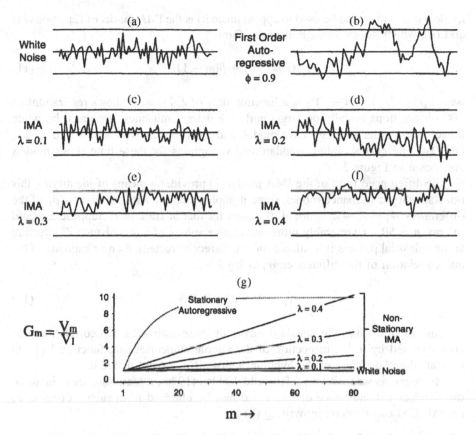

Figure 3. Examples of time series of length 100 generated by white noise, a first-order autoregressive process with $\phi = .9$, and IMA processes for various choices of the nonstationarity parameter λ with associated standardized variograms.

approach an asymptote. That this can happen quite quickly even when successive observations are highly correlated is illustrated by the variogram shown in Figure 3 for the first-order stationary autoregressive time series model $Z_t = 0.9Z_{t-1} + a_t$. In this example, although successive deviations have autocorrelation 0.9, $G(m)$ is already within 5% of its asymptotic value after only 20 lags. To use this model implies, therefore, that for an *uncontrolled* disturbance we should expect observations 100 steps apart to differ little more than those 20 steps apart. But as we have argued previously, an uncontrolled system is subject to influences that would be expected to *continuously* increase its entropy. For such a system, we might expect the variogram to increase monotonically. In Appendix A, we consider some properties of what we shall call the *sticky innovation* model of a type due to Barnard (1959), which has such a variogram that increases linearly with m. A special case of this model (which may also be used to approximate it) is the IMA model of Equations (5) and (6), which have a variogram of the form

$$G(m) = 1 + \beta(m - 1) \tag{11}$$

where $\beta = \lambda^2/\{1 + (1 - \lambda)^2\}$ is a function only of λ. Figure 3 shows realizations of 100 observations initially on target and with $\pm 3\sigma_a$ limit lines, generated by white noise, a first-order autoregressive model, and IMA models with $\lambda = 0.1$, 0.2, 0.3, and 0.4. The corresponding standardized variograms for these time series models are shown in Figure 3.

The differenced form of the IMA model (7) provides a means of identifying this nonstationary disturbance model, since it implies that the autocorrelations ρ_k of the differences $W_t = Z_t - Z_{t-1}$ are all 0 except for that at lag $k = 1$. Suppose a record of, say, $n = 50$, or preferably more, successive values of a disturbance Z_t affecting some industrial process is available (or can be reconstructed); then an estimate of the autocorrelation of the differences W_t at lag k is

$$r_k = \Sigma W_t W_{t-k}/\Sigma W_t^2 \tag{12}$$

Bearing in mind that the standard errors of these estimated autocorrelations are approximated by $n^{-1/2}$, inspection of the sample autocorrelation function $\{r_k\}$ can indicate the plausibility of the hypothesis that all ρ_k except ρ_1 are 0.

Moreover, as was shown by Box and Jenkins (1970), a least squares estimate of the parameter θ and hence of $\lambda = 1 - \theta$ may be obtained from such a disturbance record. This can be done by writing (7) as

$$a_t = \theta a_{t-1} + W_t \tag{13}$$

using it to recursively compute sets of $\{a_t$'s$\}$ and hence of minimizing their sum of squares by making such calculations for a number of trial values of θ. Details of the procedure for the selection of starting values for this procedure and of the relationship of the least squares value of θ to the maximum likelihood estimate will also be found in the above reference.

A Dynamic Model Describing Process Inertia

In the feedback scheme shown in Figure 2, control is achieved by changing catalyst formulation X_t, which in turn changes viscosity Y_t. In modeling such a relation between input and output of a process, we may need to allow for the system inertia. A useful way to do this uses linear difference equations. In particular, a simple *first-order* dynamic model that can approximate the behavior of a number of processes is characterized by the difference equation

$$Y_t = \text{constant} + \delta Y_{t-1} + g(1 - \delta)X_{t-1}, \quad 0 \le \delta < 1 \tag{14}$$

The inertial properties of the equation can be appreciated from the consideration that, t time periods after a unit step change is made in X, the change in Y will be $g(1 - \delta^t)$. Thus for this dynamic model the output change asymptotically approaches g units, where g is called the system *gain*.

The MMSE Controller

If now we suppose that the drifting disturbance is represented by the nonstationary IMA model of Equation (4) [or equivalently (7) and (8)] and that the process dynamics are represented by Equation (14), it is easily shown that PI adjustment of the form of Equation (2) produces MMSE about the target value provided that the proportional and integral constants k_P and k_I are set to the values

$$k_P = \frac{\lambda \delta}{g(1 - \delta)} \quad \text{and} \quad k_I = \frac{\lambda}{g} \tag{15}$$

Equivalently, if we suppose that the cost of being off target is the *only* cost and is a quadratic function of the output deviation from the target, then this scheme would minimize overall cost. As we show later, however, minimum-cost feedback control is of a quite different kind if we need to take account of the cost of adjusting the system and/or of obtaining observations.

The control equation (15) is of practical use only if δ is fairly small. As δ becomes larger and, in particular, as it approaches unity, the MMSE scheme requires excessive control action. Modified schemes can be employed, however, in which reduced control action can be achieved at a cost of small increases in the mean squared error at the output. See, for example, Åström (1970), Box and Jenkins (1970), Wilson (1970), MacGregor (1972), and Åström and Wittenmark (1984). We briefly discuss this point later.

A Further Simplification of Dynamics

For the purpose of the discussion that follows, we will adopt a simple dynamic model in which it can be assumed that essentially all of the change induced by a step change in X will occur in a single time interval. This corresponds to setting $\delta = 0$ in Equation (14), and the dynamic model then becomes

$$Y_t = \text{constant} + gX_{t-1} \tag{16}$$

The MMSE feedback equation is then

$$-X_t = k_0 + k_I \sum_{i=1}^{t} e_i \quad \text{with} \quad k_I = \lambda g \tag{17}$$

the discrete analogue of integral control. Although the word "automatic" is usually associated with the methods of process regulation used by the control engineer, it is the concepts that are important and not the manner in which they are implemented. Once the principles are understood, the quality practitioner can greatly benefit by adding some expertise on simple feedback control to his/her tool kit. He/she can put this to use whether adjustments are *determined* by the computer or by a chart, and whether they are *put into effect* by transducers or manually. In particular, (see Chapter D.3), *manual* adjustment charts that produce the integral or proportional integral compensation defined by Equations (17) and (2) can be employed and are almost as easy to use as Shewhart charts. They do not seem to have been widely known, however.

RESPONSE TO SOME CRITICISMS OF SPC AND APC

Having presented elementary examples of statistical process monitoring and of feedback control and the methods sometimes used to justify them, we now have a framework for discussion of the controversy referred to earlier.

SPC practitioners have sometimes criticized feedback controllers (a) for over-compensating disturbances (in particular, we have heard stories from the parts industries of improvement when automatic controllers are *disconnected*), (b) for compensating disturbances rather than removing them, and (c) for concealing information that might be used for quality improvement. To these might be added the further point discussed later that linear controllers (such as PI controllers) are not of the appropriate form when there is a fixed cost associated with the periodic adjustment of the process. On the other hand, process monitoring methods, used for process adjustment can be extremely inefficient.

We now discuss these questions using the simple models discussed previously, and we begin by considering criticisms of feedback control.

Overcompensation by Feedback Control?

Consider the feedback scheme of Equation (17). Although for the assumptions made, MMSE σ_a^2 would be produced if we set $k_I = \lambda/g$, the control could be poor or even unstable if the constant k_I were wrongly chosen. For illustration consider an extreme example in which this type of feedback adjustment is applied to a process for which the "disturbance" is white noise so that λ is 0. In the case from Equation (17), k_I should be set equal to 0 resulting in a variance about the target value of σ_a^2. But suppose instead that k_I was set equal to some positive value $\dot{k}_I > 0$. Then if e_t is the error at the output having variance σ_e^2, it is shown in Appendix B that the inflation of the output variance would be

$$\sigma_e^2/\sigma_a^2 = (1 - \tfrac{1}{2}g\dot{k}_I)^{-1}, \quad 0 \le g\dot{k}_I < 2$$
$$= \infty, \quad 2 < g\dot{k}_I < \infty$$

(18)

For illustration, when $g = 1$,

k_I	0	0.5	1	1.5	≥ 2
σ_e^2/σ_a^2	1	1.33	2	4	∞

Thus, although it is true that, for a system already in a state of iid control, feedback could increase the variance or even render the process unstable, nevertheless, if that system (for which $\lambda = 0$) had been correctly identified, then $k_I = \lambda/g$ would have been set equal to 0 and the appropriate control action of Equation (17) would then be $-X_t = k_0$—that is, *no control.*

This concept was discussed earlier by Deming (1950, 1986) in relation to an experiment in which a marble is dropped through a funnel onto a target. Deming pointed out that attempts to adjust for *random* deviations from target by moving the funnel will merely increase the dispersion. He considered a number of different "control rules," and a comprehensive critique of these was given by MacGregor (1990). Here we consider only Deming's rule (2) in which it is supposed that an adjustment is made to "compensate" the last error. This corresponds, in our notation, to making an adjustment $X_t - X_{t-1} = -e_t$ or equivalently to using the control equation $-X_t = k_0 + \sum_{i=1}^{t} e_i$ when $k_0 = -X_0$. This then is equivalent to setting $g = 1$ and $k_I = 1$ in Equation (18), and from the table we confirm that the variance is doubled, as was found earlier by Deming (see also Figure 4 in Chapter D3).

Notice, however, that if successive errors had not been white noise but had followed a nonstationary IMA model with parameter λ, then MMSE control would be obtained with $X_t - X_{t-1} = -\lambda e_t$ or $-X_t = k_0 + \lambda \sum_{i=1}^{t} e_t$. Thus the appropriate adjustment to achieve MMSE for that case would compensate only a specific fraction λ of e_t, and taking no action would increase the mean squared error. We were reminded by Alwan and Roberts (1988) that Shewhart (1931) defined a state of control as "when, through the use of past experience we can *predict* within limits how the phenomenon may be expected to vary in the future" (p. 6). Notice that this definition includes stationary autocorrelated models and also non-stationary models.

For a process in the state of iid control characterized by Equation (3), the best prediction is the target value, and no action should be taken. But this is no longer true for dependent data in which a better prediction (forecast) than the target value exists. For example, for a system with the simple dynamics of Equation (16) suppose that it is possible *on some basis or other* to make such a better prediction $\hat{Z}_{t+1}$ of Z_{t+1} using data up to time t which has prediction error $Z_{t+1} - \hat{Z}_{t+1} = e'_{t+1}$. Then using Equations (1) and (16) and adjusting location so that the arbitrary constant is set to 0, the error at the output for a process controlled by manipulating X is $e_{t+1} = e'_{t+1} + \hat{Z}_{t+1} + gX_t$. Therefore, if we set $-X_t = \hat{Z}_{t+1}/g$, then $e_{t+1} = e'_{t+1}$ and the error at the output for the adjusted process will be the one-step-ahead forecast error. This will apply whether the forecast uses past values of the output series (feedback control) or present and past values of other predictive input variables (feedforward control) or both.

In particular, in the feedback case discussed in Section 2.4, if the IMA model were precisely true, then the forecast error $e'_{t+1} = a_{t+1}$ will be white noise and the feedback scheme would have MMSE. But notice that even if this were not so, the control action would still be determined by an EWMA forecast, which would frequently provide a reasonably good estimate of the moving location of the series even if it was not precisely represented by an IMA model.

The state sometimes achieved without process adjustment by very careful standardization of raw materials, operator procedures, and so forth is probably more realistically represented by a mildly nonstationary model. For this reason (Box, 1991b) unless cost considerations ruled out such a course, even for a process nominally in control, the running of a feedback scheme like that of Equation (17) with an empirically chosen value of, say, $\lambda = 0.2$ might be wise. The reason is that, if in reality λ were equal to 0 and the process were in a perfect state of control, very little increase in the standard deviation of the output would result from this small amount of feedback (about 5% assuming the models discussed previously), whereas mild nonstationarity, not easy to detect with a Shewhart chart, could cause considerably greater increases.

It is the occasional misapplication of APC that has resulted in overcompensation and not inherent problems with the technique itself. A standard controller has often been hooked up to a system, the characteristics of which have not been identified. Two problems that can arise are (1) that the controller is of approximately the right design but is mistuned and (2) that the basic design of controller is inappropriate. So far as mistuning is concerned, formal identification of the disturbance and of the dynamics by the fitting of appropriate models could avoid this problem but might be too tedious for routine use. In modern practice self-tuning controllers are sometimes employed, but when such options were not available, considerable improvement is possible simply by using an experimental approach to tune the controllers. For example, using the output mean squared error, or preferably its logarithm, as the objective function and the settings of the control constants (e.g., k_P and k_I for PI controller) as *factor levels*, experimental design can be used to improve and optimize the controller settings on line. Such an experimental approach is particularly valuable for systems that included many controllers whose effects might interact with each other. The procedure would, of course, lead to no control action should the process be already in a state of control. To avoid upsetting the system, experimental runs might be made in the *evolutionary operation* mode (Box, 1957a). For systems containing large numbers of individual controllers fractional designs could, for example, be used initially to determine a steepest descent path leading to better settings. This could be followed, if necessary, by second-order response surface methods to explore and optimize the more important factors and by canonical analysis to exploit ridge systems. If desired, criteria other than mean squared error could be used as the objective function.

All of the preceding supposes, however, that reasonably good control would be possible with off-the-shelf controllers such as PI controllers. But, as we shall see later, consideration of costs associated with adjusting the process and in making an observation can change the form of the feedback controller design, and in particular PI controllers could be inadequate in *any* tuning.

Some Disturbances Cannot Be Removed

When a process is affected by a disturbance, we can try to eliminate it and/or we can compensate for it. Thus a switch to an alternative supplier may result in a more uniform feedstock. But if a feedstock is a naturally occurring one such as a metallic ore, crude oil, or lumber from a forest, this approach may not be possible or may be too expensive. For example, if we find the temperature variation in Wisconsin from winter to summer too extreme, we can move to California, but if we decide to stay in Wisconsin, we can, like everybody else, compensate for the cold weather by using a furnace controlled by appropriate feedback adjustment.

Information Need Not Be Concealed by Feedback Control

As often practiced, feedback control can conceal the nature of the compensated disturbance. But this need not happen. Figure 4 illustrates the point. This shows the behavior of a simulated feedback scheme in which the disturbance is an IMA process with the parameter measuring the nonstationarity of the disturbance given by $\lambda = 0.2$, the dynamic parameter $\delta = 0.5$, and $g = 1$. The calculations were made assuming that the system is controlled by the PI controller

$$-X_t = \text{constant} + 0.20e_t + 0.20\sum_{i=1}^{t}e_i \qquad (19)$$

which produces MMSE for these parameter values.

Although this is not usually done, the control action X_t in Figure 4(b), as well as the deviation from target e_t in Figure 4(d), could be charted or better still displayed on the screen of a process computer. [If the dynamics were known, the exact compensation Y_t shown in Figure 4(c) could also be computed and hence the original disturbance Z_t of Figure 4(a) could be reconstructed.] Examination of these displays could be conducted using a generalized concept of common and special causes. The disturbance model and the dynamic model together define the *common-cause* system, which is taken account of in designing the controller and could be changed by management. Suppose, for example, it was found that the pattern of control action X_t mirrored that of a particular impurity in the feedstock. If this correlation checked out as a causative relation, management might decide to change the control system either by removing the impurity from a feedstock before it reached the process, or, if this were impossible or too expensive, by measuring it and compensating for it by appropriate feedforward control.

In addition, a *special cause* producing a temporary deviation from the underlying system model, induced perhaps by misoperation of the controller or a mistake by the

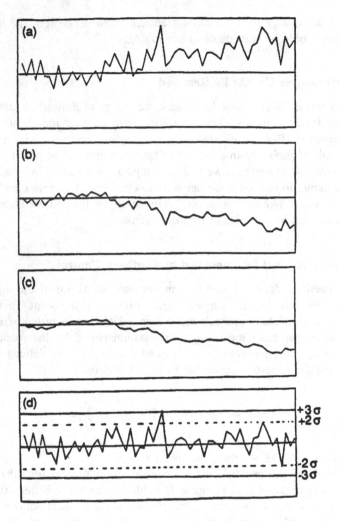

Figure 4. (a) Disturbance Z_t for a simulated process subject to feedback control; (b) feedback-control action X_t; (c) compensation of the disturbance Y_t; (d) deviations from the target value e_t.

operator, could be evidenced by an outlier in the sequence $\{e_t\}$ and then lead to remedial action. To illustrate this, we added a deviation of size $3\sigma_a$ to the 30th value of the disturbance Z_t in Figure 4(a). After the disturbance has been subjected to feedback control, this outlier is clearly visible in the record of the deviations e_t from target that can usefully be plotted as a Shewhart chart as in Figure 4(d). The control limits can be calculated directly from the models used to design the controller or from the record of the e_t's during stable operation. Notice, however, that for a process that is improperly tuned deviations from the model will be blurred and special causes more difficult to detect. Moreover, for more complicated optimal schemes the error signal may need to be appropriately filtered to ensure good detectability.

Thus our conclusion is that, although feedback control could conceal features of a process that might be used for its improvement, this need not happen.

The Efficiency of Monitoring Charts

The purpose of the Shewhart chart, the Page–Barnard CUSUM chart, and the Roberts EWMA chart is, of course, *not* to regulate the process but to monitor stable operation due to common causes and to reveal special causes. For this purpose it makes sense to react to apparent process changes only if they are, in some sense, *statistically significant*. Not to do so could result in continually hunting for phantom phenomena. But if our purpose is process regulation (adjustment), then we should use a scheme specifically designed to do this. In particular, waiting for some monitoring criteria to become statistically significant before we adjust can produce a much larger mean squared error than would be obtained from a suitable adjustment scheme.

So far as the *relative* efficiency of Shewhart charts, CUSUM charts, and EWMA charts for the purpose of *monitoring* is concerned, conclusions will of course depend on the out-of-control alternatives that are being considered. For example, the CUSUM chart is associated with a likelihood ratio test that is optimally sensitive to small changes in the mean. It will therefore behave well on the assumption that a step change in mean is what we expect may happen. Cumulative statistics and corresponding charts based on other likelihood ratio tests, or equivalently on Fisher's score statistic, are easily devised, however, for other anticipated alternatives (e.g., see Bagshaw and Johnson, 1977; Box and Jenkins, 1966; Segan and Sanderson, 1980). Recently, Ramírez (1989) and Box and Ramírez (1992) discussed "Cuscore" statistics and presented charts for detection of process cycling, for detection of changes in the rate of tool wear, and for detecting the onset of nonstationarity. For example, in the last case the null model for Z_t corresponds to the white-noise model of Equation (3) and the alternative to the nonstationary IMA model with some small value of λ (say, $\lambda = 0.2$). The Cuscore statistic is of the form $Q = \Sigma e_t \hat{e}_t$, where e_t is the deviation from target and $\hat{e}_t$ is an EWMA of past e_t's with smoothing constant $\theta = 1 - \lambda$. This is eminently sensible, since on the null model e_t is unpredictable from previous e's.

Such procedures are of great value *if we know what we are looking for*. In particular, they can continually check on the assumptions underlying a control scheme. It would be unwise, however, to rely exclusively on such specialized tests, since they may be useless to detect *unexpected* forms of disturbance. One great virtue of the Shewhart chart is that it is a plot of the actual data and so can expose types of nonrandomness of a *totally unexpected* kind (e.g., that every fifth point is unexpectedly high). It acts, therefore, partly as an exploratory data-analysis tool, which should be used in combination with other directionally sensitive procedures.

MINIMUM-COST ADJUSTMENT: SOME SIMPLE SCHEMES

The process adjustment (or regulation) schemes so far considered have the property that they minimize the mean squared error of the output quality characteristic about the target value T. In particular, the elementary assumptions of Equations (4) and (16) about the nature of the disturbance and the process dynamics lead to an MMSE controller given by Equation (17) that is the discrete equivalent of a properly tuned integral controller. This form of controller would thus also minimize the mean overall cost of adjustment *if* it could be assumed that the cost of being off target was proportional to the square of the deviation from target and that other variable costs were negligible. If, however, other costs, such as that of adjusting a process and/or of taking an observation, had to be taken account of, then the resulting minimum-cost feedback adjustment schemes take on a different design configuration.

To explore some simple possibilities of this kind, suppose that (a) the cost of being off target is assumed proportional to the square of the deviation from target and that C_T is a constant representing the cost of being off target for one time interval by an amount σ_a; (b) there is a fixed cost C_A incurred by adjusting the process; and (c) there is a cost C_M incurred each time the process is observed.

Sampling Interval Fixed. When the cost of observing the process is negligible, then only C_T and C_A need be considered. For this special case with sampling interval fixed at some value that we can define as a single *unit of time*, Box and Jenkins (1963) showed that the minimal-cost scheme could be represented by a bounded adjustment chart like that introduced in Chapter D.4 and shown again in Figure 5. In this chart, the deviation from the target Z is indicated by a circle and the exponentially smoothed (forecasted) value $\dot{Z}$ is indicated by a dot. An adjustment

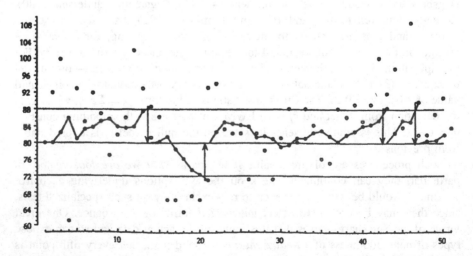

Figure 5. Illustration of a feedback-control scheme to minimize overall cost. The values of the disturbance Z are indicated by circles, the smoothed values $\dot{Z}$ are indicated by dots. The data are from an industrial series for which λ is about 0.2.

$-X_{t+} = \dot{Z}_{t-1}g$ calculated at time t is made as soon as the forecasted value $\dot{Z}_{t+1}$ falls outside one or other of the limit lines $T \pm L$. The plus sign on the subscript of X_{t+} implies that the adjustment is made during the interval between t and $t + 1$. In Figure 5 the situation is shown where $\dot{Z}$ calculated from data obtained up to time $t = 13$ is zeroed at some time between $t = 13$ and $t = 14$ by making an adjustment $-X_{13+} = \dot{Z}_{14}g$. The EWMA forecasts can be conveniently calculated from either of the recursive formulas (11), and the process can be initiated, for example, by setting $\dot{Z}_2$ equal to Z_1. Notice, however, that reinitiation is not needed after every adjustment and that after adjustment at time $t+$ the relevant value of $\dot{Z}_{t-1}$ is 0.

This *bounded adjustment chart* resembles superficially an EWMA monitoring chart. The position of the limit lines at $T \pm L$, however, are not determined by questions of statistical significance but rather depend on the ratio C_A/C_T of the adjustment cost to the cost of being off target. Simple tables that allow the value of L to be calculated given C_A/C_T, λ, and σ_a and that provide corresponding average run lengths (ARLs) between adjustments are given in the preceding references, and a more extensive table was given by Kramer (1989).

The appropriate choice of C_T that determines the off-target cost is sometimes difficult. An argument used, for example, by Taguchi (1981) is that, assuming that the cost is a quadratic function of the off-target deviation with a minimum at T, to determine the whole cost curve we need to know only one additional point. To obtain such a point it is argued that as the deviations from target increase it will reach a point $T \pm \Delta$, say, at which the manufactured material must be discarded or reprocessed at a cost c_0. Then, from the preceding definition,

$$C_T = \frac{c_0 \sigma_a^2}{\Delta^2} \quad . \tag{20}$$

Since it is sometimes hard to judge costs directly, as an alternative Box (1991b) introduced a table that provided a list of options from which the choice among minimum-cost schemes could be made empirically by balancing the advantage of longer average adjustment intervals (ARLs) against the consequential increase in the mean squared error about the target. As a further simplification, an easily used interpolation chart for manual adjustment was introduced to update the EWMA and to indicate when, and by how much, the input X should be changed.

Different Modes of Feedback Control Consequent on Cost Functions for Adjustment and for Sampling Frequency.

Before discussing further the details of feedback schemes that take account of costs of adjustment and sampling, it is important to emphasize the more general implications of such schemes. The simple feedback control loop illustrated in Figure 2 supposes (a) that the control action is applied repeatedly (at every opportunity) and (b) that the adjustment $X_t = f(e_t, e_{t-1}, \ldots)$ can be written as a linear function of past deviations from target. It is appropriate if the costs of adjustment and of sampling are negligible. Notice, however, that as soon as it is necessary to take into account the cost of adjustment a different form of feedback control becomes necessary. In that case adjustment action is triggered when some function of the e_t's exceeds certain boundary values.

Although such triggering action resembles that employed for monitoring charts such as Shewhart charts, CUSUM charts, and EWMA charts, the nature of the function considered and the position of the boundary lines is decided by relative costs and not by significance testing-like procedures. The discontinuous triggering action occurs essentially because the function describing the cost of adjustment is discontinuous—for example, the machine is either stopped to make a tool change at a cost of $\$C_A$ or it is not.

Different kinds of cost models may be appropriate for different circumstances. In particular, we mentioned earlier that although the optimal PI controller based on the IMA disturbance model and the first-order dynamic model of Equation (14) would produce MMSE at the output, this form of adjustment could require excessive control action if δ was not small. If we write $x_t = X_t - X_{t-1}$ for the adjustment made at time t, then instead of seeking a scheme that minimized mean squared error $E(e_t^2)$ or equivalently a quadratic cost function, one could minimize $E(e_t^2 + \alpha x_t^2)$, where α is a suitably chosen constant (e.g., see Åström and Wittenmark, 1984; Bergh and MacGregor, 1987; Box and Jenkins, 1970; MacGregor, 1972; and Wilson, 1970). This can represent minimization of $E(e_t^2)$ subject to a constraint on $E(x_t^2)$ or alternatively minimization of an overall cost function in which the constant α allows for the relative costs of excessive control action and of being off target. Using this approach, it is typically true, that, by appropriately changing the control equation, much smaller values of $E(x_t^2)$ are possible with relatively little increase in $E(e_t^2)$. Since the problem arises because the sampling interval is small in relation to the inertial response of the system as measured by δ, an alternative but usually less satisfactory procedure is simply to lengthen the sampling interval. A detailed comparison of these approaches was given by Kramer (1989). The resulting control schemes using this type of cost model are once again linear in the sense that the level of control action X_t can be written as linear aggregate of past deviations e_t from target. It is, of course, the nature of the problem that determines which form of cost function is appropriate or indeed whether some quite different form should be considered.

Sampling Interval Not Fixed. When the cost of taking an observation is included, the minimal cost scheme is still of the form of the EWMA bounded adjustment chart of Figure 5, but with a sampling interval of m units. The cost C_M of taking an observation is presumed to include every cost involved in obtaining the final numerical reading. This could include, among other things, the cost of sampling the product and of making a physical or chemical analysis of the sample. The sampling interval will refer to the length of time between making such observations.

To obtain minimum cost schemes when C_M is not negligible, it may be necessary to lengthen the sampling interval in comparison with schemes when the sampling cost is negligible. We can then use the following result due to Box and Jenkins (1970). If an IMA process observed at unit intervals with smoothing constant θ and innovation variance $\sigma_a^2 = \sigma_1^2$ is sampled at intervals of m units, where m is an integer, then the sampled process is also an IMA process but with smoothing constant $\theta_m = 1 - \lambda_m$ and the innovation variance σ_m^2, where

$$\theta_m = \sqrt{a_m^2 - 1}, \quad \text{where } A_m = 1 + \frac{m(1 - \theta)^2}{2\theta} \quad \text{and} \quad \sigma_m^2 = \sigma_1^2 \theta/\theta_m \quad (21)$$

Then from Equation (11) the exponentially smoothed value may be calculated from the recursive relation $\hat{Z}_{u+1} = \lambda_m Z_u + \theta_m \hat{Z}_u$ with the index $(u+1, u, u-1, \ldots)$ applied to observations made m steps apart.

Using this result, Kramer (1989) showed that, with the cost of adjustment, the monitoring cost, and the cost of being off target all included, the overall cost C per unit of time is

$$C = \frac{C_A}{\text{ARL}} + \frac{C_M}{m} + \frac{C_T}{\sigma_1^2} \frac{E\left\{\sum_{j=0}^{n-1} \sum_{k=1}^{m} e_{mj+k}^2\right\}}{\text{ARL}} \quad (22)$$

where the ARL is the average run length between adjustments and the expected value is taken with respect to both the e's and the run lengths.

By minimizing this overall cost C, it is possible to determine (a) when the process should be adjusted, (b) what size of adjustment should be made, (c) how often the process should be sampled and data collected, and (d) the average interval between adjustments (ARL) associated with each scheme.

An outline of the argument whereby this minimization is achieved is presented in Appendix C. The limit lines yielding minimum overall cost are set at $T \pm L$, where $L = l\lambda\sigma_1$ and λ and $\sigma_1 = \sigma_a$ are parameters of the original disturbance process *monitored at unit intervals*. To an adequate approximation, values of l and m can be found using Charts A and B of Figure 6. The charts require values of three quantities, the nonstationarity measure λ and two ratios R_A and R_M, given by

$$R_A = \frac{C_A/C_T}{\lambda^2}, \qquad R_M = \frac{C_M/C_T}{\lambda^2} \quad (23)$$

Strictly speaking, both l and m are functions of all three of these quantities, but the chosen parameterization has the property that l is nearly independent of λ and m is nearly independent of R_A. Thus, for most practical purposes, the simple two-dimensional representations of Charts A and B can be used. Notice that this does not imply that l and m are independent of the nonstationary measure λ but that this dependence is very nearly taken account of by the inclusion of λ^2 in the denominators of both R_M and R_A.

In those cases in which allocation of cost may be difficult, the approach described before may be adopted, where a choice between minimum-cost options is made empirically by balancing the sampling interval, the average time between adjustments, and the size of the mean squared error of the output. Tables for computing the latter two quantities are given in the earlier references.

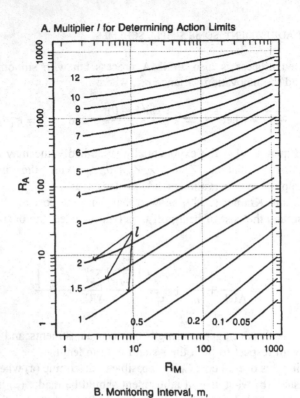

A. Multiplier *l* for Determining Action Limits

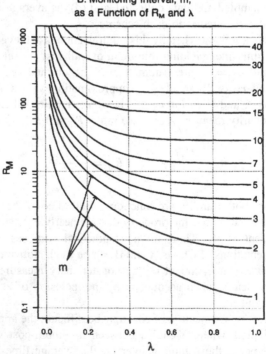

B. Monitoring Interval, m,
as a Function of R_M and λ

Figure 6. Chart A shows contours of the standardized action limit *l* for various values of R_M and R_A; Chart B shows contours of the monitoring interval *m* for various values of λ and R_M.

An Example of the Use of Charts A and B of Figure 6

Suppose that we need to control a quality characteristic subject to a disturbance that, when reconstructed from past hourly records, was identified and estimated as an IMA process with $\lambda = 1 - \theta = 0.3$ and $\sigma_a = \sigma_1 = 3.0$. Suppose that the target value is $T = 340$ and the specification limits at $T \pm \Delta$ are 325 and 355 so that $\Delta = 15$. Suppose also that material that fell just outside a specification limit would need to be reprocessed at a cost of $c_0 = \$1.350$ per hour's worth of material. Hence, from Equation (20),

$$C_T = \frac{c_0 \sigma_1^2}{\Delta^2} = \frac{1.350 \times 9}{15^2} = 54$$

The present operating procedure is to take a sample every hour and make an analysis that costs $C_M = \$200$, from the result of which the operator may decide to make an adjustment at a cost of $C_A = \$600$. Then the relative adjustment cost is $R_A = (C_A/C_T)/\lambda^2 = (600/54)/0.09 = 123.5$ and the relative monitoring cost is $R_M = (C_M/C_T)/\lambda^2 = (200/54)/0.09 = 41.2$. From Chart A we see that l is about 3.2 so that $L = l\lambda\sigma_1 = 3.2 \times 0.3 \times 3 = 2.9$ and the limit lines should be placed at 340 ± 2.9. Moreover, from Chart B with $\lambda = 0.3$ and $R_M = 41.2$, m is about 10 units, so the monitoring interval should be about 10 hours.

For this example using Equation (21) with $\theta = 0.7$ and $m = 10$, we find that $\theta_m = 0.34$. Thus the smoothed value of the disturbance monitored at the interval $m = 10$ computed from $\hat{Z}_{u+1} - 0.66 Z_u + 0.34 \hat{Z}_u$ is plotted on the chart, and the process is adjusted as soon as this smoothed value crosses either limit line.

Special Cases

1. $\lambda = 0$. In this case, the disturbance is a white-noise stationary process. The standardized limit λ and the monitoring interval m are then both infinite. This says that the control action to be taken for a process known to be in a perfect state of control is *no action*.

2. C_A Negligible. In this case, for any fixed m (say, $m = 1$) l tends to 0 and the limits $T \pm L$ converge on the target value. Adjustments must therefore be made as *each new value becomes available*. Each adjustment is made to just cancel the deviation of the exponentially smoothed value from the target value. Hence, after the ith observation, the effect of the adjustment is $\hat{Z}_{t+1} - \hat{Z}_t$. These adjustments are made by manipulating X so that the total adjustment at time t is

$$-X + t = k_0 + \frac{1}{g}\sum_{i=1}^{t}(\hat{Z}_{t+1} - \hat{Z}_t)$$

$$= k_0 + \frac{\lambda}{g}\sum_{i=1}^{t}e_t = k_0 + k_1\sum_{i=1}^{t}e_t$$

which is the MMSE feedback equation (17). Notice that it provides minimum cost of regulation only if the cost of adjustment C_A is negligible. Control action depends on the EWMA of the disturbance Z or equivalently the cumulative sum of the output errors e. This is not, of course, the same as adjustment based on a CUSUM of the disturbance Z itself as has sometimes been proposed.

3. C_A *Is Negligible*, C_M *Is Not, and m Is Not Fixed.* In this case a feedback scheme is obtained in which limit lines are on the target so that adjustments are made after each observation, but these observations may be made less frequently. Consider again the numerical example we used earlier with $C_M = \$200$, but suppose now that C_A is 0. The value of l would now be 0, but the value of m obtained from Chart B would still be about 10 so that adjustments would be made after each observation taken 10 unit intervals apart.

4. C_A *Is Not Negligible, m Is Fixed.* In this case, action limits $T \pm l\lambda\sigma$ are determined directly by R_A. This was the special case considered by Box and Jenkins (1963). Adjustments, when needed, are made to bring the predicted value to the target value.

5. $\lambda = 1$. This random-walk case could theoretically occur, and *only* in this case could adjustment action based on Z_t rather than $\hat{Z}_t$ be justified. This degree of nonstationarity is so extreme, however (meaning, for example, that the variance of the uncontrolled process will double after only two intervals of time), that it can hardly be regarded as describing any control situation likely to be met in practice. Indeed Adams and Woodall (1989) showed that Taguchi's approximations derived from behavior associated with a random walk are inadequate and can lead to suboptimal schemes.

DISCUSSION AND CONCLUSIONS

It has not been our intention to discuss in any degree of generality the various techniques we have touched on but rather to employ the simplest possible models to illustrate the philosophical points we wished to make. Although excellent works explaining some of the statistical issues to control engineers are available (e.g., see MacGregor, 1987, 1988), there still seems to be a need for further discussion with statistical-quality-control practitioners.

Some of the points we have illustrated are:

1. Standard process control charts are tools whose primary uses are to repeatedly *monitor* the common-cause system and to *detect* significant deviations possibly pointing to special (assignable) causes. They appropriately employ considerations paralleling statistical significance tests to trigger action and so reduce the chance of fruitless pursuit of phenomena produced by chance alone. Such devices are not appropriate for process adjustment and when so used may unnecessarily increase variation about the target value.

2. When the quality-control practitioner needs to *adjust* a process he/she should understand some of the simple principles of feedback control (and possibly also of feedforward control). Standard SPC ideas are not enough.

3. Process control using properly designed feedforward and feedback systems is a tool properly designed for process *adjustment* to compensate for a disturbance the cause of which is unknown or it is impossible or too expensive to remove. When such control systems are poorly *designed* or poorly *tuned* however, it is possible for them to increase rather than reduce variation about the target value.

4. Multivariate feedback schemes can be designed by modeling the process dynamics and the disturbance at the output, but this approach can be tedious. Alternatively an existing control system—involving, for example, several PI controllers—can be tuned by using experimental design.

5. Feedback-control schemes are often operated so that the nature of the disturbance that is being compensated is concealed and unusual deviations from the target cannot be taken account of. It is pointed out that this need not happen. In this connection it is useful to extend the idea of common causes and special causes to such systems.

6. One reason that "off the shelf" controllers may not work very well is that they are mistuned. Another is that the design of the controller may be wrong. For example, when costs of adjustment and/or of sampling are nonnegligible, continuous feedback may be inappropriate.

7. Methods are discussed whereby simple minimum-cost feedback-control schemes can be designed if appropriate cost constants—for being off target, for making an adjustment, and for taking an observation—are approximately known. When these constants are difficult to determine, an alternative is to choose among a sample of alternative optimal schemes by empirically balancing the increase in the average time between needed adjustments and a desirable sampling interval against the inflation of the mean squared error at the output.

ACKNOWLEDGMENT

We are indebted to an associate editor and a number of referees and to M. S. Srivastava for their constructive comments. Special thanks are due to Alberto Luceño, whose assistance has been invaluable.

APPENDIX A: A STICKY INNOVATION MODEL WITH A LINEARLY INCREASING VARIOGRAM

In Section 2.1 it was argued that an uncontrolled system is subject to influences that steadily increase its entropy or disorder. We could represent the output Z_t from such an uncontrolled system by $Z_t = b_t + S_t$, where the $\{b_t\}$ are transitory errors induced by measurement and sampling. There is also a nontransitory part $\{S_t\}$, which may also be called the signal representing the accumulated effect of "sticky" innovations that enter the system and remain there. Examples of such sticky innovations are a small increment of *permanent* damage sustained when an automobile tire hits a

Table A1. Relationship of λ to the Parameters of the Noisy Random Walk and to the Inflation Factor

σ_a/σ_b	λ	σ_a/σ_b
0.11	0.1	1.05
0.22	0.2	1.12
0.36	0.3	1.20
0.52	0.4	1.29
0.71	0.5	1.41

sharp stone, a tiny crater on the surface of a driving shaft caused by corrosion that *remains* there, and a certain detail in the standard hospital procedure for taking a blood pressure that is inadvertently omitted or changed *from that point onward*. Then (see also Barnard, 1959) for discrete data we suppose that in any time interval the probability of occurrence of a sticky innovation that changes the level of disturbance is p and also that the effect of the sticky innovation that occurs at time t is C_t. We further suppose that b_t and c_t are represented by independent white-noise innovations with variance σ_b^2 and σ_c^2, respectively. The variogram for such a process is

$$G_m = \frac{V_m}{V_1} = \frac{2\sigma_b^2 + mp\sigma_c^2}{2\sigma_b^2 + p\sigma_c^2} = 1 + \beta(m-1)$$

where

$$\beta = \frac{p\sigma_c^2}{2\sigma_b^2 + p\sigma_c^2}$$

If we define $\{d_t\}$ as a series of white-noise innovations with variance $p\sigma_c^2 = \sigma_d^2$, then we can approximate Z_t by the *noisy random walk* $Z_t = b_t + \sum_{i=1}^{t} d_i = b_t + S_t$. The feedback control problem we have considered can be thought of as that of estimating the *signal* S_t and cancelling its effect in the presence of the noise b_t. Such an estimate is supplied by $\hat{Z}_t$ of Equation (5) with error σ_a^2.

By equating moments, it is easy to show that $\sigma_d/\sigma_b = \lambda/\theta^{1/2}$. Moreover, $\sigma_a/\sigma_b = \theta^{-1/2}$ is an *inflation factor*, which measures how good the filtering process for separating the signal from the noise is. It is the ratio of the standard deviations of the white-noise *forecast errors* a_t of the filtered series and the white noise of the generating series. In particular, note from Table A1 that if λ is not larger than, say, 0.5, it is possible to filter out a great deal of the signal corresponding to the cumulated sticky innovations.

APPENDIX B: INFLATION OF VARIANCE DUE TO MISTUNING OF A SIMPLE FEEDBACK CONTROLLER

For the feedback scheme of Equation (17) with the disturbance represented by an IMA time series model and the process dynamics by Equation (16), to achieve MMSE, we need to set the controller constant $k_I = \lambda/g$, in which case $\sigma_e^2/\sigma_a^2 = 1$. But suppose that k_I is set equal to some value $\dot{k}_I$, which is not necessarily equal to k_I; then

$$e_t = Z_t + Y_t$$

$$= \text{constant} + a_t + \lambda \sum_{i=1}^{t-1} a_i - g\dot{k}_I \sum_{i=1}^{t-1} e_i \tag{B1}$$

After setting $\varphi = 1 - g\dot{k}_I$ and differencing, Equation (B1) implies that

$$e_t - \varphi e_{t-1} = a_t - \theta a_{t-1}, \quad \text{where } \theta = 1 - \lambda; \ 0 \le \theta \le 1 \tag{B2}$$

For this autoregressive moving average process,

$$\frac{\sigma_e^2}{\sigma_a^2} = \frac{1 + \theta^2 - 2\varphi\theta}{1 - \varphi^2}, \quad -1 < \varphi \le 1 = \infty, \quad \text{otherwise} \tag{B3}$$

This shows the degree of inflation of the variance due to mistuning for any desired values of $\lambda = 1 - \theta$, g, and $\dot{k}_I = (1 - \varphi)/g$.

In the special case discussed in Section 3.1, Z_t is white noise so that $\lambda = 0$, $\theta = 1$, and

$$\frac{\sigma_e^2}{\sigma_a^2} = \frac{2}{1 + \varphi}, \quad -1 < \varphi \le 1$$
$$= \infty, \quad \text{otherwise} \tag{B4}$$

From which Equation (18) follows.

APPENDIX C: MINIMAL COST REGULATION: AN OUTLINE OF THE ARGUMENT

Suppose that a process that is subject to an IMA disturbance is observed at regular intervals. Suppose further that adjustments can be made at a cost C_A, which can bring the observed process to a desired target, T, within one interval. Finally, suppose that there is a cost associated with being off target that is proportional to the square of the deviation with proportionality constant k_T.

Now, if the process is going to make just one additional unit of material, then the expected cost when an adjustment is made is

$$
k_T E(e_{t+1}^2) + C_A
$$
$$
= k_T E(a_{t+1}^2) = k_T (\hat{Z}_{t+1} + Y_{t+1})^2 + C_A \tag{C1}
$$

which is minimized by setting the compensating variable, Y_{t+1}, so that it negates the predicted disturbance, $\hat{Z}_{t+1}$. When this is done, the expected cost for adjustment is $k_T \sigma_a^2 + C_A$. If no adjustment is made, then the expected cost is

$$
k_T E(Z_{t+1} + Y_{t+1})^2 = k_T \sigma_a^2 + k_T (\hat{Z}_{t+1} + Y_{t+1})^2 \tag{C2}
$$

Hence an adjustment should be made if

$$
|\hat{Z}_{t+1} + Y_{t+1}| \geq \left(\frac{C_A}{k_T} \right)^{1/2} \equiv L_1 \tag{C3}
$$

One can show that, if N additional items are to be made, then the process should be adjusted if

$$
|\hat{Z}_{t+1} + Y_{t+1}| \geq L_N \tag{C4}
$$

where L_N is a function of the adjustment cost, the off-target proportionality constant, and the disturbance parameters. The values L_1, L_2, ... form a sequence that has a limiting value L. Therefore, the optimal strategy for an infinite horizon is to adjust the process when the forecasted observation deviates from target by an amount L.

Suppose now that the process is observed at a multiple m of the unit interval and associated with each observation is a cost C_M. The preceding logic holds once again, and the minimal cost strategy is to again adjust when the forecasted observation deviates from the target by an amount $L(m)$ in which the notation indicates that the limit will depend on the monitoring interval as well as the costs and disturbance parameters. If ARL is the average run length, it can be shown that the expected cost per run (interval between adjustments) is then

$$
C_A + \text{ARL} \frac{C_M}{m} + k_T E \left\{ \sum_{j=0}^{n-1} \sum_{k=1}^{m} e_{mj+k}^2 \right\} \tag{C5}
$$

The expectation is taken over both the deviations from target, e_i, and the run length, n, from which the expected cost per interval is

$$
\frac{C_A}{\text{ARL}} + \frac{C_M}{m} + \frac{k_T}{\text{ARL}} E \left\{ \sum_{j=0}^{n-1} \sum_{k=1}^{m} e_{mj+k}^2 \right\} \tag{C6}
$$

For a disturbance with parameters (λ, σ_a^2), monitoring interval m, and limits at $\pm L = l\lambda\sigma_a$, this cost can be expressed as

$$\frac{C_A}{mh(l/\sqrt{m})} + \frac{C_M}{m} + k_T \frac{\theta}{\theta_m}\sigma_a^2$$
$$+ mk_T\lambda^2\sigma_a^2 g(l/\sqrt{m}) - \frac{k_T(m-1)\lambda^2\sigma_a^2}{2} \tag{C7}$$

where $g(l)$ and $h(l)$ are both functions solely of l that relate to the expected squared deviation from target and ARL, respectively, for a standardized random walk prior to crossing a boundary.

This last equation may be scaled by dividing by $k_T\lambda^2\sigma_a^2$ to give

$$\frac{R_A}{mh(l/\sqrt{m})} + \frac{R_M}{m} + \frac{\theta}{\lambda^2\theta_m} + mg(l/\sqrt{m}) - \frac{(m-1)}{2} \tag{C8}$$

This form shows that the choice of l and m that minimizes the total expected cost can be obtained as functions of the three quantities R_a, R_m, and λ. Methods for obtaining approximations for $g(l)$ and $h(l)$ were given by Box and Jenkins (1963), Crowder (1987), and Lucas and Crosier (1982); see also Srivastava and Wu (1991). The approximations used in this chapter were obtained by modeling extensive simulations. These approximations were then used to determine values l and m, which minimized the preceding equation.

A difficulty in putting this result to practical use is that triple-entry tables would be required. As explained in the text, by using the preceding parameterization values of l and m that give approximately minimal cost can be obtained from two double-entry tables or alternatively from Figures 6A and 6B.

APPENDIX D: GLOSSARY OF SOME TERMS USED IN THE CHAPTER

Output—quality characteristic to be controlled.

Input—compensating variable that can change the level of the output.

EWMA—exponentially weighted moving average of past data.

Adjustment—change in input level required to compensate for output deviation.

Disturbance—course followed by output if no compensatory adjustments are made.

Process monitoring—verifying that a constant common-cause system remains in place and detecting deviations from the common-cause system pointing to possible special (assignable) causes.

Common-cause system—a probability model that describes the normal stable operation of the system.

State of control—broadly speaking, a process state in which future behavior can be predicted within probability limits determined by the common-cause system; often more narrowly interpreted as stationary or iid variation about a target value.

Forecasting—use of past data to estimate a future value.

Feedback control—using past output deviations from target to determine a process adjustment.

White-noise innovations $\{a_t\}$—a series of independent, approximately normally distributed, random variables having mean 0 and (innovation) variance σ_a forming a white-noise sequence.

ACKNOWLEDGMENT

This research was sponsored by the National Science Foundation under Grant No. DDM-8808138.

CHAPTER D.6

Discrete Proportional-Integral Control with Constrained Adjustment

SUMMARY

It is well known that discrete feedback control schemes chosen to produce minimum mean-squared error at the output can require excessive manipulation of the compensating variable. Also very large reductions in the manipulation variance can be obtained at the expense of minor increases in the output variance by using constrained schemes. Unfortunately, however, both the form and the derivation of such schemes are somewhat complicated. The purpose of this paper is to show that suitable "tuned" proportional-integral (PI) schemes in which the required adjustment is merely a linear combination of the two last observed errors can do almost as well as the more complicated optimal constrained schemes. If desired, these PI schemes can be applied manually by using a feedback adjustment chart which is no more difficult to use than a Shewhart chart. Several examples are given and tables are provided that allow the calculation of the optimal constrained PI scheme and the resulting adjustment variance and output variance. Methods of tuning such controllers by using evolutionary operation and experimental design are briefly discussed.

INTRODUCTION

In the last chapter two complementary types of process control were discussed: *process monitoring* for the elimination of removable causes of variation from the

From Box, G. E. P. and Luceño, A. (1995), *The Statistician, JRSS Series D*, **44**(4), 479–495. Copyright © 1995 Royal Statistical Society.

system and *process adjustment* for the regulation of the system. The latter involves in particular feedback control. It was emphasized that, to control a process, it is important first to standardize materials and techniques and to eliminate systematically other assignable causes of variation with the help of process monitoring charts such as Shewhart, cumulative sum, and Roberts's exponentially weighted moving average (EWMA) charts.

Often, however, disturbances produced, for example, by varying ambient temperatures and feedstock cannot be eliminated in this way but can be compensated by feedback control. This paper concerns simple ways of applying such feedback control while reducing as far as possible the amount of manipulation needed to do this.

Suppose, for some operating process, that observations of an output quality characteristic are made at equally spaced intervals of time. The length of such intervals would depend on the rate of response of the system; for a waste water treatment plant it might be a day, for a chemical process an hour, and for an atomic reactor a second. Suppose that in each such interval an adjustment to an input compensating variable X can be made in an attempt to bring the output as close as possible to some target value.

A simple scheme of feedback control to achieve such compensation is shown in Figure 1 where, for illustration, we consider the problem of controlling a dyeing process by manipulating the input dye addition rate X to maintain the output colour index of the dyed material as closely as possible to the desired target value. [More detailed discussions of discrete feedback control in relation to time series and dynamic models will be found in Box and Jenkins (1970, 1976) or in Box, Jenkins, and Reinsel (1994) and in Box and Luceño (1997) to which frequent references will be made. Other basic references are Aström (1970), Aström and Wittenmark (1984), and MacGregor (1972, 1987).] Continual adjustment of this kind may be needed, for

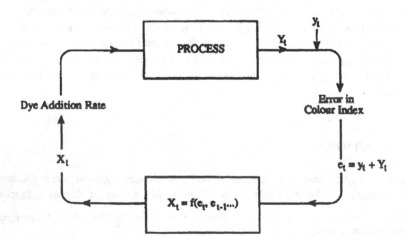

Figure 1. Feedback control loop illustrated for a dyeing operation: Y_t is the net effect at the output of all past adjustments in X_t and y_t is the net effect at the output of disturbances in the system.

example, because of drifts in the ambient temperature or in the quality of raw materials or in other "noise" variables which can neither be eliminated nor allowed for in advance. The next effect of such noise on the output of the process *when no compensatory action* is taken will be called the *disturbance* y_t. For simplicity we shall suppose that y_t is measured as a deviation from a fixed output target value T. Also we suppose that at time 0 the adjustment variable X was set so that the output was on target. Then, ideally, at time t, the net effect Y_t at the output of all the previous compensatory adjustments $X_{t-1} - X_{t-2}$, $X_{t-2} - X_{t-3}, \ldots$ would just cancel the difference y_t (so that $Y_t = -y_t$). In practice, this would not be possible to do exactly and an error (deviation from target) of $e_t = y_t + Y_t$ would occur. These errors e_t, e_{t-1}, $e_{t-2}, \ldots$ may then be fed back to decide the next appropriate adjustment $X_t - X_{t-1}$. Thus in the dyeing example the level X_t at time t of the adjustment (input) variable is the dye addition rate and this is adjusted in the light of current and past deviations e_t, $e_{t-1}, \ldots$ from target of the (output) colour index. This paper is about the choice of the *control equation*

$$X_t = f(e_t, e_{t-1}, \ldots)$$

so as to reduce as much as possible both the output error and the amount of adjustment which is needed. It will be noted from the above that the values $y_t, y_{t-1}, \ldots$ of the disturbance represent deviations from target that would occur if no adjustments were applied (i.e., if $X_t = X_{t-1} = X_{t-2} = \cdots = X_0$). The errors $e_t, e_{t-1}, \ldots$ represent such deviations when some kind of feedback control, involving manipulation of X_t, is operating.

Perhaps the most intuitive form of feedback control is such that will make the *adjustment* $x_i = X_i - X_{i-1}$ of the input variable at time i proportional to e_i, the last error at the output, so that

$$x_i = c_1 e_i \qquad (1)$$

where c_1 is a constant whose choice is considered later. By summing Equation (1) from 1 to t, we obtain a control equation of the form

$$X_t = k_0 + k_1 \sum_{i=1}^{t} e_i \qquad (2)$$

where $k_1 = c_1$ and $k_0 = X_0$ is a constant defining the initial settings of X. Thus, the control produced by Equation (1) is such that the level of the adjustment variable X_t depends linearly on the cumulative sum of past errors. The control equation (2) is the discrete analogue of the control engineer's continuous *integral control* in which the adjustment depends linearly on the integral of past errors.

An aspect of the simple form of control defined by Equations (1) and (2) is its relationship to exponential smoothing. Suppose that g is the *steady state gain* of the system, that is, a unit adjustment in the input eventually will produce g units of change at the output. Suppose also that this change becomes fully effective in one time interval so that $Y_t = gX_{t-1}$. Then (as shown below) the control equation (2) can alternatively be written

$$X_t = -\frac{1}{g}\hat{y}_{t+1} \tag{3}$$

where, with $\tau = 1 + c_1 g$,

$$\hat{y}_{t+1} = (1 - \tau)(y_t + \tau y_{t-1} + \tau^2 y_{t-2} + \cdots), \quad 0 < \tau < 1 \tag{4}$$

and τ will be referred to as the smoothing constant. Thus the control equation (2) adjusts X_t so that it just cancels an EWMA of present and past values $y_t, y_{t-1}, \ldots$ of the disturbance.

To see this, from Figure 1 and Equation (1),

$$X_t - X_{t-1} = c_1 e_t = c_1(y_t + gX_{t-1})$$

and so, defining τ as above,

$$X_t - \tau X_{t-1} = -\frac{1 - \tau}{g} y_t \tag{5}$$

which is a first-order difference equation with solution (3).

Now EWMAs are often used empirically to estimate (forecast) a future value in a non-stationary time series. The control equation is thus equivalent to continually adjusting X_t to cancel the forecast $\hat{y}_{t+1}$, made at time t, of the disturbance at time $t + 1$.

A slight elaboration of Equation (1) would make the adjustment depend linearly on the last two errors:

$$x_i = c_1 e_i + c_2 e_{i-1} \tag{6}$$

Summing as before we obtain

$$X_t = k_0 + k_P e_t + k_1 \sum_{i=1}^{t} e_i \tag{7}$$

where $k_0 = X_0 + c_2 e_0$, $k_P = -c_2$, and $k_1 = c_1 + c_2$. Thus the control produced by Equation (7) is such that X_t is a linear combination of the last error e_t and the cumulative sum of previous errors $\sum_{i=1}^{t} e_i$.

The simple adjustment equation (6) thus produces control which is a discrete analogue of continuous *proportional-integral* (PI) control widely used in the process industries. (Automatic "PID" controllers include a third "differential" term which takes account of the rate of change of the error. If in the corresponding discrete analogue the derivative is replaced by a finite difference, the adjustment $x_t =$

$X_t - X_{t-1}$ becomes a linear function of the last *three* errors.) These continuous controllers and their corresponding discrete counterparts have, when appropriately tuned, been remarkably successful. See, for example, Mayr (1970), Fearn and Maris (1991) and Box (1991b). They are also easy to apply. Indeed, the discrete controllers can be put into operation manually by using *feedback adjustment charts* that are no more difficult to use than Shewhart monitoring charts. Alternatively, it may be shown that this is equivalent to setting

$$\hat{X}_t = -\frac{1}{g}\hat{y}_t \qquad (8)$$

where $\hat{X}_t$ is an exponentially weighted mean of past values $X_{t-1}, X_{t-2}, \dots$ with smoothing constant $-c_2/c_1$. The main topic of this paper concerns the choice of the constants that determine the behavior of the control schemes and consequently the design of the appropriate charts when manual control is employed.

FEEDBACK ADJUSTMENT CHARTS

Feedback control is often thought of as exclusively a part of "automatic" process control or "computer" control. However, this form of control is also of value to the quality practitioner when manual adjustment of the process is required.

Integral Action

For illustration consider again the dyeing process referred to earlier in which, by manipulating the dye addition rate, the colour index of the dyed material is kept close to a target value ($T = 9$). Suppose at first that the desired discrete control was of the "integral" type of Equation (2) with $k_1 = c_1 = -5$. The control equation is then

$$X_t = k_0 - 5\sum_{i=1}^{t} e_i$$

and the corresponding adjustment equation is

$$x_t = -5e_t$$

A simple chart that the operator can use to achieve this action is shown in Figure 2. The input adjustment scale shown on the left-hand side and the output scale shown on the right-hand side are such that a unit deviation from target in the colour index calls for minus five units of adjustment in the dye addition rate. The operator records on the chart the current value of the colour index and immediately reads off on the left-hand scale the required adjustment to the dye addition rate. Thus at time

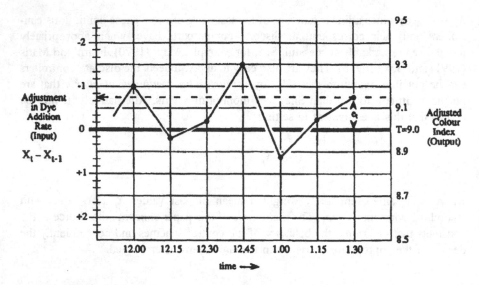

Figure 2. Feedback control chart to achieve integral action.

1.30 p.m. the colour index was 9.14, calling for a reduction of 0.7 units in the dye addition rate.

Proportional-Integral Action

If instead PI action was required (say, with $k_P = -1.25$ and $k_I = -5$), then to achieve the required action

$$X_t = k_0 - 1.25 e_t - 5 \sum_{i=1}^{t} e_i$$

A feedback control chart like that of Figure 3 could be employed with dotted lines placed at a fraction $k_P/k_I = 0.25$ of each interval. The operator would then read off the desired control action at each stage by drawing a line through the last two points and extrapolating 0.25 intervals ahead (as indicated by the arrows). The adjustment would be obtained by reading off the extrapolated value on the adjustment scale. Thus in Figure 3 the last two readings at 1.15 p.m. and 1.30 p.m. were 9.06 and 9.14. The projected value of 9.16 requires a reduction of the dye addition rate of 0.8 units. Note that the use of these charts requires no calculation, that is, they are entirely graphical. They are also very robust so that for most practical schemes visual inspection of the extrapolation would be sufficiently good.

If ∇ is the back difference operator so that $\nabla e_t = e_t - e_{t-1}$, then the adjustment equation (6) can be written

$$x_t = -G(1 + P\nabla)e_t$$

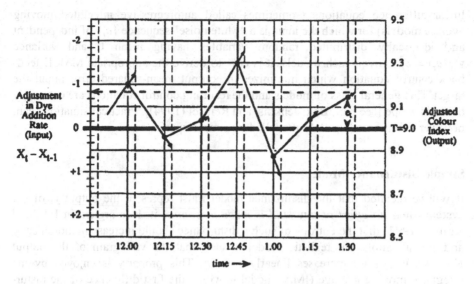

Figure 3. Feedback control chart to achieve PI action.

where $G = -k_I$ and $P = k_P/k_I$. In general then, in terms of the manual adjustment chart, the meaning of G and P is as follows.

(a) The constant G determines the relative size of the units on the scales of A_t and e_t.

(b) The constant P determines the required degree of interpolation or extrapolation of the last two points on the chart. If P is positive, the required action is obtained by *extrapolating* P units along the line joining e_t to e_{t-1}. If P is negative, a corresponding *interpolation* between e_t and e_{t-1} is called for.

SIMPLE TIME SERIES AND DYNAMIC MODELS AND MINIMUM MEAN-SQUARED ERROR CONTROL

So far we have discussed discrete PI control from an empirical point of view. No model or theoretical justification has yet been introduced. This is deliberate for it is important to understand that this form of control can be (and historically has been) developed from a purely commonsense standpoint. with this understanding it is hardly surprising that PI control is useful over a much wider field than that covered, for example, by the minimum mean-squared error (MMSE) justification briefly discussed below.

Processes sometimes contain inertia such that some time elapses before a change at the process input becomes fully effective at the output. Useful discrete models to represent such dynamic behavior are provided by linear difference equations. Correspondingly, models to describe the disturbance y_t are provided by stochastic linear difference equations (sometimes called autoregressive integrated moving average models) for which the input is a "white noise" sequence $\{a_t\}$ of independent and identically distributed random variables having mean 0 and variance $\mathrm{var}(a_t) = \sigma_a^2$. Given such models, it is easy to derive the appropriate MMSE feedback control equation which minimizes the output mean-squared error about the target. This general class of models and the resulting control equations are extremely broad in scope [see e.g., Box, Jenkins, and Reinsel (1994)]. Practical situations can, however, often be adequately modelled by simple special cases.

Simple Disturbance Model

It will be recalled that the disturbance model must represent the output y_t of the system when *no effort of any kind is made to control* it. It is argued in Box and Kramer (1992) that the entropy of such a disturbance should increase monotonically and that a simple but realistic model is such that the variogram of the output $V_m = \mathrm{var}(y_{t+m} - y_t)$ increases linearly with m. This property is enjoyed by the integrated moving average (IMA) model in which the first difference of the disturbance follows a first-order moving average time series

$$y_t - y_{t-1} = a_t - \theta a_{t-1} \tag{9}$$

From an empirical standpoint, identification of actually occurring disturbances has often led to this model or some close approximation to it. As is well known, the MMSE forecast at time t for all future time of a series generated by Equation (9) is given by $\hat{y}_{t+1}^{(0)}$, the EWMA defined by Equation (4) with $\tau = \theta$. If the system is such that corrective action has its full effect in one time interval so that $Y_t = gX_{t-1}$, it is easy to show that, for a disturbance represented by Equation (9), integral control action defined by Equations (1) and (2) yields MMSE at the output *if* the constant k_1 is chosen so that

$$k_1 = c_1 = -\lambda/g, \quad 0 < \lambda < 1$$

where $\lambda = 1 - \theta$.

More generally, when the process has inertial characteristics such that a change at time t would not be realized at the output during one interval of time, it is frequently possible to represent such dynamic behavior by a "linear filter." Specifically, the output from the system is modeled by $Y_t = g\tilde{X}_t$, where $\tilde{X}_t$ is a weighted average of previous levels of the input $X_{t-1}, X_{t-2}, \ldots$. A simple dynamic model which provides such a filter is such that the change $\tilde{X}_t - \tilde{X}_{t-1}$, which occurs at the (scaled) output in the time interval between t and $t-1$, is proportional to the deviation $X_{t-1} - \tilde{X}_{t-1}$ between the input and the (scaled) output at time $t-1$ so that

$$\tilde{X}_t - \tilde{X}_{t-1} = (1 - \delta)(X_{t-1} - \tilde{X}_{t-1}), \quad 0 < \delta < 1$$

where the constant $1 - \delta$ can be interpreted as the proportion of the eventual output change that will occur in one time interval after a "step" change has been made in the input. Thus the closer δ is to 1 the slower the system will react. Equivalently,

$$\tilde{X}_t - \delta \tilde{X}_{t-1} = (1 - \delta)X_{t-1} \tag{10}$$

This first-order difference equation provides a discrete analogue of a first-order dynamic system typified, for example, by Newton's law of cooling. If the system dynamics can be represented in this way, then it is easy to show that PI action defined by Equations (6) and (7) will yield an MMSE if the constants k_I and k_P are chosen such that

$$
\begin{aligned}
k_I &= -\lambda/g \\
k_P &= -\lambda\delta/g(1 - \delta)
\end{aligned} \tag{11}
$$

The solution of Equation (10) [compare Equation (5)] is

$$\tilde{X}_t = \hat{X}_t^{(\delta)} = (1 - \delta)(X_{t-1} + \delta X_{t-2} + \delta^2 X_{t-3} + \cdots)$$

Thus the linear filter associated with the difference equation (10) is an EWMA with smoothing constant δ. It is then easy to show that the PI control equation with constants given by Equations (11) is equivalent to Equation (8) with $\delta = -c_2/c_1$, that is, to continuously arranging that

$$\hat{X}_t^{(\delta)} = -\frac{1}{g}\hat{y}_t^{(\theta)}$$

That is, an EWMA with smoothing constant δ of past values of the input variable is chosen to cancel an EWMA with smoothing constant θ of the disturbance.

Sometimes the dynamic model must allow, in addition, for "dead time," that is, pure delay of, say, f units in response to adjustment. The equation for a first-order system with f units of delay [see Equation (10)] is

$$\tilde{X}_t - \delta \tilde{X}_{t-1} = (1 - \delta)X_{t-1-f} \tag{12}$$

In that case the MMSE is not given by PI control but by a scheme for which the adjustment equation is of the form

$$x_t = d_1 x_{t-1} + d_2 x_{t-2} + \cdots + d_f x_{t-1-f} + d(e_t - \delta e_{t-1})$$

where the d's are suitable coefficients.

From the above it will be seen that one rationalization for the use of PI control might be that, for the case where $f = 0$, it produces MMSE at the output for perhaps the simplest models for disturbance and for dynamics which can approximate reality.

However, from Equation (11) it is readily seen that the coefficients G and P of the feedback chart that then yields MMSE must be chosen such that

$$G = \lambda/g$$

$$P = \delta/(1 - \delta)$$

We see, therefore, that with $0 < \delta < 1$ the chart will always require extrapolation rather than interpolation. Furthermore, a chart thus constructed for a slow acting system in which δ is close to 1 can require very extensive extrapolation and, consequently, can lead to instability. For example, for $\delta = 0.9$, $P = 9$ and nine whole units of extrapolation would be called for. This can result in highly unstable schemes. Suppose, for instance, that X_t is a reaction temperature. We may be told at time t to increase the temperature by 2000 °C and at time $t + 1$ to reduce it by 3000 °C.

CONSTRAINED ADJUSTMENT SCHEMES

Thus although, at first sight, the criterion that the control scheme should produce MMSE at the output seems eminently sensible, such MMSE control can require excessive and impractical manipulation of the compensating variable X_t. Various schemes have been devised, therefore, in which the range of compensatory manipulation is constrained (Box and Jenkins, 1970, Åström (1970, 1976); Box, Jenkins, and Reinsel, 1994; Wilson, 1970; MacGregor, 1972; Åström and Wittenmark (1984)). In particular, such constrained schemes can be obtained by finding an unconstrained minimum of the expression

$$\text{var}(x_t) + \beta \, \text{var}(e_t) \tag{13}$$

where β can be regarded as an undetermined multiplier or as a constant which allocates the relative quadratic costs of the departures of e_t and x_t from zero. Options may be explored by making calculations for various choices of β.

Major Reductions in Manipulation Variance var(x$_t$) Possible with Constrained Schemes

A characteristic of constrained schemes obtained in this way is that, frequently, extremely large reductions in the manipulation variance $\text{var}(x_t)$ can be obtained at the expense of very minor increases in the output variance $\text{var}(e_t)$. To illustrate this point consider the following example. Suppose that the disturbance model is of the form of Equation (9) with $\lambda = 1 - \theta = 0.4$ and $\sigma_a^2 = 1.0$, and also that the dynamics can be represented by a first-order system with $g = 0.4$ and $\delta = 0.5$ and with dead time characterized by $f = 1$.

The MMSE scheme is

$$x_t = -0.40x_{t-1} - 2.00e_t + 1.00e_{t-1}$$

giving variances $\text{var}(e_t) = 1.16$ and $\text{var}(x_t) = 5.00$.

Two examples of optimal constrained schemes obtained by minimizing expression (13) for different values of β are

(a) $x_t = 0.29x_{t-1} - 0.11x_{t-2} - 0.98e_t + 0.49e_{t-1}$, giving $\text{var}(e_t) = 1.21$ and $\text{var}(x_t) = 1.00$, and

(b) $x_t = 0.75x_{t-1} - 0.21x_{t-2} - 0.39e_t + 0.19e_{t-1}$, giving $\text{var}(e_t) = 1.36$ and $\text{var}(x_t) = 0.25$.

Thus for scheme (a), at the expense of an increase of about 4% in the output variance, an 80% reduction in the input variance is possible. Scheme (b) produces an increase of 17% in the output variance and a reduction of 95% in the adjustment variance.

Substitution of Proportional-Integral Schemes for Optimal Constrained Schemes

Although optimal constrained schemes such as those illustrated above can produce dramatic reductions in the adjustment variance $\text{var}(x_t)$ at the expense of only minor increases in the output variance $\text{var}(e_t)$, they can be somewhat complicated in form and in their derivation. The main purpose of this chapter is to show that, for a simple and useful class of models, simple PI schemes when suitably tuned can do almost as well. Thus in what follows we shall assume that we are to use a PI adjustment equation

$$-x_t = G(1 + P\nabla)e_t$$

or correspondingly a control equation

$$-X_t = -k_0 + GPe_t + G\sum_{i=1}^{t} e_i$$

and we shall consider the appropriate choices of G and P for various constrained schemes.

That this approach is useful is illustrated by considering again the example in the previous subsection. Constrained PI schemes matching (a) and (b) in that they produce essentialy the same reduction in $\text{var}(x_t)$ are

(a') $x_t = -0.61(1 + 0.62\nabla)e_t$, giving $\text{var}(e_t) = 1.21$ and $\text{var}(x_t) = 1.00$, and

(b') $x_t = -0.45(1 - 0.11\nabla)e_t$, giving $\text{var}(e_t) = 1.37$ and $\text{var}(x_t) = 0.25$.

Since (a) and (b) are *optimal* constrained solutions, (a′) and (b′) are necessarily not optimal, but we see that to an accuracy of two decimal places the PI schemes do as well as the much more complicated constrained schemes. In Appendix A we further demonstrate that PI schemes doing almost as well as optimal constrained schemes can be found for a wide range of situations.

Note that for the more "heavily damped" scheme (b′) the constant $P = -0.11$ is negative. This calls for a linear *interpolation* at a point 89% of the way between e_{t-1} and e_t, rather than for an extrapolation, which accounts intuitively for the greater stability of the scheme and the smaller variance var(x_t). A study of the tables given later shows that this behaviour is typical. Notice also that the values $P = 0.62$ and $P = -0.11$ for schemes (a′) and (b′) imply that an intermediate scheme exists for which $P = 0$. This is

(c′) $x_t = -0.49e_t$, giving var(e_t) = 1.33 and var(x_t) = 0.32.

We see that this extremely simple scheme requiring only integral action produces a 94% reduction in the input variance for an increase of only 14% in the output variance.

CONSTRAINED PROPORTIONAL-INTEGRAL SCHEMES

Thus although MMSE schemes based on the simple time series and dynamic models (9) and (12) are often unsatisfactory, this is principally the fault of an inappropriate choice of the *criterion* (output MMSE) rather than of these models' structure. As we have already mentioned, there is both theoretical and empirical evidence that supports the use of the IMA time series model (9) to approximate the behavior of an uncontrolled disturbance. Furthermore, there is much empirical evidence that the first-order system, possibly with delay as represented by Equation (12), frequently provides a useful approximation to the dynamic behavior of industrial processes. In what follows we shall, therefore, retain these *models* to derive optimally constrained PI schemes. In practice the sampling interval for a process will be chosen taking into account the rapidity with which the system reacts to adjustment so that schemes for $f = 0$ and $f = 1$ are of the most practical importance.

The problem of finding the values of G and P for a PI scheme that, for a given value of var(e_t), minimizes var(x_t) is equivalent to the problem of finding the unconstrained minimum of the linear aggregate (13) of the two variances. For later convenience we write this as

$$L_\alpha = \{g(1 - \delta)\}^2 \operatorname{var}(x_t) + \alpha \operatorname{var}(e_t) \tag{14}$$

The procedure that we have used for doing this is given in Appendix A where it is shown that the resulting values of gG, P, var(e_t)/σ_a^2 and var(gx_t)/σ_a^2 are functions of λ, δ, f, and α. The results are tabled in Box and Luceño 1995 for frequently encountered ranges of the parameters. Charts prepared from these tables are given in the following Chapter D.7 from which appropriate adjustment schemes can be designed. Methods for estimating λ, δ, f, and g from actual operating data are given in time series texts and, in particular, in Box and Jenkins (1976).

Example

As an example we consider an application, presented by Fearn and Maris (1991), of feedback adjustment for the control of gluten additions in a flour mill.

In this example extensive preliminary runs provided data yielding estimates of $\lambda = 0.25$ and $\sigma_a = 0.12$. Also the dynamic response of the system was modelled by a simple delay of one time unit so that $f = 1$ and the units were chosen so that $g = 1$. On the basis of these parameters values, Fearn and Maris employed the MMSE scheme

$$-x_t = \lambda(x_{t-1} + e_t)/g$$

or

$$-x_t = 0.25(x_{t-1} + e_t)$$

Fearn and Maris found this unconstrained scheme (which we note is *not* a PI scheme) to be entirely satisfactory. However, the results of the present paper show that constrained PI schemes might have been used which for rather small increases in var(e_t) could have given dramatic reductions in var(x_t). Examples are given in Table 1. The even simpler scheme

$$-x_t = 0.19e_t$$

obtained by interpolation to make P zero would give an increase in var(e_t) of only 0.6% with a decrease in var(x_t) of 38%.

Table 1. Alternative Schemes for the Fern and Maris Example

G	P	% Increase in var(e_t)	% Decrease in var(x_t)
0.19	0.16	0.1	16
0.18	−0.13	1.0	55
0.12	−0.39	6.7	83

TUNING PROPORTIONAL-INTEGRAL CONTROLLERS WITH EVOLUTIONARY OPERATION AND EXPERIMENTAL DESIGN

The ease with which estimates of the coefficients λ, g, δ, and f can be obtained varies with different applications. Thus, for some systems, it will be known that an adjustment at the input is fully realized at the output in one time interval and, therefore, that δ and f are both 0. Also the steady state gain constant g is often known from a physical understanding of the process. Finally, when past records are available, λ and other needed constants can be estimated by direct calculation (see Box, Jenkins, and Reinsel, 1994).

Sometimes, however, especially at the initiation of an entirely new scheme, little information is available. In such a case, the PI scheme based on best guesses could be run initially and improved on by using evolutionary operation (Box, 1957a). The "variables" studied in the evolutionary operation scheme would be the two constants G and P of the PI controller.

When, as sometimes happens, there is a number m of PI control schemes operating simultaneously on the same process, multifactor designs can be used in which (G_1, P_1), (G_2, P_2), ..., (G_m, P_m) are the design variables. Initially fractional factorial designs or other orthogonal arrays can be used to determine a direction of steepest descent. If necessary, this can be followed later by a response surface study in some or all of the control variables. This approach has the advantage that it can take account of interaction between the different controllers.

CONCLUSION

MMSE feedback control schemes can require excessive manipulation of the adjustment variable. Constrained PI schemes may be used to reduce significantly the variance of the adjustments required at the expense of a small increase in the variance of the output. The performance of simple optimal constrained PI schemes appears to be almost as good as that of general linear constrained schemes for a wide range of practical situations. PI schemes, however, are much easier to implement, since the required adjustment is simply a linear combination of the two last observed errors.

ACKNOWLEDGMENTS

We are grateful to the Editor and to three referees for helpful comments. This research was supported in part by the National Science Foundation under grant DMI-9414765, the Sloan Foundation and the Spanish Dirección General de Investigación Científica y Técnica under grant PB92-0502.

APPENDIX A: DERIVATION OF CONSTRAINED PROPORTIONAL-INTEGRAL SCHEMES

In what follows we denote the back shift operator by B so that $Be_t = e_{t-1}$. Then the difference operator is $\nabla = 1 - B$ so that

$$\nabla e_t = (1 - B)e_t = e_t - e_{t-1}$$

The PI adjustment equation $x_t = -G(1 + P\nabla)e_t$ implies that x_t depends only on the last two errors e_t and e_{t-1} according to the linear relationship

$$x_t = c_1 e_t + c_2 e_{t-1} \qquad (A1)$$

with $G = -(c_1 + c_2)$ and $P = -c_2/(c_1 + c_2)$. Now let B be the backward shift operator so that, for example, $Be_t = e_{t-1}$. Then the IMA time series model (9) may be written

$$(1 - B)y_t = (1 - \theta B)a_t \qquad (A2)$$

with parameters $\lambda = 1 - \theta$ and σ_a^2. Also the delayed first-order dynamic model may be written

$$(1 - \delta B)\tilde{X}_t = (1 - \delta)X_{t-1-f} \qquad (A3)$$

Now, introducing Equations (A1)–(A3) into the equation resulting from multiplying $e_t = y_t + g\tilde{X}_t$ by $1 - B$, we find that e_t follows an ARMA(p, 2) model of the form $(1 - \phi_1 B - \cdots - \psi_p D^n)e_t = [1 - (\delta + \theta)B + \delta\theta B^2]a_t$, where the order p of the autoregressive part of the model is $f + 2$.

For $f = 0$, the autoregressive parameters are

$$\phi_1 = 1 + \delta + c_1 g(1 - \delta)$$

and

$$\phi_2 = -\delta + c_2 g(1 - \delta)$$

Similarly, for $f = 1$, we have

$$\phi_1 = 1 + \delta$$
$$\phi_2 = -\delta + c_1 g(1 - \delta)$$

and

$$\phi_3 = c_2 g(1 - \delta)$$

Using this ARMA(p, 2) model for e_t, it is possible to calculate var(e_t) and cov(e_t, e_{t-1}) as functions of the parameters $\lambda = 1 - \theta$, σ_a^2, $c_1 g(1 - \delta)$, $c_2 g(1 - \delta)$, δ, and f. Using Equation (A1), we can now calculate

$$\{g(1 - \delta)\}^2 \, \text{var}(x_t) = \{g(1 - \delta)\}^2\{(c_1^2 + c_2^2) \, \text{var}(e_t) + 2c_1 c_2 \, \text{cov}(e_t, \, e_{t-1})\}$$

as a function of the same set of parameters.

The minimization of L_α [Equation (14)] with respect to c_1 and c_2 can be performed numerically for the set of values of c_1 and c_2 that makes the ARMA$(p, 2)$ model for e_t stationary. Because $\{g(1 - \delta)\}^2 \text{var}(x_t)$ and $\text{var}(e_t)$ are functions of λ, σ_a^2, $c_1g(1 - \delta)$, $c_2g(1 - \delta)$, δ, and f, and σ_a^2 does not affect the point where the minimum of L_α is attained, the values of $c_1g(1 - \delta)$ and $c_2g(1 - \delta)$ that minimize L_α and provide the optimal constrained PI scheme are functions of λ, δ, f, and α. [Note that the minimization of L_α is carried out with respect to c_1 and c_2 or, equivalently, with respect to $c_1g(1 - \delta)$ and $c_2g(1 - \delta)$.] Then gG, P, $\text{var}(e_t)/\sigma_a^2$, and $\text{var}(gx_t)/\sigma_a^2$ are also functions of λ, δ, f, and α.

APPENDIX B

Tables 2 and 3 further illustrate the similarity of the results obtained from suitable tuned PI schemes and optimal linear constrained schemes. In what follows we use the subscripts C and U to indicate the optimal linear constrained scheme and the unconstrained scheme, respectively.

When $f = 0$ it is possible to show that

$$\text{var}_C(e_t) = \sigma_a^2(1 + \lambda^2 Q), \qquad \text{var}_C(x_t) = \text{var}_U(x_t)W$$

Table 2. Values of $100Q$ and $100W$ that Provide the Output and Input Variances for Some Optimal Constrained PI Schemes[a] and General Linear Constrained Schemes According to the Formulas $\text{var}(e_t) = \sigma_a^2(1 + \lambda^2 Q)$ and $\text{var}(gx_t) = \text{var}_U(gx_t)W$ for $f = 0$, $\delta = 0.5$ and Several Values of λ and α

α		Results for the Optimal Constrained PI Schemes and the Following Values of:				Optimal Constrained Scheme for Any λ
		0.20	0.40	0.60	0.80	
1.0742900	$100Q$	23.504	22.084	21.111	20.480	20.000
	$100W$	26.660	27.040	26.989	26.706	26.179
0.4153830	$100Q$	45.557	44.035	42.468	41.210	40.000
	$100W$	14.663	15.220	15.542	15.616	15.489
0.2152021	$100Q$	63.522	64.103	63.043	61.676	60.000
	$100W$	10.319	10.389	10.600	10.706	10.694
0.1288560	$100Q$	79.904	83.133	83.057	81.902	80.000
	$100W$	8.130	7.846	7.926	8.006	8.025
0.0844903	$100Q$	96.511	101.857	102.770	101.963	100.000
	$100W$	6.744	6.281	6.278	6.329	6.354

[a] The percentage increase in $\text{var}(e_t)$ and the percentage decrease in $\text{var}(gx_t)$ with respect to the optimal unconstrained scheme, are $100\lambda^2 Q$ and $100(1 - W)$, respectively.

Table 3. Percentage Increase in var(e_t) and Percentage Decrease in var(gx_t) for Some Optimal Constrained PI Schemes,[a] with Respect to the Optimal Unconstrained Scheme, for $f = 1$, $\delta = 0.5$, $\lambda = 0.4$, and Several Values of α

	Optimal Constrained PI Scheme		Optimal Constrained Scheme
α	% Increase in var(e_t)/σ_a^2	% Decrease in var(gx_t)/σ_a^2	% Increase in var(e_t)/σ_a^2 for the Same Decrease in var(gx_t)/σ_a^2
∞	0.64	39.08	0.38
100	0.64	39.82	0.40
1	2.72	72.13	2.50
0.01	38.6	97.93	38.32
0.0001	356.1	99.80	350.41

[a] Also shown is the percentage increase in var(e_t) for the general linear constrained schemes that have the same percentage decrease in var(gx_t) with respect to the optimal unconstrained scheme as the corresponding constrained PI schemes.

where

$$\frac{\text{var}_U(x_t)}{\sigma_a^2} = \frac{\lambda^2(1 + \delta^2)}{g^2(1 - \delta)^2} \quad \text{(for any } f = 0, 1, \ldots) \tag{B1}$$

and Q and W only depend on δ and α (see Box and Jenkins, 1970, Chap. 13). Because of this, we have used the same representation for the optimal constrained PI schemes, that is, var(e_t) = $\sigma_a^2(1 + \lambda^2 Q)$ and var(x_t) = var$_U(x_t)W$, although here Q and W depend on δ, α, and also on λ. The increase in var(e_t) with respect to the corresponding optimal unconstrained schemes is then $100\lambda^2 Q\%$ and the decrease in var(x_t) is $100(1 - W)\%$.

Table 2 is for $f = 0$ with $\delta = 0.5$ and various values of λ. It compares values of $100Q$ and $100W$ for schemes of the two types and shows that optimal PI schemes give results that are always close to those of optimal linear schemes in terms of the decrease in var(gx_t) and increase in var(e_t) and can be used without any significant loss of efficiency.

A similar conclusion is reached from Table 3. For $f = 1$, $\delta = 0.5$, $\lambda = 0.4$, and several values of α, a comparison is made of the decrease in var(gx_t) and the corresponding increase in var(e_t) that may be attained by using optimal linear constrained schemes and optimal constrained PI schemes. In this case, var$_U(e_t)/\sigma_a^2 = 1 + \lambda^2 f = 1.16$, and var$_U(gx_t)/\sigma_a^2 = 0.8$ because of Equation (B1). We note that minimizing L_α/α when $\alpha = \infty$ is equivalent to minimizing var(e_t) without paying any attention to var(x_t) and that the PI scheme for $\alpha = \infty$ produces a small increase in var(e_t) in compensation for a much bigger decrease in var(gx_t).

CHAPTER D.7

Discrete Proportional-Integral Adjustment and Statistical Process Control

This paper explains the nature and importance of proportional-integral control and shows how it may be adapted to statistical process control. The relation of this type of control to exponential smoothing, minimum mean squared error control, and optimal constrained schemes is discussed. Robustness properties which simplify considerably the practical application of this type of control are demonstrated.

INTRODUCTION

Proportional-integral (PI) control* is a form of continuous adjustment originating in the process industries and usually considered part of (automatic) engineering process control (EPC). Its application for adjustment of processes in the context of statistical process control (SPC) is less well-known. In that case the data are usually not continuous but, instead, observations and opportunities for adjustment occur at equally spaced times..., $t-1$, t, $t+1$, 7.... The object of this paper is to explain in simple terms the nature and importance of PI control and its application to SPC. We also discuss its relation to exponential smoothing and describe some robustness properties which considerably simplify its practical application.

From Box, G. E. P. and Luceño, A. (1997), *Journal of Quality Technology*, **29**(3), 248–260.

*A slightly more elaborate form is proportional-integral-derivative (PID) control in which there is an added derivative term, but we do not consider such schemes in the present paper.

DISCRETE PI CONTROL

Consider a dyeing operation where the object is to keep as small as possible deviations e_t of the color index of dyed yarn (the output variable) from the target value $T = 9.0$ by manipulation of the dye addition rate X_t (the input variable). Also suppose that one unit of change in X_t produces $g = 0.08$ units of change in the color index. The deviations $\ldots, e_{t-1}, e_t, e_{t+1}, \ldots$ will be referred to as "errors" at the output. In a PI scheme, the setting of the input variable X_t as each new observation comes to hand is a linear combination of the last error e_t and the cumulative sum of current and previous errors $e_t + \cdots + e_1$ so that the control equation is

$$gX_t = k_0 + k_1 e_t + k_2 \sum_{i=1}^{t} e_i \qquad (1)$$

where the constant k_0 is the initial level of the adjustment variable (which may as well be taken as its origin so that $k_0 = 0$), the constant k_1 determines the amount of proportional adjustment, and k_2 the amount of cumulative (integral) adjustment. For illustration, a simple scheme of this kind that might be used for the dyeing example is

$$0.08X_t = -0.4e_t - 0.4 \sum_{i=1}^{t} e_i$$

or

$$X_t = -5e_t - 5 \sum_{i=1}^{t} e_i \qquad (2)$$

in which $g = 0.08$, $k_1 = -0.4$, and $k_2 = -0.4$.

ANOTHER WAY OF LOOKING AT PI CONTROL

Instead of considering the level X_t of the input variable at time t, it is often more convenient to think in terms of the *adjustment* $x_t = X_t - X_{t-1}$ to be made at time t. Now subtracting Equation (1) applied at time $t - 1$ from the same equation at time t, we find that

$$gx_t = c_1 e_t + c_2 e_{t-1} \qquad (3)$$

with

$$c_1 = k_1 + k_2, \qquad c_2 = -k_1$$

For the dyeing example this leads to

$$0.08x_t = -0.8e_t + 0.4e_{t-1}$$

or

$$x_t = -10e_t + 5e_{t-1} \tag{4}$$

Thus, discrete PI control simply consists of making an adjustment in the input variable which is a linear combination of the last two output errors. (The discrete analog of PID control mentioned earlier is of the form $gx_t = c_1 e_t + c_2 e_{t-1} + c_3 e_{t-2}$ and uses the last three errors.)

A PI CONTROL CHART

PI control can readily be put into effect with automatic equipment. However, in the SPC context, manual process adjustment may be needed using a suitable feedback control chart. See, for example, Box, Jenkins, and Reinsel (1994, Chap. 13). Construction of such a chart also greatly assists in the understanding of this form of control.

Equation (3) may be written

$$gx_t = -G\{e_t + P(e_t - e_{t-1})\}$$

or

$$x_t = -(G/g)\{e_t + P(e_t - e_{t-1})\} \tag{5}$$

where the constants G and P are related to the previous constants as follows:

$$G = -c_1 - c_2$$
$$= -k_2$$
$$P = -c_2/(c_1 + c_2)$$
$$= k_1/k_2$$

A PI adjustment chart for the control equation (2), or equivalently for the adjustment equation (4), is shown for the dyeing example in Figure 1. For this example $G/g = 0.4/0.08 = 5$ and $P = 1$ so that

$$x_t = -5\{e_t + (e_t - e_{t-1})\} \tag{6}$$

This chart requires two scales: that on the right shows the output variable (color index) with the target value ($T = 9.0$) at the center of the scale, so that e_t is the

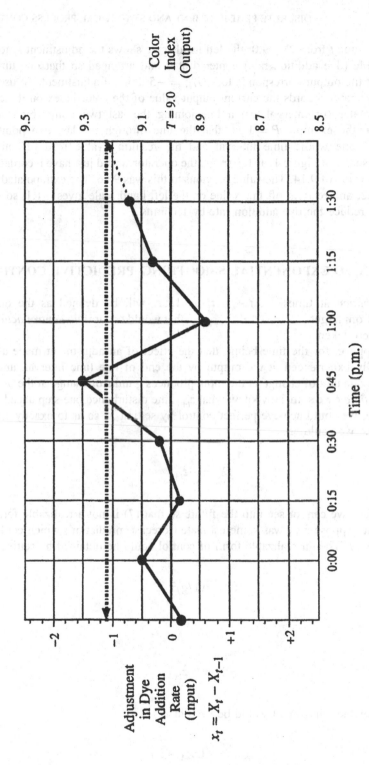

Figure 1. A feedback adjustment chart to apply PI control in SPC.

deviation at time t from $T = 9.0$; the left hand scale shows the adjustment x_t to the input variable (dye addition rate) centered at 0 and arranged so that one unit of deviation at the output corresponds to $-G/g = -5$ units of adjustment. To use the chart, the operator records the current output value of the color index on the chart and extrapolates (or interpolates) a line joining the last two points by P time intervals. In the example $P = 1$ so that the line through the last two points is extrapolated one whole time interval and the resulting value read off on the adjustment scale. In Figure 1, at 1:30 p.m., the operator would just have recorded the color index value of 9.14. The value previous to this was 9.06. The extrapolated line falls on 9.22, and reading off this value on the left hand scale gives -1.1, so s/he would now reduce the dye addition rate by 1.1 units.

RELATION TO EXPONENTIAL SMOOTHING–PREDICTIVE CONTROL

The *disturbance* at times $\dots$, $t-1$, t, $t+1$, $\dots$ will be defined as the output deviations from target $\dots$, z_{t-1}, z_t, z_{t+1}, $\dots$ that would occur if *no control action of any kind* were taken.

Now suppose, for the time being, that the effect of an adjustment made at the input is fully experienced at the output by the end of one time interval, and, as before, that one unit of change in the input produces g units of change at the output. Then, if *at time t* we somehow knew what z_{t+1}, the disturbance one step ahead, was going to be, we could achieve perfect control by setting X_t so as to exactly cancel z_{t+1}. That is, we could set

$$gX_t = -z_{t+1} \tag{7}$$

Unfortunately, we cannot see into the future so that (7) is not a realizable form of control. But suppose $\hat{z}_{t+1}$ was some estimate (forecast, prediction) which could be made at time t. Then a realizable form of control could be obtained by setting

$$X_t = -(1/g)\hat{z}_{t+1}$$

or

$$gX_t = -\hat{z}_{t+1}$$

Equivalently, the adjustment would be such that

$$gx_t = -(\hat{z}_{t+1} - \hat{z}_t) \tag{8}$$

This form of control would ensure that the error at the output, say, at time $t + 1$, was not the value of the disturbance z_{t+1}, but the error in predicting that disturbance. We denote such a "forecasting error" by

$$e_{t+1} = z_{t+1} - \hat{z}_{t+1}$$

Now as we saw earlier a quantity frequently employed as a forecast of the next value in a series is the exponentially (geometrically) discounted average of present and past values of the series. Such a quantity is called an *exponentially weighted moving average* or EWMA. An EWMA with smoothing constant (discount factor) θ will be denoted by

$$\tilde{z}_t = (1 - \theta)\{z_t + \theta z_{t-1} + \theta^2 z_{t-2} + \cdots\} \tag{9}$$

where θ has some value between -1 and 1 but is usually between 0 and 1. An equivalent useful quantity is $\lambda = 1 - \theta$ which then lies between 0 and 1. The closer θ is to 1 (λ is to 0), the slower the discounting process and the more $\tilde{z}$ will act like an ordinary arithmetic average of past data. The closer θ is to 0 (λ is to 1) the greater the weight placed on recent history. If we replace $\hat{z}_{t+1}$ by the EWMA $\tilde{z}_t$, we must set

$$gX_t = -\tilde{z}_t \tag{10}$$

Now a well-known relationship sometimes used to update an EWMA and following from (9) is

$$\tilde{z}_t - \tilde{z}_{t-1} = \lambda(z_t - \tilde{z}_{t-1})$$
$$= \lambda e_t \tag{11}$$

where e_t is now the error $e_t = z_t - \tilde{z}_{t-1}$ made by using $\tilde{z}_{t-1}$ to forecast z_t. Thus the adjustment equation (8) becomes

$$gx_t = -\lambda e_t \tag{12}$$

This adjustment equation is the same as (3) with $c_1 = -\lambda$ and $c_2 = 0$.

So, by the rather roundabout route, we have reached a form of control suggested by common sense: we should make a compensating adjustment proportional to the last error. It is rather surprising that this turns out to be equivalent to setting X_t at each stage so as to just cancel an EWMA forecast of the disturbance, as shown in (9) and (10). This is particularly so since when we use this form of control we never actually see values of the disturbance z_t itself (although knowing the control actions that have been taken this may be reconstructed). Equivalently the PI scheme in its original form (1) becomes

$$gX_t = -\lambda \sum_{i=1}^{t} e_i$$

in which the level of adjustment depends only on the cumulative sum of the errors. This is the discrete analog of pure integral control in which the proportional term k_1 in Equation (1) is zero. It is formally shown in Box and Luceño (1997) that the use of such integral control *requires* that the forecast of z_{t+1} in (7) must be the EWMA $\tilde{z}_t$ of (9) with smoothing constant $\theta = 1 + k_2$ with $-1 \le k_2 \le 0$.

Finally, $G = \lambda$ and $P = 0$ in Equation (5). So that in the adjustment equation for pure "integral" control no extrapolation or interpolation between points is needed; we simply read off the adjustment opposite the last point on the chart.

PROCESS INERTIA

So far, we have supposed that the effect of a change in the input is fully realized at the output in one time interval, implying that it is only the last adjustment of the input that is "remembered" at the output and is available to help compensate the disturbance. If the process has inertia it would be more reasonable to suppose that not only the adjustment x_t but previous adjustments might need to be taken into account. One reasonable assumption might be that past adjustments should be discounted exponentially by some discount factor δ. The effects of current and previous adjustments experienced at the output would then be given by a second EWMA

$$\tilde{x}_t = (1 - \delta)(x_t + \delta x_{t-1} + \delta^2 x_{t-2} + \cdots)$$

Thus, to take account of process inertia in this way, $\tilde{x}_t$ would replace x_t in Equation (12) to give

$$g\tilde{x}_t = -\lambda e_t \tag{13}$$

Now since $\tilde{x}_t$ is an EWMA,

$$(1 - \delta)x_t = \tilde{x}_t - \delta \tilde{x}_{t-1}$$

so Equation (13) implies that the adjustment x_t should be

$$gx_t = -\frac{\lambda}{1 - \delta}(e_t - \delta e_{t-1})$$

which is the same as (3) with

$$c_1 = -\frac{\lambda}{1 - \delta}, \qquad c_2 = \frac{\lambda\delta}{1 - \delta}$$

Equivalently, such adjustment yields discrete PI control in the form of (1) with

$$k_1 = -\frac{\lambda\delta}{1 - \delta}$$
$$k_2 = -\lambda \tag{14}$$

Finally, the adjustment chart would use

$$G = \lambda, \quad P = \delta/(1 - \delta)$$

We see that *any* PI scheme is equivalent to setting an EWMA $\tilde{X}_t$ of past input levels with discount factor δ so as to just cancel an EWMA $\tilde{z}_t$ of past disturbances with discount factor θ, so that

$$g\tilde{X}_t = -\tilde{z}_t \tag{15}$$

When δ is positive, P is positive implying that we should *extrapolate* a line joining the last two points on the chart.

Notice that everything we have said so far comes solely out of the original PI equation (1). Starting only with that equation, we have shown that this form of control can be represented in three other distinct ways, described by Equations (3), (5), and (15). Notice also that each of these four forms is intuitively reasonable and could have provided an alternative basis from which the other three could be derived. So far, however, we have said nothing about how such control might relate to the characteristics of the process to be controlled. To make further progress, therefore, we need to consider some simple models for process disturbances and process dynamics.

A TIME-SERIES MODEL FOR A DISTURBANCE

As shown by Muth (1960), the time-series model for the disturbance z_t which is such that an EWMA with smoothing constant θ_0 provides a forecast with minimum mean squared error (MMSE) is

$$z_t - z_{t-1} = a_t - \theta_0 a_{t-1} \tag{16}$$

In this expression, $\ldots, a_{t-1}, a_t, a_{t+1}, \ldots$ (sometimes referred to as shocks or innovations) are a series of statistically independent errors each with mean 0 and standard deviation σ_a. Such a series may be called "white noise." The time series model (16) is called an integrated moving average (IMA).

To see the logic of Muth's result, suppose we are using an EWMA with smoothing constant θ to forecast a series..., z_{t-1}, z_t, z_{t+1}, ... and we produce a series of forecast errors..., e_{t-1}, e_t, e_{t+1}, ... From Equation (11), an updating formula for these EWMA forecasts will be

$$\tilde{z}_t - \tilde{z}_{t-1} = \lambda e_t \tag{17}$$

and if we add $e_{t+1} - e_t$ to both sides of (17) we get

$$z_{t+1} - z_t = e_{t+1} - \theta e_t$$

This formula relates *any* series ..., z_{t-1}, z_t, z_{t+1}, ... to the forecast errors obtained when an EWMA with *any* smoothing constant θ is used to forecast it. In general, however, these forecast errors will not be statistically independent and the present error e_t will be correlated with previous errors. Consequently, at some time t, current and past errors ..., e_{t-2}, e_{t-1}, e_t, ... can be used in some way to forecast e_{t+1} and hence to improve the forecast $\tilde{z}_t$. This will always be possible unless the model is such that an EWMA with some specific smoothing constant, say, θ_0, can produce forecasts for that particular series with forecast errors that are independent and so unforecastable. Only then could the forecast not be improved upon. This model is that given by (16).

Figure 2 shows a number of disturbance series generated from (16) for $\lambda_0 = 0$, 0.1, 0.2, 0.3, and 0.4 using the same shocks..., a_{t-1}, a_t, a_{t+1}, ... We see that the parameter $\lambda_0 = 1 - \theta_0$ can be regarded as a measure of non-stationarity. It indicates the degree of departure from the ideal "state of control" that is achieved when $\lambda_0 = 0$, in which case the disturbance $z_t = a_t$ and hence is a white noise series.

The usefulness of the IMA model (16) for the representation of process disturbances is supported both on practical and theoretical grounds. A number of authors have found that this model was selected by time-series analysis from a much wider class (e.g., of ARIMA models) for the representation of process data. Also, they have usually found values for λ_0 between 0.2 and 0.5. From a theoretical point of view, (16) has been justified as a "sticky innovation model" (Box and Kramer, 1992) of the kind first suggested by Barnard (1959).

A MODEL FOR PROCESS DYNAMICS

We have supposed above that the geometrically discounted weights of an EWMA might be used to represent the inertial memory of a process. Thus the effect of a

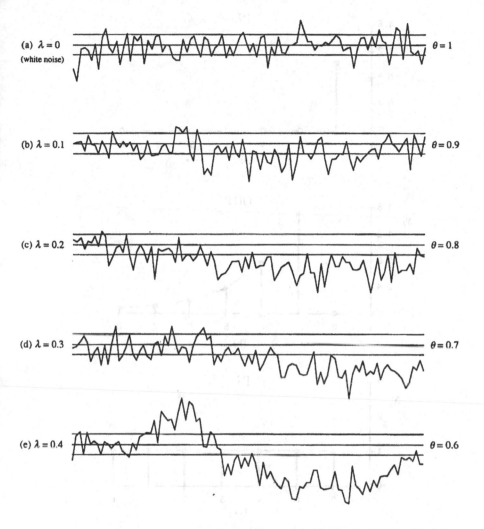

Figure 2. Disturbance series z_t generated by the IMA model with $\lambda_0 = 0$, 0.1, 0.2, 0.3, 0.4.

series of input adjustments..., x_{t-2}, x_{t-1}, x_t, ... would be experienced at the output at time $t + 1$ as

$$y_{t+1} = g\tilde{x}_t$$
$$= g(1 - \delta_0)(x_t + \delta_0 x_{t-1} + \delta_0^2 x_{t-2} + \cdots) \tag{18}$$

where δ_0 was suitably chosen for the particular process.

To see what is implied, suppose as in Figure 3a that over an extended period of time the only adjustment made to the input was a unit change at time $t = 2$. Figure 3b shows how this unit change would be experienced at the output using for

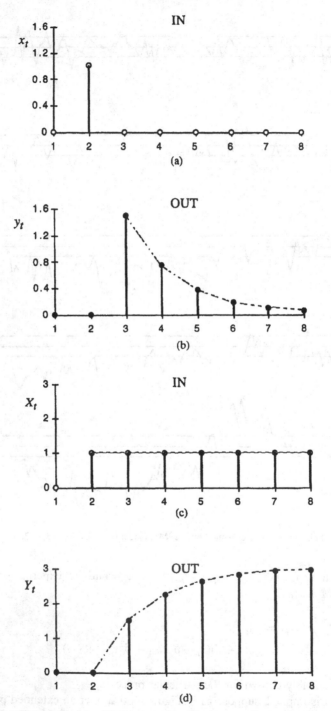

Figure 3. (a) Change in input x_t; (b) change produced in output y_t; (c) consequent level of input X_t; and (d) consequent level of output Y_t.

illustration the case $g = 3$ and $\delta_0 = 0.5$. The responses y_t at times $t = 3, 4, 5, \ldots$ would be $g(1 - \delta_0) = 1.5$, $g(1 - \delta_0)\delta_0 = 0.75$, $g(1 - \delta_0)\delta_0^2 = 0.375$, and so on. This unit adjustment to the input x_t will result, as shown in Figure 3c, in a unit change in level X_t of the input from time $t = 2$ onward. The corresponding levels $Y_t = y_t + \cdots + y_1$ (accumulated changes) at the output, shown in Figure 3d, will be $g(1 - \delta_0) = 1.5$, $g(1 - \delta_0)(1 + \delta_0) = 2.25$, $g(1 - \delta_0)(1 + \delta_0 + \delta_0^2) = 2.625$, and so on, quickly building up to the full change of $g = 3$ units. A system described by (18) is often referred to as a "first order dynamic system." Notice that $1 - \delta_0$ is the proportion of the ultimate change that occurs in the first time interval. So for a fast acting system δ_0 is close to 0; for a very slow acting system it is close to 1. In addition, sometimes there may be one or more time periods of pure delay (dead time) before the process reacts at all.

Experience has shown that first order systems with δ_0 suitably chosen, and occasionally with one (or sometimes more) units of delay, are often able to approximate the inertial behavior of systems that are in reality much more complex.

MINIMUM MEAN SQUARED ERROR CONTROL

The choice of the parameters of a control scheme would be expected to depend on the nature of the disturbance and of the dynamics of the process. Suppose then, as would frequently be the case, that the IMA time series model of (16) with some suitably chosen smoothing constant $\theta = \theta_0$ can approximately represent the process disturbance and that the first order dynamic model (18) with $\delta = \delta_0$ can approximately represent the process inertia. Then it might be supposed that, when these values θ_0 and δ_0 were substituted in (14), the corresponding control scheme would in some sense be optimal. It can, in fact, be shown that such a scheme *does* minimize the output mean squared error. Unfortunately, however, except when the process is only slightly nonstationary (λ is small) and has rapid dynamic response (δ is not far from 0), such minimum mean squared error (MMSE) schemes may not be of much practical interest. This is because, to achieve minimal error at the output, they may require excessive manipulation at the input.

For illustration, suppose for the dyeing example that $g = 0.08$ and that the disturbance affecting the system and the dynamics of the process could be adequately represented by the above models with $\lambda_0 = 0.4$ and $\delta_0 = 0.5$. The substitution of these values in (14) and (1) yields the PI scheme of equation (2) and its equivalent manifestations in (4) and (6) and in Figure 1 that we have used throughout for illustration. To demonstrate its behavior, Figure 4 shows a disturbance series z_t generated by the assumed IMA time series model with $\lambda_0 = 0.4$ and $\sigma_a = 1$. The actual value of σ_a in the dyeing example was about 0.25. However, to facilitate comparison all the diagrams in Figure 5 are based on the value $\sigma_a = 1$. Figure 5a shows the output series e_t obtained using the MMSE control scheme referred to above. The scheme produces excellent control totally eliminating the trends in the level of the disturbance z_t. However, from the input adjustment series x_t

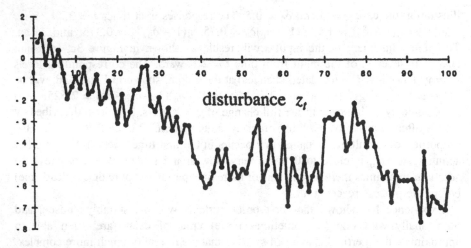

Figure 4. A typical disturbance series generated by an IMA time-series model with $\lambda_0 = 0.4$ and $\sigma_a = 1$.

shown immediately below the output, we see that excessive manipulation is needed to bring this about.

One way to see why this happens is to consider the corresponding adjustment chart of Figure 1. Recall that P is the degree of extrapolation along a line joining the last two points, which for an MMSE scheme must be $P = \delta_0/(1 - \delta_0)$. Thus for the dyeing example $P = 1$ and one whole unit of extrapolation is needed. (This is for a system in which a proportion $1 - \delta_0 = 0.5$ of the eventual change occurs in one time interval. For slower dynamics, corresponding to larger values of δ_0, even greater extrapolation would be needed.) Such extrapolation produces large adjustments which tend to be partly canceled in the next time period resulting in excessive input variance.

This phenomenon is well known and many authors such as Aström (1970), Aström and Wittenmark (1984), Box and Jenkins (1970), Box, Jenkins, and Reinsel (1994), MacGregor (1972), Wilson (1970), and Whittle (1963) (among others) have considered the problem of constraining the degree of manipulation needed at the input while still producing good control of the output. One way to do this is to minimize a combination

$$\sigma_x^2 + \alpha\sigma_e^2 \tag{19}$$

of the input and output variances instead of minimizing σ_e^2 alone. Different choices of α produce an "envelope" of schemes which can be used to minimize σ_x^2 at the expense of a small increase in variation at the output. The quantity $\sigma_x^2 + \alpha\sigma_e^2$ could also be thought of as a measure of the total cost, with α measuring the relative costs of deviations from target and input manipulations as measured by σ_e^2 and σ_x^2, respectively. [Note that minimizing (19) is equivalent to minimizing $\beta\sigma_x^2 + \sigma_e^2$ where the constants α and β are such that $\alpha\beta = 1$.]

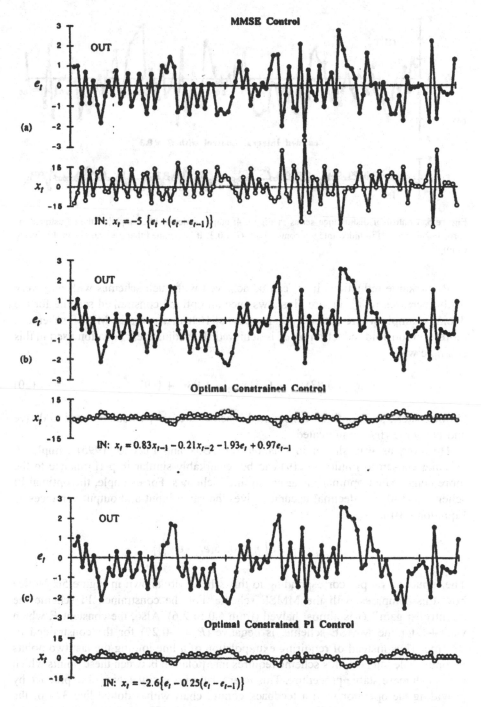

Figure 5. Control of disturbance series in Figure 4: output series e_t after applying the adjustments x_t corresponding to (a) MMSE control; (b) optimal constrained control; (c) optimal constrained PI control.

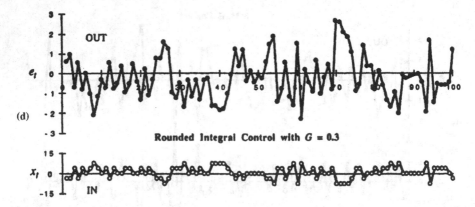

Figure 5. Control of disturbance series in Figure 4: output series e_t after applying the adjustments x_t corresponding to (d) rounded integral control with $G = 0.3$. It is assumed that $g = 0.08$ as in the dyeing example.

Remarkable reductions in σ_x^2 can be achieved with such schemes with only very slight increases in σ_e^2. Figure 5b shows such an optimal constrained scheme for the dyeing example where with only an increase of 10% in σ_e over the MMSE scheme, a more than fourfold reduction on σ_x is achieved. The adjustment equation used in this example was

$$x_t = 0.83x_{t-1} - 0.21x_{t-2} - 1.93e_t + 0.97e_{t-1} \qquad (20)$$

Unfortunately, such optimal constrained schemes are not particularly easy to derive and are somewhat complicated.

However, as was shown in Chapter D.6 (Box and Luceño, 1995), simple *PI schemes* chosen to minimize (19) can be remarkably similar in performance to the more complicated optimal linear constrained schemes. For example, the optimal PI scheme that to two decimal accuracy gives the same input and output variances as Equation (20) is

$$x_t = -2.6\{e_t - 0.25(e_t - e_{t-1})\}$$

The input and output corresponding to this scheme are shown in Figure 5c. Notice how this compares with the MMSE scheme. For the constrained PI scheme the "controller gain" G is almost halved (from 5.0 to 2.6). Also, the constant P, which was $+1$ for the MMSE scheme, is negative ($P = -0.25$) for the constrained PI scheme. Thus instead of requiring extrapolation of a line joining the last two points by one whole interval, this scheme requires interpolation between these points which is a much more stable procedure. This interpolation is easily achieved in practice by providing the operator with a feedback control chart with a dotted line 3/4 of the way between each time line, as shown in Figure 6. The operator reads off the adjustment opposite the intersection with the dotted line.

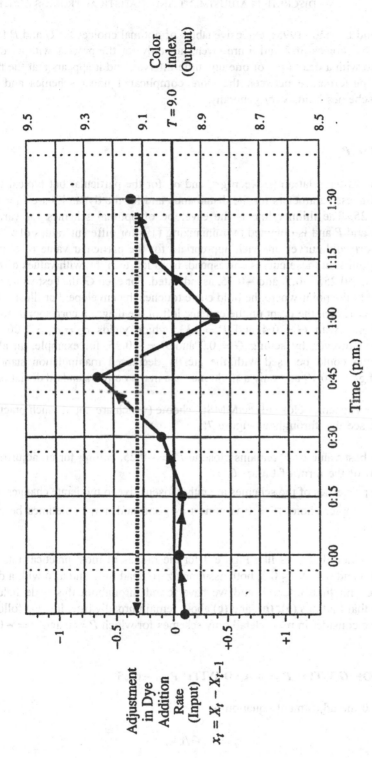

Figure 6. A feedback adjustment chart to apply constrained PI control in SPC.

In Box and Luceño (1995), extensive tables of optimal choices for G and P for a wide range of choices of λ and δ are given, not only for the process without dead time but also with a dead time of one unit time interval, and it appears that the near equality in performance between the more complicated linear schemes and the simpler PI schemes holds very generally.

CHOICE OF P

Figure 7 shows the relation between $g\sigma_x$ and σ_e for the particular but typical case where the process parameters for the disturbance and for the dynamics are $\lambda_0 = 0.4$ and $\delta_0 = 0.25$. The thinner line is the envelope of optimal schemes for various values of G and P and is obtained by minimizing (19) for different values of α. The series of short bold curves are each appropriate for the particular value of G indicated. The points on these curves correspond, from left to right, with values of P of $0.33, 0.25, 0, -0.25, -0.5$, and -0.75, as indicated. For each G, the best value of P corresponds to the point where the bold curve touches the envelope. For illustration, suppose $\sigma_a = 1$, then the point on the extreme left of the diagram corresponds to the MMSE scheme with $G = 0.4$ and $P = 0.33$, which yields $\sigma_e = \sigma_a = 1.00$ and $g\sigma_x = 0.55$. However, by setting $G = 0.2$ and $P = -0.25$, for example, an alternative scheme could be used with the greatly decreased manipulation standard deviation of $g\sigma_x = 0.19$ and only a slight increase in the output standard deviation to $\sigma_e = 1.08$.

Apart from schemes close to the MMSE scheme (which are not of much practical interest) we see that throughout Figure 7:

(a) The best value of P remains close to $P = -0.25$, calling for an adjustment chart of the form of Figure 6.

(b) The properties of the schemes are rather insensitive to moderate changes in P.

(c) The simpler choice $P = 0$ is almost as good and may sometimes be even better.

We have made diagrams like Figure 7 for the ranges of most practical interest: $0 < \lambda_0 \leq 0.6$ and $0 < \delta_0 \leq 0.5$, both assuming zero dead time and also with a dead time of one unit time interval, and we have found throughout this wide field of alternatives that findings (a), (b), and (c) above remain broadly true. In what follows, therefore, we consider in more detail only schemes for which $P = 0$ and $P = -0.25$

CHOICE OF G WITH $P = 0$ AND WITH $P = -0.25$

When $P = 0$ the adjustment equation is

$$x_t = (-G/g)e_t$$

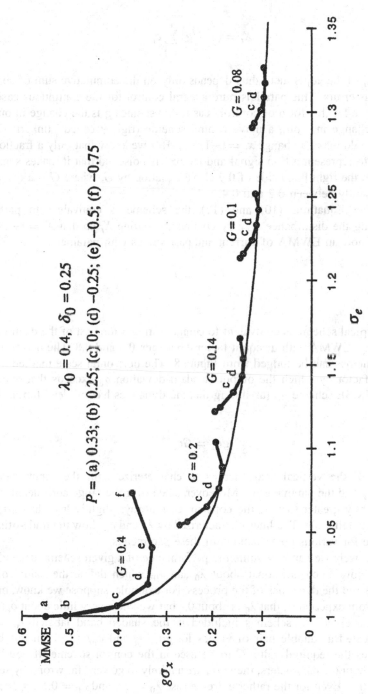

Figure 7. Relation between $g\sigma_x$ and σ_e when the process parameters for the disturbance and dynamics are $\lambda_0 = 0.4$ and $\delta_0 = 0.25$.

$$\lambda_0 = 0.4, \quad \delta_0 = 0.25$$

$$P = \text{(a) } 0.33; \text{ (b) } 0.25; \text{ (c) } 0; \text{ (d) } -0.25; \text{ (e) } -0.5; \text{ (f) } -0.75$$

and the level of the adjustment is provided by

$$X_t = (-G/g) \sum_{i=1}^{t} e_i$$

The level X_t of the adjustment thus depends only on the cumulative sum of current and previous errors. This parallels pure integral control for the continuous case.

A rationale for this form of control is as follows: since g is the change in output for a unit change in input, a naive control scheme (rightly called "tinkering" by Deming) would make a change $x_t = (-1/g)e_t$. But we know that only a fraction of e_t is likely to represent a true signal and the rest is noise, so that it makes sense to replace 1 on the right hand side of the above equation by G, where G is a damping factor frequently between 0.2 and 0.4.

Also from Equations (10) and (12), the scheme is equivalent to partially compensating the disturbance z_{t+1} by continually setting X_t so that $X_t = (-1/g)\tilde{z}_t$, where $\tilde{z}_t$ is now an EWMA of current and past values with parameter

$$\lambda = 1 - \theta$$
$$= G \tag{21}$$

Thus the typical scheme is equivalent to compensating a forecast of the disturbance made with an EWMA with discount factor θ between 0.5 and 0.8. The performance of these schemes can be judged from Figure 8. The horizontal scale marked σ_e/σ_a shows the factor by which the output standard deviation σ_e exceeds that obtained from the MMSE scheme σ_a (assuming that the dynamics has no dead time). Also, since

$$g\sigma_x = G\sigma_e$$

when $P = 0$, the vertical scale serves to characterize both the input standard deviation σ_x and the parameter G. Moreover, since over the range considered σ_e/σ_a is only slightly greater than 1, the constant G is always slightly less than $g\sigma_x/\sigma_a$, shown on the left scale. The lines characterized by λ_0 and δ_0 show the trade-offs that are possible for various combinations of these parameters.

How precisely we can determine the performance of a given scheme depends, of course, on how much is known about λ_0 and δ_0, which define the nature of the disturbance and the dynamics of the process. For example, suppose we know, or can speculate from experience, that λ_0 is about 0.4 but we know very little about δ_0; then we have the choice of schemes included in the shaded band labeled $\lambda_0 = 0.4$. Having selected a suitable pair of values for $g\sigma_x/\sigma_a$ and σ_e/σ_a within this band, $g\sigma_x/\sigma_e$ gives the required value G to be used in the control scheme. If you know very little about the parameters, then you are unlikely to go very far wrong by setting G at about 0.3. Even for the rather extreme case $\lambda_0 = 0.6$ and $\delta_0 = 0.5$, σ_e/σ_a will be only about 1.16. Figure 9 shows the corresponding performance chart for

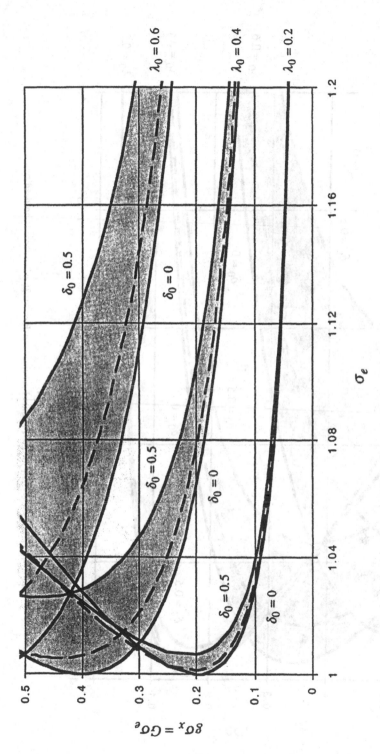

Figure 8. Performance chart for the integral control scheme ($P = 0$) when there is no dead time.

389

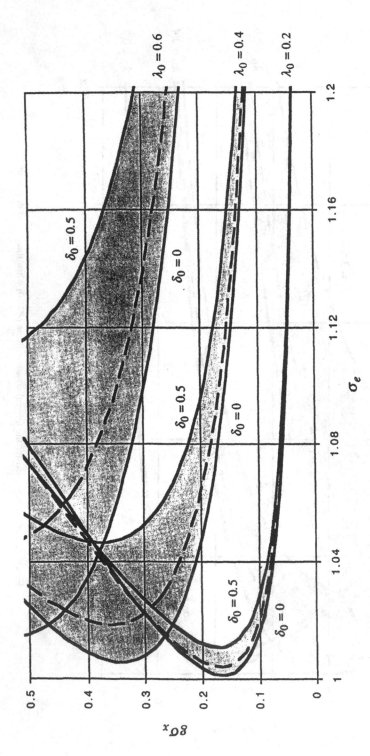

Figure 9. Performance chart for the PI control scheme with $P = -0.25$ when there is no dead time.

$P = -0.25$. The difference appears to be slight and we think that the slight loss of efficiency incurred by setting $P = 0$ will usually be compensated in practice by the added simplicity of the corresponding feedback schemes and charts it yields. Charts such as these are very easy for the operator to use. To attempt even greater simplicity Box and Jenkins (1970) suggested, as a possible modification, the use of rounded charts. See also Box, Jenkins, and Reinsel (1994, Chap. 13). In particular, with these charts action is not required after each observation but only from time to time. Figure 5d shows the input and output corresponding to a rounded scheme in which adjustments are computed using $G = 0.3$ and $P = 0$, and then rounded to the nearest number in the set $(-20, -10, 0, 10, 20)$.

Figures 10 and 11 show the performance charts corresponding to $P = 0$ and $P = -0.25$ when the dynamics has a pure delay (dead time) of one unit time interval in response to adjustment. In this case the output standard deviation corresponding to MMSE control is $\sigma_a\sqrt{1 + \lambda^2}$ rather than σ_a (i.e., $1.02\sigma_a$, $1.08\sigma_a$, or $1.17\sigma_a$ for $\lambda = 0.2$, 0.4, or 0.6, respectively). Thus while the values of σ_e/σ_a are larger when the dynamics has a dead time (Figures 10 and 11) than when there is no dead time (Figures 8 and 9), the reverse may sometimes be true for the factor $\sigma_e/(\sigma_a\sqrt{1 + \lambda^2})$ by which the output standard deviation σ_e exceeds that obtained from the MMSE scheme.

ROUNDED AND BOUNDED ADJUSTMENT SCHEMES

We mentioned above the use of rounded adjustment schemes. Although such schemes are sometimes useful, reluctance to apply adjustment at each time interval implies that, directly or indirectly, there is some cost associated with adjusting the process. When adjustment is associated with additional cost, it is better to use the bounded adjustment charts of Chapter D.4.

At first sight these charts appear similar to the EWMA charts of Roberts (1959) in that an EWMA of the data is plotted between two parallel lines. However, the motivation (feedback adjustment, rather than monitoring) and the basis on which the limit lines and the parameter of the EWMA are chosen are quite different.

The bounded adjustment charts were originally justified (Box and Jenkins, 1963). They produce minimum overall cost assuming that the disturbance is represented by an IMA model, that the loss from being off target is proportional to the square of the off-target deviation, and that a fixed cost is incurred each time the process is adjusted. We motivate them here in a somewhat different way.

We saw earlier that the discrete analog of integral control is

$$gX_t = k_0 + k_2 \sum_{i=1}^{t} e_i \qquad (22)$$

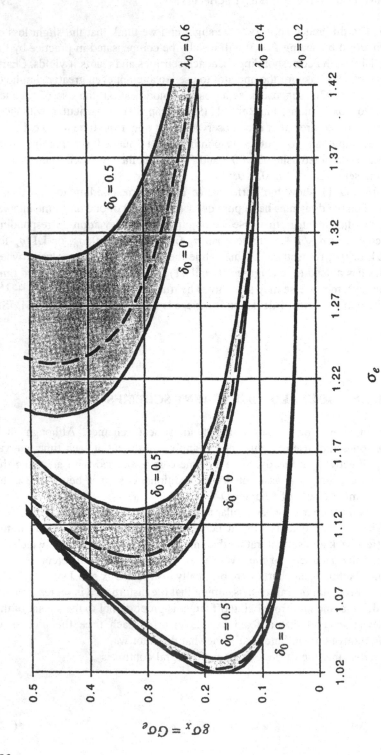

Figure 10. Performance chart for the integral control scheme with $P = 0$ when dynamics has a dead time of 1 unit time interval.

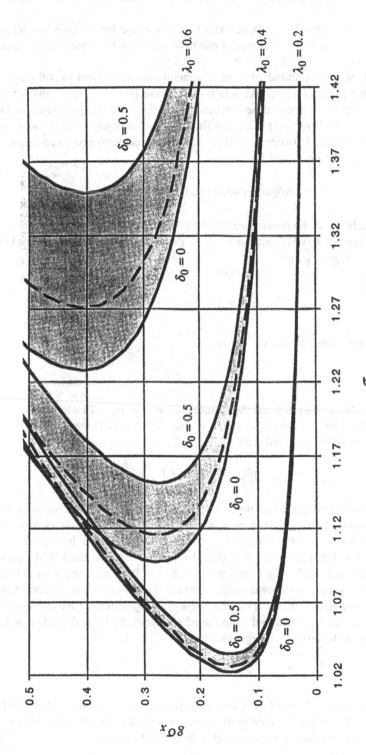

Figure 11. Performance chart for the PI control scheme with $P = -0.25$ when the dynamics has a dead time of 1 unit time interval.

393

Suppose that the process disturbance can be represented by an IMA model with parameter $\theta = 1 - \lambda$. Then this control equation will give minimum mean squared error if $k_2 = -\lambda$.

We now show that bounded adjustment schemes are equivalent to still using the control equation (22) but not to take action at each time period. Suppose that you last made an adjustment at some time r when the level of the compensating variable stood at X_r, then to follow this policy you should not adjust again until at some time t the absolute difference between X_t and X_r reaches a predetermined fixed value, say, k. Thus the control rule would be:

$$\text{Adjust as soon as } |X_t - X_r| \geq k$$

We showed earlier that the control achieved by Equation (22) is equivalent to setting X_t so as to cancel an EWMA estimate $\hat{z}_{t+1}$ of the next value z_{t+1} of the disturbance so that $X_t = -(1/g)\hat{z}_{t+1}$ and

$$X_t - X_r = -\frac{1}{g}(\hat{z}_{t+1} - \hat{z}_{r+1})$$

Thus action should be taken as soon as

$$|\hat{z}_{t+1} - \hat{z}_{r+1}| \geq L \tag{23}$$

where $L = kg$. We earlier defined the disturbance as $z_t = w_t - T$, where the w_t's are the original data and T is the target value, so that Equation (23) may equally well be written in terms of the original data w_t:

$$|\hat{w}_{t+1} - \hat{w}_{r+1}| \geq L$$

A *bounded* adjustment chart that puts this policy into effect is illustrated once again in Figure 12 using the first 50 values w_t of uncontrolled thickness series considered in Chapter D.3. We suppose that the process was controlled by adjusting the deposition rate and that $g = 1.2$, $\delta = 0$, and $T = 80$. Also, that the EWMA used in the calculations had smoothing constant $\theta = 0.8$ ($\lambda = 0.2$), a value found to best fit the series. The open circles shown on the chart are the values w_t', say. The addition of a prime (') to w indicates that it is the thickness *after adjustments have been applied*. The EWMA's $\bar{w}_t' = \hat{w}_{t+1}'$ obtained from the w_t' are shown by filled dots. The updated EWMA is best calculated using the formula

$$\hat{w}_{t+1}' = \lambda w_t' + \theta \hat{w}_t'$$

which is an alternative form of (11). In this particular case, therefore, we obtain $\hat{w}_{t+1}'$ from $\hat{w}_{t+1}' = 0.2w_t' + 0.8\hat{w}_t'$. This corresponds to an interpolation two-tenths of the way between the previous forecast $\hat{w}_t'$ and the present value w_t'.

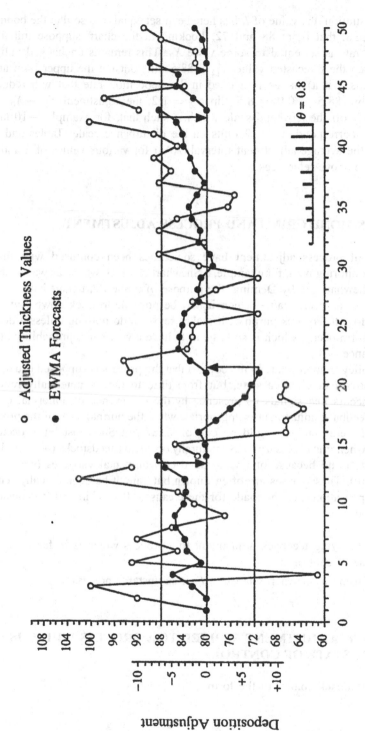

Figure 12. A plot for a bounded adjustment feedback scheme with action boundaries at 80 ± 8.

395

For illustration, the value of L has here been set equal to 8 so that the boundaries are at 80 ± 8, that is, at 88 and 72. Looking at the chart, suppose initially the deposition rate is set equal to some value X_0. This remains unchanged until time $t = 13$ when the forecasted value $\hat{w}'_{14} = 88.5$ falls outside the upper limit and the chart signals that a change is needed in the deposition rate that will reduce the thickness by $(88.5 - 80.0) = 8.5$. Since $g = 1.2$, the adjustment $X_{13} - X_0$ can be read directly off the left hand scale which is such that, for example, -10 units of adjustment corresponds to $+12$ units on the disturbance scale. Tables and charts providing the average adjustment interval (AAI) for various values of λ and L are given in the above references.

PROCESS MONITORING AND PROCESS ADJUSTMENT

Problems of process adjustment have sometimes been confused with those of process monitoring with, for example, a Shewhart chart. However, as was made clear both by Shewhart and by Deming, the purpose of a Shewhart chart is to signal the possibility of a special cause which it may be possible to track down and remove once and for all. For this purpose, it makes sense to do nothing unless a deviation from the norm occurs which is so large as to have a very small probability of being due to chance.

This policy is appropriate in the context that the process is supposed usually to be in an approximately *stable* state, but from time to time a potentially removable problem occurs that signals its presence by the occurrence of abnormal data. By contrast, feedback adjustment is appropriate when the normal state of the process, if no control were applied, would be one of *instability*. Such instability occurs, for example, when it is necessary to use naturally occurring feedstocks (oil, wood, wool, sewage, etc.) and because of uncontrolled environmental variables (e.g., ambient temperature). These causes are often known but cannot be economically removed: thus, compensation must be made for these causes. It is of interest to consider the consequences:

(a) of applying feedback adjustment to a process which is in fact in a perfect state of control; or

(b) of using a Shewhart chart to *adjust* an unstable process.

FEEDBACK ADJUSTMENT APPLIED TO A PROCESS WHICH IS IN A PERFECT STATE OF CONTROL

Suppose feedback control of the form

$$gx_t = -Ge_t$$

is applied to a process which is in a perfect state of control with the deviations e_t from target forming a random series exemplified by the white noise series a_t mentioned earlier (or, what is equivalent, represented by the IMA model with the non-stationarity parameter λ_0 exactly equal to 0). Then it is easy to show that because of this feedback control the process standard deviation would be increased by a factor $\sqrt{2/(2 - G)}$. For example, if as was recommended above, in the absence of special knowledge about the process, we used a scheme with $G = 0.3$, then the output standard deviation would be increased by a factor of 1.085. That is by less than 10%. For this modest premium we would gain the insurance that if, in fact, the system were not in perfect control, then it would be adjusted by the feedback scheme.

USING A SHEWHART CHART TO ADJUST AN UNSTABLE PROCESS

Although the purpose of a Shewhart chart is to signal the possibility of a special cause, this chart is sometimes inappropriately used to adjust an unstable process. This results in a type of control that at first sight resembles that produced by bounded adjustment schemes. Although these two types of control are similar in that they require adjusting the process only from time to time, bounded adjustment schemes are considerably more efficient. One reason for this is that, when Shewhart

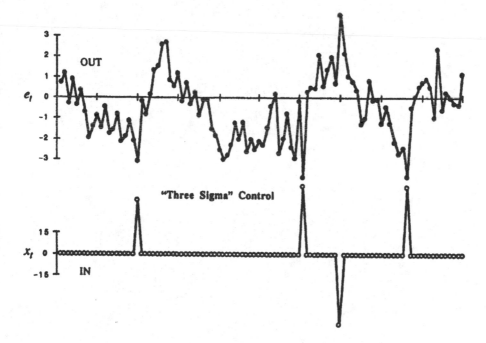

Figure 13. Output series e_t consequent to applying the adjustments x_t corresponding to "three-sigma" control with $G = 1$.

charts are used to adjust a process, action is delayed until the process is almost certainly out of control, for example, when $|e_t|/\sigma_a > 3$. However, bounded adjustment schemes of Chapters D.4 and D.8 call for action at considerably smaller values of $|e_t|$. (See also Luceño, González, and Puig-Pey, 1996.) Another reason is that the magnitude of the adjustment called for by a bounded adjustment scheme should be based on an estimate $\hat{z}_{t+1}$ of the current level of the process which takes account of recent history, rather than on the last value z_t, which has a much larger error (see also Luceño, 1993) and corresponds to taking $\theta = 0$ in (9) and $G = 1$ in (21). Thus, for example, while the variances of the outputs in Figures 5a, 5b, 5c, and 5d are 1.01, 1.24, 1.25, and 1.25, respectively, the control scheme in Figure 13, which is based on a Shewhart chart, gives an output variance of 2.35 for the same disturbance.

ACKNOWLEDGMENT

This research was partially supported by the National Science Foundation under Grant DMI-9414765. Alberto Luceño also acknowledges the support of the Spanish DGICYT under Grant PB 95-0583.

CHAPTER D.8

Selection of Sampling Interval and Action Limit for Discrete Feedback Adjustment

An important problem in process adjustment using feedback is how often to sample the process and when to apply an adjustment. Schemes designed to minimize the overall cost were developed by Box and Kramer, but unfortunately it is not always easy to assign values to the individual costs required to define such schemes. These are the costs of making an adjustment, of taking an observation, and of being off target. In this chapter, charts are provided in which the same schemes are alternatively characterized by the mean squared deviation from target they produce, the frequency with which they require observations to be made, or equivalently the overall length of time between adjustments. This characterization allows a particular scheme to be chosen by judging the advantages and disadvantages of alternative options in the light of the special circumstances of the application. The schemes are derived on certain assumptions relating, in particular, to the model for the disturbance affecting the process. An investigation is undertaken of the effect of two important kinds of failure of this model. We conclude that the procedures we discuss are reasonably robust against such failures.

INTRODUCTION

Two complementary approaches to process control may be called *process monitoring* and *process adjustment*. The former is part of what is called statistical process

From Box, G. E. P. and Luceño, A. (1994), *Technometrics*, 36(4), 369–378. Reprinted with permission from TECHNOMETRICS. Copyright © 1994 by the American Statistical Association. All rights reserved.

control (SPC) and uses, for example, Shewhart, cumulative sum (CUSUM), and exponentially weighted moving average (EWMA) charts to verify that a process remains in a stable "state of control" and to seek out and eliminate sources of disturbance. The latter, sometimes called engineering process control (EPC), uses, in the simplest case, manipulation of some compensating variable X to adjust a quality characteristic to be close to target employing feedback control, feedforward control, or sometimes a combination of them. In the process industries, measurements and adjustments are often made continuously. In this article, we consider only feedback control, and we assume that observations and opportunities for adjustment occur at discrete times $\dots t-1, \, t, \, t+1 \dots$ equispaced at what we define as *unit* intervals. Such a unit interval might be a second of time if we were controlling a nuclear reactor or a day if we were controlling a water treatment plant. Moreover, depending on circumstances, it might be more feasible, convenient, or economic to apply a particular scheme automatically using a computer and transducers or manually using a suitable chart. For clarity, in what follows we illustrate the schemes assuming that a manual feedback adjustment chart is to be used. We assume that an adjustment $x_t = X_t - X_{t-1}$ in the "input" compensatory variable will produce gx_t units of change in the "output" quality characteristic that has its full effect in one unit time interval. Moreover, we define the disturbance z_t at time t as the *deviation from target* of the output quality characteristic that would occur if *no* compensatory action were taken. Thus at time $t+1$, the deviation from target with the adjustment variable still set at the level X_t is

$$\varepsilon_{t+1} = z_{t+1} + gX_t \tag{1}$$

We refer to this deviation ε_{t+1} as the *error* at time $t+1$. The most intuitive form of feedback control would make the adjustment $x_t = X_t - X_{t-1}$ proportional to the last error so that

$$x_t = -\frac{\lambda}{g}\varepsilon_t, \quad 0 \le \lambda \le 1 \tag{2}$$

where λ is a "damping factor" to prevent overadjustment. By summing Equation (2), we see that, at time t, the level X_t of compensatory action would then be a linear function of the CUSUM of past errors

$$X_t = X_0 - \frac{\lambda}{g}\sum_{i=1}^{t}\varepsilon_i \tag{3}$$

This form of control is thus the discrete analog of *integral* control, a basic form of adjustment employed in EPC. It is also easily shown to be equivalent to setting X_t so as just to cancel an EWMA forecast $\hat{z}_{t+1}$ of z_{t+1} computed at time t (e.g., see Box, Jenkins, and Reinsel, 1994) so that

$$X_t = -\hat{z}_{t+1}/g \tag{4}$$

with

$$\hat{z}_{t+1} = \lambda(z_t + \theta z_{t-1} + \theta^2 z_{t-2} + \cdots) \tag{5}$$

where $\theta = 1 - \lambda$. If then the forecast error is

$$z_{t+1} - \hat{z}_{t+1} = e_{t+1} \tag{6}$$

substitution of (6) and (4) in (1) shows that $\varepsilon_{t+1} = e_{t+1}$. So the effect of this simple form of feedback control is that the *error* in forecasting the deviation from target is substituted for the deviation itself. Notice that if the process were already in a perfect state of control with z_t independently and identically distributed (iid) about 0, then the best forecast of z_{t+1} would be 0 and *no* control action would be called for.

The preceding paragraph makes no assumptions concerning the precise form of the disturbance, but it was shown by Muth (1960) that the EWMA (5) produces a minimum mean squared error forecast if the disturbance is generated by the (integrated moving average) IMA(0,1,1) time series model

$$z_{t+1} - z_t = a_{t+1} - \theta a_t \tag{7}$$

where the shocks a_t are independent random variables with mean 0 and variance σ_a^2. It follows then that $\varepsilon_t = a_t$ and the deviations from target of this control scheme have minimum variance σ_a^2. For brevity we will refer to (7) as the IMA model, and unless otherwise stated it is assumed that the shocks are normally distributed.

A well-known aphorism is that "no model is true but some models are useful." The fact that the simple form of control equation (3) is often very effective over a wide range of conditions suggests that (7) might be a model useful for approximating process disturbances more generally. This possibility is further supported by the frequent success of the EWMA as an approximate forecast and by the fact (see Box and Kramer, 1992) that the model implies that *in the absence of any control* the variance of the difference $z_{t+m} - z_t$ will increase linearly with m.

If the operating cost of being off target is a quadratic function of ε_t and is the *only* control cost, then the control models (1) and (3) in which repeated adjustments are made in each time interval define a minimum cost scheme. We will call this a *repeated adjustment* scheme. In this article, we use for illustration 100 successive measurements of the thickness of a metallic film for which the estimated values for (λ, g, σ_a) were (.2, 1.2, 11.0). The data were originally used to illustrate a scheme of feedback control employing repeated adjustment of the deposition rate X (Box, 1991b).

Now, Box and Jenkins (1963) showed that such a repeated adjustment scheme no longer gives minimum cost feedback if it is assumed that there is a fixed adjustment cost (e.g., a cost of stopping a machine and changing a tool). They showed that in that case the scheme yielding minimum overall expected cost would require that action $X_t = -\hat{z}_{t+1}/g$ such as to produce a compensation $-\hat{z}_{t+1}$ should be taken only when $\hat{z}_{t+1}$ crossed either of the lines defined by $|\hat{z}_{t+1}| = L$. A "bounded adjustment" scheme of this kind, again illustrated with the metallic-film data, was given by Box (1991c, 1993), who described a simple interpolation chart to aid the operator in putting it into effect. See also Baxley (1990). Figure 1(a) again shows the bounded adjustment scheme for the first 50 values of the metallic-thickness example with limits $\pm L$ set at ± 8. With the deposition rate initially set at some value X_0, X remains unchanged until time $t = 13$, when the forecasted value 88.7 falls outside the upper limit and the chart signals that a change $X_{13} - X_0 = -8.7/1.2$ is needed in the deposition rate calculated to reduce the thickness by -8.7. Notice that such an adjustment does not upset the calculation of the next EWMA. For example, the forecasted thickness at time $t = 14$ is $(.2 \times 81.3) + (.8 \times 80.0) = 80.3$, where 80.0 is the appropriate previous forecasted value *after the adjustment has been made to bring the process on target.*

The value $L = 8$ in Figure 1(a) was originally chosen to minimize the overall expected cost assuming the accuracy of guesses of the relative costs of being off target and of making a change. For the 100 data available it is interesting to compare this scheme with the repeated adjustment scheme that makes changes in each time interval and for which $\tilde{\sigma}_a = 11.0$. The bounded adjustment scheme but, has an estimated average adjustment interval of 14 but increases the estimated root mean squared error about the target to a value of only $\tilde{\sigma}_\varepsilon = 11.9$. This is an increase of only about 8%. These are purely empirical calculations, but assuming the applicability of the IMA model, one can easily calculate values of the average adjustment interval (AAI) and the percent increase in the standard deviation for any choice of L/σ_a.

The term *average run length* has been widely used to specify and compare the properties of process-monitoring schemes using Shewhart charts, CUSUM charts, and EWMA charts. We prefer to use here instead the term *average adjustment interval* to emphasize that the schemes we discuss are intended to determine when (and by how much) an *adjustment* should be made.

This approach can be generalized by considering also the *sampling* cost—that is, the cost of taking each observation. Supposing that the process was initially sampled at unit intervals, Box and Kramer (1992) considered the possibility of sampling at intervals m units apart, where m is an integer. The sampled process is then also an IMA(0,1,1) time series model like (7) but with parameters λ_m and σ_m, where

$$\lambda_m^2 \sigma_m^2 = m\lambda^2 \sigma_a^2 \quad \text{and} \quad \theta_m \sigma_m^2 = \theta \sigma_a^2 \tag{8}$$

and the overall expected cost C per unit interval is

$$C = \frac{C_A}{\text{AAI}} + \frac{C_M}{m} + \frac{C_T}{\sigma_a^2}(\text{MSD}) \tag{9}$$

where C_T is the cost of being off target for one time interval by an amount of σ_a, C_A and C_M are, respectively, the fixed costs incurred each time the process is adjusted

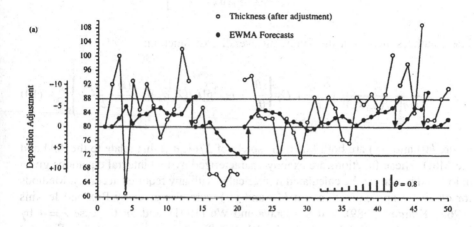

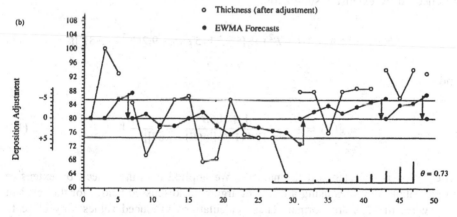

Figure 1. (1) Scheme O applied to the first 50 values of the thickness data with action boundaries at 80 ± 8 and unit sampling interval ($S = 1$). (b) Scheme E applied with $S = 2$ using only the odd values of the data and boundaries at 80 ± 5.55. In both cases the open circles are the adjusted levels of thickness, the solid circles are the EWMA forecasts.

and observed the AAI is the average interval between adjustments measured in terms of *unit* intervals, and the MSD is the mean squared deviation defined by

$$\text{MSD} = \frac{1}{\text{AAI}} E \left[\sum_{j=0}^{n-1} \sum_{k=1}^{m} \varepsilon_{mj+k}^2 \right]$$

These authors arrived at the following overall cost function:

$$C = \frac{C_A}{mh[L/(\lambda_m \sigma_m)]} + \frac{C_M}{m} + C_T \left\{ \frac{\theta}{\theta_m} + m\lambda^2 g[L/(\lambda_m \sigma_m)] - \frac{(m-1)\lambda^2}{2} \right\} \quad (10)$$

where $h(\cdot)$ and $g(\cdot)$ are both functions solely of $L/(\lambda_m \sigma_m)$ that relate to the AAI and the MSD. These functions are exactly characterized by two integral equations given below and hence can be calculated numerically with any required accuracy. Methods for obtaining approximations for $h(\cdot)$ and $g(\cdot)$ were also given by Box and Jenkins (1963), Kramer (1989), and Srivastava and Wu (1991), and for the case $\lambda = 1$ by Crowder (1986) and Adams and Woodall (1989). The charts presented in Figure 2 have been obtained by using the following approximations of Kramer (1989), which he checked by extensive simulation:

$$h(B) = (1 + 1.1B + B^2) \times \{1 - .115 \exp[-9.2(B^3 - .88)^2]\} \quad (11)$$

and

$$g(B) = \frac{1 + .06B^2}{1 - .647\Phi\{1.35[\ln(B) - .67]\}} - 1, \quad (12)$$

where $\Phi(\cdot)$ is the standard normal cdf. We applied a further check by extensive recalculation of values using a different argument, discussed below, for the relevant case when the a_t's are normal. These calculations produced values very close to those given by Equations (11) and (12).

The main purpose of this chapter is to provide charts from which one can choose L and m for these schemes on the basis of how much the root mean squared error would need to increase to achieve the advantage of taking samples and making adjustments less frequently. This approach avoids the direct assignment of values to the costs C_A, C_M, and C_T. A second purpose of this chapter is to explore the sensitivity of the procedure to certain aspects of the assumed disturbance model. Our results suggest that the schemes are reasonably robust to the types of departure we studied. These calculations employ two integral equations that characterize the functions $h(\cdot)$ and $g(\cdot)$ exactly for different choices of the pdf of the shocks.

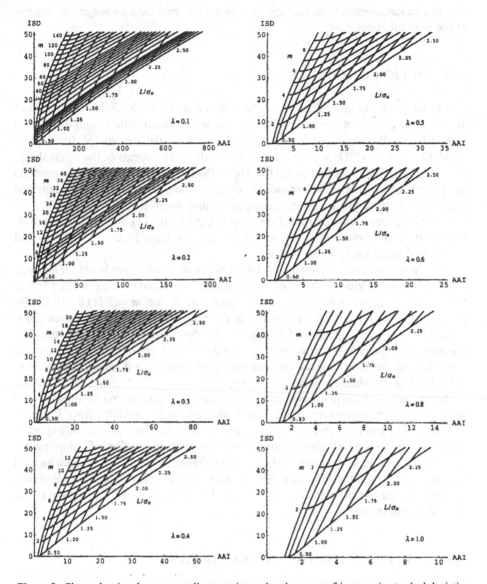

Figure 2. Charts showing the average adjustment interval and percent of increase in standard deviation with respect to σ_a for $\lambda = .1(.1).6(.2) 1.0$, $L/\sigma_a = .0(.25) 2.5$, and several values of m from 1 to 140. The AAI is in terms of the original unit intervals and the ISD= $100 \times (MSD^{1/2}/\sigma_a - 1)$.

Other writers, including Taguchi (1981) and Srivastava and Wu (1991), have considered a similar problem, but they used charts in which action is based on the actual value z_t instead of the EWMA $\hat{z}_{t+1}$. On the assumptions mentioned previously, their schemes are not optimal when the nonstationarity measure λ is different from 1 (see also Adams and Woodall 1989). Recently, Luceño (1993) showed that the main loss of efficiency arose from making an adjustment producing a change $-z_t$ instead of the optimal change $-\hat{z}_{t+1}$.

DESCRIPTION OF CHARTS

Charts are given in Figure 2 for the eight values $\lambda = .1$ (.1) .6, .8, and 1.0. Each of these charts gives, for various values of the standardized action limit $L/\sigma_a = .0$ (.25) 2.5 and the sampling interval $m = 1, \ldots, 140$, the percent of increase in standard deviation (ISD) with respect to σ_a and the AAI in terms of the original unit intervals. The ISD is defined as $\text{ISD} = 100 \times (\text{MSD}^{1/2}/\sigma_a - 1)$ and thus represents the percent of inflation of the standard deviation compared with that of a repeated adjustment scheme for the process sampled at unit intervals.

The charts were obtained from Equations (8), (9), (10), (11), and (12), noting that, since θ/θ_m is a function of λ and m and $L/(\lambda_m\sigma_m) = (L/\sigma_a)/(\lambda m^{1/2})$, the AAI and ISD only depend on λ, L/σ_a, and m.

The charts cover the interesting regions where only small to moderate ISDs occur. For this reason the greater values of m only appear combined with the smaller values of λ. For example, when $\lambda = .1$, we give charts for $m = 1$ (1) 6, 10 (5) 50, 60 (10) 140, but for $\lambda = .6$ we cover only the range $m = 1$ (1) 7. Less attention is given to values of λ larger than .6 because such values occur only for highly unstable processes less likely to be of interest. For instance, $\lambda = 1$ implies that the variance of the difference between observations two unit intervals apart is double the variance of the difference between neighboring observations.

An interesting particular case occurs when the cost C_A of making an adjustment is 0. In this case minimization of the overall expected cost calculated from Equation (10) with (11) and (12) gives $L = 0$ and then $\text{AAI} = m$. Thus, as is to be expected, we return to a repeated adjustment scheme but with the option of sampling the process at intervals m units apart. The charts of Figure 2 with $L = 0$ can then be used to produce values of the ISD for different values of m and λ. It should be noted that, in this case, the overall expected cost regarded as a function of L and m is very flat around its minimum.

AN EXAMPLE OF THE USE OF THE CHARTS

The metallic-thickness data referred to earlier were taken at a particular stage of manufacture of a computer chip when standard SPC techniques had been only partially successful in stabilizing the process. To preserve proprietary information, the data have been coded. A sequence of 100 observations from the uncontrolled process was used in this example to obtain the estimated values $\hat{\lambda} = .2$ and $\hat{\sigma}_a = 11.0$ using standard time series analysis. More frequently, records are available only when some attempt at control has already been applied. So long as the effect of such control actions can be allowed for, the parameters λ and σ_a can still be estimated if necessary by "reconstructing" the disturbance. For clarity in this

Table 1. Some Alternative Schemes for the Metallic-Thickness Example

L m	.0			5.5			11.0		
	Scheme	AAI	ISD%	Scheme	AAI	ISD%	Scheme	AAI	ISD%
1	A	1	0	B	10	3	C	32	10
2	D	2	4	E	12	6	F	35	13
5	G	5	11	H	16	13	I	41	20
L/σ_a	.0			.5			1.0		

example, let us suppose that the unit interval is one minute so that m becomes the interval *in minutes* between observations and the AAI becomes the average interval *in minutes* between adjustments.

Then Table 1 shows the AAI in minutes and the corresponding ISD for nine alternative schemes using all combinations of $L = .0$, 5.5, and 11.0 with $m = 1$, 2, 5. These schemes are denoted by A, B, ..., I and are indicated by circles on the chart for $\lambda = .2$ in Figure 2.

The scheme illustrated in Figure 1(a) with $L = 8$ and $m = 1$ was chosen in a particular situation in which the sampling interval was *fixed* at one minute. We see that this scheme is an interpolation between schemes B and C and yields theoretical values for the AAI and ISD of 20 minutes and 6%, which compare reasonably well with the empirical values mentioned earlier. Table 1 shows alternative possibilities.

Thus one alternative (scheme E) would be to set $L = 5.5$ and observe the process at 2-minute intervals ($m = 2$) with an AAI and ISD of 12 minutes and 6%, respectively. The operation of this scheme is illustrated in Figure 1(b) by using only the odd values of the thickness series as data.

The choice of schemes would depend, for example, on "how capable" the *controlled* process was of providing product within specifications. If the capability index were sufficiently high, then a moderate increase in the standard deviation might be tolerated if this resulted in major savings in sampling and/or adjustment costs.

THE EXPECTED COST FUNCTION UNDER ALTERNATIVE MODELS

For the charts to be useful in practice, one needs to know how sensitive the results are to the assumptions made. For example, we found it at first surprising that very small values of the ISD could be associated with large increments in m and in the AAI. One needs to be reassured, therefore, that these results will hold up at least approximately in a more general context. (Extreme emphasis on precision is, however, not justified because in any case the models are only approximations to real processes.) In this section then we consider how the AAI and ISD are affected by certain specific failures of the disturbance model.

Departures of the Parameter λ from Its True Value

Concerning this first kind of failure, Box (1991a) showed that, under the repeated adjustment scheme, the consequence of choosing a value $\tilde{\lambda}$ moderately different from the true value of λ is to slightly increase the output variance by an amount of $100 \times (\tilde{\lambda} - \lambda)^2/(2\tilde{\lambda} - \tilde{\lambda}^2)\%$. Moreover, Luceño (1993) showed, for $\tilde{\lambda} = 1$ and $m \geq 1$, that for any pair (L, λ), where $L \geq 0$, one can choose a pair $(\tilde{L}, \tilde{\lambda})$ such that their corresponding adjustment schemes have the same AAI and that the MSD is larger for the latter pair than for the former by an amount not greater than $100 \times (\tilde{\lambda} - \lambda)^2\%$. Although these results tend to show that moderate departures $(\tilde{\lambda} - \lambda)$ from the true value of λ have small consequences, we believe that this cannot justify the routine use of the extreme value $\tilde{\lambda} = 1$, which Taguchi (1981) and Srivastava and Wu (1991) employed.

Nonnormal Shocks

Table 2 compares values of the AAI and the quantity $1 + g(\cdot)$ for the IMA model (a) with normally distributed errors and (b) with uniformly distributed errors, where the errors $e_t = z_t - \hat{z}_t = \varepsilon_t - \hat{\varepsilon}_t$ are assumed to have mean 0 and variance σ_e^2. We shall show later in this section how the AAI and MSD corresponding to these models can be computed using a numerical method based on two integral equations. It should be remembered that when $m = 1$ we can express the MSD in terms of $g(\cdot) = (\text{MSD}/\sigma_e^2 - 1)/\lambda^2$, so that $1 + g(\cdot)$ is equal to the MSD when $\lambda = 1$, $m = 1$, and $\sigma_e^2 = 1$. Given $L/(\lambda\sigma_e)$ and $m > 1$ also, the AAI is m times the AAI for $m = 1$ and $L/(\lambda_m\sigma_{em})$, and the MSD$= \sigma_e^2[(\theta/\theta_m)\{1 + \lambda_m^2[g(L/(\lambda_m\sigma_{em})) - (m - 1)/(2m)]\}]$, where $\lambda_m^2\sigma_{em}^2 = m\lambda^2\sigma_e^2$ and $\theta_m\sigma_{em}^2 = \theta\sigma_e^2$.

Table 2. Average Adjustment Interval and $1 + g(\cdot)$ for (a) Normal and (b) Uniform Errors

	IMA Model			
	(a) Normal Errors		(b) Uniform Errors	
$L/(\lambda\sigma_e)$	AAI	$1 + g(\cdot)$	AAI	$1 + g(\cdot)$
.00	1.000	1.000	1.000	1.000
.25	1.245	1.004	1.169	1.003
.50	1.607	1.031	1.406	1.024
.75	2.115	1.092	1.764	1.081

Note: By definition $g(\cdot) = (\text{MSD}/\sigma_e^2 - 1)/\lambda^2$ with $m = 1$ so that $1 + g(\cdot)$ is equal to the MSD when $\lambda = 1$, $m = 1$, and $\sigma_e^2 = 1$.

Barnard Model

A somewhat different structure for the nonstationary distribution model was suggested by Barnard (1959). He assumed that the mean μ_t makes random jumps occurring at intervals whose lengths are distributed as a Poisson distribution with mean δ. The size b of each jump is normally distributed with mean 0 and variance σ_b^2, and on this is superimposed a random noise c, normally distributed with variance σ_c^2. Barnard's model is closely related to the "sticky innovation model" used in Chapter D.5 to justify the IMA disturbance model.

Some results are shown in Table 3. The fourth and fifth columns show for the normal IMA model, with $m = 1$ and for various values of $L/(\lambda\sigma_e)$, the AAI and the quantity $1 + g(\cdot)$. The next two columns show corresponding simulation results for the Barnard model for various values of δ and σ_b^2. The last two columns are for an IMA model with the distribution of e_t a mixture of two normal distributions. Particularly if $L/(\lambda\sigma_e)$ is small, this normal-mixture model is related to the Barnard model, as is explained more fully in Appendix A.

Such results suggest that moderate departures of the kind we consider produce only moderate changes in the AAI and MSD or, equivalently, in the functions $h(\cdot)$ and $g(\cdot)$ that have been used to construct Figure 2.

Evaluation of the AAI and MSD

As a general rule, whether the IMA model with normal errors or a different disturbance model is used, the overall expected cost function can be computed if the AAI and MSD can be calculated. We now give a numerical method that may be used to compute the AAI and MSD for a wide class of disturbance models containing the IMA model.

The AAI can be evaluated using an approach similar to that used by Crowder (1987) in the context of SPC, provided that the action criterion is $|\hat{\varepsilon}_t| \geq L$, where $\hat{\varepsilon}_t = \hat{z}_t + gX_{t-1}$; the *computed* one-step-ahead forecast is updated with $\hat{z}_t = \hat{z}_{t-1} + \lambda e_{t-1}$, and the pdf of the next *computed* shock, or forecast error, $e_{t-1} = z_{t-1} - \hat{z}_{t-1} = \varepsilon_{t-1} - \hat{\varepsilon}_{t-1}$, is known, where $\varepsilon_t = z_t + gX_{t-1}$. The result is also given by an integral equation, although the disturbance is not assumed here to be a series of iid observations. Furthermore, the numerator of the MSD is also given by a similar integral equation.

Let $A(u, L)$ be the average remaining adjustment interval after $t - 2$, given that the computed one-step-ahead forecast is u and the action limit is L. Then, at time $t - 1$, we will conclude that the process needs an adjustment if $|\hat{\varepsilon}_t| \geq L$. In this case the remaining adjustment interval after $t - 2$ will be equal to 1. If $|\hat{\varepsilon}_t| < L$, the process will not need adjustment for at least one more unit of time, the computed

Table 3. Average Adjustment Interval and Mean Squared Deviation Under the Normal IMA, Barnard, and Normal-Mixture IMA Models with the Same Variance of the Next Computed Shock, for $\sigma_c^2 = 1$ and Several Values of $L/(\lambda\sigma_e)$, δ, and σ_b^2

			IMA Model (Normal Errors): Numerical		Barnard Model: Simulation		IMA Model (Normal-Mixture Errors): Numerical	
$L/(\lambda\sigma_e)$	δ^{-1}	$\sigma_b^2 (\sigma_c^2 = 1)$	AAI	$1 + g(\cdot)$	AAI	$1 + g(\cdot)$	AAI	$1 + g(\cdot)$
.5	64	4.00			1.64	1.06	1.64	1.03
		1.00			1.61	1.04	1.62	1.03
		.25			1.61	0.96	1.61	1.03
	16	4.00			1.66	1.05	1.68	1.03
		1.00	1.61	1.03	1.62	1.09	1.62	1.03
		.25			1.61	1.06	1.61	1.03
	4	4.00			1.74	1.01	1.73	1.03
		1.00			1.64	1.01	1.63	1.03
		.25			1.60	1.13	1.61	1.03
1.0	64	4.00			2.90	1.02	2.89	1.19
		1.00			2.82	1.05	2.80	1.19
		.25			2.79	1.36	2.78	1.19
	16	4.00			3.04	1.12	2.99	1.20
		1.00	2.78	1.19	2.83	1.11	2.82	1.19
		.25			2.79	1.29	2.78	1.19
	4	4.00			3.22	1.13	3.09	1.19
		1.00			2.88	1.13	2.84	1.19
		.25			2.80	1.10	2.78	1.19
2.0	64	4.00			7.79	1.48	7.12	1.86
		1.00			7.17	1.26	6.91	1.87
		.25			6.92	2.09	6.88	1.87
	16	4.00			8.46	1.45	7.28	1.85
		1.00	6.90	1.86	7.22	1.79	6.94	1.86
		.25			7.00	1.81	6.88	1.87
	4	4.00			8.44	1.63	7.32	1.84
		1.00			7.42	1.71	6.95	1.86
		.25			7.01	1.92	6.88	1.87
4.0	64	4.00			29.42	3.04	21.42	4.23
		1.00			24.11	3.22	21.10	4.24
		.25			21.81	3.87	21.03	4.24
	16	4.00			28.14	3.38	21.65	4.22
		1.00	21.02	4.27	23.86	3.83	21.11	4.24
		.25			21.96	3.97	21.04	4.24
	4	4.00			24.35	3.96	21.67	4.22
		1.00			22.44	4.10	21.15	4.24
		.25			21.96	4.02	21.03	4.24

Note: The corresponding values of λ, σ_e^2, α, s, and σ_a^2 satisfy $\delta\sigma_b^2/\sigma_c^2 = \lambda^2/\theta$, $\sigma_e^2 = \sigma_a^2 = \sigma_c^2/\theta$, $\alpha = \delta/(1 - \theta^2)$, and $s = 2/(1 + \theta)$.

one-step-ahead forecast will change from $\hat{\varepsilon}_{t-1}$ to $\hat{\varepsilon}_t = \hat{\varepsilon}_{t-1} + \lambda e_{t-1}$, and the remaining adjustment interval after $t-2$ will be $1 + A(\hat{\varepsilon}_t, L)$. Therefore

$$A(u, L) = 1 + \int_{|u+\lambda e|<L} A(u + \lambda e, L)f(e)\, de$$

$$= 1 + \frac{1}{\lambda}\int_{-L}^{L} A(x, L)f\left(\frac{x-u}{\lambda}\right) dx \tag{13}$$

where $f(e)$ is the pdf of the computed shock e_{t-1}.

Analogously, the numerator of the MSD satisfies the integral equation

$$M(u, L) = u^2 + 2uE(e) + E(e^2) + \frac{1}{\lambda}\int_{-L}^{L} M(x, L)f\left(\frac{x-u}{\lambda}\right) dx \tag{14}$$

where $E(e)$ and $E(e^2)$ are the first and second moments of e_{t-1} about 0. Equations (13) and (14) are valid provided that $f(e)$ does not depend on the past of the process or, otherwise, that the whole influence of the past is contained in the pdf of e_{t-1} conditioned to $\hat{\varepsilon}_{t-1}$, which does not depend on t. We will, however, assume that $f(e)$ does not depend on u and also that $E(e) = 0$. Hence

$$M(u, L) = u^2 + \sigma_e^2 + \frac{1}{\lambda}\int_{-L}^{L} M(x, L)f\left(\frac{x-u}{\lambda}\right) dx \tag{15}$$

where σ_e^2 is the variance of e_{t-1}.

Equations (13) and (15) are Friedholm integral equations of the second kind for $A(u, L)$ and $M(u, L)$, respectively, that can be solved analytically in simple cases. In more complex cases, $A(u, L)$ and $M(u, L)$ can be obtained numerically for each L using a method of discretization of (13) and (15). One such method, which is very convenient and accurate, is the collocation method (see Appendix B). A detailed discussion of methods used to obtain numerical solutions to integral equations was given by Baker (1977).

For the cases being considered, the solutions $A(u, L)$ and $M(u, L)$ of (13) and (15) can be written in the form

$$A(u, L) = \tilde{h}[u/(\lambda\sigma_e), L/(\lambda\sigma_e)]$$

$$\frac{M(u, L)}{A(u, L)} = \sigma_e^2 + \lambda^2\sigma_e^2\tilde{g}[u/(\lambda\sigma_e), L/(\lambda\sigma_e)]$$

When the IMA model is true, $f(\cdot)$ is the normal density function with mean 0 and variance σ_a^2. In this case, the numerical results found for $\tilde{h}[0, .]$ and $\tilde{g}[0, .]$ coincide very closely with the approximations to the functions $h(\cdot)$ and $g(\cdot)$ given by Kramer (1989), which are based on smoothed computer simulations.

Once $\tilde{h}[0, .]$ and $\tilde{g}[0, .]$ have been calculated, the results can easily be extended to cover the case in which the process is sampled every m unit intervals if the sequentially computed shocks are uncorrelated both for the process sampled every unit e_t and for the process sampled every m unit intervals. Using a reasoning similar to that of Kramer (1989) or Box and Kramer (1992) one gets

$$AAI = m\tilde{h}[0, \ L/(\lambda_m \sigma_{em})]$$
$$MSD = \sigma_{em}^2 + \lambda_m^2 \sigma_{em}^2 \{\bar{g}[0, \ L/(\lambda_m \sigma_{em})] - (m-1)/(2m)\}$$

where $\lambda_m^2 \sigma_{em}^2 = m \lambda^2 \sigma_e^2$ and $\theta_m \sigma_{em}^2 = \theta \sigma_e^2$.

Additional details on the examples of Tables 2 and 3 are given in Appendix A.

CONCLUSIONS

The charts of Figure 2 avoid the direct assignment of values to the costs of making an adjustment, of taking a sample, and of being off target. They are, however, calculated making specific assumptions concerning the dynamic and disturbance models that are used to calculate the AAI and ISD.

It is shown that two integral equations characterize exactly the AAI and MSD for a class of disturbance models containing the normal IMA model assumed here. By using these equations and by computer simulation, some alternative models for the disturbance have been explored. The results suggest that the robustness of the proposed procedure to these alternative models is satisfactory.

ACKNOWLEDGMENTS

We thank Kuo-Tsung Wu for performing the simulations. This research was supported by the Sloan Foundation and by the Spanish DG1CYT under Grants BE91-150 and PB92-0502. We thank the editor, an associate editor, and two referees for very helpful comments.

APPENDIX A: ADDITIONAL DETAILS ON THE EXAMPLES OF TABLES 2 AND 3

A simple example of the use of (13) and (15) can be shown if $f(\cdot)$ is the uniform distribution in the range $[-(3\sigma_e^2)^{1/2}, (3\sigma_e^2)^{1/2}]$. In this case, the solution of the integral equations is particularly simple if $L/(\lambda\sigma_e) \leq 3^{1/2}/2$, and the AAI and MSD are then given, respectively, by

$$\text{AAI} = \left[1 - \left(\frac{L}{\lambda\sigma_e}\right)\frac{1}{\sqrt{3}}\right]^{-1}$$

$$\text{MSD} = \sigma_e^2 + \lambda^2\sigma_e^2\left(\frac{L}{\lambda\sigma_e}\right)^3\frac{1}{3\sqrt{3}}$$

Table 2 shows some values of the AAI and MSD that have been computed solving (13) and (15) under the IMA and the uniform models.

In Barnard's model, the disturbance is assumed to be generated by the process $z_t = \mu_t + c_t$, where the shocks c_t have iid $N(0, \sigma_c^2)$ distributions, $\mu_t = \mu_{t-1} + \omega_t$, and ω_t is the sum of k independent $N(0, \sigma_b^2)$ random variables, k being an independent Poisson random variable with mean δ for each time t. Clearly, the variance of ω_t is $\sigma_\omega^2 = \delta\sigma_b^2$.

If a process generated using the Barnard model is adjusted according to the action criteria $|\hat{\varepsilon}_t| \geq L$, where $\hat{\varepsilon}_t = \hat{z}_t + gX_{t-1}$, and the computed one-step-ahead forecast is updated with the formula $\hat{z}_t = \hat{z}_{t-1} + \lambda e_{t-1}$, where $\lambda = 1 - \theta$, then the next computed shock $e_{t-1} = z_{t-1} - \hat{z}_{t-1} = \varepsilon_{t-1} - \hat{\varepsilon}_{t-1}$ has mean 0 and variance

$$\sigma_e^2 = \frac{2\sigma_c^2}{1+\theta} + \frac{\delta\sigma_b^2}{1-\theta^2} \tag{A1}$$

It can be easily shown that, if $\delta\sigma_b^2/\sigma_c^2 = \lambda^2/\theta$, the autocorrelation structures of the IMA with smoothing constant θ and the Barnard's disturbances are the same. The variance of the noise in this IMA model is $\sigma_a^2 = \sigma_c^2/\theta$, which coincides with the variance of the next computed shock given by (A1) when $\delta\sigma_b^2/\sigma_c^2 = \lambda^2/\theta$.

The distribution of the next computed shock for the Barnard's model is a mixture of an infinite number of normal distributions with mean 0 and increasing variances. For this distribution, Equations (13) and (15) become very cumbersome to solve.

A somewhat related but easier model can be found assuming that the distribution of the next computed shock is a mixture of an $N(0, s\sigma_c^2)$ and an $N(0, s\sigma_c^2 + \sigma_b^2)$, with probabilities $1 - \alpha$ and α, respectively. The variance of the next computed shock is then $s\sigma_c^2 + \alpha\sigma_b^2$. For δ small, if $\alpha = \delta/(1 - \theta^2)$ and $s = 2/(1 + \theta)$, the variance of the next computed shock is the same for the Barnard and for this normal-mixture models.

Table 3 gives the AAI and MSD obtained by simulation with the Barnard's model and the AAI and MSD obtained solving (13) and (15) numerically using the normal-mixture and the IMA models with corresponding values of λ, σ_e^2, α, s, and σ_a^2 for $\sigma_c^2 = 1$ and several values of $L/(\lambda\sigma_e)$, and δ, and σ_b^2.

APPENDIX B: A NUMERICAL METHOD TO SOLVE THE INTEGRAL EQUATIONS (13) AND (15)

To obtain approximate solutions for $A(u, L)$ and $M(u, L)$ using the method of collection, let

$$A_n(u, L) = \sum_{i=1}^{n} \alpha_i p_i(u, L)$$

$$M_n(u, L) = \sum_{i=1}^{n} \gamma_i p_i(u, L) \tag{B1}$$

where α_i and γ_i are constants to be determined, and $p_i(u, L)$ for $i = 1, \ldots, n$ are linearly independent functions of u for $-L \leq u \leq L$.

Introducing (B1) into (13), one has the remainder

$$R[A_n(u, L), u] = 1 - \sum_{i=1}^{n} \alpha_i[p_i(u, L) - I_i(u, L)]$$

where

$$I_i(u, L) = \frac{1}{\lambda} \int_{-L}^{L} p_i(u, L) f\left(\frac{x - u}{\lambda}\right) dx$$

If $R[A_n(u, L), u]$ vanishes at n given points $\{u_1, \ldots, u_n\}$, which are within the domain of interest, the constants α_i can be calculated by solving the linear system

$$\sum_{i=1}^{n} \alpha_i[p_i(u_j, L) - I_i(u_j, L)] = 1 \tag{B2}$$

for $j = 1, \ldots, n$.

Analogously, the constants γ_i can be calculated solving the linear system

$$\sum_{i=1}^{n} \gamma_i[p_i(u_j, L) - I_i(u_j, L)] = u_j^2 + \sigma_e^2 \tag{B3}$$

for $j = 1, \ldots, n$, where we note that (B2) and (B3) have the same coefficients matrix.

The numerical results shown in Table 3 have been obtained solving (B2) and (B3) for $n = 3$ (2) 15, evaluating $A_n(0, L)$ and $M_n(0, L)$ for $n = 3$ (2) 15 and using the Shanks's transformation to accelerate the convergence to the limit when n tends to infinity (see Bender and Orszag, 1984). Besides, each function $p_i(u, L)$ has been chosen as a piecewise continuous linear function of u that takes the value 1 at u_i and 0 at u_j for $j \neq i$, where $u_i = 2L[i - (n + 1)/2]/(n - 1)$ for $i = 1, \ldots, n$.

For the cases being considered, the approximate solutions $A_n(u, L)$ and $M_n(u, L)$ for $n = 3, 4, \ldots$ can be written in the form

$$A_n(u, L) = \tilde{h}_n[u/(\lambda\sigma_e), \ L/(\lambda\sigma_e)]$$

$$\frac{M_n(u, L)}{A_n(u, L)} = \sigma_e^2 + \lambda^2\sigma_e^2\tilde{g}_n[u/(\lambda\sigma_e), \ L/(\lambda\sigma_e)]$$

Hence the same expression applies to the limits of $A_n(u, L)$ and $M_n(u, L)$ when n tends to infinity, provided that functions $\tilde{h}_n[\cdot]$ and $\tilde{g}_n[\cdot]$ are substituted for their limits $\tilde{h}[\cdot]$ and $\tilde{g}[\cdot]$.

Use of Cusum Statistics in the Analysis of Data and in Process Monitoring

INTRODUCTION

Cusums were originally devised by Page (1954) and Barnard (1959) and put to extensive use both in England and the United States, particularly for the monitoring of synthetic fiber manufacture (see, e.g., Goldsmith and Whitfield, 1961; Lucas and Crosier, 1982). They provide an extremely sensitive on-line procedure for detecting small shifts in the process mean; that is, for the detection of "step changes." In its simplest form, the Cusum S is an accumulated sum of the deviations $y_1 - T$, $y_2 - T, \ldots$ from the process target value T. Thus, $S_t = (y_1 - T) + (y_2 - T) + \cdots + (y_t - T)$, and a Cusum chart is obtained by plotting S_t against t. If the mean is on target, then S_t will simply be the sum of a series of deviations having zero mean, some of which are positive and some negative, but as soon as a shift in the mean away from the target occurs, the Cusum plot exhibits a change in *slope*. This is so because as soon if an increase in the mean of d units occurs, then as each new observation is obtained, the quantities d, $2d$, $3d, \ldots$ are added to successive values of the Cusum.

For example, Figure 1(a) shows observations plotted on a Shewhart run chart for which the target value is $T = 30$. Figure 1(b)shows the Cusum chart in which, for the same observations, the successive sum S_t of the deviations from target is plotted against t. The marked change in slope that occurs in the Cusum plot strongly suggests that a shift in mean has occurred close to observation number 16. Also, over the last 15 observations, a total increase of about 32 occurs in the Cusum so that the amount of shift can be estimated to be about $32/15 = 2.1$ units. It is true that by

From Box, G. E. P. (1999), *Quality Engineering*, **11**(3), 495–498.
Copyright © 1999 by Marcel Dekker, Inc

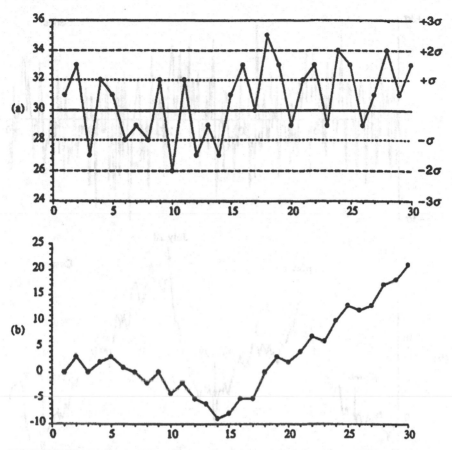

Figure 1. A series with a one-standard-deviation step change occurring at observation 16 plotted (a) as a Shewhart chart and (b) as a Cusum chart.

using only the Shewhart chart, with extended Western Electric rules, a change can also be detected. However, the Cusum characterizes the point of change and its magnitude more distinctly.

Although, over the last few decades, there has been extensive development of the formal theory of Cusum tests, the usefulness and simplicity of Cusums in the analysis of *past* data have sometimes been overlooked. In particular, their ability to determine when and by how much the process mean has changed is often of great use in tracking down assignable causes of trouble.

An interesting example showing the Cusum used as a diagnostic tool occurred in the analysis of data from one stage of a large chemical process in which two gases were brought together at very high temperatures to form a desired product. In order to avoid the production of impurities, it was necessary that the gas flows were maintained as close as possible to specified constant levels. Deviations from target for daily flow readings of one of the gases over an 8-month period are shown in Figure 2(a), together with 2σ and 3σ limits. This run chart does not bring to

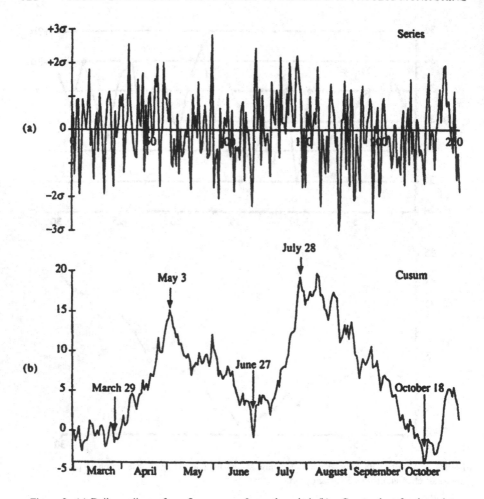

Figure 2. (a) Daily readings of gas flow over an 8-month period; (b) a Cusum chart for these data.

attention any features of particular interest. However, in the Cusum chart for the same data, shown in Figure 2(b), a series of slope changes are seen clearly, indicating possible changes in the mean in the original data. The chart suggests that these changes occurred close to the following dates: March 29, May 3, June 27, July 28, and October 18. After some thought, it was realized that these times were close to those when the meters measuring the gas flow were recalibrated. As a result of this discovery, the calibration system was drastically modified and the problem eliminated.

THE CUSUM AS AN EXAMPLE OF A CUSCORE STATISTIC

The Cusum is a particular example of what Box and Ramírez (1992) called Cuscore statistics. They regarded Shewhart and other quality control charts used for process

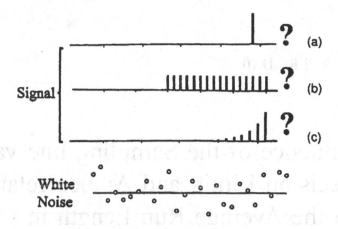

Figure 3. Three kinds of signal: (a) spike; (b) step-function; (c) exponential increase.

monitoring as particular devices for detecting a *signal* buried in *noise*. The signal represents a particular kind of pattern in the data, characterizing a malfunction, which we need to detect. The noise is the background variation in which the signal is hidden. This background variation might, for example, be characterized by "white noise": that is, by *independently* distributed deviations from the mean drawn from an approximately Normal distribution. Alternatively, it might be characterized by dependent deviations generated by some kind of time-series model and possibly by distributions other than the Normal.

In any case, given (a) the kind of signal you are looking for, and (b) the kind of noise in which it is buried, these authors showed how to determine the most sensitive statistic for signal detection by using the efficient score statistic of Fisher (1925b). As an illustration, Figure 3 shows three different kinds of signals that might be of interest: (a) a "spike," (b) a step-function, and (c) an exponential increase. If such signals are concealed in white noise, then the plots of the appropriate Cuscore statistics can be shown to be, respectively, (a) the Shewhart chart, (b) a Cusum chart, and (c) the exponentially weighted moving average chart (Box and Luceño, 1997).

The Cuscore principle can be used quite generally for finding the most efficient detector for virtually any kind of signal in any kind of noise. In particular, it may be used (Box and Luceño, 1997) to detect a signal in a process which is subject to feedback control.

CHAPTER D.10

Influence of the Sampling Interval, Decision Limit, and Autocorrelation on the Average Run Length in Cusum Charts

This chapter shows how the average run length (ARL) for a one-sided Cusum chart varies as a function of the length of the sampling interval between consecutive observations, the decision limit for the Cusum statistic, and the amount of autocorrelation between successive observations. It is shown that the rate of false alarms can be decreased considerably, without modifying the rate of valid alarms, by decreasing the sampling interval and appropriately increasing the decision interval. Also, this can be done even when the shorter sampling interval induces moderately autocorrelation between successive observations.

INTRODUCTION

By process monitoring we mean sequential surveillance of a system to detect a change of state that can characterize a malfunction. The general problem can also be stated as that of serially collecting and analyzing data to quickly detect a signal buried in noise. We suppose that observations y_1, y_2, ... of some characteristic of interest are made at equispaced times τS, where $\tau = 1$, 2, ... and S is the *sampling interval*, that is, the time interval between successive observations, and that monitoring is performed by continually updating the value of some quantity Q calculated from the incoming data and declaring detection of a malfunction as soon as Q

From Luceño, A. and Box, G. E. P. (2000), *Journal of Applied Statistics*, **27**(2) 172–183.

changes in some specifically defined manner. Correct detection of a malfunction will be called a *valid alarm*; incorrect detection a *false alarm*. Because of the system noise, some detection errors will occur. An efficient monitoring scheme is one that produces alarms, the largest proportion of which are valid and the smallest proportion false.

Such "change point" problems have been considered by a number of authors (see, e.g., Basseville and Nikiforov, 1993; Crowder, Hawkins, Reynolds, and Yashchin, 1997; Ferger, 1995; Grodon and Pollak, 1995; Lai, 1995; Lorden, 1971; and Roberts, 1966). In particular, a cumulative Fisher score statistic (see Box and Ramírez, 1992; Box and Luceño, 1997) or Cuscore has been proposed for the quantity Q.

In this paper we suppose that the system is in a satisfactory* state when

$$y_t = \mu_0 + a_t \qquad (1)$$

where $\{a_t\}$ is a series of normally, independently, and identically distributed random variables with mean zero and variance $\sigma_a^2 = \sigma_y^2$, often referred to as white noise. We suppose also that a malfunction is characterized by an increase $d = D\sigma_a$ in the mean from μ_0 to $\mu_0 + d = \mu_1$, where the quantity D is called the *standardized increase in the mean*. The monitoring problem is then one of detecting a step change signal buried in white noise and the appropriate detection statistic (Barnard, 1959; Page, 1954, 1957, 1961) is the Cusum

$$Q = \sum (y_i - \mu_0)$$

Comprehensive descriptions and applications of Cusums may be found, for example, in Box and Luceño (1997, Chap. 3), Duncan (1974, Chap. 22), or Montgomery (1991, Chap. 7).

For formal detection, it is convenient to use the centered Cusum

$$Q^* = \sum (y_i - \bar{\mu})$$

where $\bar{\mu} = (\mu_0 + \mu_1)/2 = \mu_0 + d/2$. This centered Cusum is essentially a handicapped score. When the mean is equal to μ_0, the quantity $d/2$ will be subtracted from each deviation, producing a downward trend in the centered Cusum. But if the mean level is greater than $\mu_0 + d/2$, the centered Cusum will tend to increase. The Cusum detection procedure requires that we accept that a genuine increase in the mean above μ_0 has occurred as soon as the centered Cusum has risen from its previous minimum by more than some quantity h called the *decision interval*.

For a given standardized increase in the mean D and the sampling interval S, the standardized decision interval $H = h/\sigma_a$ will determine the characteristics of the

---

* The assumption that an operating system can remain, even approximately, in a stationary state for long periods of time is certainly invalid (see, e.g., Chapter D.1). However as an approximation, we make this assumption here.

Cusum chart. A quantity describing the performance of such schemes is the *run length* (RL), that is, the elapsed time before an alarm is triggered. This is a random variable varying from one run to another, and its mean is called the *average run length* (ARL). When the ARL is large, the probability distribution for the RL is close to a geometric distribution (Ewan and Kemp, 1960; Brook and Evans, 1972) so that it can be completely determined from its ARL. On this basis, it has become customary to define the Cusum scheme in terms of the ARL for false alarms (L_a) when the process is in the acceptable condition $\mu = \mu_0$, and the ARL for valid alarms (L_r) when the process is in the rejectable condition $\mu = \mu_1$.

As a specific example, suppose that the data were measurements of the temperature of a fluid in a reactor taken every $S = 6$ minutes. Then if μ_0 were the acceptable temperature, our monitoring system would be required to indicate an alarm if the temperature rose to the unacceptable level μ_1. If $D = 3$ and $H = 2.64$, the methods in Goel and Wu (1971) give $L_r = 2.435$ and $L_a = 16.666$. Thus if the temperature stayed at the satisfactory level μ_0, the average elapsed time before a false alarm was triggered would be $T_a = L_a S = 100,000$ minutes (69.4 days). However, with the temperature at the unsatisfactory level μ_1, an alarm will occur on the average after an elapsed time of $T_r = L_r S = 14.61$ minutes.

If we characterize the Cusum in terms of these averages, the above scheme seems very satisfactory. Unfortunately, however, as is well known, RLs are subject to very large variations about their ARLs. This may be particularly disturbing in the case of the RL distribution for satisfactory performance, when the ARL is very large, because the standard deviation of the RL is then almost equal to its ARL and hence is very large. For illustration, Table 1 shows nine quantiles of the RL distribution for the acceptable state with $\mu = \mu_0$ in the example above. Although it would be true that when the reactor is in the acceptable state, the RL distribution would have its mean at 100,000 minutes (69.4 days), we see that more than 5% of false alarms would occur in the first 4 days and about 25% in the first 20 days. It is then necessary:

(1) to find ways of increasing the smaller quantiles of the RL distribution (expressed as elapsed time) corresponding to satisfactory performance;

(2) to characterize potential performance in such a way that a fairer picture of the behavior of the monitoring scheme is gained.

Table 1. Some Quantiles of the RL Distribution for False Alarms Having $T_a = 100,000$ minutes (equivalent to 69.4 days), Scheme A

Percent of False Alarms	1	5	10	25	50	75	90	95	99
Minutes	1,014	5,136	10,542	28,776	69,318	138,624	230,238	299,544	460,470
Days	0.70	3.57	7.32	19.98	48.14	96.27	159.89	208.02	319.77

These quantiles have been computed numerically using the true RL distribution, rather than its geometric approximation.

In the remainder of this chapter we consider the effect on the ARL and quantiles of the RL distribution of reducing the sampling interval and simultaneously increasing the decision interval and how the resulting ARL and quantiles change when there is a moderate autocorrelation between consecutive observations.

EFFECT OF REDUCING THE SAMPLING INTERVAL AND AUGMENTING THE DECISION INTERVAL

Given the assumptions, efficient design of a Cusum chart may be accomplished by observing that the ARL (expressed as a multiple of the sampling interval S) increases approximately *exponentially* with the decision interval when the process behaves satisfactorily with mean μ_0, whereas this increase is approximately *linear* when the process mean is μ_1. In addition, the ARL (expressed as elapsed time) is a linear function of S irrespective of whether the process behavior is satisfactory or not. Thus suppose as in scheme B the length of the sampling interval was divided by 3 so that $S = 2$ minutes instead of 6 minutes and that simultaneously the standardized decision limit was multiplied by 3 so that $H = 7.92$. Then if we could assume that model (1) remained essentially valid with the shorter sampling interval, then compared with scheme A, the ARL and upper quantiles ($q_{0.95}$ and $q_{0.99}$) for the RL distribution corresponding to valid alarms would decrease slightly, whereas the ARL and lower quantiles ($q_{0.01}$ and $q_{0.05}$) for the RL distribution of false alarms would increase drastically. In particular, scheme A produces 5% of false alarms by 5136 minutes

Table 2. ARLs and Quantiles in Minutes for Different Schemes and Temperature Conditions

Scheme	False Alarms Acceptable Condition			Valid Alarms Rejectable Condition		
	$q_{0.01}$	$q_{0.05}$	ARL (T_a)	ARL (T_r)	$q_{0.95}$	$q_{0.99}$
A	1014	5136	100,000	14.61	24.00	30.00
$S = 6$ minutes						
$H = 2.64$, $D = 3$						
B	2.56×10^9	13.1×10^9	255×10^9	11.91	18.00	22.00
$S = 2$ minutes						
$H = 7.902$, $D = 3$						
C	0.556×10^6	2.84×10^6	55.3×10^6	17.63	28.00	36.00
$S = 2$ minutes				(15.14)	(26.00)	(32.00)
$H = 7.92$,						
$D = 1.964$						
autocorrelation						
$\phi = 0.4$						

(3.6 days), whereas theoretically the same percent of false alarms would be reached only by 13.1 billion minutes (24,924 years) with scheme B.

An explanation is as follows. By dividing the sampling interval by 3, the ARL and quantiles for the RL distribution (expressed in minutes) is divided exactly by 3 no matter whether the temperature in the reactor is in the acceptable or rejectable condition. Now on the one hand, when the temperature is in the rejectable condition, the ARL and upper quantiles for the RL distribution increase approximately *linearly* with the standardized decision interval so that, when this is multiplied by 3, the ARL and upper quantiles for valid alarms are roughly multiplied by 3 and therefore remain approximately unchanged with respect to the initial scheme. On the other hand, when the temperature is in the acceptable condition, the ARL and lower quantiles for false alarms increase approximately *exponentially* with the standardized decision interval and hence are much larger for the new scheme than for the initial scheme. This rationalization is in agreement with Siegmund's approximate formula for evaluating the ARLs for false and valid alarms (see Siegmund, 1985, p. 27; or Woodall and Adams, 1993).

EFFECT OF THE AUTOCORRELATION BETWEEN SUCCESSIVE OBSERVATIONS

We now suppose that the system stays in a stationary state but that when the length of the sampling interval is reduced to $S = 2$ minutes, model (1) does not remain valid but is replaced by an autoregressive time series model of order one, AR (1), such as

$$(y_t - \mu_0) = \phi(y_{t-1} - \mu_0) + b_t \tag{2}$$

where ϕ is the autocorrelation coefficient between consecutive observations and b_t is a white noise with variance σ_b^2. Obviously, for $\phi = 0$, model (2) coincides with model (1). If model (2) is assumed to be valid for $S = 2$ minutes, the corresponding model for $S = 6$ minutes would be

$$(y_t - \mu_0) = \phi^3(y_{t-3} - \mu_0) + a_t \tag{3}$$

and, assuming that σ_y^2 remains constant, we have

$$\sigma_y^2 = \frac{\sigma_a^2}{1 - \phi^6} = \frac{\sigma_b^2}{1 - \phi^2}$$

which relates σ_y with σ_a and σ_b.

Suppose now that we want to look for the temperature change (step signal) from μ_0 to μ_1 hidden in the AR (1) noise of Equation (2) with $\phi = 0.4$. Then application of the results in Box and Luceño (1997, Appendix 10B) shows that this problem is equivalent to looking in white noise for a signal that has a first spike of magnitude

$$D = \frac{d}{\sigma_b} = \frac{3\sigma_y}{\sigma_b} = 3(1 - \phi^2)^{-1/2} = 3.2733 \tag{4a}$$

and the remaining spikes of magnitude

$$D = \frac{d(1 - \phi)}{\sigma_b} = \frac{3(1 - \phi)\sigma_y}{\sigma_b} = 3(1 - \phi)(1 - \phi^2)^{-1/2} = 1.9640 \tag{4b}$$

Let us suppose, for simplicity, that all these spikes including the first one may be taken equal to 1.9640 so that looking for a change of the mean temperature of magnitude $d = 3\sigma_y$ in the AR (1) noise of Equation (2) with $\phi = 0.4$ is approximately equivalent to looking for a standardized change in the mean temperature of magnitude $D = 1.9640$ in white noise. Then scheme C of Table 2 gives the ARL and upper quantiles for the RL distribution corresponding to the rejectable condition along with the ARL and lower quantiles for the RL distribution for the acceptable condition. We see that the parameters for valid alarms are close to those obtained previously for the scheme A, whereas the parameters for the false alarms remain considerably larger than for the original scheme (although not as large as those of scheme B). In particular, whereas the original scheme produces a 5% of false alarms by 5136 minutes (3.6 days), this percent of false alarms is reached only by 2.84 million minutes (5.4 years) for scheme C when the autocorrelation coefficient is $\phi = 0.4$.

In the computations for scheme C it has been assumed that the first spike of magnitude 3.2733 may be replaced by a smaller spike of magnitude 1.9640 without introducing important errors. Consequently, the signal used to compute the scheme C is a step signal of magnitude 1.9640. If this simplification is not made, that is, if our scheme is looking for a step signal of magnitude 1.9640 but the true signal is given by Equation (4b), then the resulting ARL and upper quantiles are those shown in parentheses in the third line of Table 2. We see that these values are even closer than their approximations to the values for scheme A. The ARL and quantiles for the false alarms remain unchanged because these are computed assuming that the true signal is null.

CONCLUDING REMARKS

The performance of Cusum charts has been usually analyzed in the literature under the assumption that the noise sequence in model (1) is stationary and normally distributed. In particular, Goel and Wu (1971) provided a chart to evaluate the ARL for one-sided Cusum charts under the normal assumption. This assumption is often justified in practice because the values y_t in Equation (1) are obtained as an average of a few individual observations of the quality characteristic. If the standardized decision interval H is very large, however, the resulting ARLs are somewhat sensitive to the assumption that the tails of the distribution are not very different from those corresponding to the normal assumption. This problem is one of lack of knowledge of the exact tails of the distribution (because data in the far tail are only very rarely observed), so that no alternative to the normal distribution is likely to be considered more appropriate. However, the assumption of stationarity over long periods of time is likely to be much more important; consequently, the actual performance of any tentative scheme chosen on the guidelines provided by theoretical studies should always be checked in practice.

ACKNOWLEDGMENT

This research was partially supported by the National Science Foundation Grant DMI-9812839 and the Spanish DGESIC Grant SAB1995-0766. Alberto Luceño also acknowledges the support of the Spanish DGESIC Grant PB97-0555.

Cumulative Score Charts

SUMMARY

Shewhart charts are direct plots of the data and they have the potential to detect departures from statistical stability of unanticipated kinds. However, when one can identify in advance a kind of departure specifically feared, then a more sensitive detection statistic can be developed for that specific possibility.

In this chapter Cuscore statistics are developed for this purpose which can be used as an adjunct to the Shewhart chart. These statistics use an idea due to Box and Jenkins (1966), which is in turn an application of Fisher's score statistic. This article shows how the resulting procedures relate to Wald–Barnard sequential tests and to Cusum statistics which are special cases of Cuscore statistics. The ideas are illustrated by examples.

INTRODUCTION

Process control can have different objectives and consequently can take different forms. One important objective is *process regulation* achieved, for example, by automatic feedback and feedforward control schemes designed to maintain a quality characteristic close to target. Statistical considerations associated with such schemes have been discussed, for example, by Box and Jenkins (1966), Aström (1970), Aström and Wittenmark (1984), MacGregor (1972), and Box and Kramer (1989).

From Box, G. E. P. and Ramírez, J. (1992), *Quality and Reliability Engineering International*, **8**, 17–27. Copyright © 1992 by John Wiley & Sons, Ltd.

This paper concerns the situation where the object is not primarily to regulate but to monitor the process. Such monitoring provides sequential verification of the continuous stability of the process, once a state of statistical control has been achieved, and allows detection of deviations from the state leading to an appropriate search for assignable causes. Such *process monitoring* may be achieved, for example, by using a Shewhart chart.

One of the many virtues of the Shewhart chart is that it is a direct plot of the actual data and so can expose types of deviations from statistical stability, of a totally *unexpected* kind. Thus it can serve in an *inductive* role. Inspection of such a chart might, for example, show that every fifth observation was low as compared to the mean of the previous four which might in turn lead to the discovery of an assignable cause previously not even thought of. However, this property of potential response to global alternatives also ensures that the Shewhart chart will not be as sensitive to some *specific* deviation from randomness as a correspondingly specially chosen procedure. When such a specific kind of deviation is feared, therefore, it is appropriate to employ a procedure especially sensitive to that possibility and to use it *in addition* to the overall Shewhart chart. The problem has been likened by Box (1980) to that faced by a small country wishing to install an early warning radar system against air attack. In addition to a multi-directional screen it would be wise to add very sensitive radar beams in certain directions known to be likely sources of aggression.

In the monitoring of quality, one such additional specific procedure employs the Page–Barnard cumulative sum (Cusum) chart (Page, 1954; Barnard, 1959). Such a chart is particularly sensitive to small changes in the mean level of a process characteristic, indicated by a change of slope of the Cusum. The sensitivity of this procedure arises from its similarity to a Wald–Barnard sequential likelihood test for change in mean. Equivalently, it can be regarded as a test based on the accumulated Fisher score statistic. Such charts can provide the basis for formal tests of statistical significance but have often been used informally to track down assignable causes evidenced by small changes in mean level. For example, in Chapter D.9 we saw how a Cusum chart for gas flow showed marked changes of slope indicating small shifts in mean on specific days. Careful investigation revealed that these changes occurred close to times at which the gas flowmeters had been recalibrated. Used in this diagnostic way Cusum charts are a valuable adjunct to, for example, Ishikawa's "seven tools" and similar exploratory methods for tracking down the causes of problems and so improving quality.

Now, occasions occur when some specific kind of deviation other than a change in mean is feared as a likely possibility. To cope with this kind of problem, in an unpublished report, Box and Jenkins (1976) proposed a general *cumulative score* statistic. Although the immediate objective of these authors was to warn of possible changes in the parameters of time series models, the concept has much broader applicability.

THE CUSCORE STATISTIC

Consider a statistical model written in the form

$$a_t = a_t(y_t, x_t, \theta), \quad t = 1, 2, \ldots, n \tag{1}$$

where y_t are observations, θ is some unknown parameter, and the x_t's are known independent variables. Then standard normal theory models assume that when θ is the true value of the unknown parameter, the resulting a_t's are a sequence of independently identically normally distributed random variables with mean zero and variance $\sigma_a^2 = \sigma^2$, which we will call a white noise sequence. Thought of in this way, statistical models are recipes for reducing data to "white noise."

Apart from a constant, the log likelihood for $\theta = \theta_0$ is then

$$l = -\frac{1}{2\sigma^2} \sum a_{t0}^2 \tag{2}$$

where the a_{t0}s are obtained by setting $\theta = \theta_0$ in Equation (1) and unless otherwise stated, $\sum$ will indicate summation from $t = 1, 2, \ldots, n$. If we now write

$$-\frac{\partial a_t}{\partial \theta}\Big|_{\theta=\theta_0} = d_{t0}$$

then

$$\frac{\partial l}{\partial \theta} = \frac{1}{\sigma^2} \sum a_{t0} d_{t0} = \frac{1}{\sigma^2} Q_0 \tag{3}$$

and we shall refer to

$$Q_0 = \sum a_{t0} d_{t0} \tag{4}$$

as the *Cuscore* associated with the parameter value $\theta = \theta_0$.

Now expanding a_t about a_{t0} the following formula is exact if the model is linear in θ and approximate otherwise:

$$a_{t0} = (\theta - \theta_0)d_{t0} + a_t \tag{5}$$

Hence we see that if the value of the parameter is not equal to θ_0 an increment of the discrepancy vector d_{t0} is added to the vector a_t. The Cuscore statistic equation (4) which sequentially correlates a_{t0} with d_{t0} is thus continuously searching for the presence of that particular discrepancy vector.

Equivalently, since $(\hat{\theta} - \theta_0) = \sum a_{t0} d_{t0} / \sum d_{t0}^2$ is the least squares estimator of $\theta - \theta_0$,

$$Q = (\hat{\theta} - \theta_0) \sum d_{t0}^2 \qquad (6)$$

When plotted against n the Cuscore can thus be expected to provide a very sensitive check for changes in θ, and such changes will be indicated by a change in slope of the plot just as changes in the mean are indicated by a change in slope of the Cusum plot. By noticing at which specific times changes in slope have occurred, the Cuscore may thus be used to provide clues as to the time of occurrence, and hence possibly of the identity of specific problems.

Detecting a Sine Wave Buried in Noise

As a first illustration, consider an industrial process in which trouble has been experienced with a characteristic harmonic cycling about a target value T at a particular period and phase, and suppose that although action has been taken which seems to have solved the problem it is feared that it might recur. A test procedure is needed which will give the earliest possible warning of its recurrence.

Suppose the sine wave has amplitude θ and period p, then the model is

$$y_t = T + \theta \, \sin(2\pi t/p) + a_t$$

or

$$a_t = y_t - T - \theta \, \sin(2\pi t/p) \qquad (7)$$

When the process is operating correctly, $\theta = \theta_0 = 0$, and there will be no sine component in $y_t - T$. Since $d_{t0} = \sin(2\pi t/p)$ the Cuscore statistic which is looking for a recurrence of the sine component is

$$Q_0 = \sum (y_t - T) \sin(2\pi t/p) \qquad (8)$$

A sine wave with period 12 and amplitude $\theta = \frac{1}{2}$ starting at the 48th observation and stopping after the 96th observation is shown in Figure 1(a). Figure 1(b) shows a Shewhart chart for the *composite* series obtained by adding the sine wave to the white noise series consisting of 144 random normal deviates with standard deviation $\sigma = 1$. It will be seen that the sine wave is effectively hidden in the noise. The appearance and disappearance of the sine wave is, however, quite evident in the Cuscore chart of Figure 1(c) in which

$$Q = \sum (y_t - T) \sin(\pi t/6) \qquad (9)$$

is plotted against t.

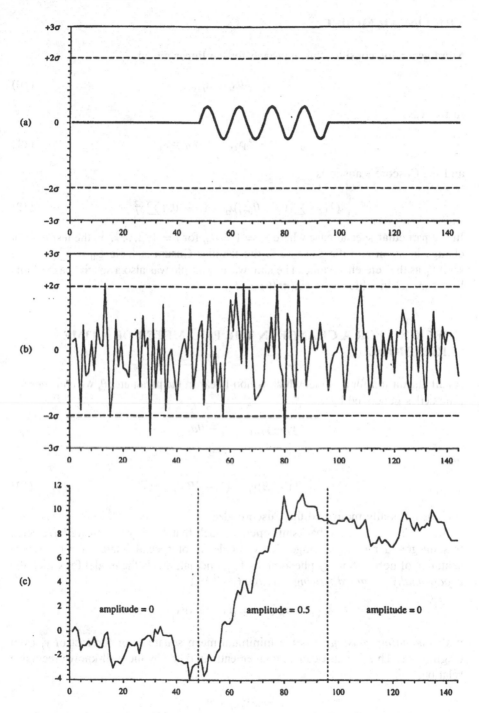

Figure 1. Detection of a sine wave: (a) an intermittent sine wave with amplitude 0.5σ; (b) the sum of the sine wave and white noise with $\sigma = 1$; and (c) the Cuscore statistic applied to the data of (b).

THE LINEAR MODEL

More generally, consider this approach for the linear model

$$y_t = \theta x_t + a_t \tag{10}$$

In this case

$$a_{t0} = y_t - \theta_0 x_t, \qquad d_{t0} = x_t \tag{11}$$

and the Cuscore statistic is

$$Q_0 = \sum (y_t - \theta_0 x_t) x_t = (\hat{\theta} - \theta_0) \sum x_t^2 \tag{12}$$

In the particular special case where $x_t = 1 = d_{t0}$ for $t = 1, 2, \ldots, n$, the test is for a change in mean and the Cuscore is the familiar Cusum statistic $Q_0 = \sum (y_t - \theta_0)$ with θ_0 as the reference value. The sine wave example was also a special case of this linear model in which $x_t = \sin(\pi t/6) = d_{t0}$.

CHECKING FOR A CHANGE IN THE PARAMETER OF A TIME SERIES MODEL

As an illustration for a model which is non-linear in the parameter θ, we consider the time series generated by

$$y_t - y_{t-1} = a_t - \theta a_{t-1}$$

or

$$(1 - B)y_t = (1 - \theta B)a_t \tag{13}$$

such as originally motivated the Cuscore idea.

In this model B is the backshift operator such that $By_t = y_{t-1}$. As we have seen, this integrated moving average (IMA) model is of special interest for the representation of non-stationary phenomena. In particular, this is the model for which the *exponentially weighted moving average* (EWMA)

$$\hat{y}_t = (1 - \theta)(y_{t-1} + \theta y_{t-2} + \theta^2 y_{t-3} + \cdots)$$

with *smoothing constant* θ is the minimum mean square error forecast of y_t from origin $t - 1$. This forecast can be conveniently updated by the well-known recursive relation,

$$\hat{y}_t = \lambda y_{t-1} + \theta \hat{y}_{t-1}$$

where $\lambda = 1 - \theta$.

For this model then

$$a_t(\theta) = \frac{(1 - B)}{1 - \theta B} y_t \tag{14}$$

and

$$d_t = -\frac{\partial a_t}{\partial \theta} = -\frac{(1 - B)}{(1 - \theta B)^2} y_{t-1} = \frac{-a_{t-1}}{1 - \theta B} = \frac{-\hat{a}_t}{\lambda} \tag{15}$$

where $\hat{a}_t$ is the exponentially weighted moving average of previous a_t's with smoothing constant θ, and again is conveniently updated using

$$\hat{a}_t = \lambda a_{t-1} + \theta \hat{a}_{t-1}$$

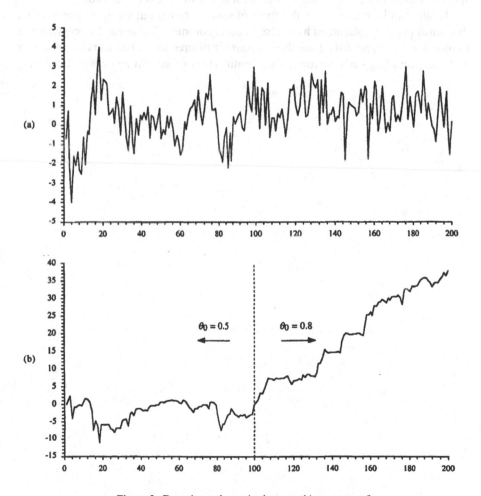

Figure 2. Detecting a change in the smoothing constant θ.

The Cuscore statistic appropriate for $\theta = \theta_0$ is $Q_0 = -1/\lambda \sum a_{t0} \hat{a}_{t0}$. This statistic makes excellent sense since if θ_0 is the true value of θ, the a_0's would be independent and, in particular, the exponentially weighted average $\hat{a}_{t0}$ of previous a_{t0}'s would not be correlated with a_{t0}.

For illustration with the model (14) we used unit random normal deviates to generate the series of 200 values, shown in Figure 2(a) where the value of θ was changed after 100 values from 0.5 to 0.8. Although it is hard to see any difference in the series itself, the Q statistic does well in quickly reacting to the change.

CONCLUSIONS

In this chapter it is pointed out that very sensitive sequential checks to monitor specific feared deviations from a specific model can be easily derived.

Results can be presented in the form of suitable charts but we also have in mind that small process computers have also become common. These can be programmed to monitor simultaneously a number of possible discrepancies specific to a particular process, providing early warning and identification of several sources of trouble.

PART E

Variance Reduction and Robustness

CHAPTER E.0

Introduction

In 1904 Guinness's decided to build a pilot plant to help them learn how to further improve their beer. W. S. Gosset was put in charge of the project. He found, however, that it was difficult to draw reliable conclusions from his experiments because his data were extremely variable. To try to deal with this problem he decided to learn more about statistics and in 1906 was given a year's leave of absence to study statistics at University College London. The head of the department—Karl Pearson, who then dominated the subject—had developed a system of frequency curves; these had been fitted to the very large samples of data that he and his followers collected and they had found that often their distributions were very far from normal.

But Gosset's samples were small—typically he might want to compare the mean values from two samples of just four or eight observations. He reluctantly concluded that the large sample methods that he had learned about were not of much help. So to find a technique more applicable to his work, he developed what was later called the t statistic and was able to derive its distribution by making the assumption of distributional normality. This work intrigued the young student R. A. Fisher, who some years later, as a result of his experiences at Rothamsted experimental station, derived a series of new statistical methods for use by experimenters.

Publication of Fisher's book, *Statistical Methods for Research Workers*, in 1925 began a new epoch in the development of statistics, but of the dozens of reviews it received, not one was favorable. His methods were for small samples, not large ones, and assumed normality. They were revolutionary and, as every author knows, the most disturbing thing that you can set before a reviewer or a referee is something new.

But perhaps they had a point. If distributions were non-normal then how could Fisher's techniques be valid? Even Gosset had doubts. So a number of investigations were begun to find out what would happen to Fisher's methods if the distributions were non-normal. Fortune favors the brave, and, by and large, the investigators found

Improving Almost Anything: Ideas and Essays, Revised Edition. By George Box and Friends
Copyright © 2006 John Wiley & Sons, Inc.

that his techniques held up remarkably well for the kind of comparative experiments he had discussed.

So was it really true that non-normality didn't matter? Well, no. It was later demonstrated, for example, that the *same* deviations from normality that were fairly innocuous in tests for comparing means could have disastrous effects in tests for comparing variances. We would now say that the tests to compare means were *robust* to these departures from normality, but that those to compare variances were not.

In the early part of the twentieth century a different kind of robustness had been sought by the brewers at Guinness's. They wanted to find a variety of barley that, while not necessarily giving the absolutely best results for beer making, would give good and reasonably *uniform* results even though the barley was grown under widely different farming conditions. The trials they ran were early examples of experiments to design a product (the barley seed) that was insensitive to environmental variation. Later, an industrial statistician (Michaels, 1964) took this concept further and demonstrated how factorial arrangements with split plots could be used to design a product (he used a washing detergent as an example) that was robust to changes in a number of environmental variables (temperature of wash, hardness of the water, etc.). Earlier, Shewhart* (1931) and Morrison (1957) had discussed and solved a different problem in robust product design—that of minimizing the effect on an assembled product of variation in its components.[†] At the time neither in Britain, where they first discovered, nor elsewhere were the important implications of these ideas of robust design appreciated.

In the 1980s the U.S. automobile industry was astonished and embarrassed by the clever designs and narrow tolerances of their Japanese competitors. A number of senior executives from the Ford Motor Company and other industries visited Japan and discovered that one reason for the superiority of their products was the wide application of experimental design. In particular, they found that an engineer, Dr. Genuchi Taguchi, had been instrumental in independently developing and widely disseminating his solutions to both of the problems of robust design mentioned earlier. He referred to his methods as "parameter design." Thereafter there were vigorous campaigns to use "Taguchi methods" in the United States and Britain. We owe Dr. Taguchi a great debt, particularly for his industrial examples, which are of enormous value. Unfortunately, although some of his techniques were, faulty and unnecessarily complicated, certain of his followers insisted that strict adherence to his precepts must be regarded as an issue of doctrinal purity. For a time this seemed to threaten the healthy development applied statistics. Some of the research described in later chapters in this section was conducted to help clarify the situation.

To properly appreciate the ideas of variance reduction and robustness it is necessary to consider a number of different topics. In many industrial problems we must contend with not one but a number of different sources of variation—

*Shewhart solved the slightly more general problem of minimizing such error transmission while maintaining the mean value of the product characteristics at a fixed value.
[†] In 1999, in his lecture to the Royal Academy of Engineering on "Engineering for Corporate Success in the New Millennium," Richard Parry Jones, Group Vice President, Ford Motor Company, emphasized the great importance of this idea and correctly attributed it to Morrison.

measurement variation, sampling variation, process variation, and so forth. Chapter E.1 is about estimating and using these different components. Another issue that arises is the role of data transformation. This is discussed in Chapter E.2.

Chapter E.3 is a critique of methods for reducing the transmission of variation of components to variation in the assembled product. Chapter E.4 shows how split plot arrangements can be used more generally to reduce the cost and difficulty of experimental investigation.

The important connection between robust statistical analysis and robust product design is clarified in Chapter E.5. Of particular importance are the coefficients that appear in two sets of equations. It is noteworthy that none of the coefficients that define a robust product coincide with those that define an optimal product. Thus, as is otherwise obvious, we could have a robust product that was uniformly bad, or an "optimal" product that was disastrously nonrobust, so compromise is necessary. In Chapter E.6, a locus of compromises between robust and optimal solutions is found using a different approach to environmental robust design (see Figure 3 of that Chapter).

Split plot arrangements play an important role in robust design and in Chapter E.7 the relative merits of various arrangements are considered. The *cost* of performing different operations is, in practice, often the deciding factor.

Two of Taguchi's concepts that I think are unhelpful are *accumulation analysis* and performance criteria based on so-called *signal-to-noise* ratios. These are discussed in Chapters E.8 and E.9.

CHAPTER E.1

Multiple Sources of Variation: Variance Components

Industrial processes frequently contain multiple sources of variation, and in planning investigations and analyzing the results, these must be properly taken into account (see Daniels, 1938). On the following page three sets of data (a), (b), and (c) from an operating batch process are plotted, listed, and analyzed in Figure 1. The data represent (a) 10 repeat tests performed on a single sample, (b) 10 individual tests made on 10 different samples from the same batch, and (c) 10 individual tests made on individual samples from 10 different batches. We see that the variation associated with each of the procedures (a), (b), and (c) is very different and illustrate below how mistakes are sometimes made because the wrong error estimate has been used.

USING THE WRONG ESTIMATE OF ERROR

When you wish to compare things, it is important to collect the right data and to use the appropriate estimate of error. The kind of mistake that is easily made is illustrated by the following example. The figures were supposed to show that a process modification denoted by B gave a larger mean than the standard process denoted by A.

Standard Process A	Modified Process B
58.3	63.2
57.1	64.1
59.7	62.4
59.0	62.7
58.6	63.6
$\bar{y}_A = 58.54$	$\bar{y}_B = 63.20$

From Box, G. E. P. (1998), *Quality Engineering*, **11**(1), 171–174.
Copyright © 1998 by Marcel Dekker, Inc.

a) Ten single sample tests from a single sample.

b) Ten single batch samples from a single batch.

c) Ten batches from production

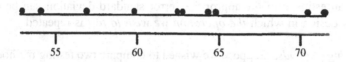

	(a) Ten tests on one sample	(b) Ten samples each tested once	(c) Ten batches each sampled and tested once
	59.6	57.0	62.2
	59.2	63.2	56.8
	61.3	61.4	64.5
	60.4	64.9	70.3
	59.7	57.9	54.1
	60.8	62.5	53.3
	58.8	59.1	64.2
	59.9	61.3	59.7
	60.4	60.5	62.4
	60.1	60.2	71.5
Averages	60.0	60.8	61.9
Variances	$V_T = 0.55$	$V_S = 5.81$	$V_P = 37.82$
Standard Deviations	$s_T = 0.74$	$s_S = 2.41$	$s_P = 6.15$

Figure 1. Repeated tests (a) from a single sample, (b) from a single batch, and (c) from 10 different batches.

In support of this assertion, it was pointed out that a test to compare the means of these two sets of data produced a value of Student's t of 8.8, which, if there were no real difference, would occur less than once in a thousand times by chance.

So it is certainly true that the difference in means for these data is over-whelmingly statistically significant. However, this does not demonstrate, as was supposed, that Process B gives a higher mean value than Process A. It turned out that these results were obtained by taking *one sample* from a batch manufactured by Process A and one sample from a batch manufactured by Process B and *testing* each sample five times. The fact that this difference is significant tells us only that the difference between the two averages is almost certainly not due to *testing error*. This is because the standard deviations estimated from the variation in these results measure only the variance due to testing. To determine the effect of a change in the process, it would be necessary to test a series of *batches* made by Process A and a series made by Process B. Only from such an experiment would it be possible to make a judgment about the effectiveness of the *change* from Process A to Process B.

SOME COMPARISONS YOU MIGHT WISH TO MAKE

In general, an estimate of the appropriate error standard deviation can be obtained only from a design in which *the operation we wish to test* is repeated.

> *Two Testing Methods.* Suppose we wished to compare two testing methods A and B; then, we must compare,* let us say, five *determinations* using test method A with five determinations using test method B with all tests made on the *same sample.*
>
> *Two Sampling Methods.* To compare two different sampling methods, we might similarly compare determinations made on five different *samples* using sampling method A with those made on five samples using method B with all samples coming from the *same batch.*
>
> *Two Processing Methods.* By the same principle, to compare a standard and modified method of processing, we might compare determinations from five *batches* taken from the standard Process A with determinations from five batches made by the modified Process B.

VARIANCE COMPONENTS

A diagrammatic representation of how testing error, sampling error, and process error contribute to the total deviation $y - \mu$ of an observation y from the process mean μ is shown in Figure 2. Variation resulting *only* from the testing method is measured by the standard deviation σ_T of the test error e_T—the deviation of a particular test result from the true mean for that sample. Similarly, variation due *only* to sampling is measured by σ_S, which is the standard deviation of e_S—the deviation of the mean of this particular sample from the true batch mean. Finally, variation due *only* to the process is measured by σ_P—which is the standard deviation of e_P—the deviation of that particular batch mean from the process mean.

Now look again at Figure 1. The estimated variance $V_T = 0.55$ obtained from data set (a) provides an estimate of σ_T^2 of the testing variance. The variance $V_S = 5.81$ obtained from data set (b) does not estimate σ_S^2 alone, however, but rather $\sigma_S^2 + \sigma_T^2$. This is because each sample value includes not only the deviation e_S due to the sample but also the deviation e_T due to the test. The variance of $e_S + e_T$ is

---

* In each of the three cases discussed, comparisons might be made not only of the two means but also of the standard deviations (and, for that matter, of any other characteristics of interest).

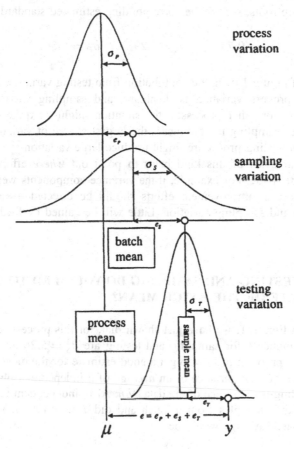

process
variation

sampling
variation

batch
mean

testing
variation

process
mean

sample mean

Figure 2. Contributions of testing, sampling, and process variation to the overall variation of data coming from the process.

$\sigma_S^2 + \sigma_T^2$ because, for independent sources of variation, the variance of the sum is simply the sum of the variances. In a similar way, the variance $V_P = 37.82$ obtained from the third column of data is an estimate of $\sigma_P^2 + \sigma_S^2 + \sigma_T^2$. Thus, if we use an arrow to mean "is an estimate of," we obtain

$$V_T = 0.55 \rightarrow \sigma_T^2, \qquad V_S = 5.81 \rightarrow \sigma_T^2 + \sigma_S^2$$
$$V_P = 37.82 \rightarrow \sigma_T^2 + \sigma_S^2 + \sigma_P^2$$

Estimates of the separate *variance components* σ_P^2, σ_S^2, and σ_T^2 can now be obtained by subtraction. Using a "hat" notation to indicate an estimate, we find

$$\hat{\sigma}_T^2 = 0.55, \qquad \hat{\sigma}_S^2 = 5.81 - 0.55 = 5.26$$

$$\hat{\sigma}_P^2 = 37.82 - 5.81 = 32.01$$

After taking square roots, we get the corresponding estimated standard deviations

$$\hat{\sigma}_T = 0.74, \qquad \hat{\sigma}_S = 2.29, \qquad \hat{\sigma}_P = 5.66$$

For the data of Figure 1 then, the contribution from testing variation is seen to be relatively small, process variation is dominant, and sampling variation is large relative to testing. For other processes, the situation might be quite different, of course. Sometimes sampling the product is the largest source of variation. In other circumstances, the testing procedure might produce large variation.

Note that an analysis of this kind helps to point out *where* efforts to reduce variance should be made. For example, if the variance components were like those obtained for Figure 1, improvement efforts should be directed toward reducing process variation and sampling variation. Little will be gained by further reducing testing variation.

HOW MUCH TESTING AND SAMPLING DO YOU NEED TO GET A GOOD ESTIMATE OF THE BATCH MEAN?

Using the data in Figure 1, we have just shown that, for this process, estimates for the variance components for sampling and testing are $\hat{\sigma}_S^2 = 5.26$ and $\hat{\sigma}_T^2 = 0.55$. Now, consider the problem of how to get a good estimate for the mean of a *particular* batch. Remember the variance of an *average* of n independent observations is obtained by dividing the variance of the data by n. Also, independent variance s add up. So, if we take S samples from a batch and did T tests *on each sample*, the variance of the batch average would be

$$\sigma^2 = \frac{\sigma_S^2}{S} + \frac{\sigma_T^2}{TS}$$

In this formula, the reason σ_T^2 is divided by TS and not by T is that, in this procedure, a total of TS tests would be averaged. Let us experiment with this formula using the above estimates of σ_S^2 and σ_T^2.

If we took just one sample and tested it once, then $S = 1$, $T = 1$, and $\hat{\sigma}^2 = 5.26/1 + 0.55/1 = 5.81$ and $\hat{\sigma} = 2.41$.

If, instead, we tested this one sample 10 times, then $S = 1$, $T = 10$, and $\hat{\sigma}^2 = 5.261 + 0.55/10 = 5.32$ and $\hat{\sigma} = 2.31$. With this arrangement, nine extra tests would have produced hardly any worthwhile improvement!

If, however, we tested each of 10 samples once, then $S = 10$, $T = 1$, and $\hat{\sigma}^2 = 5.26/10 + 0.55/10 = 0.58$ and $\hat{\sigma} = 0.76$—a very worthwhile reduction.

In the case of a chemical product, you could instead take S samples, mix them together, and then only do a *single* test on the mixture (you let the mixing do the averaging). Then, by making a *single* test on the mixture of 10 samples, you would get $\hat{\sigma}^2 = 5.26/10 + 0.55/1 = 1.08$ and $\hat{\sigma} = 1.04$.

For any given application, you can play with the formula to yield a number of alternative schemes. If you know the cost of taking a sample and the cost of making an analysis, you can calculate the cost of each individual scheme and choose the one that best meets your needs.

ASSIGNABLE CAUSES

When a point (or points) is found outside the control limits of a Shewhart or some other control chart, it may be too quickly assumed that we must look to *process* operation for an assignable cause. However, it should be remembered (see Fig. 2) that the outlying point(s) could be due to an upset in the system of sampling, or in the system of testing.

SPLIT PLOT DESIGNS AND ROBUST EXPERIMENTATION

Multiple components of error may often be used to economize on experimental effort by using split plot designs discussed in Chapter E.4. These split plot arrangements correspond to the so-called inner and outer array designs of Taguchi (1987). However, this split plot structure must be taken account of to give a correct analysis. Ways to do this were discussed in Fisher (1925a) and Yates (1937) and, for example, by Michaels (1964) and Box (1996), where you will find other references. Also a discussion of the correct use of split plot designs for robust product and process experimentation will be found in Chapter E.6.

A NESTED DESIGN TO ESTIMATE VARIANCE COMPONENTS

A more common way to *estimate* variance components is by means of a "nested" design such as that shown in Table 1. In this example, five batches were run, two samples taken from each batch, and two tests made on each sample. Thus, 20 tests were performed in all.

A simple way to analyze these 20 data values is illustrated in Table 1. The basic data consist of the 20 test results from which 10 sample averages and 5 batch averages can be calculated. Remembering that a variance estimated from the sample of two by taking half the squared difference, an estimate $\hat{V}_T = 0.10$ can be obtained by averaging the estimates obtained from the pairs. The estimate $\hat{V}_T$ has 10 degrees of freedom because each pair produces an estimate having one degree of freedom. Similarly, an estimate $\hat{V}_S = 23.40$ having five degrees of freedom is obtained from the average of the five separate estimates obtained from the pairs of sample averages. Finally, an estimate $\hat{V}_P = 49.48$ having four degrees of freedom is obtained in the usual way from the sum of squares of deviations from the overall sample average of

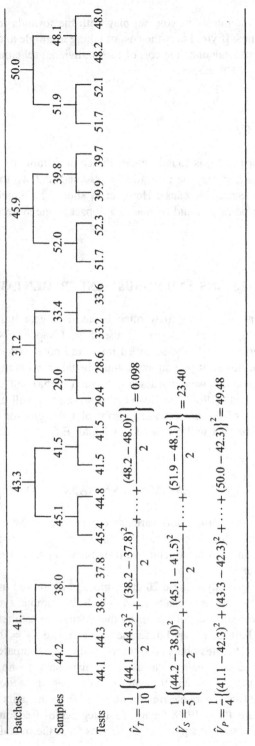

Table 1. An Illustration of Calculations Using a Nested Design

Batches, Samples, Tests (nested design tree diagram with values)

$$\hat{V}_T = \frac{1}{10}\left\{\frac{(44.1-44.3)^2}{2} + \frac{(38.2-37.8)^2}{2} + \cdots + \frac{(48.2-48.0)^2}{2}\right\} = 0.098$$

$$\hat{V}_S = \frac{1}{5}\left\{\frac{(44.2-38.0)^2}{2} + \frac{(45.1-41.5)^2}{2} + \cdots + \frac{(51.9-48.1)^2}{2}\right\} = 23.40$$

$$\hat{V}_P = \frac{1}{4}\left\{(41.1-42.3)^2 + (43.3-42.3)^2 + \cdots + (50.0-42.3)^2\right\} = 49.48$$

42.3. Bearing in mind that each sample is analyzed twice and each batch is analyzed four times, the variance components can now be obtained by subtraction as follows:

$$\hat{\sigma}_T^2 = 0.10, \qquad \hat{\sigma}_S^2 = \hat{V}_S - \tfrac{1}{2}\hat{V}_T = 23.35$$

$$\hat{\sigma}_P^2 = \hat{V}_P - \tfrac{1}{2}\hat{V}_S - \tfrac{1}{4}\hat{V}_T = 37.73$$

You can find more about these designs and about an alternative method for analyzing the data using an analysis of variance table in, for example, BHH II.

CHAPTER E.2

The Importance of Data Transformation in Designed Experiments for Life Testing

In science and engineering, data are often measured on "transformed" scales. Two common examples are the *Richter scale* for measuring earthquakes and the *decibel scale* for measuring noise. Both of these scales are logarithmic—they are such that *multiplying* the intensity (of an earthquake or of noise) by a fixed amount *adds* a fixed amount to the scale. As a further example, in an experiment to compare the effect of different diets on the performance of athletes in a race, one investigator might measure the *time* it took to run a fixed distance, whereas another might look at them their average *speed*. Each of these measures is the *reciprocal* of the other. Which is better? Does it matter? As has been noted (see e.g., BHH II; Bisgaard, 1993; Bisgaard and Fuller, 1994), the careful choosing of an appropriate data transformation can sometimes be very important.

The transformations given in the above examples can be considered as particular cases of *power transformations* in which the data value y is raised to some power λ—instead of analyzing y, you analyze y^λ. Power transformations with λ usually taking some value between -1 and $+1$ are appropriate for data which are essentially positive, such as the weight, length, or time to failure of a sample. For example, if y is the measured time in hours it takes an athlete to run a 1-mile course, then if we decide to analyze his speed in miles per hour, this is given by y^{-1} and $\lambda = -1$. Taking $\lambda = 1$ implies that the original data values y are to be used, and $\lambda = \frac{1}{2}$ implies the square root of the data. In practice, it is often better to work not with y^λ itself but equivalently with $(y^\lambda - 1)/\lambda$, which we denote by $y^{(\lambda)}$. The reasons, explained more

From Box, G. E. P. and Fung, C. A. (1995), *Quality Engineering*, **7**(3), 625–638.
Copyright © 1995 by Marcel Dekker, Inc.

Table 1. Cycles to Failure of Samples of Woolen Thread in a 3^3 Factorial Experiment

Factor Levels			Cycles to Failure
x_1	x_2	x_3	y
−1	−1	−1	674
−1	−1	0	370
−1	−1	+1	292
−1	0	−1	338
−1	0	0	266
−1	0	+1	210
−1	+1	−1	170
−1	+1	0	118
−1	+1	+1	90
0	−1	−1	1414
0	−1	0	1198
0	−1	+1	634
0	0	−1	1022
0	0	0	620
0	0	+1	438
0	+1	−1	442
0	+1	0	332
0	+1	+1	220
+1	1	−1	3636
+1	−1	0	3184
+1	−1	+1	2000
+1	0	−1	1568
+1	0	0	1070
+1	0	+1	566
+1	+1	−1	1140
+1	+1	0	884
+1	+1	+1	360

x_1: length of test specimen
x_2: amplitude of loading cycle
x_3: load

Best-fitting equation obtained by least squares:

$$\hat{y} = 551 + 660x_1 - 536x_2 - 311x_3 + 239x_1^2 + 276x_2^2 - 48x_3^2 - 457x_1x_2 - 236x_1x_3 + 143x_2x_3$$

fully in Box and Cox (1964), are (a) that the transformed data do not reverse their order when negative values of λ are used and (b) that the value $\lambda = 0$ then corresponds to the log transformation. The logarithm is, thus, a power transformation which fits naturally into the transformation sequence.* The purpose of this article is

---

* A third reason (see Box and Cox, 1964) is that one way of *choosing* a transformation is to carry out analyses of variance of the data transformed to $y^{(\lambda)}/\dot{y}^{\lambda-1}$ for different values of λ (where $\dot{y}$ is the geometric mean of the data) and pick the value of λ that gives the smallest error sum of squares. This usually gives similar results to the simpler procedure we discuss here.

to illustrate a simple and instructive way of choosing an appropriate transformation of this kind and is best introduced by looking at an actual example.

The data y shown in Table 1 are the "lifetimes" of test specimens of woolen thread subjected to repeated loadings under 27 different test conditions (Box and Cox, 1964). The data were obtained from a full three-level factorial experiment in three factors x_1, x_2, and x_3. The original investigators fitted the second degree model (with 10 parameters!) shown at the bottom of Table 1. Although they found that *all but one* of the fitted parameters were "statistically significant," we might ask, "Is such a complicated model really necessary, and does it explain the relationship most efficiently?"

A graphical way to answer such questions is to fit the full second degree model to the transformed data $y^{(\lambda)}$ for a range of values of λ and, then, make a "*lambda*" *plot*" as shown in Figure 1 (Fung, 1986; Box, 1988). This lambda plot is a graph of the t-value for each coefficient in the second degree equation, plotted versus the transformation parameter λ. It will be recalled that this t-value is the ratio of the estimated coefficient to its standard error (its estimated standard deviation). Thus, it corresponds to the *engineer's* signal-to-noise ratio (but has little to do with the performance criteria discussed under that name by Taguchi). The necessary calculations can be done very quickly with any standard regression software.

It is easy to see from the lambda plot that the model becomes much simpler when λ approaches zero (which corresponds to the log transformation). At this value of λ, all of the squared terms and interaction terms essentially disappear, leaving a first degree model with only the linear effects of the three variables x_1, x_2, and x_3. Thus, for these data, the log transformation provides a linear prediction equation which is much easier to interpret. Furthermore, the size of the t-values (the signal-to-noise

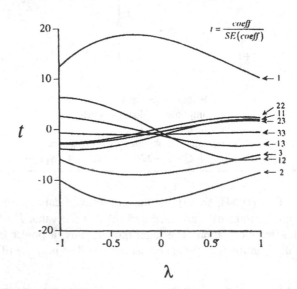

Figure 1. Lambda plot of t-values for the woolen thread example. The label "1" indicates the graph of the t-values for the coefficient of the x_1 term: "22" indicates that for the x_2^2 term; "23" indicates that for the $x_2 x_3$ term; etc.

ratios) for the linear effects are much larger when the log transformation is used. A third advantage is that, other things being equal, a "parsimonious" model (a prediction equation with fewer parameters) provides more precise estimation of the response. Specifically, the average variance of the response estimated by the equation in the immediate neighborhood of the experimental points is proportional to p/n, where p is the number of parameters and n is the number of observations. For this example, p is reduced from 10 to 4 so that this factor alone reduces the variance of the estimated response by a factor of 2.5×.

An auxilliary plot which measures the overall effectiveness of the various power transformations in producing the simpler *first degree model* is obtained by calculating the F-ratio, F = (Regression mean square for first degree model)/(Residual mean square), for each of the λ values. This is shown for the thread testing data in Figure 2. It will be seen again that this lambda plot produces a maximum value of F close to the value $\lambda = 0$. Specifically the F value for the logged data is about ten times that for the untransformed data, signaling once more that the log transformation should be used.

As a further example, look at the data in Table 2. This is an animal study of four different antidotes to guard against three different poisons (Box and Cox, 1964). Four randomly chosen animals were allocated to each of the 12 "cells" in Table 2. Survival times (measured in units of 10 hr) for each of the 48 animals are shown in the table. We can apply the same technique as before to these data. In particular, plots of F against λ, for poisons, treatments, and interactions, all shown in Figure 3, indicate a value of λ close to -1, pointing to the need for a reciprocal transfor-

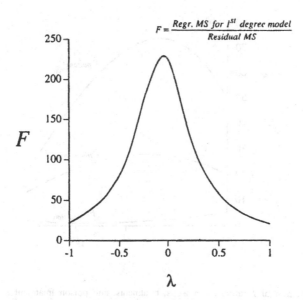

$$F = \frac{Regr.\ MS\ for\ 1^{st}\ degree\ model}{Residual\ MS}$$

Figure 2. Lambda plot of F-ratios for a first degree model for the woolen thread example.

Table 2. Survival Times (in Units of 10 hr) of Four Animals per Cell, Dosed with One of Three Poisons and Treated with One of Four Antidotes

	Treatments			
Poison	A	B	C	D
I	0.31 [0.41]	0.82 [0.88]	0.43 [0.57]	0.45 [0.61]
	0.45 (0.07)	1.10 (0.16)	0.45 (0.16)	0.71 (0.11)
	0.46 {0.50}	0.88 {0.20}	0.63 {0.49}	0.66 {0.36}
	0.43	0.72	0.76	0.62
II	0.36 [0.32]	0.92 [0.82]	0.44 [0.38]	0.56 [0.67]
	0.29 (0.08)	0.61 (0.34)	0.35 (0.06)	1.02 (0.27)
	0.40 {0.82}	0.49 {0.55}	0.31 {0.42}	0.71 {0.70}
	0.23	1.24	0.40	0.38
III	0.22 [0.21]	0.30 [0.34]	0.23 [0.24]	0.30 [0.32]
	0.21 (0.02)	0.37 (0.05)	0.25 (0.01)	0.36 (0.03)
	0.18 {0.53}	0.38 {0.42}	0.24 {0.23}	0.31 {0.24}
	0.23	0.29	0.22	0.33

The average, standard deviation, and the standard deviation of the reciprocals of the data in each cell are shown in square, round, and curly brackets, respectively.

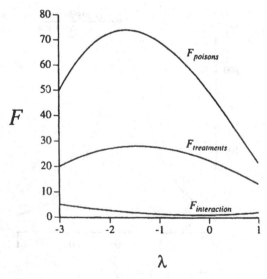

Figure 3. Lambda plot of F-ratios for poisons, treatments, and poison–treatment interaction, for the animal survival study.

mation—which would be looking at the *rate* of "failure" of the animals rather than the *time* to "failure."

RATIONALE FOR THE USE OF RECIPROCAL TRANSFORMATIONS IN ANALYZING WEAR DATA

To understand why a reciprocal transformation might make sense for life testing data (whether it is the "life" of an animal, or of a tool, or of a piece of thread under tension), look at Figure 4, which shows a *hypothetical* situation where the amount of "wear" w of a specimen (tool, animal, or thread) is a linear function of time y. The value w_0 is imagined to be the critical level of wear w at which point failure occurs. Now suppose that the specimens are treated in three different ways, A, B, and C, yielding three different rates of wear. In the examples we have quoted, the actual rates of wear cannot be measured. All we know is the time to failure y. But if our hypothesis is roughly correct, the rate of wear r (say) will be proportional to the *reciprocal* of the time taken to reach the critical level w_0. So $r = w_0/y$ and by taking the reciprocal of time to failure data in this example, we would, in effect, be analyzing the *rate* of wear. Further supposing, as indicated in Figure 4, that the standard deviations of the *rates* of wear for tests A, B, and C are the same, we see that the standard deviations of times to failure represented by the distributions A′, B′, and C′ will be very different, and treatments yielding longer life will also show much greater variability.

Now look again at the data in Table 2. By comparing the averages and standard deviations of the data in each cell you will see that, for the original data, the cells with longer lifetimes also have greater variability. However, after reciprocal transformation and allowance for sampling variation, the standard deviations are

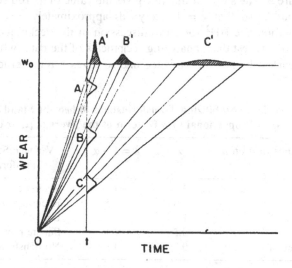

Figure 4. Illustration of the reciprocal relationship between time to failure and rate of wear.

essentially constant across the cells regardless of the mean. Thus, a reciprocal transformation "breaks" the dependence of the standard deviation on the mean.

Most statistical analysis depends on the assumption that the standard deviation of each of the data values is approximately constant—in other words, that the variance is reasonably *homogeneous*. Now there are two kinds of variance inhomogeneity: (a) *inherent* inhomogeneity—an example would be the smaller variance in a particular manufacturing operation achieved by an experienced rather than an inexperienced operator; and (b) *transformable* inhomogeneity, which comes about only because the untransformed observations give rise to a needlessly complicated model which induces unnecessary variance inhomogeneity (and often unnecessary interactions). A vital clue that tells us that transformation is needed to eliminate variance inhomogeneity is when the standard deviation depends on the mean. Moreover, when, as in Table 2, the data cover a wide range, with some data values many times larger than others, an analysis that disregards the possible need for transformation can result in grossly misleading conclusions (Box and Hill, 1979).

Now different variance stabilizing transformations are appropriate in different situations. One way to discover what is an appropriate transformation when there are replicated observations in each cell is to find out how the standard deviation in each cell depends on the average in that cell. In particular, when there appears to be no such dependence, no transformation is needed. When the standard deviation is proportional to the average, the appropriate transformation is the log, and when the standard deviation is proportional to the square of the average, a reciprocal transformation is required. These are examples of the more general rule (see Table 3) that if the standard deviation is proportional to some power α of the average, the transformation $y^{(\lambda)}$ needed to stabilize the variance is $\lambda = 1 - \alpha$. To get an idea of what α should be for data arranged in a two-way table of this kind, you can plot the log of the standard deviation in each cell against the log of the average in that cell, as is done in Figure 5. The slope of this line gives the value of α. You see that for these data, α is about 2, so that $\lambda = 1 - \alpha$ yields approximately $\lambda = -1$, giving the reciprocal as before. See BH^2 for more discussion of this technique.

For this set of life test data, analyzing reciprocals of the data (which, in effect, is the *rate* of wear) stabilizes the variance and so gives a more efficient and easily

Table 3. Variance Stabilizing Transformations When the Standard Deviation σ Is Proportional to a Power α of the Average μ, $\sigma \propto \mu^{\alpha}$

Dependence of σ on μ	α	$\lambda = 1 - \alpha$	Variance Stabilizing Transformation
$\sigma \propto \mu^2$	2	-1	Reciprocal
$\sigma \propto \mu^{3/2}$	1.5	-0.5	Reciprocal square root
$\sigma \propto \mu$	1	0	Log
$\sigma \propto \mu^{1/2}$	0.5	0.5	Square root
$\sigma \propto$ constant	0	1	No transformation

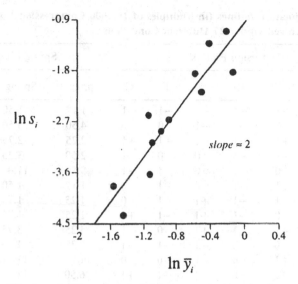

Figure 5. Plot of the logged cell standard deviations versus the logged cell averages for the animal survival study.

interpreted analysis. However, the reciprocal transformation is not appropriate for all lifetime data. To get a plot that looked like Figure 4, we might need to transform the *wear itself.* In particular, this would imply that some transformation was needed, say, w^ϕ to obtain a constant variance for the rate of wear. The required variance stabilizing transformation $y^{(\lambda)}$ for the survival time y would then be obtained when $\lambda = -\phi$. In particular, if the logarithm of rate of wear standard deviation was constant, then the variance stabilizing transformation for the time to failure would also be the logarithm.

LIFE TEST OF CLUTCH SPRINGS

An interesting experiment involving the survival time of 81 automobile clutch springs was described by Taguchi and Wu (1980). We use their data below to illustrate our methods of analysis. The analysis by the original authors was different. Clutch springs were manufactured under 27 different conditions involving 7 experimental factors* and were subjected to an accelerated test to see how many compressions they could survive before breaking. Table 4 shows the lifetimes of the springs (in multiples of 100,000 compressions). The original data do not give actual lifetimes of the individual springs but rather the number of springs in each group still

---

* The design factor (BC) in Table 4 is a constructed variable whose -1, 0, and $+1$ levels correspond to the combinations B_-C_-, B_-C_+, and B_+C_-, respectively.

Table 4. Approximate Lifetimes (in Multiples of 100,000 Compressions) of 81 Clutch Springs Manufactured Under 27 Different Conditions

Group									

Group	A	(BC)	D	E	F	G	Spring 1	Spring 2	Spring 3
			Design Factors[a]					Spring Lifetimes	
1	−1	−1	−1	−1	−1	−1	1.17	1.50	1.83
2	0	0	−1	−1	0	0	4.50	5.50	11.+[b]
3	+1	+1	−1	−1	+1	+1	2.25	2.75	11.+
4	−1	−1	−1	0	0	+1	2.50	3.25	3.75
5	0	0	−1	0	+1	−1	5.50	11.+	11.+
6	+1	+1	−1	0	−1	0	1.17	1.50	1.83
7	−1	−1	−1	+1	+1	0	1.25	1.75	3.50
8	0	0	−1	+1	−1	+1	1.25	1.75	2.50
9	+1	+1	−1	+1	0	−1	3.25	3.75	4.50
10	−1	0	0	−1	−1	−1	1.25	1.75	2.50
11	0	+1	0	−1	0	0	11.+	11.+	11.+
12	+1	−1	0	−1	+1	+1	6.50	11.+	11.+
13	−1	0	0	0	0	+1	11.+	11.+	11.+
14	0	+1	0	0	+1	−1	2.17	2.50	2.83
15	+1	−1	0	0	−1	0	1.50	2.25	2.75
16	−1	0	0	+1	+1	0	2.50	3.50	4.50
17	0	+1	0	+1	−1	+1	2.17	2.50	2.83
18	+1	−1	0	+1	0	−1	11.+	11.+	11.+
19	−1	+1	+1	−1	−1	−1	3.50	4.25	4.75
20	0	−1	+1	−1	0	0	11.+	11.+	11.+
21	+1	0	+1	−1	+1	+1	11.+	11.+	11.+
22	−1	+1	+1	0	0	+1	11.+	11.+	11.+
23	0	−1	+1	0	+1	−1	11.+	11.+	11.+
24	+1	0	+1	0	−1	0	5.50	11.+	11.+
25	−1	+1	+1	+1	+1	0	4.25	4.75	6.50
26	0	−1	+1	+1	−1	+1	2.25	2.75	3.50
27	+1	0	+1	+1	0	−1	11.+	11.+	11.+

[a] A: shape; B: hole ratio (a two-level factor); C: coining (a two-level factor); D: stress (t); E: Stress (c); F: Shot peening; G: Outside perimeter planning.
[b] A data value of "11+" indicates a spring that survived the experiment.

surviving at the end of each of 11 periods comprising 100,000, 200,000, 300,000, . . . , 1,100,000 compressions. We have obtained approximate individual survival times for each spring by allocating the lifetimes of failed springs evenly in the interval in which they failed. However, there is an added complication in this example because some of the springs had not yet failed by the end of the test, which was terminated after 1,100,000 compressions. An approximate way to deal with these survivors is sketched below.

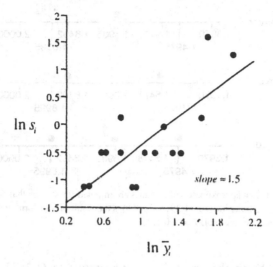

Figure 6. Plot of the logged group standard deviations versus the logged group averages for the clutch spring experiment (employing the 16 groups having 2 or more observed failure times).

AN APPROXIMATE ANALYSIS: ESTIMATING DATA VALUES FOR THE "LIFETIMES" OF THE SURVIVORS

First, we need to estimate an appropriate variance stabilizing transformation. As there were no survivors or one survivor in 16 of 27 groups, at least 2 of the 3 springs in each of these groups have *observed* failure times. Thus, for these particular groups, a plot can be made of the logs of the standard deviations against the logs of the averages.* See Figure 6. The graph has a slope of about 1.5, which implies a variance stabilizing transformation $y^{(\lambda)}$ with $\lambda = -0.5$ the reciprocal square root transformation. Thus $y^{(-0.5)} = (y^{-0.5} - 1)/ - 0.5 = 2(1 - y^{-0.5})$. Assuming that this transformation also applies to the springs that had not yet failed when the test was ended at period 11, you will see that a failure time in the infinite interval (11, ∞) transforms to a value in the finite interval (1.3970, 2.000); that is, $11^{(\lambda)} = 1.3970$ and $\infty^{(\lambda)} = 2.0$ when $\lambda = -0.5$. We can now impute approximate failure times for springs that survived past the end of the test by allocating "transformed" lifetimes evenly in the interval (1.3970, 2.000). A single spring surviving the end of the test would be assigned a value of 1.6985 on the transformed scale (the midpoint of the final interval). If two springs in a given group survived beyond the end of the test, they would be assigned the values 1.5478 and 1.8492, respectively. (These are the midpoints of the two intervals that result from dividing the final interval in half.) If all three springs in a group survived, they would be assigned the values 1.4975, 1.6985, and 1.8995, respectively. (These are the midpoints of the three intervals that result from dividing the final interval into thirds.) This procedure is illustrated schematically in Figure 7. The same principle can be applied to give transformed lifetimes for springs that failed in earlier intervals as well; for example,

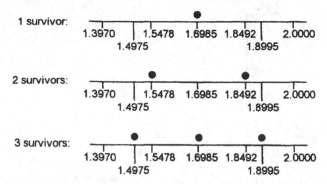

Figure 7. Illustration of how approximate failure times are imputed for springs that survived beyond the end of the experiment. The finite interval (1.3970, 2.0000) is the result of transforming the infinite interval $(11, \infty)$ by the reciprocal square root, $\lambda = -0.5$.

the interval (10, 11) on the original scale transforms to (1.3675, 1.3970) when $\lambda = -0.5$. Applying this procedure to all the clutch spring data yields a full set of transformed lifetimes, displayed in Table 5, that can be used to give an approximate analysis.

Analysis of the imputed data identifies key factors that influence the lifetimes of the springs: D (stress), which has a linear effect, and F (shot peening), which has linear and quadratic effects. By employing the high level of factor D and the middle level of factor F, the manufacturer can produce the longest lasting springs.

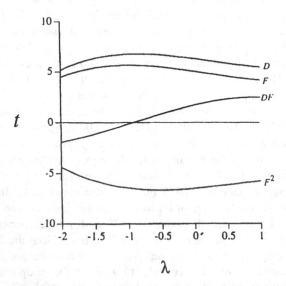

Figure 8. Lambda plot of t-values for the clutch spring experiment. Only the four contrasts shown are "significant" in this range of λ.

Table 5. Imputed Clutch Spring Lifetimes on the Transformed Scale, $\lambda = -0.5$

Group	Spring 1	Spring 2	Spring 3
1	0.0976	0.2929	0.4882
2	1.0528	1.1445	1.6985
3	0.6507	0.7804	1.6985
4	0.7156	0.8840	0.9613
5	1.1445	1.5478	1.8492
6	0.0976	0.2929	0.4882
7	0.1464	0.4393	0.9226
8	0.1464	0.4393	0.7156
9	0.8840	0.9613	1.0528
10	0.1464	0.4393	0.7156
11	1.4975	1.6985	1.8995
12	1.2138	1.5478	1.8492
13	1.4975	1.6985	1.8995
14	0.6290	0.7156	0.8020
15	0.2929	0.6507	0.7804
16	0.7156	0.9226	1.0528
17	0.6290	0.7156	0.8020
18	1.4975	1.6985	1.8995
19	0.9226	1.0264	1.0792
20	1.4975	1.6985	1.8995
21	1.4975	1.6985	1.8995
22	1.4975	1.6985	1.8995
23	1.4975	1.6985	1.8995
24	1.1445	1.5478	1.8492
25	1.0264	1.0792	1.2138
26	0.6507	0.7804	0.9226
27	1.4975	1.6985	1.8995

The cost of *not* transforming the data can be seen in the lambda plot in Figure 8. The t-values for the D, F, and F^2 terms are larger for the transformed data with $\lambda = -0.5$, so that the need for these terms can be discovered more easily. Furthermore, a simpler model without the DF interaction results when the data are appropriately transformed.

The clutch spring example is discussed in more detail in Fung (1986). Also see Delozier (1994) for an application using data from a tool life experiment.

Two Goals to Keep in Mind

We have pointed out two frequent consequences of data transformation: variance stabilization and model simplification. However, transforming *just to simplify* the model can be dangerous. For example, if you had a theoretical model like

$y = 1/(a + bx)$, you might be tempted to use a reciprocal transformation of y, because $y^{-1} = a + bx$ is an easy model to fit. But remember that if the original data y have constant variance and the range of the data is R, then the variance for y^{-1} would vary by a factor of R^4 from one end of the scale of measurement to the other. For example, if your data covered a range where the largest observation was 10 times larger than the smallest, then analyzing the reciprocal of the data would wrongly place 10,000 times as much emphasis on one end of the scale as the other. The data would act as if you had 10,000 observations at one point and only one at the other! So model simplification for its own sake can be a very dangerous undertaking.

ACKNOWLEDGMENTS

This work was supported by a grant from the Alfred P. Sloan Foundation. The authors are grateful to Howard T. Fuller for help in preparing the figures.

CHAPTER E.3

Is Your Robust Design Procedure Robust?

Robustness means *insensitivity to variation.* (a) A "robust" photocopying machine works well in both hot and cold and in dry and damp climates. (b) A "robust" television set gives stable performance even when there is variation in its component transistors. What these examples have in common is that the designer can use his control over the *specifications* of the product to make it insensitive to variation of one kind or another. Taguchi (Taguchi and Wu, 1980; Kackar, 1985) called this process of "robustifying," or of minimizing sensitivity by appropriate choice of specifications, *parameter design.* The examples above illustrate *two* distinct problems of this kind: (a) minimizing sensitivity to environmental or "use" factors (hot–cold, dry–damp) and (b) minimizing sensitivity to manufacturing variation in components (variability of transistor characteristics). The first problem has been discussed extensively (Box and Jones, 1992a; Nair, 1992; Shoemaker, Tsui, and Wu, 1991).

In this chapter we consider only the second problem: that of providing a highly uniform product, such as a television set, by minimizing variation transmitted by its components. See also Shewhart (1931, p. 259) Morrison (1957), Taguchi and Phadke (1984), Box and Fung (1986), Taylor (1991), and Parry-Jones (1999). We describe what we think is the best way to do this following Morrison who emphasized that whatever method we use, the choice of product design can be extremely sensitive to certain assumptions which we must be careful to check out. Sometimes, tacit assumptions that seem innocuous turn out to be perilous. Thus, we need to consider the robustness to assumptions of the *robust design procedure* itself.

From Box, G. E. P. and Fung, C. A. (1994), *Quality Engineering*, 6(3), 503–514.
Copyright © 1994 by Marcel Dekker, Inc.

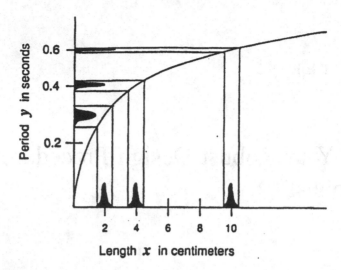

Figure 1. The period of a pendulum's swing as a function of its length. When the error variation in x is constant over the entire range of x, the longest pendulum produces the least transmitted variation in y.

BACKGROUND: MINIMIZING TRANSMITTED VARIATION—ONE VARIABLE

To fix ideas, we begin with a very simple example. Suppose we are designing a pendulum for inclusion in some kind of timing mechanism. The period of swing y depends on the pendulum's length x so that small manufacturing variations in the length x will be transmitted to variation in y. If we want the percentage error in y to be as small as possible, should we design our timing mechanism with a long pendulum, or a short one? We suppose:

(a) The relationship* plotted in Figure 1 between the length x of the pendulum and its period y is

$$y = 0.2\sqrt{x}$$

(b) To accommodate the pendulum in the timing device, it must not be shorter than 2 cm nor longer than 10 cm.

(c) Manufacturing variation which occurs in the length x of the pendulum has a constant standard deviation $\sigma = 0.01$ cm.

(d) Because the actual timing transmitted to the timing device can readily be scaled up or down by suitable gears, we need to minimize the *percentage* (proportional) variation in y. This is the same thing as saying we want to minimize the standard deviation of log y, so in Figure 1, y is shown on a log scale.

---

* This is a close approximation to the formula for a "perfect" pendulum, where y is measured in seconds and x is measured in centimeters.

It is easy to see that because of the nonlinear relationship shown in Figure 1, transmission of manufacturing variation in pendulum length is minimized when the pendulum is made as long as possible. On this basis, we might recommend, therefore, that the timing mechanism should be designed to have a 10-cm pendulum.

MINIMIZING TRANSMITTED VARIATION—MORE THAN ONE VARIABLE

When there are more variables involved, the problem is the same in principle but is more complicated. For illustration, Taguchi and Wu (1980) considered the problem of designing a Wheatstone Bridge whose circuit is shown in Figure 2. This instrument allows us to measure a *response*, in this case an *unknown resistance y*. To use it, you adjust the variable resistance B until no current flows through the ammeter X. The unknown resistance can then be calculated from the formula

$$y = B\frac{D}{C} \tag{1}$$

where B is the value read from the variable resistor, and C and D are fixed resistances whose nominal values and error tolerances are specified in the product's design. Thus, a specified product design might require that C have a nominal value, say, of $10\,\Omega \pm 2\%$ and that D have a nominal value of $30\,\Omega \pm 1\%$. The problem might be solved by making the error tolerances for *all* the components as small as possible, but this solution would usually be unnecessarily costly because parts manufactured to closer tolerances are usually more expensive. If we design the circuit right, however, it might be possible to obtain a very accurate instrument using inexpensive parts.

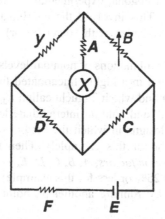

Figure 2. Schematic diagram of a Wheatstone Bridge. Nominal levels are to be chosen for the design variables A, C, D, E, and F.

Now for any given instrument and any given determination, the estimate y will be slightly in error: first, because manufacturing variation will cause the components $B, C, \ldots$ etc for that particular instrument to deviate slightly from their nominal values and, second, because slight inaccuracies in the ammeter will mean that a small current X may, in fact, be flowing in the ammeter even though we believe the circuit to be in balance. When this is the case, the actual value of y would be given by

$$y = \frac{BD}{C} - \frac{X}{C^2 E}[A(C + D) + D(B + C)][B(C + D) + F(B + C)] \qquad (2)$$

and using the standard formula (1) we would be in error. In formula (2), A, B, C, D, and F are the actual resistances, E is the actual battery voltage, and X is the current actually flowing through the ammeter.

Now the variables whose levels define the design of the instrument are the five nominal values of the factors, A, C, D, E, and F, we call *design variables* and are distinguished by bold type in Figure 2. The object of parameter design is to choose target values for these five design variables to minimize error transmission from all seven error variables.

Taguchi and Wu characterized precision of the response y in terms of a "signal-to-noise ratio"

$$SN = \frac{\bar{y}^2}{s^2}$$

which was to be maximized. Maximizing the SN ratio is of course equivalent to minimizing the coefficient of variation $s/\bar{y}$ (which means minimizing the *percentage* variation in y), which, in turn, is essentially equivalent to minimizing the standard deviation of $\ln y$ (Box, 1988; Nair and Pregibon, 1988).

To carry out the maximization of SN, Taguchi and Wu used a numerical simulation method employing orthogonal experimental layouts* in a manner illustrated schematically in Figure 3. This involved the combination of an "inner (design) array" for the five design variables, with "outer (error) arrays" for the seven error variables.

In the *design array*, 36 combinations of nominal levels of the *design factors A, C, D, E*, and *F* were considered in a highly fractionated three-level factorial arrangement due to Seiden (1954) and which Taguchi called L_{36}, with low and high levels indicated by "−" and "+," and with an intermediate level indicated by "0." This produced 36 trial circuit designs in which the factors were varied over very wide ranges (25-to-1, or 2500% for this example!). Then for each of these 36 trial circuit designs, the seven *error factors, A, B, C, D, E, F*, and *X*, were varied at three levels over small ranges (0.25% or less for this example) to represent manufacturing variation. This was done by running another 36-point *error array* for *each* trial

---

* Also known as "orthogonal arrays," these are selected subsets of all possible combinations of the levels of the experimental variables.

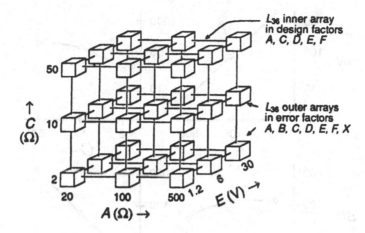

50

↑
C 10
(Ω)

2

L_{36} inner array
in design factors
A, C, D, E, F

L_{36} outer arrays
in error factors
A, B, C, D, E, F, X

30

20   100   500 1.2 6   E (V) →

A (Ω) →

Figure 3. Conceptual illustration of how an outer ("error") array is arranged at each inner ("design") array point.

circuit design. Thus, Equation (2) needed to be evaluated $36 \times 36 = 1296$ times in all.

For each trial circuit design, the 36 values of the corresponding error array set where then used to calculate a single SN ratio, thus characterizing each of the 36 trial circuit design's tendency to transmit variation from the components to the response y. Taguchi and Wu then plotted the marginal averages of these SN ratios[*] as shown in Figure 4. By inspection of these averages, they reached the conclusion that SN was maximized for the levels $A(-)$, $C(+)$, $D(0)$, $E(+)$, and $F(-)$. This is not, however, the combination of factor levels that maximizes SN. The actual maximization occurs at the values $A(-)$, $C(-0.28)$, $D(-0.47)$, $E(+)$, $F(-)$ which gives a value of SN 7.3% higher than that found by Taguchi and Wu. If the design is limited to the particular (three) levels of the components tested, then the best combination is $A(-)$, $C(0)$, $D(0)$, $E(+)$, $F(-)$, which gives a value 6.5% higher.

The situation can be understood by noting that *within the ranges explored,*[†] the extremes, $A(-)$, $E(+)$, $F(-)$, are the best values for these factors irrespective of the values of C and D. It remains, therefore, to find the best values for C and D with A, E, and F set at these extreme levels. Figure 5 shows contours of SN as a function of C and D with Taguchi's solution marked "T." Also marked is the point at which the maximum is actually obtained. The contours around the maximum are obliquely oriented, implying a large interaction between C and D. When such an interaction occurs, marginal graphs like Figure 4 will not, of course, yield the true maximum.

---

[*] Following Taguchi and Wu, SN is plotted in Figure 4 on a log scale.
[†] The most important question raised by this analysis seems to have been overlooked. The effects A, F, and E in Figure 4 are all essentially linear and are telling us to explore the possibility of using values for the resistances A and F that are *even smaller* than those tested and a battery voltage E that is *even greater*— that is, to explore a path of ascent (see, e.g., Chapter C.1).

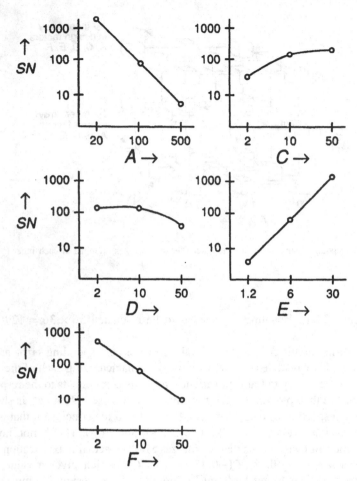

Figure 4. Marginal averages of SN for the Wheatstone Bridge problem, plotted versus each design variable.

An alternative approach which does not suffer from this deficiency and which is simpler and quicker, is as follows. As noted before, maximizing SN is equivalent to minimizing the coefficient of variation of y, or simply logging the data and minimizing the variance V of $Y = \log y$ (Box, 1988; Nair and Pregibon, 1988). Now Y depends on k variables $x_1, x_2, \ldots, x_k$ so we can write

$$Y = F(x_1, x_2, \ldots, x_k) \tag{3}$$

We want to minimize V_x, the variation transmitted to Y when the factors are at some particular settings $x = x_1, x_2, \ldots, x_k$. If we can calculate V_x for any given set of x's we will be able to minimize it by standard numerical methods. Also if desired, to clarify specific aspects, V_x may be plotted to produce diagrams like Figure 5. The problem of minimizing V_x is, in fact, an example of *constrained optimization* or

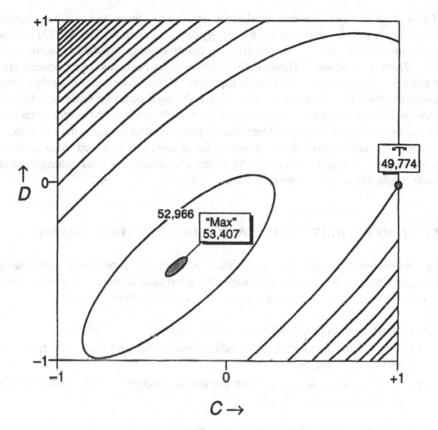

Figure 5. Contours of SN as a function of C and D for the Wheatstone Bridge problem, assuming proportional errors in all variables.

nonlinear programming for which programs are now widely available even for personal computers. The program will find the values of the x's which give a minimum value for the calculated function V_x within any stated ranges of the factors x decided to be important. The problem of *characterizing* the dependence of transmitted variation on design variables can also be studied by response surface approximation (Vining and Meyer, 1990; Tribus and Szonyi, 1989; Bailey, Chatto, Fellner, and Pfeifer, 1991).

In a paper entitled "The Study of Variability in Engineering Design," Morrison (1957) gave a simple solution to this problem over 40 years ago.* He employed the well-known error transmission formula:

$$V_x = g_1^2 \sigma_1^2 + g_2^2 \sigma_2^2 + \cdots + g_k^2 \sigma_k^2 \qquad (4)$$

where g_i is the gradient of the function F in the ith direction and σ_i is the standard deviation of the component x_i, both evaluated at the particular settings $x_1, x_2, \ldots, x_k$

*Also J. S. Hunter has recently pointed out to me that the same problem and its solution had been proposed by Shewhart in 1931, in his book *Economic Control of Quality of Manufactured Product*, p. 259.

of x. In Equation (4), you can see that the amount of transmitted variation depends on x not only via the gradients $g_1, g_2, \ldots, g_k$ obtained from the function (3) but also on the values of $\sigma_1, \sigma_2, \ldots, \sigma_k$ representing manufacturing variation associated with the different components. Though close attention has been paid to the specification or empirical approximation of $F(x)$ in Equation (3), the literature has largely ignored questions about the values of $\sigma_1, \sigma_2, \ldots, \sigma_k$, although common sense says that this question of how large is the variation transmitted by each of the components must clearly be of first importance. Morrison emphasized that we must have reasonably precise information about these variances, for as might be expected, what is chosen as an "optimal" design turns out to be extremely sensitive to what assumptions are made, explicitly or tacitly, about $\sigma_1, \sigma_2, \ldots, \sigma_k$.

WHAT DO WE NEED TO KNOW TO CONDUCT THE ANALYSIS?

No matter how the problem is tackled, whether by constrained optimization, orthogonal arrays, or by some other method, its formulation requires that we know or assume that we know the answers to a number of questions. These are:

(a) What is an appropriate performance criterion?
(b) What is the functional relationship between the response y and the design and error variables?
(c) What are the relations between the standard deviations of the error variables and their nominal values?

Let us consider each of these questions in turn.

What Is an Appropriate Performance Criterion?

Thoughtful choice of performance criteria (objective functions) is clearly important. In the above examples, it is assumed that a best assembly will result if the *percentage variation* of the quality characteristic y is minimized. However, in both of the examples considered, the tacit assumption is made that the cost of the components is the same irrespective of their numerical value. In practice, we would need to take account of a realistic costing of different designs.

What Is the Functional Relationship Between y and the Design and Error Variables?

In the pendulum example and in the circuit example, the relationships between the quality characteristic y and the design variables x were assumed *known*. When dealing with reasonably well-understood physical phenomena such as electrical circuitry, the necessary functional relationships may be available; more usually, however, such relationships would have to be determined experimentally.

What Are the Relations Between the Standard Deviations of the Error Variables and Their Nominal Values?

In the pendulum example, we supposed that the standard deviation of the length x of the pendulum remained the same for pendulums from 2 cm in length to 10 cm in length. An alternative assumption might have been that the *percentage* error in the length of the pendulums remained constant over this range. Such a change in what is assumed completely changes the solution. Similarly, in the Wheatstone Bridge example, Taguchi and Wu assumed constant *percentage* errors for the resistances A, C, D, F and the voltage E, or equivalently that the standard deviation of each of the design variables was proportional to its level (over the whole 25-to-1 range over which these factors were varied). As we shall see, quite different solutions are found if these assumptions are changed.

SENSITIVITY OF THE PARAMETER DESIGN SOLUTION TO THE MANNER IN WHICH THE STANDARD DEVIATION OF EACH FACTOR VARIES WITH ITS MEAN

It will be rare in practice that the answers to questions (b) and (c) will be known exactly. We need, therefore, either to reassure ourselves that the assumptions are innocuous, or we must face the fact that experimentation will be needed to characterize these more precisely. In particular, it turns out that the design we end up with can be completely different depending on what we conclude is the answer to question (c).

Consider first the pendulum example and suppose, as we did initially, that instead of assuming that the standard deviation of x is constant over the range 2–10 cm, we assumed instead, that x is subject to constant *percentage error*. This situation is graphed in Figure 6. Now the longest pendulum is not uniquely best. The error transmission is the *same* irrespective of the length of the pendulum. One pendulum length is as good as another!

Thus an important issue is the nature of the dependence of the *variation* of x on the *magnitude* of x. Is σ_x independent of x as in Figure 1, or is it proportional to x as in Figure 6? Or is it related to x in some other way? The "robustness" of parameter design solutions is thus dependent to a considerable degree on what we know about $\sigma(x)$. A sensitivity analysis should routinely be carried out to assess how the transmitted variance V_x varies over the range of assumptions that might plausibly be encountered.

Such considerations are of even greater importance when there are more variables. In the Wheatstone Bridge example, Taguchi and Wu assumed that the manufacturing variation of the design components should be represented by *percentage error*. Implied in that assumption is that the standard deviation is proportional to the nominal value over the wide ranges of 25-to-1 in all the design factors. Thus, for example, if a resistor of $1\,\Omega$ had a standard deviation of $0.01\,\Omega$, then a resistor of $25\,\Omega$ would have to have a standard deviation of $0.25\,\Omega$. As shown

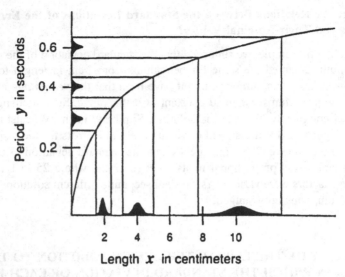

Figure 6. The period of a pendulum's swing as a function of its length. When the error variation in x is proportional to the nominal level of x, the *same* percentage variation is transmitted to y, regardless of nominal level of x.

in Figure 7, a totally different solution for the settings of the two critical factors (resistors C and D) is obtained if you assume, for example, that the standard deviations are constant over this range.*

We should first find out, therefore, whether the solution is sensitive to plausible changes in assumptions (c). When different plausible assumptions lead to radically different solutions, experiments will be needed to determine, at least approximately, the relative values of the standard deviations and/or the relation between the means and the standard deviations, for those design variables for which these issues are critical.

CONCLUSIONS

1. Vital for reliable robust product design is knowledge of the functional relationship between Y and x, the standard deviations of all of the error variables and design variables, and the *dependence of these standard deviations on the nominal levels of the variables.*

2. In particular, parameter design solutions can be extremely sensitive to what is assumed (explicitly or tacitly) about the relationship between the standard deviation and the nominal levels of the variables.

* In this calculation, the actual levels of the standard deviation σ_i were chosen to give the same variance V_x for both the constant error and proportional error cases when the design factors are set at their central levels.

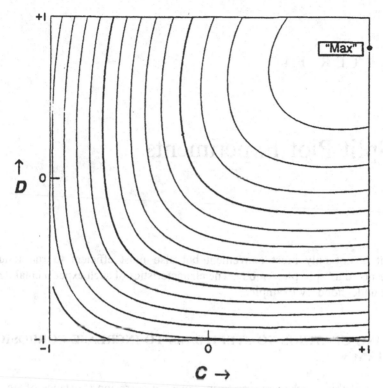

Figure 7. Contours of SN as a function of C and D for the Wheatstone Bridge problem, assuming constant errors in all variables.

ACKNOWLEDGMENTS

This work was supported by a grant from the Alfred P. Sloan Foundation and by the National Science Foundation under Grant No. DDM-8808138.

CHAPTER E.4

Split Plot Experiments

Often, it is not only most convenient but also most efficient to run industrial experiments in a "split plot" mode. The characteristics of such experimental designs are best understood by example.

APPLYING COATINGS TO STEEL BARS TO INCREASE CORROSION RESISTANCE

The object of this experiment was to improve the corrosion resistance of steel bars by applying a surface coating and then baking the bars in a furnace for a fixed time. Four different coatings C_1, C_2, C_3, and C_4 were tested at three different furnace temperatures: 360°C, 370°C, and 380°C, each of which was run twice. The experimental arrangement and the measured corrosion resistances of the 24 treated bars are shown in Table 1. Thus, in all, six furnace "heats" were run. In each such heat, the furnace was set to the desired temperature, four bars treated with coatings C_1, C_2, C_3, and C_4 and randomly positioned in the furnace were baked for the prescribed length of time, and then removed, cooled down, and tested.

Now it could be quite misleading to treat such an experiment as if it were a standard randomized design with only one kind of experimental unit and only one error source. There are, in fact, two different experimental units: "heats" called generically *whole plots* and "positions" in the furnace called generically *subplots* (or

---

* The terminology "whole plots, subplots, and split plots" comes from agricultural experimentation where these arrangements were first used. In such experimentation, it is frequently convenient to apply certain treatments (e.g., different dressings of fertilizer) to moderately large areas of land called "whole plots" and to apply other treatments (e.g., different spacings between seeds) to smaller "subplots" within these whole plots.

From Box, G. E. P. (1996), *Quality Engineering*, 8(3), 515–520.

Improving Almost Anything: Ideas and Essays, Revised Edition. By George Box and Friends

Table 1. A Split Plot Design to Study the Corrosion Resistance of Steel Bars Treated with Four Different Coatings Randomly Positioned in a Furnace and Baked at Three Different Replicated Temperatures

Heats (Whole Plots)	Positions (Subplots)			
360°C	C_2	C_3	C_1	C_4
	73	83	67	89
370°C	C_1	C_3	C_4	C_2
	65	87	86	91
380°C	C_3	C_1	C_2	C_4
	147	155	127	212
380°C	C_4	C_3	C_2	C_1
	153	90	100	108
370°C	C_4	C_1	C_3	C_2
	150	140	121	142
360°C	C_1	C_4	C_2	C_3
	33	54	8	46

sometimes split plots).* In this particular experiment, it was known that setting and maintaining the furnace at a particular temperature was a very imprecise operation. Consequently, the whole-plot standard deviation σ_w was expected to be large. This standard deviation measured the variation from one heat to another heat when an attempt was made to replicate the same conditions. However, it was believed that σ_s, measuring the subplot variation from position to position in the furnace within a given heat, was likely to be relatively small.

In this particular experiment, because of the difficulty of adjusting the whole-plot variable (temperature of the furnace) up and down, the whole plots (heats) were not run in random order. Instead, heats were successively set at 360, 370, 380, 380, 370, and 360°C as indicated in Table 1. There was, however, no difficulty in positioning the bars randomly in the furnace within each heat. In industrial experimentation, randomizing the whole-plot variable would, as in this case, often be troublesome. As we shall see, however, this need not be important in comparing the coatings.

The numerical calculation of the analysis of variance for split plot experiments—computation of the degrees of freedom, sums of squares, and mean squares—is the same as for any other design and can be performed by any of the many excellent computer programs now available. If you want to refresh your memory about the basis of analysis of variance calculations you might refer to Chapter B.10 or to BHH II, for example. Having calculated the various entries, however, experimenters sometimes encounter difficulties in identifying appropriate error terms. These difficulties are best overcome by rearranging the analysis in two parallel columns which relate, respectively, to whole plots and subplots as in Table 2 (Box, 1950). On

Table 2. Parallel Column Layout of Analysis of Variance Showing How Various Contributions Are Associated with Their Appropriate Error

Whole Plots—Heats (Averaged over Coatings)		Subplots—Within Heats (Interactions with Coatings)	
Source	df	Source	df
Mean, I	1	C	3
Temperature, T	2	$T \times C$	6
Error, E_w	3	Error, E_s	9

the left is the analysis of the whole plots (between the six heats); on the right is the analysis for the subplots (within the six heats).

In the whole-plot analysis, the contribution from the grand mean (sometimes called the correction for the mean) is denoted by I and has one degree of freedom. The contribution from differences associated with the three temperatures—the main effect T—has two degrees of freedom. The whole-plot error E_w measures the differences between replicated heats at the three different temperatures.

In the subplot analysis contributions can be conveniently thought of as "interactions" of the factor coatings (C) with the corresponding whole-plot effects shown on the left. Thus, the contribution from the differences in the corrosion resistances of the four different coatings (the main effect C) is positioned opposite I because it shows whether the *means* are the same for the different coatings and so can be thought of as the interaction $C \times I = C$. Similarly, the contribution $T \times C$ measuring the interaction between T and C is shown opposite T. It measures the extent to which temperature effects are the same or different for different coatings. Also, the subplot error E_s can be thought of as the interaction of E_w with C. It measures the extent to which the coatings give similar results within each replicated temperature when the differences between whole-plot averages are taken into account. An analysis of this kind for the corrosion resistance data is shown in Table 3.

Table 3. Analysis of Variance for Corrosion Resistance Data Showing Parallel Column Layout Identifying Appropriate Errors for Whole-Plot and Subplot Effects[a]

		Heats (Whole Plots)				Coatings (Subplots)			
Source	df	SS	MS		Source	df	SS	MS	
Mean, I	1	245,430	245,430		C	3	4,289	1,430	$F_{3,9} = 11.5$**
Temperature, T	2	26,519	13,260	$F_{2,3} = 2.8$	$T \times C$	6	3,270	545	$F_{6,9} = 4.4$*
Error$_w$	3	14,440	4,813		Error$_s$	9	1,121	125	
(Reps)					(Reps$\times C$)				

[a] We have used the convention that a single asterisk indicates significance at the 5% and two asterisks at the 1% level.

Because there are four coatings in each heat, the whole-plot error mean square 4813 is an estimate of $4\sigma_w^2 + \sigma_s^2$. The subplot error mean square 125 is an estimate of σ_s^2. Thus, estimates for the individual whole-plot and subplot error variances are

$$\hat{\sigma}_w^2 = \frac{(4813 - 125)}{4} = 1172, \qquad \hat{\sigma}_s^2 = 125$$

and

$$\hat{\sigma}_w = 34.2, \qquad \hat{\sigma}_s = 11.1$$

Thus, in this experiment, $\hat{\sigma}_w$, the standard deviation associated with furnace heats, is estimated to be over three times as large as $\hat{\sigma}_s$, the standard deviation for coatings within heats.

The whole-plot analysis for differences in temperature T shows a value of $F_{2,3} = 2.8$ with a significance probability of only about 25%. The comparison is, of course, extremely insensitive because of the large whole-plot error and also the very small numbers of degrees of freedom involved. However, the main purpose of this experiment was to compare coatings and to determine their possible interactions with temperature. For this purpose, the split plot design has produced a very sensitive experiment revealing statistically significant differences for C and for $T \times C$ and has required much less effort than a fully randomized design.

As always, an analysis of variance should be accompanied by (i) a detailed analysis of residuals to check for the possibility of *unexpected* occurrences not allowed for by the analysis of variance model and (ii) appropriate tables of means to show the *nature* of effects found to be of possible statistical significance.

For split plot data, we should look separately at whole-plot residuals and at subplot residuals. For completeness these analyses are given in Table 4. The whole-plot residuals in Table 4a are obtained by subtracting out the three temperature means from the table of whole-plot means. Making due allowance for rounding error, the sum of squares of these residuals when multiplied by 4 (the number of coatings) yields the whole-plot error sum of squares in Table 3. The subplot residuals of Table 4b are obtained from the individual two-way tables (Replicates × Coatings) for each temperature. For example, the data for 360°C are shown in Table 4c. The subplot residuals can be obtained by comparing the deviation of a data value from its row average with the deviation of the corresponding column average from the overall average. Thus, the residual 11.1 in the first row, second column of Table 4b is obtained as $(73 - 78) - (40.5 - 56.6) = 11.1$. Their sum of squares yields the subplot error sum of squares in Table 3.

The individual residuals at first look a little peculiar because of the dependence between them induced by the analysis. In particular, replicate residuals are "mirror images" of each other because they represent deviations from a mean of two items. Also, we are not surprised to see in this experiment that the whole-plot residuals are very much larger than the subplot residuals. After allowance is made for these

Table 4. Analysis of Whole-Plot and of Subplot Residuals

(a) Whole-Plot Residuals

	360°C	370°C	380°C
Rep. 1	21.4	−28.0	−23.8
Rep. 2	−21.4	28.0	23.8

(b) Subplot Residuals

		C_1	C_2	C_3	C_4
360°C	Rep. 1	−4.4	11.1	−2.9	−3.9
	Rep. 2	4.4	−11.1	2.9	3.9
370°C	Rep. 1	−9.5	2.5	11.0	−4.0
	Rep. 2	9.5	−2.5	−11.0	4.0
380°C	Rep. 1	−0.3	−10.3	4.8	5.8
	Rep. 2	0.3	10.3	−4.8	−5.8

(c) Row and Column Data and Averages at 360°C

	C_1	C_2	C_3	C_4	Avg.
Rep. 1	67	73	83	89	78
Rep. 2	33	8	46	54	35.2
Avg.	50	40.5	64.5	71.5	56.6

characteristics, nothing particularly surprising shows up, although with so few data we can expect to detect only very gross discrepancies.

FURTHER ANALYSIS OF THE COATING DIFFERENCES

Before looking at tables of means exemplifying the significant coating effects, we first discuss two factors (not previously mentioned) which characterized the four coatings: these were the base treatment (B) and the finish (F) both at two levels as shown in Figure 1.

The subplot analysis can thus be further broken down to produce Table 5, from which it will be seen that the effects B, F, BF, and TBF provide statistically significant effects whose meaning can best be understood by inspecting the appropriate two-way tables of averages in Figure 2.

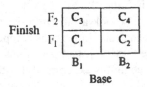

Figure 1. Base treatment (B) and finish (F) characterizing four coatings.

Table 5. Analysis of Variance for Corrosion Resistance Data with Coatings C Further Characterized by Base Treatment B and Finish F

Heats (Whole Plots)

Source	df	SS	MS	
Mean I	1	245,430	245,430	
Temperature T	2	26,519	13,260	$F_{2,3}=2.8$
Error$_w$ (Reps)	3	14,440	4,813	

Interactions with Coatings (Subplots)

Source	df	SS		df	SS	MS	
C	3	4,289	B	1	852	852	$F_{1,9}=6.9*$
			F	1	1820	1820	$F_{1,9}=14.7**$
			BF	1	1617	1617	$F_{1,9}=13.0**$
$T \times C$	6	3,270	TB	2	601	301	$F_{2,9}=2.4$
			TF	2	788	394	$F_{2,9}=3.2$
			TBF	2	1881	940	$F_{2,9}=7.5*$
Error$_s$ (Reps $\times C$)	9	1,121				125	

Note: One asterisk indicates significance at 5% level; two asterisks indicates significance at 1% level.

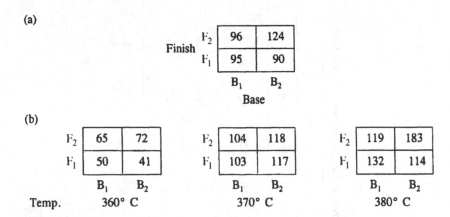

Figure 2. Tables of average corrosion resistance exemplifying interactions of base, finish, and temperature: (a) overall averages for base and finish; (b) base and finish averages for different temperatures.

Figure 2a makes the meaning of the B, F, and BF effects clear. The base and finish factors act synergistically to produce a coating having a large additional anti-corrosion effect when level 2 of finish and level 2 of base are used *together*. Figure 2b exemplifies the TBF interaction showing how the synergistic BF interaction is greatly enhanced by the use of high temperature.

CONCLUSIONS ABOUT THE COATING EXPERIMENT

The actions called for as a result of this experiment were, as always, of two kinds:

1. *"Cashing in" on the New Knowledge.* Coating C_4 (combination B_2F_2) with a temperature of 380°C should be adopted for current manufacture.
2. *Using the New Knowledge to Look for Further Possibilities of Improvement.*

   (i)   Coating C_4 should be tested at even higher temperatures.
   (ii) Because (see Figure 2b) getting the best out of coating C_4 depends critically on temperature which is, at present, very poorly controlled, considerable effort is justified in getting better control of the furnace.
   (iii) To exploit further the results obtained, attempts should be made to *understand* the possible physical, chemical, or metallurgical reasons for the synergistic BF and TBF effects by consulting expert opinion. Promising conjectures would, of course, need to be investigated by further experimentation, in the lab or on the plant scale as appropriate.

SOME GENERAL CONCLUSIONS ABOUT SPLIT PLOT EXPERIMENTS

As Cuthbert Daniel once remarked, perhaps with slight exaggeration, "All industrial experiments are split plot experiments." It is essential for industrial experimenters to be aware of the following:

1. By the deliberate use of a split plot design, it is often possible to achieve great precision in studying split plot factors of major interest even with highly variable whole plots.

2. A split plot design can produce considerable economy of effort. Thus, a fully randomized design involving 24 heats tested with one coating in each heat would, because of the large value of σ_w, have been far less efficient than the split plot design actually used. It would also have been almost impossible to perform!

3. It is important to recognize split plot structure (often unintended*) in experiments and to take account of it in subsequent data analysis. We can be misled seriously by failure to take account of "split plotting" (see Daniel, 1976).

4. Experimenters have sometimes had trouble identifying which effects should be tested against which error with designs of this kind. However, arrangement of the analysis of variance table in two parallel columns alleviates this problem.

5. The "inner and outer array" experiments popularized by Taguchi (1987) for developing robust design of processes and products are, of course, examples of split plot designs (Box and Jones, 1992b). Unfortunately, they have usually not been recognized and analysed as such. An early discussion of their correct use for this purpose will, however, be found in a remarkable paper by Michaels (1964).

Split plot experiments were first developed by R. A. Fisher and F. Yates. A general discussion will be found in Cox (1958).

ACKNOWLEDGMENTS

This work was sponsored by grants from the Alfred P. Sloan Foundation, Procter and Gamble, and NSF grant #DMI-9414765. Special thanks are due to Timothy Kramer of Hewlett Packard for his help. Thanks are also due to Elisa Santos and Howard Fuller for very helpful comments leading to improvement of this article.

---

*The quality improvement experiment in Chapter B.10 used a split plot design at first unrecognized.

Robustness in Statistics

If statistics is to be an essential catalyst to industrial experimentation, it is important to use and to fashion techniques that are suitable for everyday use. Such techniques rest on tentative assumptions (stated or unstated), and since all models are wrong, but some are useful, they must be robust to likely departures from assumption. As is indicated by the wavy lines in Figure 1, robustness concepts are important both for statistical analysis and also for the process of informed extrapolation required in order that the results of an investigation can be put to practical use.

ROBUST STATISTICAL ANALYSIS

I want to first discuss the robustness of statistical analysis. It is sometimes supposed that if the assumptions are "nearly right," then so will be a desired procedure and that if they are "badly wrong," then the desired procedure will not work. Both ideas are faulty because they take no account of robustness. As a specific example, consider the estimation of the standard deviation, σ, for a control chart for individual items. Suppose an estimate is employed based on the average moving range, $\overline{MR}$ of two successive observations (see, e.g., Duncan, 1974). If it is assumed that the data are normally, identically, and *independently* distributed, then an unbiased estimate of σ is provided by $\overline{MR}/1.128$. But for data of this kind collected in sequence it is to be expected that successive deviations from the mean may be appreciably correlated. It can be shown that even small serial correlations of this kind can seriously bias this estimate of σ (see, e.g., Box and Luceño, 1997).

In all such examples, the effect of a departure from assumption depends on two factors: (i) the magnitude of the deviation from assumption and (ii) a robustness

From Box, G. E. P. (1998), "Statistics as a Catalyst to Learning by Scientific Method, Part II, A Discussion," *Journal of Quality Technology*, 31(1), 16–29.

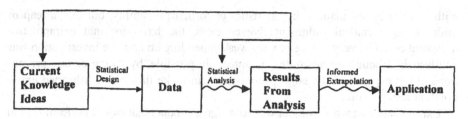

Figure 1. The role of statistics in the process of scientific learning. Wavy lines indicate the need for robust procedures.

factor which measures how sensitive the outcome is to such a deviation. The concept is completely general, but for illustration I will again use the correlation example. Suppose that some outcome of interest (the estimate of σ in the example, but in general some characteristic denoted by Y) is sensitive to an assumption about some characteristic (the zero value of the serial correlation coefficient for this example, but in general some quantity defining the assumption denoted by Z). Now let a deviation, z, from assumption produce a change, y, in the outcome. Then approximately

$$y = z \times dy/dz.$$

Thus the effect y on the outcome is obtained by multiplying z (the discrepancy from assumption) by a *robustness factor* dy/dz (the rate of change of the outcome in relation to the change in assumption). Thus, as is well known, two different procedures, even though derived from identical assumptions (such as a test to compare means by the analysis of variance and a test to compare variances by Bartlett's test), can be affected very differently by the *same* departure from assumption (the test to compare means is robust to certain likely kinds of non-normality of the error distribution, but the test to compare variances is not).

Such facts have led to the development of a plethora of robust estimators and tests. However, practitioners must be cautious in the choice of such methods. In particular, we should ask the question "Robust to what?" For example, common so-called "distribution free" tests recommended as substitutes for the t-test are robust to distributional non-normality. But they are not distribution free; in particular they are just as disastrously affected by failure of the *distributional assumption* of zero serial correlation as is the t-test. When such sensitivity occurs, inclusion of the sensitive parameter in the formulation of the original model is often necessary to obtain a robust procedure.

ROBUST DESIGN FOR INFORMED EXTRAPOLATION

Almost never is an experimental result put to use in the circumstances in which it was obtained. Thus, it might be hoped that a result obtained from a laboratory study published in a Polish journal might find application in, say, an industrial process in the United States. However, as was emphasized by Deming (1950, 1986), such a link with practice is not made using statistics or formal probability, but by "a leap of faith" using technical judgment. Nevertheless, the basis for that extrapolative judgment could be very strong or very weak depending on how the investigation was conducted; although no absolute guarantees are possible, by taking certain precautions in the experimental design process, we can make this job of informed extrapolation less perilous.

Experimentation whose object was to design a product that would perform well in the environmental conditions of the real world is a concept with a long history. An example are the field trials conducted by Guinness's at the turn of the twentieth century to find a variety of barley for brewing beer with properties that were insensitive to the many different soils, weather conditions, and farming techniques found in different parts of Ireland. The virtues of experimental design in such studies was emphasized by Fisher (1935, p. 112) who pointed out that "[extraneous factors] may be incorporated in experiments designed primarily to test other points with the real advantages, that if either general effects or interactions are detected, that there will be so much knowledge gained at no expense to the other objects of the experiments and that, in any case, there will be no reason for rejecting the experimental results on the ground that the test was made in conditions differing in one or other of these respects from those in which it is proposed to apply the results."

We owe to Taguchi (1986) our present awareness of the importance of statistics in achieving robust processes and products in industry. Many such applications of robust design fall into one of two categories:

(1) minimization of the variation in system performance transmitted by its components or
(2) minimization of the effect on system performance of variation in environmental variables which occurs in everyday use.

Overlooked solutions to both problems, which in my view are better than those later proposed by Taguchi, are due to Shewhart (1931), Morrison (1957), and to Michaels (1964) respectively. As described in Chapter E.3, Shewhart and Morrison solved the first problem directly using the classical error transmission formula. Morrison further made the critical observation that for any solution to be reliable, standard deviations of component errors must be reasonably well known. Michaels showed how the solution to the second problem is best dealt with by an application of split plot designs; see Chapter E.6. One illuminating approach to the environmental robustness problem can be understood by an extension of the earlier discussion on robust statistical analysis.

The Environmental Robustness Problem

As an example of environmental robust design consider the formulation of a washing machine detergent (see Michaels, 1964). Suppose we have an initial "prototype" formulation of the detergent (product design) for which, however, the effectiveness, Y, is unduly sensitive to the temperature, Z actually used in the domestic washing machine. (Effectiveness of a detergent can be measured by applying a "standard soil" to a sample of white cloth and making a colorimetric determination of its whiteness after washing.) To make the detergent suitable for household use we need to modify the formulation so that its effectiveness is robust to a moderate departure, z, from the ideal washing temperature. If y is the change induced by z in the measure of effectiveness, then, as before, approximately

$$y = z \times dy/dz$$

Now let $dy/dz|_p$ be the robustness factor for the initial prototype design and suppose we can find a design variable (say, the proportion of compound X in the formula) which has a substantial interaction (measured by $d^2y/dz\,dx$) with the temperature Z. If we assume that terms of third and higher orders can be ignored, then the robustness factor can be changed in accordance with the equation

$$dy/dz = dy/dz|_p + (d^2y/dz\,dx) \times x \qquad (1a)$$

where x is the deviation of the design variable X from its value in the prototype formulation. Theoretically, therefore, the robustness factor can be reduced to zero and a formulation insensitive to temperature obtained by setting $x = x^*$ such that x^* is the solution of the equation

$$dy/dz|_p + (d^2y/dz\,dx) \times x^* = 0$$

Now suppose we have a suitable experimental design centered at the prototype conditions x_p so that $dy/dz|_p$ can be estimated by the linear effect of the temperature (denoted below by c) and so that $d^2y/dz\,dx$ can be estimated by the interaction (denoted by C) of X with temperature. The value x^* required for robustness can then be estimated from the equation

$$-c = Cx^* \qquad (1b)$$

More generally, suppose that in Equation (1b), c now represents a vector of the linear effects of p environmental variables so that $\mathbf{c} = (c_1, c_2, \ldots, c_p)'$ and $\mathbf{C} = \{c_{ij}\}$ represents a $p \times q$ matrix of interactions such that the element of its ith row and jth column is the interaction between the environmental variable Z_i and the design variable X_j. Then the solution of the equations $\mathbf{x}^* = (x_1^*, x_2^*, \ldots, x_q^*)'$, if a solution exists, estimates the values of the design variables required for a robust design. If there are more design variables than environmental variables ($q > p$), then an infinity of solutions may exist. If $q = p$ and the matrix $\mathbf{C}$ is non-singular, then there will be a unique solution; if $q < p$, then no solution may exist. Also, for the solution to be of any value, $\mathbf{x}^*$ will need to be located in the immediate region where the approximations can be expected to hold and the coefficients in $\mathbf{c}$ and $\mathbf{C}$ will need to be estimated with reasonable precision. A discussion of the effects on the solution of errors in the coefficients of a linear equation is given in Box and Hunter (1954).

For illustration, suppose $q = p = 2$ and the environmental variables are Z_1, the temperature of the wash, and Z_2, its duration. Also suppose the design variables are the deviations from prototype levels of the amounts of two ingredients, X_1 and X_2. Then the conditions (x_1^*, x_2^*) which satisfy the robustness criterion are such that

$$-c_1 = c_{11}x_1^* + c_{12}x_2^*$$
$$-c_2 = c_{21}x_1^* + c_{22}x_2^*$$

(2)

where it must be remembered that the coefficients c_{11}, c_{12}, etc. are all interactions of the environmental variables with the design variables. Thus, for example, c_{11} is the interaction coefficient of Z_1 with X_1. Notice that no account of the *level* of response (the effectiveness of the detergent) is taken by these equations. The robust formulation could give *equally bad* results at different levels of the environmental variables.

Now suppose that the environmental variables are at their fixed nominal values and that locally the response is adequately represented by a second degree equation in the design variables,

$$\hat{y} = b_0 + b_1 x_1 + b_2 x_2 + b_{11}x_1^2 + b_{22}x_2^2 + b_{12}x_1 x_2$$

The coordinates (x_1^0, x_2^0) of a maximum in the immediate region of interest will then satisfy the equations

$$-b_1 = 2b_{11}x_1^0 + b_{12}x_2^0$$
$$-b_2 = b_{12}x_1^0 + 2b_{22}x_2^0$$

(3)

Note that Equations (3) have *no coefficients in common* with the robustness equations (2).

Thus you could have an "optimal" solution that was highly non-robust and a robust solution that was far from optimal. In practice, some kind of compromise is needed. This can be based on costs, the performance of competitive products, and the like. One approach, which combines considerations of robustness and optimality and provides a locus of compromise between the two solutions, was given by Box and Jones (1992a,b), who showed which coefficients needed to be estimated to achieve various objectives and who provided appropriate experimental designs. (See Chapter E.7). It is clearly important that robustness concepts and response surface ideas should be considered together (see, e.g., Vining and Myers, 1990, and Kim and Lin, 1998).

ENVIRONMENTALLY ROBUST DESIGN USING SPLIT PLOTS

As Michaels pointed out, to achieve environmental robustness, convenient and economical experimental designs are provided by split plot arrangements employing "main plots" and "subplots" within main plots. (The designs which Taguchi refers to as containing "inner" and "outer" arrays are split plot arrangements often incorrectly analyzed.)

Depending on whether the design variables or the environmental variables are applied to the subplots, design main effects or environmental main effects will be estimated with the subplot error. In either case, however, all the design × environmental interactions will be estimated with the subplot error. Because it is frequently true that the subplot error is considerably smaller than the whole plot error, the different kinds of split plot arrangements can have different theoretical efficiencies. These were given by Box and Jones (1992b) who also showed that strip-block designs can be even more efficient. They point out, however, that in practice the numbers of such operations and the difficulty and cost of carrying them out are usually of most importance in deciding the way in which an experiment is conducted. In his experiments on different detergent formulations, Michaels applied the design variables (test products) to the subplots because this produced experimental arrangements which were easier to carry out. Notice, however, that if the first and second order terms in robustness equations such as (2) are to be determined using the subplot error, then it is the environmental variables that should be applied to the subplots. These issues are discussed in greater detail in Chapter E.6, which follows.

In practice, discovering *which* environmental factors' effects need to be modified and *identifying* the design factors that can achieve these modifications are of greatest importance. The principle of parsimony is likely to apply to both kinds of factors. Fractional factorials and other orthogonal arrays of highest projectivity are thus particularly valuable to carry both the environmental and the design factors. Particularly when there are more design factors than environmental factors, different choices or different combinations of design factors may be used to attain robustness. Relative estimation errors, economics, and ease of application can help decide the best choice. Most important of all, the *nature* of the interplay between design factors and environmental factors revealed by the analysis should be studied by subject matter specialists. This can lead to an understanding of *why* the factors behave and interact in the way they do and can produce new ideas and, perhaps, even better means for obtaining robustness. Notice that these considerations require that we look at *individual* effects. Portmanteau criteria, such as signal to noise ratios that mix up these effects, are I think unhelpful.

CHAPTER E.6

Split Plots for Robust Product and Process Experimentation

Environmental robustness* and optimality are important concerns for improving current products and processes and for designing new ones. Since these two characteristics can rarely be achieved simultaneously, we must aim for a design that is *robust* and *good*: that is, a design that, although not necessarily best for any particular artificial set of environmental conditions, is reasonably effective over those conditions in which the product or process will need to operate. In this chapter we explore the use of split plot arrangements introduced in Chapter E.4 to achieve such designs. For a fuller discussion see, for example, Box and Jones (1992a,b), and the references contained therein.

A REMINDER ABOUT SPLIT PLOTS

In Chapter E.4 it was explained that for many experiments conducted in industry a completely random experimental arrangement is impractical. This is usually because certain factors (variables) are much more difficult to change than others. In such circumstances split plot experiments can be employed that are often very efficient for exploring the main issues and are much more easy to run. I used as an example an experiment to improve the corrosion resistance of steel bars by first applying a surface coating and then baking the bars in a large furnace. In the experiment, four

---

* A different problem concerns minimizing sensitivity of the performance of an assembled product to variation transmitted by its components. This problem was correctly posed and solved by Shewhart (1931) and by Morrison (1957). See also Chapter E.3 and Parry-Jones (1999).

From Box, G. E. P. and Jones, S. (1992b), *Journal of Applied Statistics*, 19(1), 3–26 and Box, G. E. P. and Jones, S. (2000), *Quality Engineering*, 13(1).

different coatings were to be compared at three different temperatures. A completely random arrangement in which the furnace temperature had to be reset repeatedly would have been impractical but a split plot experiment was easily run in the following way: a temperature condition was set and the four different kinds of coated bars were randomly positioned in the furnace and baked for the prescribed length of time. The furnace was then reset at a different temperature to bake a second set of coated bars and so on.

As mentioned in the earlier chapter split plot arrangements were first developed in agriculture, (see Fisher, 1935; Yates, 1937). Using their original terminology, repeated furnace heats would correspond to "main plots" and coatings applied to the bars within a single heat to "subplots." In such an experiment it is essential to understand that there will be two distinct sources of error that affect the results. In terms of this example, the main plot error is associated with the variation between furnace heats at the same temperature setting, and the subplot error with the variation between coatings within a given heat. It is typical of such experiments that the main plot variation (between heats) can be much larger than the subplot variation (between coatings within heats). Consequently, by using a split plot arrangement it would be possible to compare the effectiveness of the different coatings with greater accuracy than with a fully randomized arrangement.

DESIGNING PRODUCTS ROBUST TO ENVIRONMENTAL FACTORS USING SPLIT PLOTS

A concept described by Michaels (1964), which went largely unnoticed at that time, was the use of split plot experiments to design products robust to environmental variation. The value of statistically planned experiments to achieve such robustness was later brought to the attention of many engineers and statisticians by Genichi Taguchi (see, e.g., Taguchi and Wu, 1980) with many cogent industrial illustrations. However, the importance of exploiting split plot structure to minimize experimental effort and increase statistical efficiency and also the profound effect that such structure has on the correct analysis of the data were not discussed.

In Michael's example of a split plot design to achieve environmental robustness, the product was a detergent to be designed (formulated) so that its washing ability was not much affected by the hardness of the water, the temperature of the wash, and so forth. The product could equally well be an ignition system to be designed so that its performance was insensitive to the humidity and temperature of the intake air. Also, for several decades the U.S. food industry has used experiments such as is shown in Table 1 to obtain recipes that were robust to departures from recommended cooking instructions. In the discussion that follows whether the objective is to obtain a robust *formulation* for a detergent, a robust *design* for an ignition system, or a robust *recipe* for a boxed cake mix, we will in every case describe this activity by the single word *design*. Also, to avoid confusion the structure of the data collection strategy will be described by the words *experimental arrangement*.

Table 1. Data for the Cake Mix Example

Recipe	Design Variables F	S	E	T: − t: −	+ −	− +	+ +	Average	Range	Added Cost	Standard Deviation	Quadratic Loss
(1)	−	−	−	1.1	1.4	1.0	2.9	1.6	1.9	0	0.9	119.0
(2)	+	−	−	1.8	5.1	2.8	6.1	4.0	4.3	2	2.0	49.1
(3)	−	+	−	1.7	1.6	1.9	2.1	1.8	0.5	3	0.2	107.3
(4)	+	+	−	3.9	3.7	4.0	4.4	4.0	0.7	5	0.3	36.3
(5)	−	−	+	1.9	3.8	2.6	4.7	3.2	2.8	4	1.2	60.9
(6)	+	−	+	4.4	6.4	6.2	6.6	5.6	2.2	6	1.0	10.5
(7)	−	+	+	1.6	2.1	2.3	1.9	2.0	0.7	7	0.3	101.3
(8)	+	+	+	4.9	5.5	5.2	5.7	5.3	0.8	9	0.4	11.6

F, flour; S, shortening; E, egg; T, temperature; t, time.

In the somewhat simplified example for which the data are shown in Table 1, the manufacturer was seeking a design for a boxed cake mix that would be good and also robust to deviations from the recommended baking temperature (T) and the time of cooking (t). Thus, in this example, T and t were the environmental factors and were arranged in a 2^2 factorial. The design factors that were under the control of the manufacturer—the amounts of flour (F), shortening (S), and egg powder (E) included in the mixture—were arranged in a 2^3 factorial. All 32 combinations of the 8 cake recipes (designs) in the 4 environmental conditions of temperature and time were run and the results shown are average "hedonic" scores, on a scale of 1–7, obtained from a taste panel and measuring "how good" the cakes tasted.

As is later discussed, the efficiency of the experiment and the analysis of the subsequent data depend critically on how the experiment is actually carried out. In this preliminary discussion we will suppose that each of the eight separate product designs (cake mixes) was individually prepared and then subdivided into four batches, which were tested at the four environmental conditions of baking time and temperature. In this arrangement, therefore, the product designs were the main plots and the environmental combinations were the subplots.

A first appraisal of the data is gained from Table 1 by looking at the *Average*, *Range*, and *Additional Cost* for each of the eight product designs.* The average supplies a measure of the overall goodness of the design and the range is a measure of its robustness over the combinations of the environmental factors tested. The numbers in the "Additional Cost" column are in fractions of a penny, treating design (1) as a base.

---

* In addition to the average $(\bar{y})$ and the range, the estimated standard deviation $(\hat{\sigma})$ is also shown and may be used instead of the range if preferred. The results are essentially the same. Also shown is the "Quadratic Loss" $L = \sum (7 - y)^2$. This is a somewhat arbitrary combination of $\bar{y}$ and $\hat{\sigma}$, (for this example it is $L = 4[(7 - \bar{y})^2 + \frac{3}{4}\hat{\sigma}^2])$ and we think you will be better off looking at $\bar{y}$ and $\hat{\sigma}$ separately.

We see, for example, that design number (3) appears to be very robust (the standard deviation for the four cases made at these conditions is 0.2) but uniformly inedible (the average hedonmic index is only 1.8); design number (6) is good on the average but is not particularly robust; design (8) is essentially as good as (6) and is robust (but is the most expensive).

On (more substantial) data and analysis of this kind, management could make informed choices between alternatives that in addition took account of many subjective factors (the perceived quality of the competition, the relative profitability of this particular product, etc.).

A CLOSER ANALYSIS

The main effects and interactions for the design and the environmental factors can be obtained by analyzing in the usual way the data of Table 1 as a 2^5 factorial in the factors F, S, E, T, and t. These calculated effects may then be conveniently arranged in the 8×4 table as shown in Table 2, in which the symbol I is used to indicate a mean value.

We have seen that an important characteristic of split plot arrangements is that the subplot effects and *all their interactions* with the main plot effects are estimated with the same smaller subplot error. Only the main plot effects have the main plot error. Consequently, if the data from a split plot arrangement are analyzed graphically, *two* separate normal plots are needed. In the cake experiment the design factors are associated with the main plots and the environmental factors with subplots. The seven design effects F, S, E, FS, FE, SE, and FSE, which have the main plot error, have therefore been graphed in Figure 1a. The remaining 24 effects associated with the environmental factors and their interactions, which have the subplot error, are graphed separately in Figure 1b. From the gradients of the "error lines" in these two

Table 2. Calculated Effects for the Cake Mix Data

		Environments			
		I	T	t	Tt
	I	3.45	1.11	0.53	0.07
	F	2.57	0.31	0.01	−0.08
	S	−0.33	−0.92	−0.22	−0.08
Designs	E	1.21	−0.03	−0.08	−0.24
	FS	0.19	−0.17	0.01	0.22
	FE	0.31	−0.26	−0.08	0.08
	SE	−0.43	0.14	0.02	0.01
	FSE	0.28	0.37	0.07	−0.02

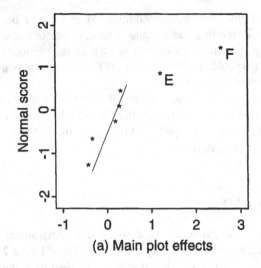

(a) Main plot effects

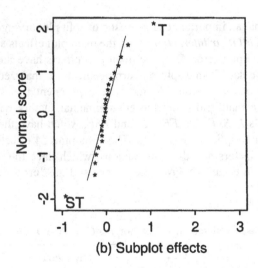

(b) Subplot effects

Figure 1. Normal plots (a) for main plot effects and (b) for subplot effects.

normal plots we see that, as expected, in this experiment the subplot error is considerably smaller than the main plot error. With this arrangement in which main plots are associated with designs, Figure 1a tells us which factors control how *good*, on the average, are the various designs. Thus we see that higher values for flour (F) and egg powder (E) give cakes that, on the whole, are better tasting. Figure 1b shows which factors affect robustness. Significant effects T, t and Tt indicate sensitivity to the temperature and time of baking. If a robust design can be achieved it will be by using the interactions between the design and environmental factors as "levers" to reduce such sensitivity. Suppose the factors are scaled so that minus and plus signs correspond to -1 and $+1$ in the coded variables. Suppose, in general, that there is an environmental effect E that we wish to modify and a design factor D that has an interaction DE with E. Then by setting* the design factor D to a value d, we can change E to a modified value E_m in accordance with the equation

$$E_m = E + (DE \times d) \tag{1}$$

which may be compared with equation 1a in Chapter E.5.

In the cake example, we would like the effects of T, t, and Tt to be as small as possible. But from Figure 1b we see that T is not small, indicating that the taste of the cake is highly sensitive to baking temperature—an effect clearly seen in the original data. Fortunately, there is also a large ST interaction. This can be employed to modify sensitivity to temperature to a value T_m by setting the design variable S (shortening) to a value S_0 in accordance with the equation

$$\begin{aligned} T_m &= T + (ST \times S_0) \\ &= 1.11 + (-0.92 \times S_0) \end{aligned} \tag{2}$$

where s denotes the amount of shortening on the coded scale.

Ideally then, and if there were no experimental error, sensitivity to temperature, as measured by T_m, could be reduced to zero by setting the design variable S to the value

$$S_0 = \frac{1.11}{0.92} = 1.2$$

The value $S_0 = 1.2$ is close enough to $s = 1$ and by comparing results in the rows of Table 1 we see that indeed the sawtooth pattern associated with temperature sensitivity in runs 1, 2, 5, and 6 (when S is at its minus level -1) essentially disappears in runs 3, 4, 7, and 8 (when S is at its plus level $+1$). Finally then, it is very logical that by setting F and E at their plus levels to get high response, and S at its plus level to minimize sensitivity to temperature, we obtain a cake mix (8) that is both *robust* and *good*.

In this example the environmentally sensitive factor T was not affected by the design variables E and F that changed the average response. When this happy separation does not occur some compromise solution will be necessary.

An estimate of the subplot standard deviation can be obtained from the twelve design $\times$ environment interactions between three factors or more shown below the dotted line in Table 2. The estimate obtained from the square root of the average value of the squares of these elements is 0.16. This provides an alternative means of

* For the argument, necessary assumptions, and generalizations associated with this concept, see Box and Jones (1992a,b) and Chapters E.5 and E.7.

assessing those effects shown above the dotted line in relation to the noise. The conclusions are as before.

WHAT? HOW? AND WHY?

As always, when considering the results of an experiment, we should have in mind the questions: *What* happened? *How* did it happen? *Why* did it happen?

What the various tested cake designs do when exposed to different environmental conditions is seen in Table 1.

How, for example, sensitivity to baking temperature may be reduced is shown by Equation (2).

It would remain to check with the subject matter specialist (the food technologist in this case) *why* there might be such effects such as those found. Good questions are: Would you have expected this? If yes, please explain why. If no, then what do you think might be going on? If your speculation is right and might lead to new ideas for developing the product, might it be a good idea to check it out and perhaps move on from there?

CHOOSING THE BEST EXPERIMENTAL ARRANGEMENT

Using the cake mix experiment for illustration we have so far supposed that the experiment was conducted with the ($n = 8$) designs as main plots and the ($m = 4$) environments as subplots. We will call this arrangement (a). The types and numbers of operations with this and other possible arrangements are summarized in Table 3.

In a fully randomized design 32 cake mixes would have been prepared independently and each one baked independently at the prescribed conditions of temperature and time, resulting in 32 separate bakes.

In the split plot arrangement of type (a) already described, 8 separate mixes would be prepared for each of the various combinations of flour (F), shortening (S), and egg (E). Each mix would then be divided into four and the resulting 32 mixes baked independently. In this arrangement the oven is operated 32 times.

Table 3. Four Possible Arrangements for the Cake Mix Example

| | | | Example | | General | |
			Mixes	Bakes	Design Models	Environments
	Fully randomized		32	32	$n \times m$	$n \times m$
Split	Bakes are subplots	(a)	8	32	n	$n \times m$
plots	Designs (recipes) are subplots	(b)	32	4	$n \times m$	m
	Strip block	(c)	8	4	n	m

In a split plot design arrangement of type (b), 32 cake mixes would be made up independently. These would then be separated into 4 sets each containing all 8 combinations of F, S, and E. The first set of 8 would then all be baked together, with the bake temperature (T) and bake time (t) chosen randomly from the four combinations of levels. The next set of 8 would be run at a second randomly selected time and temperature, and so on. So in this arrangement the oven is operated only four times.

Another type of cross product array is the strip block arrangement (c) due to Yates (1937). Using the arrangement a batch large enough to make four cakes would be prepared for each of the 8 combinations of F, S, and E and a complete set of these eight combinations would be baked at each of the four conditions of temperature and time. Again in this arrangement the oven is operated only four times.

It is not necessary, of course, that the design array or the environmental array should themselves be factorials. You could, for example, just test *any* set of product designs in *any* set of environments in a cross-product array.

CONSIDERATIONS IN CHOOSING THE EXPERIMENTAL CONFIGURATION

Statistical Considerations

It is important to be able to study the design × environment interactions individually and it should be noted that *whichever* cross product configuration is used these will all be estimated with the subplot error, which is often relatively small. Split plot arrangement (a) used for the cake mix experiment has the additional advantage that the main effects of all the environmental variables are also estimated with the subplot error. Thus, for example, in Equation (2) the coefficients T and ST are both estimated with the smaller subplot error. This advantage is of even greater importance when there is more than one important environmental effect to be minimized and/or more than one design variable that can be adjusted. In such cases Equation (1) is replaced by a set of linear equations (see equations (2) of Chapter E.5 also Box and Jones, 1992a; Box, 1999). If arrangement (a) is used, then all the coefficients in these equations will be estimated with the same subplot error.

We emphasize that it is the manner in which the experiment is actually carried out that decides the appropriate analysis. However, the *numerical* quantities that are needed for *any* analysis of the cake mix data are always those in Table 2. If the experiment had been conducted with split plot arrangement (b), the 7 design effects and the 21 design × environment interactions would have the subplot error and should appear on the same normal plot. The three environmental effects would have the main plot error. If the strip block of arrangement (c) had been used, there would have been three separate sources of error to consider, and if normal plots were used, there would be three different plots needed for the analysis: one for the design effects, one for the environmental effects, and one for the design × environment interactions. The last of these would usually have the smallest error term.

In the paper by Box and Jones (1992b) there is a discussion of detailed models, analyses of variance, and the statistical efficiency for each of the various arrangements.

Experimental Effort and Cost

In practice, as well as statistical considerations, the cost and difficulty of experimentation must be taken into account in deciding whether subplots are associated with the environmental conditions or with the product design factors.

For example, if we were developing a copying machine, arrangement (a) would almost certainly be best from a practical point of view because the production of a number of different types of copier (product designs) would be expensive; but running these copiers at different environmental conditions (ambient temperature, humidity, type of paper, etc.) would be easy and inexpensive. Furthermore if factors such as temperature and humidity were the only environmental variables, these might be changed en masse in a specially fitted room (a biotron) in which all the copiers could be tested together at each environmental condition in a strip block design. This arrangement is usually so easy to run that a number of repeats in which the environmental conditions are replicated can be made inexpensively, thus giving increased precision for the estimated effects and the design × environment interactions and allowing better estimation of the sources of error.

On the other hand, if the experiment referred to earlier to test the corrosion resistance of coated steel bars was part of a robustness study, then it would be the environmental factors (the furnace temperatures) that were difficult and expensive to change. These would therefore be chosen as main plot factors and the product designs (the various coatings applied to the steel bars) would be the subplots; and arrangement (b) would be the preferred arrangement.

The same principles apply to robust *process* design. In a robust process experiment using arrangement (a), experiment parts were treated in a heated chamber that had gas pumped into it. The problem was to minimize processing differences arising from different orientations and different locations of the parts within the chamber. The design variables included hold temperature, dwell time, gas flow rate, and removal temperature. The environmental variables, which were easy to change, were the orientation of the part and the location of the part within the chamber and formed the subplots in the arrangement. The robust process design finally arrived at allowed the chamber to be run at its maximum capacity.

Extreme Conditions

One technique for reducing the number of environments to be tested that has sometimes been used is to run experiments only at what are judged to be some of the severest environmental conditions. For example, for the cake mix experiment it is reasonable to assume that the effect of baking time and baking temperature would to some extent be mutually compensating. Thus, a mistake in using too high a baking temperature might be roughly compensated by using a shorter baking time. Thus by running the environmental factors only at the combinations $(T, t) = (-, -)$, and $(+, +)$ representing the severest conditions, the robustness of the various product designs might be assessed with fewer runs. The same kind of reasoning can be used in other technologies, but the disadvantage of this technique is that it is necessary for the food technologist, engineer, or chemist to be able to guess in advance what the extreme environmental combinations will be—a task that becomes more and more difficult as the number of environmental factors increases. Thus, this approach has serious limitations.

Finding Active Factors

In some instances one of the most difficult problems is to find out *which* of a large number of environmental effects are important and which of a large number of design factors might eliminate them by the use of suitable interactions.

For this purpose, fractional factorials or other orthogonal arrays may be used to carry both the design variables and the environmental variables. For example, using the same overall number of runs as in Table 1, seven design factors might have been accommodated by using a saturated 2^{7-4} fractional factorial for the design array, and three environmental variables by using a saturated 2^{3-1} fractional factorial for the environmental array. Notice that on the assumption that interactions of three or more factors can be ignored, all 21 of the design × environment interactions could still be estimated without bias. The main effects for the design variables would be aliased with their two factor interactions, and similarly for the environmental variables; however, in two replications, the signs for the environmental variables could be switched. Then making use of the foldover principle described e.g. in Chapter B.6, and on the same assumptions concerning higher order interactions, all main effects for the environmental variables would be free of aliases (see, also BH^2, and the references therein). Bear in mind, however, that environmental sensitivity might occur because of an interaction between environmental variables; such possible interactions may therefore need to be isolated. Also, while in principle the design variables could similarly be switched in the second replication, this might not be easy to do since it would involve using a further set of different designs that were independently built. The use of fractional factorials and orthogonal arrays would be particularly attractive when the more easily performed strip block arrangements could be used.*

Finding the Critical Factors

Perhaps most important is to *find* the environmental factors that need to be desensitized and to discover design factors that will interact with them to achieve desensitization. When fractional factorials and Plackett–Burman designs are employed to carry the environmental and design factors, they can be used as screens

---

*If these designs are replicated, "partial confounding" could be employed to better isolate interactions (see, e.g., Fisher, 1935; also Bisgaard, 1992, 2000; and Kulahci, 2000).

to search both for environmentally sensitive effects and for appropriate design variables that can neutralize them. The projective properties of these designs can then be particularly useful to isolate main effects and their interactions for small subgroups of the important environmental and design factors (see e.g., Chapters B.2, B.3, B.4 and B.5).

When there is one or more environmental factors that have not been neutralized with design factors so far tested, it will be important to make clear to the subject matter specialist the nature of the problems that remain. This is vital to produce new ideas.

A FURTHER EXAMPLE

Murphy never sleeps and planned experimentation provides him with an almost irresistible temptation. Fortunately, however, sometimes even an imperfect experiment can yield valuable results. This was true in a study conducted by John Shuerman and Todd Algrim, engineers with DRG Medical Packaging Company in Madison, Wisconsin, presented and discussed at my Monday night sessions.

In this investigation, the company sought to design a package that would give a good seal over the wide range of sealing process conditions used by its customers. This packaging manufacturer identified five product design factors that might affect the quality of the seal. These were the paper supplier (PS), the solids content (SC), the cylinder type (CY), the oven temperature (OT), and the line speed (LS), each to be tested at the two levels shown in Table 4. The customer had identified two environmental factors—the sealing temperature (ST), which was set at three levels, and the sealing dwell time (SD), at two levels.

The experiment was conducted in the following manner. An eight-run fractional factorial experiment was run using the five packaging factors, and large quantities of the packaging material were manufactured at each of the eight conditions. The eight types of material were then shipped to each customer's plant and each batch was subdivided into six subbatches. The subbatches were then used to run a 2×3

Table 4. Design Factors and Their Levels for the Medical Packaging Example

Factors	Levels	
	−	+
Paper supplier (PS)	U.K.	U.S.A.
Contents of solids (SC)	28%	30%
Cylinder (CY)	Fine	Coarse
Oven temperature (OT)	180°C	215°C
Line speed (LS)	Low	High

Table 5. Number of Defects in Packaging Experiment

Product Design	Design factors					Environmental Factors[a]					
	PS	SC	CY	OT	LS	ST − 0 + − 0 +					
						SD − − − + + +					
(1)	+	−	−	+	+	8	4	*	0	*	7
(2)	+	−	−	+	−	*	*	*	*	*	*
(3)	−	−	+	−	+	3	2	1	1	0	0
(4)	−	−	+	−	−	7	1	1	2	0	4
(5)	+	+	+	−	+	0	0	0	0	0	1
(6)	−	+	+	+	+	4	0	0	1	1	1
(7)	−	+	−	+	−	9	1	0	4	5	1
(8)	+	+	−	−	−	2	1	3	1	0	0

[a] Asterisk (*) indicates missing observations.

experiment in the two sealing process factors. This was, therefore, a split plot arrangement of type (a) in which the designs were main plots and the environments subplots.

Table 5 shows the observed number of defective seals from a sample of 15. This experiment was imperfectly carried out because the two levels of one of the factors were inadvertently switched so that the eight-run experiment as actually carried out was unbalanced. Also, there were a number of missing observations in the first run in Table 5 and a complete loss of information for the second run.

Nevertheless, it is easy to see from visual examination of the data that product design number (5) behaved almost perfectly under all environmental conditions.

The complete analysis indicated how the manufacturer could produce packaging material that would give a sterile seal under a range of sealing process conditions. This and other experiments led to discoveries that turned out to be worth several million dollars.

ACKNOWLEDGMENT

This research was supported by NSF Grant Number DMI-9812839.

CHAPTER E.7

Designing Products that Are Robust to the Environment—A Response Surface Approach

INTRODUCTION

In Chapter E.6, we discussed the use of split plot and related arrangements for obtaining products that were good and also robust to environmental factors. In this chapter we consider an approach using response surface methods.

A geometric understanding of the environmental robustness problem can be gained by considering Figure 1. We have some measured *response y* (the hedonic index for the cake example), we have a set of product *design* variables $\mathbf{x}$ (flour, shortening, and egg for the cake example), which we suppose can be experimentally varied within some region of interest R_x, and we have a set of *environmental* variables $\mathbf{z}$ (baking temperature and baking time for the cake example), which we suppose can be experimentally varied within some region of interest R_z. We are looking for some particular combination of the product design variables, $\mathbf{x}_0$ within R_x, which continues to yield a high response y as the environmental variables $\mathbf{z}$ are varied over R_z. We have illustrated this geometrically in Figure 1(a), for a single product design variable x and a single environmental variable z.

A particular choice of product design corresponds to taking a section as in Figure 1(b), through the environmental space at some particular point x. We are looking for a choice of x that will give a section showing a uniformly high response for y over the whole range of interest of z. Suppose now that τ is an "ideal" value for the

From Box, G. E. P. and Jones, S. (1992a), *Total Quality Management*, 3(3), 265–282 and Box, G. E. P. and Jones, S. (1992b) *Journal of Aplied Statistics*, **19**(1), 3–26.

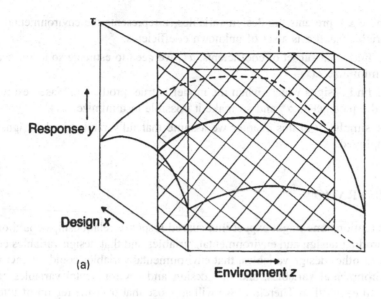

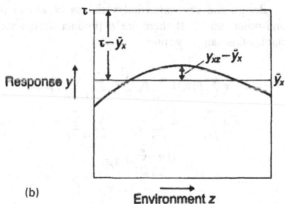

Figure 1. (a) Geometric representation of the environmental robustness problem. (b) A section of part (a) for a particular choice of **x**.

response. One measure of how far we fall short of this ideal is the integrated squared error loss

$$L(x) = k \int_{R_z} [\tau - y_{xz}]^2 \, dz$$

The approach that we adopt is as follows.

(1) We entertain a class of models which we suppose, tentatively at least, approximates the underlying system

$$y = g(\mathbf{x}, \mathbf{z}, \alpha) + e$$

where **x** represents the design variables, **z** represents the environmental variables, and α is a set of unknown coefficients.

(2) We find what subset of coefficients, α^*, we need to estimate so as to be able to minimize $L(x)$.

(3) We find designs which, when the model is true, produce unbiased estimates of the parameters α^* and so make it possible to minimize $L(x)$.

(4) For simplicity in this chapter we assume that fully randomized designs are employed.

AN ILLUSTRATION

In some situations one might expect that curved response relationships could occur both for product design and environmental variables and that design variables could interact with other design variables, that environmental variables could interact with other environmental variables, and that design and environmental variables could interact with each other. Therefore, we will suppose that over the region of interest the behavior of the design and environmental variables $(\mathbf{x}, \mathbf{z})$ can be represented by a general second-order model. If there are n product design variables and m environmental variables this can be written

$$
\begin{aligned}
y_{xz} = \beta_0 &+ \sum_{i=1}^{n} \beta_i x_i + \sum_{i=1}^{n} \beta_{ii} x_i^2 + \sum_{i=1}^{n-1} \sum_{k=i+1}^{n} \beta_{ik} x_i x_k \\
&+ \sum_{j=1}^{m} \gamma_j z_j + \sum_{j=1}^{m} \gamma_{jj} z_j^2 + \sum_{j=1}^{m-1} \sum_{k=j+1}^{m} \gamma_{jk} z_j z_k \\
&+ \sum_{i=1}^{n} \sum_{j=1}^{m} \delta_{ji} x_i z_j
\end{aligned}
\tag{1}
$$

or, in matrix notation,

$$
y_{xz} = \beta_0 + \mathbf{x}'\boldsymbol{\beta} + \mathbf{x}'\mathbf{B}\mathbf{x} + \mathbf{z}'\boldsymbol{\gamma} + \mathbf{z}'\mathbf{C}\mathbf{z} + \mathbf{z}'\mathbf{D}\mathbf{x}
$$

To minimize $L(x)$ over **x** we need unbiased estimates of β_0, $\boldsymbol{\beta}$, **B**, $\boldsymbol{\gamma}$, tr **C**, and **D**. Some experimental plans that satisfy these conditions are given in the fuller account in Jones (1990).

For illustration suppose we require an arrangement with $n = 3$ product design variables, x_1, x_2, x_3, and $m = 4$ environmental variables, z_1, z_2, z_3, z_4, with the response, y_{xz}, represented by the full second-order model given in Equation (1) with $n = 3$ and $m = 4$.

One useful class of designs is somewhat similar to the standard central composite designs (see, e.g. Box and Wilson, 1951; Box and Hunter, 1957; Box and Draper, 1987). The cube portion of the experimental arrangement is chosen to be a resolution IV design which confounds the interactions of the environmental factors in strings with each other, but not with the main effects nor the other two-factor interactions, and star points are added not for all the factors but only for the n product design factors.

For our example, an appropriate experimental arrangement can be obtained by using, for the cube portion of the design, a resolution IV design with only 32 runs by writing out the 32-run full factorial design in 5 variables, A, B, C, D, and E, and assigning A, B, and C as product design factors, and D, E, and the interactions ABCD and ABCE as the environmental factors. The complete experimental design would be obtained by adding two star points for each of the three product design factors and a certain number (say, four) center points. Thus the complete experimental plan would require $32 + 6 + 4 = 42$ runs. The following table (Table 1) gives the factor levels for this design and a set of hypothetical data. Table 2 shows the estimates of the effects that can be calculated from the data set.

Now it can be shown that

$$\min L(x) = \min\{[\tau - (\beta_0 + x'\beta + x'Bx + \tfrac{1}{3}\operatorname{tr} C)]^2 + \tfrac{1}{3}(\gamma + Dx)'(\gamma + Dx)\}$$

Let us assume that the ideal value, τ, is taken to be 100. A FORTRAN 77 computer program using the IMSL subroutines DU4INF and DUMINF was written to conduct the minimization by a quasi-Newton method. Using the computer program for this example we conclude that $L(x)$ is minimized at

$$x = \begin{bmatrix} -0.324 \\ -0.909 \\ -0.010 \end{bmatrix}$$

Contour plots of the function

$$\{[\tau - (\beta_0 + x'\beta + x'Bx + \tfrac{1}{3}\operatorname{tr} C)]^2 + \tfrac{1}{3}(\gamma + Dx)'(\gamma + Dx)\}$$

are shown in Figure 2.

The calculations for this fictitious example have been repeated using errors simulated from alternative normal distributions. It has been concluded that if the coefficient of variation is of the order of 10% or less then the value of x that minimizes $L(x)$ is reasonably consistent for different error samples.

From the minimization that has been conducted the investigator has an indication of the region of the design space, R_x, in which the value of the loss function is minimized. This experimental design and analysis could be regarded as a stage in a sequential investigation where the investigator would now use the information gained from this experiment to guide the choice of where further experimentation should take place.

Table 1. Experimental Design and Response

Run	y	x_1	x_2	x_3	z_1	z_2	z_3	z_4
Cube Points								
1	36.55	−1	−1	−1	−1	−1	1	1
2	41.10	1	−1	−1	−1	−1	−1	−1
3	9.26	−1	1	−1	−1	−1	−1	−1
4	21.63	1	1	−1	−1	−1	1	1
5	46.29	−1	−1	1	−1	−1	−1	−1
6	39.95	1	−1	1	−1	−1	1	1
7	49.37	−1	1	1	−1	−1	1	1
8	18.56	1	1	1	−1	−1	−1	−1
9	63.13	−1	−1	−1	1	−1	−1	1
10	60.17	1	−1	−1	1	−1	1	−1
11	24.25	−1	1	−1	1	−1	1	−1
12	17.36	1	1	−1	1	−1	−1	1
13	42.03	−1	−1	1	1	−1	1	−1
14	70.55	1	−1	1	1	−1	−1	1
15	19.19	−1	1	1	1	−1	−1	1
16	53.25	1	1	1	1	−1	1	−1
17	13.85	−1	−1	−1	−1	1	1	−1
18	15.95	1	−1	−1	−1	1	−1	1
19	5.71	−1	1	−1	−1	1	−1	1
20	15.00	1	1	−1	−1	1	1	−1
21	22.71	−1	−1	1	−1	1	−1	1
22	18.62	1	−1	1	−1	1	1	−1
23	42.00	−1	1	1	−1	1	−1	1
24	16.16	1	1	1	−1	1	−1	1
25	38.76	−1	−1	−1	1	1	−1	−1
26	37.41	1	−1	−1	1	1	1	1
27	19.29	−1	1	−1	1	1	1	1
28	12.12	1	1	−1	1	1	−1	−1
29	16.05	−1	−1	1	1	1	1	1
30	36.03	1	−1	1	1	1	−1	−1
31	10.14	−1	1	1	1	1	−1	−1
32	44.48	1	1	1	1	1	1	1
Star and Center Points								
33	69.12	−1.68	0	0	0	0	0	0
34	70.19	1.68	0	0	0	0	0	0
35	77.58	0	−1.68	0	0	0	0	0
36	60.49	0	1.68	0	0	0	0	0
37	63.74	0	0	−1.68	0	0	0	0
38	73.09	0	0	1.68	0	0	0	0
39	73.32	0	0	0	0	0	0	0
40	80.92	0	0	0	0	0	0	0
41	77.15	0	0	0	0	0	0	0
42	76.49	0	0	0	0	0	0	0

Table 2. Estimate of Effects from Data

Coefficient	Estimate	Coefficient	Estimate
β_0	76.970		
β_1	1.635	δ_{11}	4.291
β_2	−6.643	δ_{12}	−3.335
β_3	3.443	δ_{13}	−2.355
		δ_{21}	−0.164
γ_1	4.735	δ_{22}	4.763
γ_2	−7.762	δ_{23}	−0.551
γ_3	2.840	δ_{31}	1.078
γ_4	0.439	δ_{32}	7.208
		δ_{33}	1.291
β_{11}	−2.592	δ_{41}	0.102
β_{22}	−2.812	δ_{42}	0.100
β_{33}	−3.030	δ_{43}	0.281
β_{12}	−0.659	trC	−9.502
β_{13}	1.247		
β_{23}	4.477		

In the strategy that we have proposed we have taken the loss function to be the criterion that we desire to minimize. The loss function, a measure of how far we fall short of the theoretical ideal τ, is

$$L(x) = k \int_{R_z} [\tau - y_{xz}]^2 \, d\mathbf{z}$$

where

$$k^{-1} = \int_{R_z} d\mathbf{z}$$

There are certain difficulties associated with direct use of the squared error loss. We discuss below a somewhat modified approach.

Let us write $\bar{y}_x$ for the mean level of the response at a particular $\mathbf{x}$ averaged over the environmental variables

$$\bar{y}_x = k \int_{R_z} y_{xz} \, d\mathbf{z}$$

Then

$$L(x) = M(x) + V(x)$$

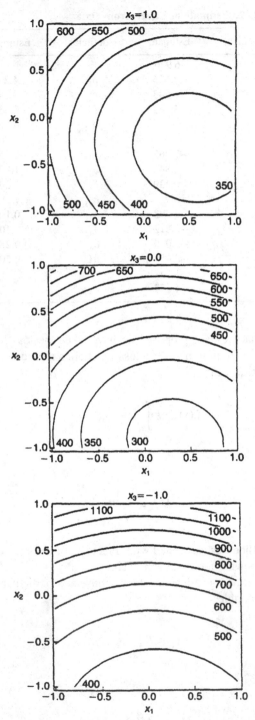

Figure 2. Contour plots of the function $\{[\tau - (\beta_0 + \mathbf{x}'\boldsymbol{\beta} + \mathbf{x}'\mathbf{Bx} + \frac{1}{3}\operatorname{tr}\mathbf{C})]^2 + \frac{1}{3}(\boldsymbol{\gamma} + \mathbf{Dx})'(\boldsymbol{\gamma} + \mathbf{Dx})\}$.

where

$$M(x) = k \int_{R_z} (\tau - \bar{y}_x)^2 \, d\mathbf{z}$$

measures the deviation of the mean response from the theoretically perfect value τ, and

$$V(x) = k \int_{R_z} (y_{xz} - \bar{y}_x)^2 \, d\mathbf{z}$$

measures the mean square variation about the average response.

The meaning of $M(x)$ and of $V(x)$ is readily understood. $M(x)$ is measuring how good the product design is on the *average* when it is exposed to different environmental conditions, and $V(x)$ is measuring how much variation there is in performance at different environmental conditions. Both of these are important qualities. Unfortunately, the relative importance placed upon them if we use the portmanteau criterion $L(x)$ is entirely dependent on how we choose τ. If we give a higher value for τ then more emphasis will be given to $M(x)$ and less to $V(x)$, and *vice versa*. This introduces an arbitrary element because τ may be unknown and unknowable. We therefore believe that $M(x)$ and $V(x)$ should be considered separately and an overall measure of robust performance, if such is needed, should be of the form

$$R(x) = \lambda V(x) + (1 - \lambda)M(x)$$

where a particular combination, determined by λ, is chosen after the available alternatives become manifest.

The situation will become clear by considering Figure 3, where we suppose that there are two product design variables, x_1 and x_2. If we were interested only in achieving a mean performance close to τ then we would choose the product design corresponding to the point x_M where $M(x)$ is minimized. If we were only interested in reducing variation about the mean we would choose the point x_V where $V(x)$ is minimized. For different values of λ $(0 \leq \lambda \leq 1)$ the combined criteria $R(x)$ will be minimized on the locus where the contours of $M(x)$ and the contours of $V(x)$ "kiss."

Referring to Equations (2) and (3) in Chapter E.5 and allowing for differences in notation, this locus will be seen to represent a series of compromises between the product design denoted by x^* and yielding maximum robustness and x° yielding maximum response.

Arguing as above, we see that reasonable experimental objectives would be: (a) to minimize $R(x)$ for any value λ_0; (b) to find the locus of minimum $R(x)$ for different λ; (c) to minimize $V(x)$; and (d) to minimize $M(x)$. In practice some understanding of the options available from (c) and (d) will be all that is needed to make an appropriate choice. It turns out that to achieve these various goals, (a), (b), (c), and (d), we need to be able to estimate only certain constants in the model. For the second-order model that we have considered above, the constants that we need to be able to

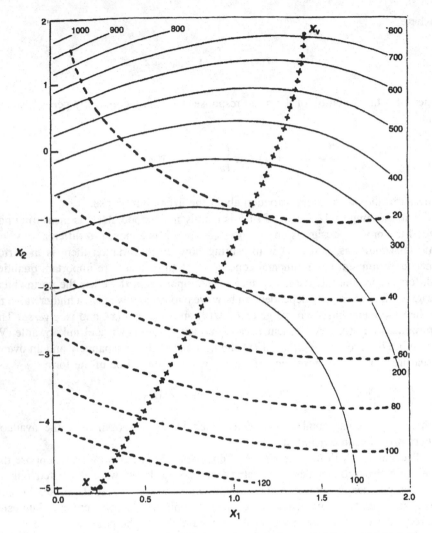

Figure 3. Contours of $M(x)$ (——), $V(x)$ (– – –), and the locus of minimum $R(x)$ (+ + +).

estimate for the various objectives of interest are indicated in Table 3. Since, for the second-order model, we know what these constants are, we can search for experimental designs which estimate them efficiently.

In Table 3, the entry "tr" under the column "**C**" indicates that only the trace of the matrix **C** needs to be known, where trace of **C** is tr $\mathbf{C} = \gamma_{11} + \gamma_{22} + \cdots + \gamma_{mm}$.

In the strategy that we outlined above our objective was to minimize the loss function, which is equivalent to minimizing $R(x)$ with $\lambda = 0.5$. Table 3 indicates that smaller designs might be possible for the other objectives since these objectives require that fewer constants be estimated.

Table 3. Estimation Table for the Full Second-Order Model

Objective	Constants						
Compute/Minimize	β_0	β	**B**	γ	**C**	**D**	Need to Know τ
$R(x) = \lambda_0 V(x) + (1 - \lambda_0)M(x)$	×	×	×	×	tr	×	×
Locus of minimum $R(x)$ for different λ		×	×	×		×	
$V(x)$				×		×	
$M(x)$		×	×				

In the full version of this paper more elaborate models are explored and a list of possible designs is given.

ACKNOWLEDGMENTS

Sponsored by the National Science Foundation Grant No. DDM-8808138, and by the Vilas Trust of the University of Wisconsin-Madison.

CHAPTER E.8

An Investigation of the Method of Accumulation Analysis

ABSTRACT

"Accumulation analysis" is a method introduced by Taguchi (1974) for the analysis of ordered categorical data. Using his original data on the efficacy of two rival drugs, we discuss the reasons Taguchi gives for substituting his accumulation statistic for Karl Pearson's χ^2 statistic. We point out the superiority and greater simplicity of a well-known procedure for decomposing the χ^2 statistic and employing this linear component to determine the efficacy of a treatment. Finally, using a result due to Yates, we show that the extraction of the linear component of χ^2 is itself equivalent to simply analysing equally spaced scores and conducting a standard analysis using the t statistic or the analysis of variance. In view of these results and the findings of earlier research by ourselves (1986b), Nair (1986), and Hamada and Wu (1986), we conclude that accumulation analysis is inefficient, as well as unnecessarily complicated, and consequently should not be taught or recommended.

INTRODUCTION

Industrial and other experimental results often appear as ordered categorical data. Such data are also frequently subjective. For example, Fisher (1963), in his book *Statistical Methods for Research Workers*, which first appeared in 1925, discussed the analysis of an experiment in which a number of samples of human blood were tested with a number of different sera and the degree of agglutination was subjectively rated in five classes of increasing activity. Data occurring in ordered categories

From Box, G. E. P. and Jones, S. (1990), *Total Quality Management*, 1(1), 101–113.

are common. In the food industry, for example, how good a cake tastes may be assessed on a subjective seven-point scale. Other examples concern how unpleasant is a particular odour, or how soft to the touch is a particular fabric. As an aid to assessment, a set of standard samples chosen by a panel are agreed to represent, in an appropriately equally graded manner, the characteristic that is being assessed. Typically $k = 5$ or 7 standards would be used.

Although more elaborate methods, some of which are mentioned in Nair (1986), are available, we have usually analysed *the raw scores themselves* with standard procedures, as was done, for example, in the book edited by Davies (1958). In particular, for analysis of unreplicated factorials and fractional factorial designs we have calculated effects from the scores and used Daniel's normal probability plots (Daniel, 1959) and Bayes plots (Box and Meyer, 1986a). For more extensive data, we have employed t statistics and the analysis of variance with such scores. If preferred, non-parametric methods may be substituted, although in practice the results obtained are usually very little different.

It may be argued that such analysis is inappropriate because (1) the scale is subjective, (2) it is discrete rather than continuous, and (3) the necessary distributional assumptions are not satisfied. We feel that the first objection does not carry much weight, provided that the scale is carefully and thoughtfully chosen by a consensus of knowledgeable people and bearing in mind that the results will be expressed on the same scale. On the second and third points, it is known that the standard statistical procedures are remarkably robust to drastic rounding and to departures from distributional assumptions. In any case, we will argue that accumulation analysis avoids none of these difficulties and on closer examination turns out to be more complicated and less efficient.

Accumulation analysis is an alternative approach for analysing ordered categorical data. In a basic reference, to which his later publications refer, Taguchi (1974) illustrated the arguments in favour of his method using a particular set of hypothetical data which we reproduce in Table 1.

To illustrate his approach, Taguchi supposed that trials were made of rival drugs, A and B. In each trial a control and treated group, each of 80 patients, were tested and were allocated to the categories not effective $(-)$, slightly effective $(+)$, effective $(++)$, and markedly effective $(+++)$.

Table 1. Taguchi's Hypothetical Data for Treatment and Control Patients with Two Therapeutic Drugs A and B, Given by Taguchi (1974)

	Drug A				Drug B			
	$-$	$+$	$++$	$+++$	$-$	$+$	$++$	$+++$
Control	40	24	10	6	40	24	10	6
Treatment	24	40	10	6	24	29	16	11

t-TEST ANALYSIS USING SCORES AS DATA

To provide a standard against which Taguchi's accumulation method may be judged, we will first carry through the calculations which would be appropriate if we simply scored the therapeutic effects $-$, $+$, $++$, $+++$ as 0, 1, 2, 3 (or employed any other equally spaced scores) and calculated the usual t statistics or equivalently conducted analysis of variance. Although the scores could be analyzed using the t statistic, to allow later comparisons to be made more easily we here employ the (equivalent) analysis of variance shown in Table 2. It will be seen that a highly significant effect is found with drug B but not for drug A. Notice that the score for each *individual* patient is regarded as an observation. Notice also that the number of degrees of freedom for error (158) is so large that the tabulated $F = t^2$ distribution would be very closely approximated by a χ^2 distribution with one degree of freedom.

CHI-SQUARE ANALYSIS

Taguchi's method of accumulation analysis is prompted by what he feels is an inadequacy of the Pearson χ^2 test. After noting the values of the Pearson χ^2 statistics and the corresponding levels of significance (given in Table 3), Taguchi presents the following motivation for introducing his alternative analysis: "the efficacy of the drug A is significant but that of drug B is not despite the fact that it is clear the efficacy of B is much greater than that of A. This example shows that the use of the χ^2 test is unsuitable in cases in which the ... data are arranged in order" (Taguchi, 1974).

Karl Pearson's test is for general discrepancies from the expected proportions and is not, of course, intended to be sensitive to the ordering of categories (of which it knows nothing). A standard way to look for a tendency toward an increase in the number of treated patients in the higher effect categories would be to isolate the linear component χ_1^2 of the overall χ^2 statistic (see, e.g., Snedecor and Cochran, 1980). If this is done we obtain the results in Table 4, which clearly indicate the

Table 2. Analysis of Variance Test for the Differences Between Treatment and Control for the Drugs A and B When Using Scores (0, 1, 2, 3)

	df	Sum of Squares	Mean Square	$F = t^2$
		Drug A		
Treatment/control difference	1	1.600	1.600	1.977
Error	158	127.900	0.809	$\alpha = 16.2\%$
Total	159	129.500		
		Drug B		
Treatment/control difference	1	6.400	6.400	6.675
Error	158	151.500	0.959	$\alpha = 1.1\%$
Total	159	157.900		

Table 3. Hypothetical Data for Treatment and Control Patients with Two Therapeutic Drugs A and B, Given by Taguchi (1974)

	Drug A				Drug B			
	−	+	++	+++	−	+	++	+++
Control	40	24	10	6	40	24	10	6
Treatment	24	40	10	6	24	29	16	11
y	−16	16	0	0	−16	5	6	5
	80	80	80	80	80	80	80	80
		$\chi_1^2 = 8.000$				$\chi_3^2 = 7.327$		
		$\alpha = 4.7\%$				$\alpha = 6.0\%$		

The (treatment − control) differences, y, are used in a later analysis.

Table 4. The Linear χ^2 Components for the Drugs A and B, Both Based on One Degree of Freedom

Drug A	Drug B
$\chi_1^2 = 1.977$	$\chi_1^2 = 6.485$
$\alpha = 16.0\%$	$\alpha = 1.1\%$

superiority of drug B over drug A in the manner desired by Taguchi. Also notice that the values of the linear components of χ^2, 1.977 and 6.485, and the associated levels of significance are essentially the same as those obtained previously for the $F = t^2$ statistics in Table 2 obtained by allocating equally spaced scores. The reason for this is discussed later.

MANTEL'S STATISTIC

A formal test for ordered categorical data based on a chi-square statistic having one degree of freedom was proposed by Mantel (1963). In the case that we are considering, again using χ_1^2 for the linear component of Pearson's χ^2, his statistic M (say) would be $M = \chi_1^2 \times ((N - 1)/N)$ with $N = 160$. Thus, for examples of this kind, the practical difference from the χ_1^2 statistic is negligible. In discussing the statistic, Mantel remarked "that a linear regression is being tested does not mean that an assumption of linearity is being made. Rather it is that test of a linear component of regression provides power for detecting any progressive association which may exist. [This] chi-square may have various interpretations, the basic one being that it provides a test with power for any progressive relation ..." (Mantel, 1963, p. 698).

Table 5. Cumulated Frequencies for the Data of Table 1

	Drug A				Drug B			
	I	II	III	IV	I	II	III	IV
Control	40	64	74	80	40	64	74	80
Treatment	24	64	74	80	24	53	69	80
$\hat{q}_i$	64	128	148	1	64	117	143	1
	160	160	160		160	160	160	

ACCUMULATION ANALYSIS

Instead of employing the simple decomposition of χ^2 presented above, Taguchi introduced accumulation analysis. For this he used the cumulated frequencies shown in Table 5. In this table the frequencies in class I are the $(-)$ frequencies, those in class II are the sum of the frequencies in the first two classes $(-)$ and $(+)$, and so on.

Analysis of variance tables were then constructed from these cumulated frequencies by calculating separate sums of squares for categories I, II, and III, scoring each patient as a 1 if they appear in the given category and a zero otherwise. The individual sums of squares for the categories I, II, and III were then weighted with the reciprocal of the binomial variance appropriate for that category. As an example, Table 6 shows such an analysis of variance table for drug B taken from Taguchi (1974). Also shown is the general structure of the degrees of freedom and sums of squares for k categories ($k = 4$ in this example) and for n patients exposed to a treatment and n patients exposed to a control ($n = 80$ in this example).

It is not our purpose to rationalize this analysis of variance table. In particular, the 474 degrees of freedom for "error" and the 477 for "total" seem impossible to justify, since the total number of patients in this trial is only $2n = 160$. For the purpose of subsequent discussion, however, note that:

(i) The entries in the table only involve the frequency data via the quantity T. In general, we will call T the *accumulation statistic*, using subscripts A and B to denote its value for the two drugs. For Taguchi's data we find $T_A = 6.667$ and $T_B = 12.160$.

(ii) The "mean square for error" $(2n(k-1) - T)/2(n-1)(k-1)$ will usually be close to unity.

Table 6. Taguchi's Accumulated Analysis of Variance Table for Drug B

Source	df	Sum of Squares	Mean Square	
Treatments	$k - 1 = 3$	$T_B = 12.160$	4.053	$F = 4.107$
Error	$2(n-1)(k-1) = 474$	$2n(k-1) - T_n = 467.840$	0.987	
Total	$(2n-1)(k-1) = 477$	$2n(k-1) = 480.000$		

(iii) The stated number of error degrees of freedom $2(n-1)(k-1)$ will usually be large, and consequently referral of the mean square ratio to an F distribution will be closely equivalent to referring the accumulation statistic T to a χ^2 distribution with $k-1$ degrees of freedom.

NATURE OF THE ACCUMULATION STATISTIC T

One way to better understand the nature of the accumulation statistic T was noted by Takeuchi and Hirotsu (1982) and is illustrated in Table 7.

From the accumulated frequencies in Table 5 we have constructed three 2×2 tables for each drug. In these tables, $\bar{J}$, $(J = I, II, III)$, means the complement of category J and refers to the number of patients not in category J and hence the number that did better than those in category J. The accumulation statistics T_A and T_B are those which appear in Taguchi' analysis of variance tables (see, e.g., Table 6 for the analysis of drug B). They may be obtained by summing the individual χ^2 values for the separate 2×2 tables in Table 7. The accumulation method does take some account of the ordering of the categories but because the components are not independent these T statistics do not of course follow χ_3^2 distributions as Taguchi supposes, nor is the F distribution appropriate in the analysis of variance of Table 6, for example.

An approximation we gave earlier (Box and Jones, 1986), based on equating the first two moments of the statistic to those of χ^2 (also mentioned by Nair, 1986), yields the result that T is approximately distributed as $g\chi_f^2$, where $f = (k-1)/g$ is the fractional degrees of freedom for chi-square and

$$g = 1 + \frac{2}{k-1} \sum_{i=1}^{k-2} \sum_{j=i+1}^{k-1} r_i/r_j$$

$$r_i = \hat{q}_i/(1 - \hat{q}_i)$$

Table 7. Individual χ^2 Values for Separate 2×2 Tables

	I	$\bar{I}$	II	$\bar{II}$	III	$\bar{III}$
			Drug A			
Control	40	40	64	16	74	6
Treatment	24	56	64	16	74	6
	$\chi^2 = 6.667$		$\chi^2 = 0$		$\chi^2 = 0$	
			$T_A = 6.667$			
			Drug B			
Control	40	40	64	16	74	6
Treatment	24	56	53	27	69	11
	$\chi^2 = 6.667$		$\chi^2 = 3.848$		$\chi^2 = 1.645$	
			$T_B = 12.160$			

and $\hat{q}_i$ are the empirical cumulated marginal probabilities as in Table 5. The accuracy of such approximations has been investigated by Box (1954b) and in this example it can be expected to be quite good. Using it we find that

$$T_A \rightarrow 1.36\chi^2_{2.2}$$

$$T_B \rightarrow 1.43\chi^2_{2.1}$$

where the arrow is used to mean "is distributed as."

For $T_A = 6.67$, $\alpha = 10.2\%$ and for $T_B = 12.16$, $\alpha = 1.6\%$. Thus the accumulation statistic when referred to an appropriate reference distribution takes some account of the ordering but with much less sensitivity than is displayed by the χ^2_1 statistic, or the $F = t^2$ statistic based on equally spaced scores. We show in the next section why this relative lack of sensitivity of the accumulation statistic occurs.

EIGENVALUE–EIGENVECTOR ANALYSIS

Suppose $z = Ty$ is distributed as the multinormal distribution $N(0, \Sigma)$, where T is a $(k - 1) \times (k - 1)$ lower triangular matrix with all non-zero elements equal to unity. Then, by an extension of a theorem due to Cochran (1934), a quadratic form $z'Mz$, where M is a positive semi-definite matrix, can be written as

$$\lambda_1\chi^2_1 + \lambda_2\chi^2_2 + \lambda_3\chi^2_3$$

where $\chi_i = u_i'z$ are independent standard normal deviates ($i = 1, 2, 3$), and the u_i's are the eigenvectors and the λ_i's eigenvalues of the matrix (ΣM)

Now consider a linear combination $l = \sum_{j=1}^k a_j y_j$. Since $\sum_{j=1}^k y_j = 0$, we will without loss of generality take $\sum_{j=1}^k a_j = 0$. Then, taking $\dot{a} = (a_1, a_2, \ldots, a_{k-1})'$, $\dot{y} = (y_1, y_2, \ldots, y_{k-1})'$, and denoting by $\underline{1}$ a $(k - 1)$ dimensional column vector of 1's, we have $-a_k = \underline{1}'\dot{a}$ and $-y_k = \underline{1}'\dot{y}$. Then

$$l = \dot{a}'\dot{y} + \dot{a}'\underline{1}\,\underline{1}'\dot{y} = \dot{a}'(I + \underline{1}\,\underline{1}')T^{-1}z$$

Now let $u' = \dot{a}'(I + \underline{1}\,\underline{1}')T^{-1}$. Then $1 = u'z = u'G$ where $G = T(I + \underline{1}\,\underline{1}')^{-1}$. We therefore have $\chi_i = a_iy(i = 1, 2, 3)$, where the first $(k - 1)$ elements of a_i are the elements of $\dot{a}_i' = u_iG$ and the kth element is chosen so that the elements of a_i sum to zero.

If we apply this for the drug data, then we obtain

$$T_A = 6.67 = 1.82\chi^2_1 + 0.78\chi^2_2 + 0.39\chi^2_3$$

with

$$\chi_1 = -9.9y_1 - 5.2y_2 + 2.4y_3 + 12.8y_4 = 0.94$$
$$\chi_2 = 4.2y_1 - 7.9y_2 - 5.5y_3 + 9.2y_4 = -2.42$$
$$\chi_3 = -0.5y_1 + 5.1y_2 - 13.8y_3 + 9.2y_4 = 1.12$$

and

$$T_B = 12.19 = 1.91\chi_1^2 + 0.72\chi_2^2 + 0.37\chi_3^2$$

with

$$\chi_1 = -9.0y_1 - 4.2y_2 + 2.4y_3 + 10.8y_4 = 2.41$$
$$\chi_2 = 4.3y_1 - 7.3y_2 - 6.2y_3 + 9.2y_4 = -1.22$$
$$\chi_3 = -1.3y_1 + 6.5y_2 - 11.5y_3 + 6.3y_4 = 0.20$$

If equal numbers (i.e., 40 patients) happened to fall in each of the four categories then the accumulation statistic would have been of the form

$$T = 2.00\chi_1^2 + 0.66\chi_2^2 + 0.33\chi_3^2$$

with

$$\chi_1 = -8.5y_1 - 2.8y_2 + 2.8y_3 + 8.5y_4$$
$$\chi_2 = 6.3y_1 - 6.3y_2 - 6.3y_3 + 6.3y_4$$
$$\chi_3 = -2.8y_1 + 8.5y_2 - 8.5y_3 + 2.8y_4$$

In this last case the coefficients in the linear aggregates are seen to be proportional to $(-3, -1, 1, 3)$, $(1, -1, -1, 1)$, and $(-1, 3, -3, 1)$; the *linear, quadratic,* and *cubic orthogonal* polynomials. Note that the coefficients in T_A and T_B also approximate these same functions.

The situation can be clarified by considering further the case where there are equal numbers of patients that fall in each of the categories. In this case we see that the weighting of the associated chi-square distributions will be as shown in Table 8.

Table 8. Weighting of the χ^2 Distributions

	χ_L^2	χ_Q^2	χ_C^2		
Pearson χ^2	1	1	1	→	χ_3^2
Accumulation statistic	1.33	0.44	0.22	→	χ_2^2
χ_1^2	1	0	0	→	χ_1^2

While it is true that the accumulation statistic places greater emphasis on the linear component, it contaminates this component with elements from the quadratic and cubic effects in an arbitrary manner. A more powerful statistic, taking account of the desired categorization, will always be provided by χ_1^2.

THE EQUIVALENCE OF THE LINEAR COMPONENT OF χ^2 AND THE t^2 SCORE STATISTIC

We saw earlier that the computation of the linear component, χ_1^2, of χ^2, and the calculation of the t statistic by analysing equally spaced scores, yielded almost the same result. This is interesting and perhaps surprising because at first sight it might be supposed that the violation of assumptions involved in the calculation of t^2 directly from the scored data are much more serious than are those involved in computing the linear component of χ^2. That the two procedures are in fact equivalent was shown by Yates (1948) and his result was extended by Cochran (1954). We will not rederive these results, but rather consider their implications for the Taguchi data.

A convenient way to compute the χ^2 statistic and to break it up into its components is due to Cochran (1950). Using Cochran's method, patients from the control and treatment groups would be indicated by two levels of a suitable indicator variable and these values of the indicator variable are treated as the basic data on which a standard analysis, such as a t analysis or an analysis of variance, is performed. This procedure has recently been referred to as "Taguchi's" minute analysis. Cochran shows that it provides an approximately valid analysis for independent data but this of course does not apply to Taguchi's application to accumulated data which are not independent.

For the drug data let us employ the indicator variables -1 for a patient in the control group and $+1$ for a patient in the treatment group. Shown in Table 9 is an analysis of variance, treating the -1's and $+1$'s as data. The total sum of squares will obviously be fixed and equal to the total number of patients, namely, 160. The sum of squares for regression yields the overall χ^2 statistic based on three degrees of freedom. As expected, the values of 8.000 and 7.327 are identical to those already given in Table 3. In the analysis of variance table the number of degrees of freedom is large and the residual mean square of $(160 - \chi^2)/(160 - 4)$ will be close to one, so that the formal F test will be closely equivalent to simply referring the regression sum of squares to a χ^2 distribution with three degrees of freedom. [It was indeed argued by Mantel (1963) that the F approximation is slightly more accurate.]

The linear, quadratic, and cubic components of χ^2 can be determined by defining three contrasts with levels of $-3, -1, 1, 3$; $1, -1, -1, 1$; and $-1, 3, -3, 1$ corresponding to the four categories $-, +, ++, +++$. The drug data give the ANOVA tables shown in Table 9 with the sums of squares for the linear components reproducing the linear chi-square components (1.977 and 6.485) obtained above, in Table 4.

Now compare this analysis of variance with that shown in Table 2 for equally spaced scores, 0, 1, 2, 3. The $F = t^2$ values obtained earlier from the scores are

Table 9. Analysis of Variance Giving the χ^2 Components for the Drug Data

		df	Sum of Squares		Mean Square
			Drug A		
	Linear	1		1.977	1.977
Between groups	Quadratic	3	1	8.000	4.573	2.667	4.573
	Cubic	1		1.450	1.450
Within groups		156	152.000		0.974
Total		159	160.000		
			Drug B		
	Linear	1		6.485	6.485
Between groups	Quadratic	3	1	7.327	0.811	2.442	0.811
	Cubic	1		0.031	0.031
Within groups		156	152.673		0.979
Total		159	160.000		

precisely what we would get, if in the analyses of Table 9, we compared the linear component with an error sum of squares which included the two degrees of freedom for the quadratic and cubic components. Specifically, for drug A

$$F = \frac{1.977}{(152.000 \mid 4.573 \mid 1.450)/158} = 1.977$$

and for drug B

$$F = \frac{6.485}{(152.673 + 0.811 + 0.031)/158} = 6.675$$

We see, therefore, that the $F = t^2$ statistic based on the equally spaced scores, and the linear component for chi-square are identical. This is in spite of the fact that they have somewhat different justifications and at first sight the necessary assumptions for the χ^2 tests look much less demanding than those for the t^2. The point that we wish to stress is that the linear component of overall chi-square, χ_1^2, which extracts the component of interest more efficiently and simply than accumulation analysis, is precisely what we get from a straightforward analysis of variance or t analysis made on the *equally spaced scores*.

CHOICE OF SCORING PROCEDURE

To investigate the sensitivity of the analysis to the method of assigning scores we have repeated the analysis using different scoring schemes. Table 10 shows the t statistics and the significance levels, α, for three different sets of scores. We see that the conclusions are essentially unchanged even though different scoring schemes are

Table 10. *t* Statistics for Different Scoring Schemes

Score				Drug A		Drug B	
−	+	++	+++	*t*	α	*t*	α
0	1	2	3	1.406	16.2%	2.584	1.1%
0	1	2	4	1.154	25.0%	2.407	1.7%
0	1	2	5	0.960	33.9%	2.246	2.6%

Table 11. The Scores for Drug B from Fisher's Method

	Drug B			
Class	−	+	++	+++
Score	0	0.63	0.88	1

used in the analysis. Notice too that if they *did* change this would point to the need either to get more data or for the experts to make up their minds about the appropriateness of alternative scoring schemes. The eigenvalue–eigenvector analysis makes clear that such questions cannot be resolved by the use of accumulation analysis or, for that matter, of any other purely statistical device.

Fisher proposed a discriminant method for automatically assigning scores for his agglutination data. The method is equivalent to choosing scores which maximize the F value in an analysis of variance table, such as Table 2. He also proposed a test for whether the discriminant scores differed significantly from equally spaced scores. If we apply his technique to the data for drug B, for example, we obtain the scores shown in Table 11.

Results using this scoring system are not very different from those using the equally spaced scores, nor, using Fisher's test, are the discriminant scores significantly different from the equally spaced scores. The logic of the discriminant method, however, seems somewhat dubious to us and we would prefer scores based on human judgment.

AN EXAMPLE FROM AN EXPERIMENT ON THE SEALING PROCESS OF A LIGHT BULB MANUFACTURE

In this section we analyse a set of ordered categorical data obtained from an experiment conducted at the Mysore Electric Bulb Plant in Mysore, India (see Taguchi, 1987, pp. 463–469). Taguchi analyzed the data by the method of accumulation analysis. In this section we show the data analysed by assigning scores and employing a standard analysis.

The objective of the investigation was to decrease the number of defective light bulbs by conducting a fractional factorial experiment on the sealing machine

Table 12. Factors for the Light Bulb Experiment

Factors	−	+
A: Diameter of stem	Large	Small
B: Speed of sealing machine	4.80 rpm	5.40 rpm
C: Gas quantity at first stage	Standard	Higher
D: Air quantity at first stage	D_1	D_2
E: Gas quantity at second stage	Standard	Higher
F: Air quantity at second stage	F_1	F_2
G: Gas quantity at third stage	Standard	Higher
H: Air quantity at third stage	H_1	H_2
I: Gas quantity at fourth stage	Standard	Higher
J: Air quantity at fourth stage	J_1	J_2

process. Prior to the experiment, the process of sealing the stem into the glass bulb was causing up to 15% of the bulbs manufactured to be defective in shape. The factors of interest, and their levels, are given in Table 12.

It was assumed that two interactions, G × H and I × J, were most likely to occur, and provision was made to estimate these on the assumption that the remaining 43 two-factor interactions were negligible. Ten electric bulbs were produced in each of the 16 runs of the experiment, each bulb classified as defective (B), not defective but with poor shape (T), and non-defective (G). Table 13 gives the number of bulbs from

Table 13. Design, Data, and Effects for Light Bulb Experiment

A	B	C	D	E	F	G	H	I	J	G×H	I×J	B	T	G	Score
−	−	−	−	−	−	−	−	−	−	−	+	1	4	5	14
−	−	+	+	+	+	−	−	−	+	−	+	0	4	6	16
+	+	−	−	+	+	−	−	+	−	−	+	2	4	4	12
+	+	+	+	−	−	−	−	+	+	−	−	2	5	3	11
+	+	+	+	−	+	−	+	−	−	+	−	5	3	2	7
+	+	−	−	+	−	−	+	−	+	+	+	1	3	6	15
−	−	+	+	+	−	−	+	+	−	+	+	7	0	3	6
−	−	−	−	+	−	+	+	+	+	+	−	6	2	2	6
−	+	−	+	+	−	+	−	−	−	+	−	2	4	4	12
−	+	+	−	−	+	+	−	−	+	+	+	0	5	5	15
+	−	−	+	−	+	+	−	+	−	+	+	2	4	4	12
+	−	+	−	+	−	+	−	+	+	+	−	4	3	3	9
+	−	+	−	+	+	+	+	−	−	−	−	4	4	2	8
+	−	−	+	−	−	+	+	−	+	−	+	0	2	8	18
−	+	+	−	−	−	+	+	+	−	−	+	5	3	2	7
−	+	−	+	+	+	+	+	+	+	−	−	7	1	2	5
Effects 1.4	−0.6	−1.9	0.1	−0.9	−1.4	−0.1	−3.4	−4.6	2.1	−1.1	3.6				

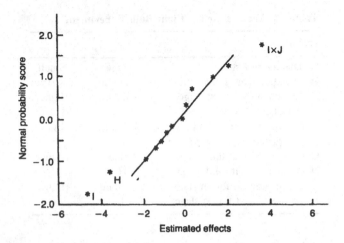

Figure 1. Normal probability plot of effects.

the sample of ten that falls into each of these categories. Taguchi analyzed these data using an accumulation analysis of considerable complexity, leading to an analysis of variance for the 160 light bulbs which resulted in a series of F tests for the factors each based on 2 and 288 degrees of freedom. The largest F values, all declared significant at the 1% level, were for factors H, I, and I × J.

For the sake of comparison we have adopted Taguchi's assumptions concerning possible interactions but have simply assigned scores of 0, 1, and 2 to the categories B, T, and G and calculated the total scores shown in the last column of Table 13. The bottom line of the table gives the estimates of the effects from this experiment, using as the response the total score. A normal probability plot of the effects is shown in Figure 1.

The use of the accumulation statistic correctly emphasizes the linear component but unnecessarily contaminates it with components of the quadratic and higher terms. The incorrect analysis of variance associated with it then overestimates the significance of the various effects. In the present example the effects are sufficiently strong to survive this process and the H, I, and I × J effects are also indicated as distinguished from the noise in our analysis.

CONCLUSIONS

We believe that the accumulation analysis is not a useful technique and to teach it to engineers would be a serious mistake. Also the distribution proposed for the accumulation statistic is incorrect. When compared with much simpler directional procedures, accumulation analysis does poorly. Although it places extra emphasis on the linear χ^2 component, it dilutes that component with unnecessary noise.

For ordered categorical data, we believe that the engineer should use his expertise to devise a simple scoring system and then analyse his results in the usual way.

ACKNOWLEDGMENTS

Sponsored by the National Science Foundation Grant No. DDM-8808138 and by the Vilas Trust of the University of Wisconsin-Madison; and was aided by access to the Statistics Department Research Computer at the University of Wisconsin-Madison.

CHAPTER E.9

Signal-to-Noise Ratios, Performance Criteria, and Transformations

For the analysis of designed experiments, Taguchi uses performance criteria that he calls signal-to-noise (SN) ratios. Three such criteria are here denoted by SN_T, SN_L, and SN_S. The criterion SN_T was to be used in preference to the standard deviation for the problem of achieving, for some quality characteristic y, the smallest mean squared error about an operating target value. León, Shoemaker, and Kackar (1987) showed how SN_T was appropriate to solve this problem only when σ_y was proportional to μ_y. On that assumption, the same result could be obtained more simply by conducting the analysis in terms of log y rather than y. A more general transformation approach is here introduced for other, commonly met kinds of dependence between σ_y and μ_y (including no dependence), and a "lambda plot" is presented that uses the data to suggest an appropriate transformation. The criteria SN_L and SN_S were for problems in which the objective was to make the response as large or as small as possible. It is argued here that these predecided "portmanteau" criteria provide an inadequate summary of data and that, regarded as measures of location, they can be extremely inefficient. In preference to such performance criteria, the merits of data analysis that can uncover information both expected and unexpected are urged. A reanalysis of an interesting experiment due to Quinlan (1985) illustrates the value of this approach and its contribution to the art of discovery. It is argued that improvement of quality will best be catalyzed by engineers using data analysis with computer graphics rather than by those trained only to employ more rigid predecided criteria.

From Box, G. E. P. (1988), *Technometrics*, **30**(1), 1–17. Reprinted with permission from TECHNO-METRICS. Copyright © 1988 by the American Statistical Association. All rights reserved.

INTRODUCTION

Three problems referred to by Taguchi (Taguchi, 1986; Taguchi and Phadke, 1984; Taguchi and Wu, 1985) under the general title of *parameter design* employ experimental design (a) to achieve minimum dispersion of some quality characteristic with its location adjusted to some desired target value, (b) to minimize product sensitivity to variation transmitted from components, and (c) to minimize product sensitivity to environmental fluctuations.

These may be called, respectively, problems of closeness to target, robustness to transmitted variation, and robustness to environmental variation. These are important problems but the statistical methods of design and analysis recommended by Taguchi are often unnecessarily inefficient and complicated and, where so, should be replaced by simpler and more efficient alternatives (see, e.g., Box and Bisgaard, 1987; Box and Fung, 1986; Box and Jones, 1986a,b; Box and Meyer, 1986a,b; Gunter, 1987; Hunter, 1987; Nair, 1986; Nair and Pregibon, 1986, 1987).

The present article and earlier work by Box (1986b) and Nair and Pregibon (1986) consider problem (a). Problems in which the largest or the smallest response is being sought are discussed later. (See also Box and Ramírez, 1986.) In the analysis of such problems, Taguchi employed performance criteria that he called *signal-to-noise (SN) ratios*. The extent to which maximization of such criteria can be linked with minimization of quadratic loss was considered by León, Shoemaker, and Kackar (1987). Typically in such problems, the data analyzed have arisen from replicated orthogonal arrays and, in particular, from replicated fractional factorials (see, e.g., Phadke, Kackar, Speeney, and Grieco, 1983; Quinlan, 1985). The discussion that follows would also apply, however, in the analysis of any design in which the analysis of dispersion as well as location was important.

I consider here three criteria, which I denote by $\{SN_T\}$, $\{SN_L\}$, and $\{SN_S\}$. The criterion

$$\{SN_T\} = 10 \log_{10}(\bar{y}^2/s^2)$$

with

$$\bar{y} = \frac{1}{n}\sum_{i=1}^{n} y_i$$

and

$$s = \left\{\frac{1}{n-1}\sum_{i=1}^{n}(y_i - \bar{y})^2\right\}^{1/2}$$

is employed when the objective is *closeness to target*. The criterion

$$\{SN_L\} = -10\log_{10}\left[\frac{1}{n}\sum_{i=1}^{n}\left(\frac{1}{y_i}\right)^2\right]$$

is employed when the objective is to make the response as *large as possible*. The criterion

$$\{SN_S\} = -10 \log_{10}\left[\frac{1}{n}\sum_{i=1}^{n} y_i^2\right]$$

is employed when the objective is to make the response as *small as possible*.

THE CRITERION SN$_T$

In all that follows, it is important to distinguish between (a) the selection of an appropriate performance criterion that may be defined in terms of parameters such as μ and σ, and (b) the efficient estimation of such criteria in terms of sample quantities such as $\bar{y}$ and s.

We first consider, therefore, the motivation for a performance criterion

$$SN_T = 10 \log_{10}(\mu^2/\sigma^2) = -10 \log_{10} \gamma^2$$

where γ is the coefficient of variation σ/μ. I later consider the corresponding sample criterion $\{SN_T\}$.

In explaining the use of $\{SN\}$, Phadke (1982) said

> Why do we work in terms of the S/N ratio rather than the standard deviation? Frequently, as the mean decreases the standard deviation also decreases and vice versa. In such cases, if we work in terms of the standard deviation, the optimization cannot be done in two steps i.e., we cannot minimize the standard deviation first and then bring the mean on target. (p. 13)

He later explained,

> Among many applications, Professor Taguchi has empirically found that the two stage optimization procedure involving the S/N ratio indeed gives the parameter level combination where the standard deviation is minimum while keeping the mean on target. This implies that the engineering systems behave in such a way that the manipulatable production factors can be divided into three categories:
>
> 1) control factors, which affect process variability as measured by the S/N ratio,
> 2) signal factors, which do not influence (or have negligible effect on) the S/N ratio but have a significant effect on the mean, and
> 3) factors which do not affect the S/N ratio or the process mean. (p. 13)

The idea is carried further in the article by León et al. (1987). In their terminology, given that the preceding division of the factors can be achieved, the SN ratio would be an example of a performance measure independent of adjustment (PerMIA). The signal factors would be called *adjustment factors*, and the control and signal factors together would be *design factors*.

In this context, consider the following argument: Suppose that we wish to minimize the mean squared error (MSE) $M = E\{y - T\}^2$ of some quality characteristic y about a target T with respect to design factors (variables) $\mathbf{x}$. Equivalently this would minimize *expected loss* if an appropriate quadratic loss function is assumed.

In general, both $\mu = E(y)$ and $\sigma^2 = E(y - \mu)^2$ will depend on $\mathbf{x}$, and for fixed $\mathbf{x}$ we may write

$$M(\mathbf{x}) = E(y(\mathbf{x}) - T)^2 = \sigma^2(\mathbf{x}) + (\mu(\mathbf{x}) - T)^2 \qquad (1)$$

A very general way to proceed would be to regard $M(\mathbf{x})$ (or preferably its logarithm) as a response and to conduct designed experiments (e.g., to allow first-order steepest ascent followed by second-order exploration) to find conditions $\mathbf{x}_0$, which gave a response close to the minimum (see, e.g., Box and Draper, 1987; Box and Fung, 1986; Box and Wilson, 1951).

I shall not pursue that possibility here but, following Phadke (1982), suppose that the system has certain special characteristics that can be used to simplify the problem. The general idea is to get rid of dispersion effects (see e.g., Box and Meyer 1986b) that arise only because of dependence of σ on μ. Specifically I assume that the standard deviation and the mean are linked in a manner such that a function $f[\mu(\mathbf{x})]$ can be found for which $\sigma^2(\mathbf{x})/\{f[\mu(\mathbf{x})]\}^2$ is a measure of dispersion $P(\mathbf{x}_1)$, which is a function of only a subset $\mathbf{x}_1$ of the design variables $\mathbf{x} = (\mathbf{x}_1, \mathbf{x}_2)$. Then $P(\mathbf{x}_1)$ is independent of μ, because for given $\mathbf{x}_1$, μ is a function of $\mathbf{x}_2$ only. But P is not a function of $\mathbf{x}_2$, and consequently $P(\mathbf{x}_1|\mu) - P(\mathbf{x}_1)$. Equivalently, if dispersion is measured in terms of $P(\mathbf{x}_1)$, then only a subset of the design factors $\mathbf{x}$ will have dispersion effects, $P(\mathbf{x}_1)$ will be a PerMIA, and $\mathbf{x}_2$ will be a vector of adjustment factors that can be changed without changing dispersion.

On these assumptions

$$M(\mathbf{x}) = \{f[\mu(\mathbf{x})]\}^2 P(\mathbf{x}_1) + (\mu(\mathbf{x}) - T)^2 \qquad (2)$$

and for fixed μ, $M(\mathbf{x})$ is minimized with respect to $\mathbf{x}_1$ when $P(\mathbf{x}_1)$ is minimized. Suppose now that $P(\mathbf{x}_1)$ is uniquely minimized when $\mathbf{x}_1 = \mathbf{x}_{10}$, an absolute minimum for $M(\mathbf{x})$ may be found if, by changing only $\mathbf{x}_2$, μ can be adjusted to its minimizing value μ_0, where

$$\mu_0 = T - f(\mu_0) f'(\mu_0) P \qquad (3)$$

Notice that in the preceding and in what follows an important further assumption needs to be made—namely, that a point on the hyperplane $\mathbf{x}_1 = \mathbf{x}_{10}$ exists that minimizes $M(\mathbf{x})$ with respect to μ within the region $R(\mathbf{x})$ of interest. With this proviso, $M(\mathbf{x}) = E(y(\mathbf{x}) - T)^2$ may be minimized in two stages as follows:

1. Adjust $\mathbf{x}_1$ to minimize $P(\mathbf{x}_1)$ or some monotonic function of $P(\mathbf{x}_1)$.
2. Adjust $\mathbf{x}_2$ so that μ_0 satisfies the equality $\mu_0 = T - f(\mu_0) f'(\mu_0) P$.

The quantity $f(\mu_0)f'(\mu_0)P$ measures the distance of μ_0 from the target T. This is usually small, and I shall call it the *aim-off factor*. For illustration, I consider some special cases assuming that the assumptions made previously are satisfied.

$f(\mu) = \mu^\alpha$. In this important special case,

$$P(\mathbf{x}_1) = \sigma^2/\mu^{2\alpha} \tag{4}$$

and the mean should be adjusted to the value

$$\mu_0 = T/(1 + \alpha\gamma_0^2) \tag{5}$$

where $\gamma_0 = $ coefficient of variation at the minimizing value.

$\alpha = 0$. If $\alpha = 0$, then σ is not a function of μ and $P(\mathbf{x}_1) = \sigma^2(\mathbf{x}_1)$ and $\mu_0 = T$. As is otherwise obvious, $M(\mathbf{x})$ is then minimized by first adjusting $\mathbf{x}_1$ to minimize $\sigma(\mathbf{x}_1)$ and then using $\mathbf{x}_2$ to adjust μ to the target value $\mu_0 = T$. In this case there is a zero aim-off factor and the PerMIA is simply the variance σ^2.

$\alpha = 1$. This is the special case that would partially justify Phadke's discussion (see also Nair and Pregibon, 1986, 1987). Here $P(\mathbf{x}_1) = \sigma^2/\mu^2 = \gamma^2$ and, on the assumptions previously made, minimizing P is the same as maximizing the SN ratio μ^2/σ^2 or, equivalently, SN_T. An overall MSE may then be obtained by setting P to its minimizing value $P(\mathbf{x}_{10}) = \gamma_0^2$ or SN_T to the value $-10\log_{10}\gamma_0$. Strictly speaking, however, the adjustment variables $\mathbf{x}_2$ should then be employed to move the mean μ not to the target T but to the minimizing value

$$\mu_0 = T/(1 + \gamma_0^2) \tag{6}$$

A Graphical Illustration

The general situation is clarified by considering the contour diagrams of Figure 1. Suppose that μ and σ depend on x_1 and x_2 in the manner indicated in Figure 1(a). Then for the target value $T = 10$, Figure 1(b) shows the contours of the MSE $M(x_1, x_2) = E(y - 10)^2 = \sigma^2 + (\mu - 10)^2$, showing a minimum at the point Q. The diagrams for this example were actually generated by supposing that there is a PerMIA $P = (\sigma/\mu^2)^2$ which we suppose is a function of x_1 only as illustrated in Figure 1(c). [All of the other diagrams were in fact generated from Figure 1(c).] The supposition is, therefore, that the unnecessary dispersion effects associated with x_2 in Figure 1(d), as well as 1(a), arise because the standard deviation is proportional not to the mean but to its square.

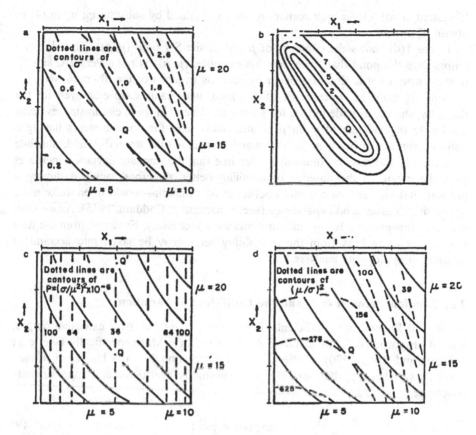

Figure 1. Simplification induced only by employment of the current PerMIA of μ^2: (a) contours of μ and σ for two design variables, x_1 and x_2; (b) a plot of mean squared error for the same situation showing a minimum at Q; (c) contours of μ and of the PerMIA $P = (\sigma/\mu^2)^2$ for the same situation (note that x_2 is now an adjustment factor); (d) contours of μ and of the signal-to-noise ratio $(\mu/\sigma)^2$.

Figure 1(c) illustrates a situation in which, by adjusting x_1 alone, $P(x_1) = (\sigma/\mu^2)^2$ can be brought to its minimizing value of 36×10^{-6}. The adjustment factor x_2 may now be manipulated to bring μ to the value

$$\mu_0 = T/(1 + 2P\mu_0^2) \tag{7}$$

This equation may be solved approximately by setting $\mu_0 = T$ in the denominator on the right side. We then find

$$\mu_0' = \frac{T}{1 + 2PT^2} = \frac{10}{1 + 2 \times (36 \times 10^{-6}) \times 100}$$

$$= \frac{10}{1.0072} = 9.93 \tag{8}$$

If desired, a still closer approximation can be obtained by substituting μ_0' in (7) to obtain μ_0'' and so on.

Figure 1(d) shows the contours of μ and of the SN ratio $(\mu/\sigma)^2$. This diagram emphasizes the point that since, for this example, $(\mu/\sigma)^2$ is not a PerMIA, simplification would *not* in this case occur by analysis in terms of the SN ratio SN_T.

Now typically, in experiments of this kind, wide ranges are employed for the factors **x**; therefore, wide ranges for y are quite likely. In these circumstances some detectable dependence of σ on μ is also likely, and for measurements having a natural origin at 0 (such as height, weight, area, tensile strength, yield, absolute temperature, and reaction time) it is often true that the standard deviation increases with the mean. In the absence of anything better, therefore, tacitly assuming a proportional increase—a constant coefficient of variation—would often make more sense than to assume no dependence (see, in particular, Gaddum, 1945). As we shall see later, however, such rigid assumptions are unnecessary. Evidence from the data themselves as well as from the probability setup may be taken into account in making an appropriate analysis.

The Relation Between $\sigma_{\ln y}$ and the Coefficient of Variation

As is well known, the coefficient of variation $\gamma = \sigma/\mu$ is, over extensive ranges, almost proportional to the standard deviation of $\ln y$. More specifically (see, e.g., Johnson and Kotz, 1970), if after a log transformation $Y = \ln y$ has mean μ_Y and constant variance σ_Y^2, then, exactly if Y is normally distributed and approximately otherwise,

$$\mu = \exp(\mu_Y + \tfrac{1}{2}\sigma_Y^2) \tag{9}$$

$$\sigma = \mu\{\exp(\sigma_Y^2) - 1\}^{1/2} \tag{10}$$

Consider Equation (10). We see that σ, the standard deviation of y, is exactly proportional to its mean μ or, equivalently, that the coefficient of variation $\sigma/\mu = \{\exp(\sigma_Y^2) - 1\}^{1/2}$ is independent of μ. Moreover, σ/μ is the monotonic function of σ_Y sketched in Figure 2. Thus analysis in terms of σ/μ (or of the SN ratio μ/σ) is essentially equivalent to the analysis of σ_Y—that is, of $\sigma_{\ln y}$. In addition, although the argument is conducted here in terms of natural logarithms indicated by the notation ln, the conclusions will apply equally for logarithms taken to base 10 indicated by the notation log.

In practice σ_Y will be replaced by the estimate $s_Y = s_{\ln y}$, which itself has a standard deviation that is proportional to its mean. (See, e.g., Bartlett and Kendall, 1946.) Thus *on the hypothesis that a log transformation will stabilize the variance* we are led to an analysis of $\ln s_{\ln y}$. This is almost equivalent to $\{SN_T\}$, since

$$\{SN_T\} = -10\log(s/\bar{y})^2 \doteq \text{const} - 20\log s_{\log y}$$

$$= \text{const} - \text{const}\ln s_{\ln y} \tag{11}$$

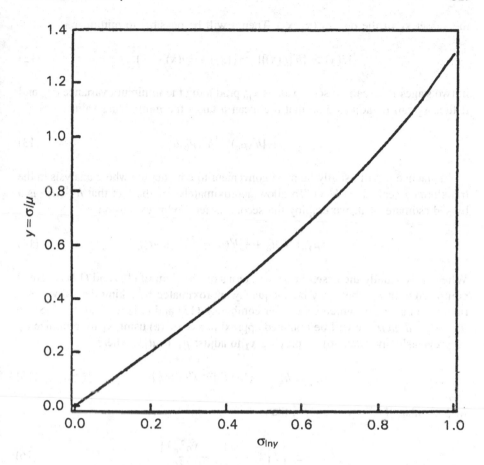

Figure 2. The relation between $\sigma_{\ln y}$ and the coefficient of variation σ/μ when $\ln y$ is normally distributed.

Relevance of General Transformation of y

Now we know both from practical experience and theoretical considerations that kinds of dependence between σ and μ other than proportionality (including absence of any dependence) are common and that, correspondingly, kinds of transformations other than the log, and sometimes no transformation, can result in variance stabilization. Both theoretical evidence for the probability setup (e.g., that y is a binomial variate) and empirical evidence from the data may be used to find an appropriate transformation. I now, therefore, broaden the concept discussed previously by considering variance stabilizing transformations other than the log (see also Nair and Pregibon, 1986).

Suppose that $Y = h(y)$ is a variance stabilizing transformation such that $\sigma_Y \simeq h'(\mu)\sigma_y$ is independent of μ, and suppose further that σ_Y^2 is a function of only the subset x_1 of the design (x_1, x_2). Then it will be possible to minimize

$$M(x) \simeq \{h'[\mu(x)]\}^{-2}\sigma_Y^2(x_1) + (\mu(x) - T)^2 \tag{12}$$

in two stages if x_1 can be set to values x_{10} producing the minimum variance σ_{Y0}^2 and if then x_2 can be adjusted so that the mean μ takes the minimizing value

$$\mu_0 = T + \{h'(\mu_0)\}^{-3}h''(\mu_0)\sigma_{Y0}^2 \tag{13}$$

In practice it will usually be most convenient to conduct the whole analysis in the transformed scale $Y = h(y)$. To allow approximately for the fact that $h^{-1}(\mu_Y)$ is a biased estimate of μ, we employ the second-order Taylor expansion

$$h^{-1}(\mu_{Y0}) \approx \mu_0 + \tfrac{1}{2}(h'(\mu_0))^{-3}h''(\mu_0)\sigma_{Y0}^2 \tag{14}$$

When, as is usually the case, the second terms on the right of (13) and (14) are small compared with μ_0, they may be adequately approximated by taking derivatives at T rather than at μ_0, in which case, after combining (13) and (14) we find that MSE *in the original metric y* will be obtained approximately by (a) using x_1 to minimize σ_Y as previously and then (b) employing x_2 to adjust μ_Y to μ_{Y0}, where

$$\mu_{Y0} = h\{T + \tfrac{3}{2}[h'(T)]^{-3}h''(T)\sigma_Y^2\} \tag{15}$$

In particular, if $Y = y^\lambda$ $(\lambda \neq 0)$, then

$$\mu_{Y0} \simeq \left\{T\left[1 - \frac{3(1-\lambda)}{2}\frac{\sigma_{Y0}^2}{\lambda^2 T^{2\lambda}}\right]\right\}^\lambda \tag{16}$$

and if $Y = \ln y$, then

$$\mu_{Y0} \simeq \ln T - \tfrac{3}{2}\sigma_{Y0}^2 \tag{17}$$

Adequacy of Approximation and Size of Aim-Off Factor

Some idea of the adequacy of the approximation that I am employing can be gained by considering the case in which the log transformation $Y = \ln y$ induces normality, for in that case the preceding minimization procedure is exact. To see this, note that, using Equation (10), the MSE in the original untransformed metric can be written exactly as

$$M(x) = \mu(x)^2\{\exp[\sigma_Y^2(x_1)] - 1\} + (\mu(x) - T)^2 \tag{18}$$

which is minimized with respect to x_1 by minimizing $\sigma_Y^2(x_1)$.

The overall minimum is now obtained if $\mathbf{x}_2$ can be adjusted so that μ has the minimizing value

$$\ln \mu_0 = \ln T - \sigma_y^2 \tag{19}$$

From (9) the exact relation to allow for "bias" of the transformed mean is

$$\mu_Y = \ln \mu - \tfrac{1}{2}\sigma_Y^2 \tag{20}$$

and, combining (19) and (20), I now obtain (17) exactly.

From this example we can also gain some appreciation of the size of the adjustments implied by (19), (20), and (17). Suppose we had a process in which the quality characteristic y had a coefficient of variation as large as 10%. Then from (19) the aim-off factor would correspond to a 1.0% reduction from target. The bias adjustment of (20) would account for a further 0.5% reduction. Thus, if we are conducting the whole analysis in the log metric [Eq. (17)], a total reduction of 1.5% from the target value would be called for.

AN EMPIRICAL PROCEDURE FOR DETERMINING A SIMPLIFYING TRANSFORMATION

León et al. (1987) showed how, when an appropriate mechanistic model for the system is known, it may be possible to derive a PerMIA mathematically. Frequently the necessary mechanistic information about the model will be lacking, however, and an empirical approach is needed.

Suppose for a particular set of experimental data there exists some monotonic transformation $Y = f(y)$, in terms of which a small set of active design variables occurs and within which there is a subset of adjustment variables. Then it is easy to see that, in some other metric, complications will occur. For example, linear models in the preferred metric will transform to nonlinear models in the alternative metric. Moreover, factors that in the optimal transformation affect location but not dispersion will in the alternative metric affect both. I shall refer to the elimination of unnecessary complication in the model form as the achievement of *parsimony* and to the elimination of transformable dependence between the mean and the standard deviation as the achievement of *separation* (or the elimination of "crosstalk" between location and dispersion effects).

A way of finding, when it exists, a transformation that achieves maximum simplicity in terms of parsimony and separation was illustrated by Box and Cox (1964, Fig. 8) and Box and Fung (1983, Fig. 2). Suppose that a class of transformations $y^{(\lambda)}$ (not necessarily power transformations) is indexed by a parameter λ. (We suppose here that λ is a scalar, but the idea could be extended to multiple parameter transformations.) Then, both for the dispersion effects and the location effects, suitable relevant statistics such as t values and F values may be plotted against the value of λ, as an aid in selecting the transformation yielding maximum simplification and separation.

The efficiency of usual analyses in terms of F and t statistics and the analysis of variance (ANOVA) is enhanced by using a data transformation for which simplicity of structure, constancy of error variance, normality, and independence of error distributions are approximately achieved. The first two of the preceding character- istics are also those required for parsimony and separation. Thus not infrequently that the metric in which the optimization discussed previously should be carried out is also a desirable metric for analysis and estimation.

A Constructed Example

Ideas are clarified by the following constructed example for eight factors A, B, C, D, E, F, G, H tested in a 2_{IV}^{8-4} fractional factorial with four replicates at each of the 16 factor combinations. It was supposed that a simple representation was possible in the reciprocal metric y^{-1} and that *in this metric* the model was such that (a) the mean was 50, (b) only three location main effects $(B = -30, D = +20, G = -20)$ occurred, and (c) a single dispersion main effect occurred for factor D so that $\sigma = 7.5$ at the plus level of D and $\sigma = 2.5$ at the minus level.

Using (a) and (b), mean values were obtained for each of the 16 factor combi- nations, and normal random deviates with the appropriate standard deviations were added to obtain the four replicates. These were then untransformed to obtain the "original data" shown in Table 1; also shown for each run are values of the sample averages $\bar{y}$ and of the logarithms of the standard deviations.

Table 1. "Original Untransformed" Data ×100 for a 2_{IV}^{8-4} Design in Four Replicates

A	B	C	D	E	F	G	H	y_1	y_2	y_3	y_4	$\bar{y}$	$\ln s$
-1	-1	-1	-1	-1	-1	-1	-1	2.50	2.85	2.80	2.92	2.77	-1.69
1	-1	-1	-1	1	1	1	-1	1.83	1.87	1.87	1.70	1.82	-2.52
-1	1	-1	-1	1	1	-1	1	1.55	1.56	1.64	1.56	1.58	-3.17
1	1	-1	-1	-1	-1	1	1	1.12	1.14	1.23	1.18	1.17	-3.08
-1	-1	1	-1	1	-1	1	1	1.67	1.65	1.83	1.89	1.76	-2.13
1	-1	1	-1	-1	1	-1	1	2.79	2.75	2.95	3.18	2.92	-1.63
-1	1	1	-1	-1	1	1	-1	1.15	1.19	1.18	1.16	1.17	-3.98
1	1	1	-1	1	-1	-1	-1	1.55	1.52	1.62	1.66	1.59	-2.78
-1	-1	-1	1	-1	1	1	1	2.95	4.05	2.73	2.13	2.96	-0.22
1	-1	-1	1	1	-1	-1	1	9.41	4.37	5.06	4.20	5.76	0.90
-1	1	-1	1	1	-1	1	-1	1.38	1.88	2.05	1.54	1.71	-1.17
1	1	-1	1	-1	1	-1	-1	2.14	2.79	2.65	1.85	2.36	-0.82
-1	-1	1	1	1	1	-1	-1	7.48	5.79	3.55	13.63	7.61	1.46
1	-1	1	1	-1	-1	1	-1	3.13	1.98	2.24	3.14	2.62	-0.50
-1	1	1	1	-1	-1	-1	1	2.48	1.87	2.92	2.21	2.37	-0.82
1	1	1	1	1	1	1	1	2.00	1.42	1.36	1.23	1.50	-1.08

Using these generated data, t values were then calculated and are plotted in Figure 3 for various values of λ. The necessary standard errors for the location effects were calculated using the within-runs sums of squares based on $3 \times 16 = 48$ df. Standard errors for the dispersion effects were obtained from the approximation $\sigma_{\ln s} \approx [1/(2(n-1))]$.

As expected, in the reciprocal metric the location effects B, D, and G and the single dispersion effect D are the only effects distinguishable from noise. As we move away from the value $\lambda = -1$, however, two things happen:

1. Induced nonlinearities produce interactions between B, D, and G. They show up in the contrasts associated with the alias strings BD $(+AF + CH + EG)$; DG $(+AC + DE + FH)$; BG $(+AH + CF + DE)$ labeled BD, DG, and BG in the location-effects plot.

2. Induced dependence of the standard deviation on the mean produces dispersion effects for B and G and magnifies the dispersion effect D as λ increases from -1 in the dispersion-effects plot.

If these were real data, these λ plots would show that the reciprocal transformation yielded greatest separation and simplification and the analysis would, therefore, have been performed in that metric. Factor D would be set at its lower level to achieve smallest variance and then, depending on technical convenience, either or both of factors B and G would be used to adjust the mean of the transformed data to the value given by Equation (16) with $\lambda = -1$.

To illustrate ideas, I have deliberately generated an example yielding wide ranges for the data. Analysis in terms of any moderate power transformation (or in terms of SN ratios) will not make much difference for data covering narrow ranges, because such transformations are then almost linear.

Fractional factorial designs and other orthogonal arrays, of course, provide us with only a very elementary idea of the nature of the relationships $\mu(\mathbf{x})$ and $\sigma(\mathbf{x})$. The "optimization" procedure used to exploit location and dispersion effects from such designs is, therefore, necessarily crude. When more is known about the functional relationships (e.g., from response-surface studies or from mechanistic modeling) the ideas of this article will allow more precise conclusions to be drawn.

PERFORMANCE MEASURES AND STATISTICAL EFFICIENCY

A first analysis of experimental results should, I believe, invariably be conducted using flexible data-analytic techniques—looking at graphs and simple statistics—that so far as possible allow the data to "speak for themselves." The unexpected phenomena that such an approach often uncovers can be of the greatest importance in shaping and sometimes redirecting the course of an ongoing investigation. At some later stage when the phenomena under study are becoming understood, we may

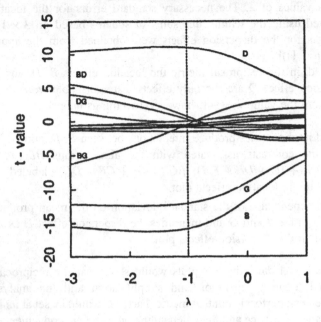

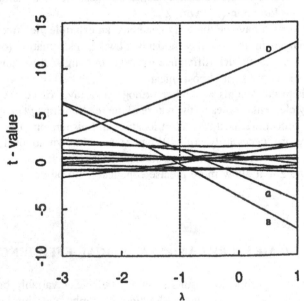

Figure 3. Lambda plots of *t* values for location and dispersion effects, constructed example.

wish to assess the results in terms of some performance measure that allows economic considerations to be taken into account.

In considering the usefulness of SN_T, SN_L, and SN_S introduced at the beginning of the article, it is important to distinguish between two issues—(a) the choice of an appropriate performance criterion and (b) the best way to estimate it.

Performance measures relate to the *objective* of a given investigation. We must consider carefully what this is and seek a measure of performance whose optimization will achieve it. But once the nature of the performance measure is agreed, it is important also to ensure that it be estimated from the available data with greatest possible efficiency. The two desiderata—the best choice of performance measure and the best way to employ the data to estimate it—are distinct and frequently attainable, and they ought not to be confused.

For example, suppose the measure of performance of interest was $1/\mu$ and we had data $y_1, y_2, \ldots, y_n$ believed to be approximately $N(\mu, \sigma^2)$. The maximum likelihood estimator of $1/\mu$ is $1/\bar{y}$, not $[\sum_{i=1}^{n}(1/y_i)]/n$. The use of the latter criterion suggested by the performance measure would, given the usual error assumptions, result in considerable loss of efficiency, equivalent to throwing away a portion of the data.

Again we saw that the performance measure SN_T was relevant when a log transformation of y validated the standard normal model. In that case, Equations (9) and (10) establish the exact functional relationship $\ln(1 + \sigma^2/\mu^2) = \sigma^2_{\ln y}$. This does not, of course, establish the equivalence of the corresponding sample functions as estimators for $\sigma^2_{\ln y}$. The statistic $s^2_{\ln y}$ would then be sufficient for $\sigma^2_{\ln y}$, and any function of $\bar{y}^2/s^2$ including $\{SN_T\}$ would be necessarily less efficient. Theoretical studies by Finney (1951) and Monte Carlo studies by J. Ramírez (1987, personal communication) have shown that the loss of efficiency involved in the use of $\{SN_T\}$ can be quite high.

Statistical Criteria

The choice of efficient criteria for the analysis of experimental data has long been the concern of statisticians. In particular, Gauss's development of least squares estimators in the early 1800s, Fisher's introduction of efficiency and sufficiency in the 1920s, Pearson and Neyman's theory of testing hypotheses in the 1930s, and the upsurge of interest in Bayesian estimators were all directed toward this end. They countered the idea that statistical criteria might be arbitrarily chosen, and they revealed the loss of information (equivalent to throwing data away) that could result from this practice. In particular, for normally distributed samples all of these ideas directed attention to the sample mean and the sample variance as efficient measures of location and scale.

Recently, new classes of useful statistics have been introduced, notably robust estimators (see, e.g., Huber, 1964, 1981) and shrinkage estimators (see, e.g., Stein, 1956). These, like their predecessors, were directed to achieve statistical efficiency but with somewhat more realistic models in mind. For robust estimators, allowance was made for the possibility of discordant observations, and for shrinkage estimators, allowance was made for the clustering expected to exist among means to be compared (see also Box, 1980).

Concerns that naturally occur, therefore, in considering proposed criteria for data analysis, such as SN ratios, are their relevance, their efficiency, their robustness, and their data-transformation implications.

THE CRITERION $SN_L = -10 \log_{10}[(1/n)\sum_{i=1}^{n}(1/y_i)^2]$

This criterion is for the analysis of experiments in which a larger response is better.

Relevance

Figure 4 shows five samples of four observations. Although each produces the same value of $\sum_{i=1}^{n}(1/y_i^2)$ and hence of SN_L, they would suggest widely different ideas about an operating process. In particular, the data of panels (a), (b), and (c) would suggest very different states of process variability, because SN_L confounds location and dispersion. Moreover, Figure 4(d) might suggest that process variability could be

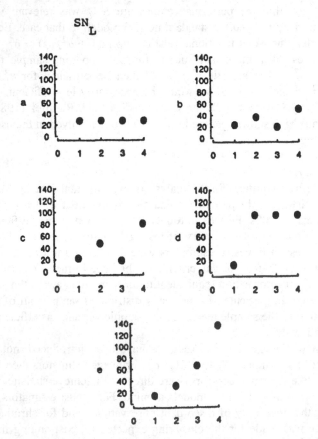

Figure 4. Five samples of four observations giving the same value for SN_L.

caused by some intermittent fault. Figure 4(e) suggests the possibility of an uncontrolled variable giving rise to a smoothly increasing change.

Of course, definite conclusions could not be arrived at on the basis of a single sample of four observations, but these examples illustrate the limitations of "portmanteau" criteria such as SN_L selected before looking at the data.

Power and Efficiency

The ideas of "the larger the better" and "the smaller the better" are ideas about location. For data that were approximately normally independent and identically distributed the sample mean is the natural choice to measure location whatever the desired direction of change might be. Figure 6 shows results from a simulation experiment motivated by a published example by Quinlan (1985), which I discuss later. The graphs compare the power of SN_L, SN_S, and $\bar{y}$ as measures of location. In this illustration we have supposed that we are interested in a possible difference in means produced by some factor tested at two levels. At the minus level of the factor, eight sets of four observations are available, from which eight values of the criteria SN_L, SN_S, and $\bar{y}$ can be calculated, and similarly at the plus level of the factor. A t test at the $\alpha = 0.05$ level for the difference in means of two sets of eight values of SN_L, SN_S, and $\bar{y}$ may then be run (or equivalently, an ANOVA with 14 df for error).

In the sampling experiment, observations were drawn from $N(40, \sigma^2)$ for the first sample and from $N(40 + \delta, \sigma^2)$ for the second sample with $\delta = 0, 1, \ldots, 20$. The random normal deviates were generated using the International Mathematical and Statistical Libraries (1979) subroutine GGNML. At each level of δ, 10,000 pairs of drawings were made to obtain each point on the curves. The experiment was repeated twice. Figure 6(a) shows results for $\sigma = 10.0$ and Figure 6(b) for $\sigma = 13.3$.

The efficiency of SN_L as a measure of location shift may be obtained by considering how much larger the sample would have to be to yield approximately the same power as $\bar{y}$. In these examples the efficiency of SN_L is about 42% when $\sigma = 10$ and about 30% when $\sigma = 13.3$. Thus for this setup the use of SN_L as a location measure is equivalent to discarding about 58% and 70% of the data, respectively.

Robustness

The possibility of occasional faulty observations, or of an error distribution with heavier tails than normal, would provide a legitimate reason for replacing $\bar{y}$ by a robust alternative. Notice, however, that for many real-life situations the function $\sum_{i=1}^{n}(1/y_i)^2$, which depends on the *square* of *reciprocals* of the data, is likely to be exceptionally *nonrobust* to the effect of outlying observations.

Transformation Considerations

Data (particularly data from life tests) may of course *require* a reciprocal transformation to induce approximate properties of constancy of variance, normality, and additivity. In this case analysis in terms of y^{-1} would be the preferred choice, and $\bar{y}$ and s would be inefficient. The choice of this performance criterion is not motivated by considerations of this kind, however.

THE CRITERION $SN_S = -10\log_{10}[(1/n)\sum_{i=1}^{n} y_i^2]$

This criterion is for the analysis of experiments in which a smaller response is better.

Relevance
The criterion can be motivated by an assumption of quadratic loss. If it is assumed that the loss arising from y_i being greater than 0 is proportional to y_i^2, the overall loss from a sample of n items is

$$L = \text{constant} \times \left(\frac{1}{n}\sum_{i=1}^{n} y_i^2\right) \tag{21}$$

and the criterion SN_S is then

$$SN_S = -10\log_{10} L \tag{22}$$

Figure 5 shows five samples of four observations that all yield the same value for $\sum y_i^2$ and hence the same quadratic loss and the same SN_S value. As before,

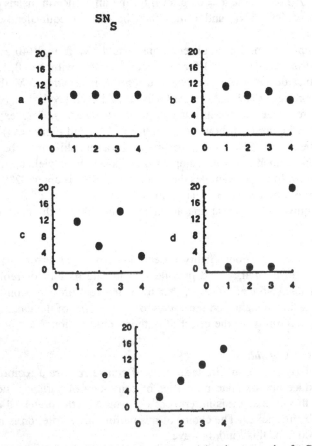

Figure 5. Five samples of four observations giving the same value for SN_S.

however, all five samples convey very different ideas about the characteristics of the process.

The "portmanteau" criterion SN_S is a function of

$$\frac{1}{n}\sum_{i=1}^{n} y_i^2 = \bar{y}^2 + \frac{n-1}{n}s^2 \qquad (23)$$

so, apart from the bias adjustment $(n-1)/n$, the criterion is the sum of the squared sample mean $\bar{y}^2$ and the sample variance s^2 and thus it confounds location effects with dispersion effects. Although it is true that in the circumstances envisaged reduction of SN_S is desirable, most investigators would want to look at the two kinds of effects separately.

Efficiency

Results shown in Figure 6 from the sampling experiment show that for both $\sigma = 10$ and $\sigma = 13.3$ the efficiency of SN_S is about 68%. This corresponds to discarding nearly one-third of the data.

AN ALTERNATIVE ANALYSIS OF A FRACTIONAL FACTORIAL DESIGN

For further illustration, consider the analysis of an experiment reported by Quinlan (1985) at a symposium on Taguchi methods organized by the American Supplier Institute. The objective was to reduce post-extrusion shrinkage of a speedometer casing.

The experiment used an L_{16} orthogonal array—that is, a two-level, fully saturated, 16-run fractional factorial design, or a 2_{III}^{15-11} fractional design. Very careful organization is needed to run such an industrial experiment, and all of those involved are to be congratulated on successfully carrying it through. Although I feel that the Quinlan paper, which was awarded first prize at the symposium, provides an excellent example of the use of such highly fractionated designs in screening experimentation, I also feel that the methods used in its analysis were unnecessarily complicated and incomplete.

The shrinkage values for four samples taken from 3,000-foot lengths of product manufactured at each set of conditions are shown in Table 2 with values of various statistics obtained from these data. In particular, in the first column are shown the 16 values of

$$SN_S = -10\log_{10}\left(\frac{1}{4}\sum_{i=1}^{4} y_i^2\right) \qquad (24)$$

Following Taguchi, Quinlan used these values as data for an ANOVA in which the error mean square was obtained by pooling the mean squares from the seven smallest effects.

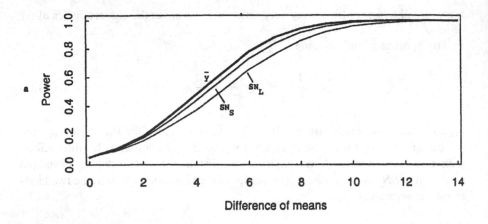

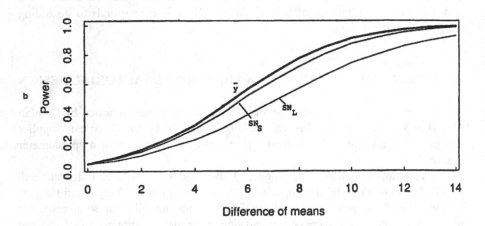

Figure 6. Power curves for t tests using $\bar{y}$, SN_L, and SN_S as data. Each of the criteria is computed from a sample of four drawings. The test compares two groups of eight such data values. Group 1 is drawn from $N(40, \sigma^2)$; group 2 is drawn from $N(40 + \delta, \sigma^2)$, with $\delta = 0, 1, 2, \ldots, 20$. (a) $\sigma = 10$; (b) $\sigma = 13.3$.

Using this procedure, it was found that all remaining eight factors (in order of absolute magnitude E, G, K, A, C, F, D, H) had significant effects. The author noted, however, that "Factors E and G are most important in terms of shrinkage. The two factors account for more than 70% of the experimental variance" (p. 14).

Bias Produced by Pooling

The extreme bias induced by this kind of pooling of mean squares is well known. For illustration I have used the first $15 \times 5 = 75$ values given by the Rand Corporation (1955) tables to supply five samples of 15 $N(0, 1)$ variates. I have treated these random numbers as if they were five sets of 15 treatment effects and conducted an ANOVA precisely as was done in this analysis. Specifically, for each sample an

Table 2. An Experiment on Speedometer-Cable Shrinkage Reported by Quinlan (1985)

Run	H	D	-L	B	-J	F	N	A	-I	-E	M	-C	K	G	-O	y_1	y_2	y_3	y_4	SN_S	$\bar{y}$	$\frac{1}{4}\sum y_i^2$	$=\bar{y}^2$	$+\frac{3}{4}s^2$
1	-1	-1	1	1	1	-1	-1	-1	1	1	-1	1	-1	-1	1	.49	.54	.46	.45	6.2626	.4850	.236450	.235225	.0012250
2	1	-1	-1	-1	-1	-1	-1	-1	-1	-1	-1	-1	-1	-1	-1	.55	.60	.57	.58	4.8024	.5750	.330950	.330625	.0003250
3	-1	1	1	-1	-1	-1	-1	1	1	-1	1	-1	1	1	-1	.07	.09	.11	.08	21.0375	.0875	.007875	.007656	.0002187
4	1	1	-1	1	1	-1	-1	1	-1	1	1	1	1	1	1	.16	.16	.19	.19	15.1074	.1750	.030850	.030625	.0002250
5	-1	-1	1	1	-1	1	-1	1	-1	1	1	-1	1	-1	-1	.13	.22	.20	.23	14.0285	.1950	.039550	.038025	.0015250
6	1	-1	-1	-1	1	1	-1	1	1	-1	1	1	1	-1	1	.16	.17	.13	.12	16.6857	.1450	.021450	.021025	.0004250
7	-1	1	1	-1	1	1	-1	-1	-1	1	-1	1	-1	1	1	.24	.22	.19	.25	12.9115	.2250	.051150	.050625	.0005250
8	1	1	-1	1	-1	1	-1	-1	1	-1	-1	-1	-1	1	-1	.13	.19	.19	.19	15.0446	.1750	.031300	.030625	.0006750
9	-1	-1	1	1	-1	-1	1	-1	-1	-1	1	1	1	1	1	.08	.10	.14	.18	17.6700	.1250	.017100	.015625	.0014750
10	1	-1	-1	-1	1	-1	1	-1	1	1	1	-1	1	1	-1	.07	.04	.19	.18	17.2700	.1200	.018750	.014400	.0043500
11	-1	1	1	-1	1	-1	1	1	-1	-1	-1	1	-1	-1	-1	.48	.49	.44	.41	6.8183	.4550	.208050	.207025	.0010250
12	1	1	-1	1	-1	-1	1	1	1	1	-1	-1	-1	-1	1	.54	.53	.53	.54	5.4325	.5350	.286250	.286225	.0000250
13	-1	-1	1	1	1	1	1	1	1	1	1	1	-1	-1	-1	.13	.17	.21	.17	15.2724	.1700	.029700	.028900	.0008000
14	1	-1	-1	-1	-1	1	1	1	-1	-1	1	-1	-1	-1	1	.28	.26	.26	.30	11.1976	.2750	.075900	.075625	.0002750
15	-1	1	1	-1	-1	1	1	-1	1	1	-1	1	1	1	1	.34	.32	.30	.41	9.2436	.3425	.119025	.117306	.0017187
16	1	1	-1	1	1	1	1	-1	-1	-1	-1	-1	1	1	1	.58	.62	.59	.54	4.6836	.5825	.340125	.339306	.0008187

*Note:#*The names of the factors are as follows: *A*, liner OD; *B*, liner die; *C*, liner material; *D*, liner line speed; *E*, wire braid type; *F*, braiding tension; *G*, wire diameter; *H*, liner tension; *I*, liner temperature; *J*, coating material; *K*, coating die type; *L*, liner material; *M*, screen pack; *N*, cooling method; *O*, line speed.

Table 3. Sampling Experiment with Random Numbers Illustrating Bias Effect Produced by Selection and Pooling

	Number of Eight Largest Mean Squares Found "Significant"				
Significance Level	Sample 1	Sample 2	Sample 3	Sample 4	Sample 5
1%–5%	1	1	1	1	2
0.1%–1%	2	3	3	3	1
<0.1%	0	1	2	2	0
Total proportion "significant"	3/8	5/8	6/8	6/8	3/8

"error mean square" was obtained from the seven smallest mean squares to test the remaining eight. The number of effects found with this random data to be "significant" at the 5%, 1%, and 0.1% levels are shown in Table 3.

It was because such methods of analysis almost guaranteed spurious conclusions that Daniel (1959) introduced normal probability plotting. This technique not only avoids invalid pooling and allows for selection but it is much easier for nonstatisticians to comprehend than is the ANOVA. If desired, Daniel plots may be supplemented with Bayes plots of posterior probabilities that factors have active effects (Box and Meyer, 1986a). Available statistical programs make this an easily used technique.

An Alternative Analysis

In Figure 7, I have used normal plots and Bayes plots to reanalyze the data of Table 2.

Figure 7(a) shows a normal plot and a Bayes plot for SN_S. These strongly indicate that *only* factors E and G (wire braid type and wire diameter) produce effects distinguishable from noise and that the remaining six factors are shown as significant by the ANOVA procedure because of selection bias.

Figure 7(b) and (c) shows separate analyses for $\bar{Y}$ and $\ln s_Y$, where $Y = \ln y$. From this analysis it is clear that the E and G effects are essentially associated with location and not with dispersion. At this point a much simpler analysis, which could have been conducted with only normal plots, has shown (a) that two rather than eight factors are distinguishable from noise and (b) that these factors affect location rather than dispersion. Although the analyses illustrated were conducted for $Y = \ln y$, very similar results are obtained in the original metric. I shall discuss the choice of metric later.

It has been argued in favor of SN_S that it combines information about changes in mean and changes in variance. Notice, however, that, even if we accept that such a combination is useful, on normal assumptions the likelihood criterion that tests

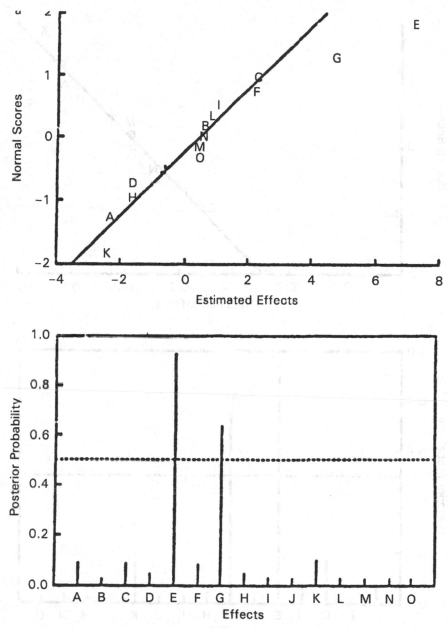

Figure 7. Normal plots and Bayes plots for the speedometer-cable experiment, using as data (a) 16 values of SN_S. In (b) and (c), $Y = \ln y$. The Bayes plots were made with $\alpha = .2$, $k = 10$ (Box and Meyer, 1986a). Varying these parameters from .1 to .3 for α and from 5 to 15 for k had little effect on the appearance of the plot.

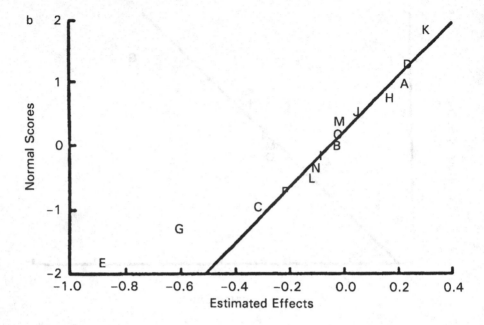

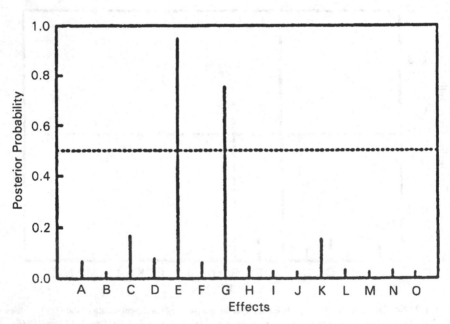

Figure 7. (*continued*) (b) 16 values of $\bar{Y}$.

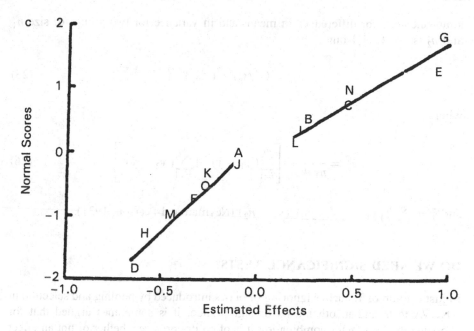

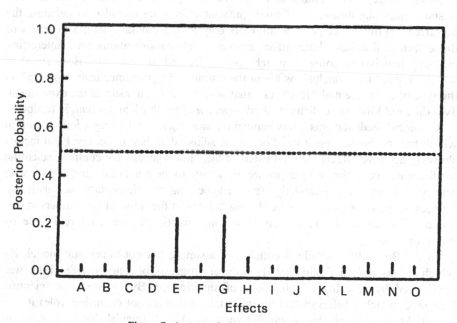

Figure 7. (*continued*) (c) 16 values of $\ln s_Y$.

simultaneously for differences in means and in variance for two groups of size n_1 and n_2 is not $\{SN_S\}$ but

$$\lambda_H = (s_1^2/s_p^2)^{n_1/2}(s_2^2 s_p^2)^{n_2/2} \tag{25}$$

where

$$S_p^2 = \frac{1}{n_1 + n_2}\left[\sum_{i=1}^{n_1}(y_{1i} - \bar{y})^2 + \sum_{j=1}^{n_2}(y_{2j} - \bar{y})^2\right] \tag{26}$$

and $\bar{y} = (\sum_{i=1}^{n_1} y_{1i} + \sum_{j=1}^{n_2} y_{2j})/(n_1 + n_2)$ (Neyman and Pearson, 1931).

DO WE NEED SIGNIFICANCE TESTS?

In justification of Taguchi's ignoring of biases introduced by pooling and selection in ANOVA tests and in other tests of significance, it is sometimes argued that for choosing the best factor combination it is of no importance whether or not an effect is statistically significant, for if nonsignificant effects are included in deciding the preferred "optimal" recipe, this amounts only to the arbitrary setting of levels of those factors that have little effect anyway. Such considerations raise interesting philosophical issues, some of which were addressed in an article (Box, 1966) in which I argued that conclusions about the efficacy of significance tests depended on the type of consequential "feedback" that would occur as a result of the experiment. Two different kinds were distinguished—*empirical* feedback and *scientific* feedback.

Empirical feedback was a mechanical reaction such as adopting a formulation in which factors showing positive effects were adjusted to their upper level and factors showing negative effects were adjusted to their lower levels. By contrast, scientific feedback was conditioned by exposure of the data to the mind of an investigator with relevant subject-matter knowledge (e.g., engineering and elementary data analysis). Suggestive patterns of changes could spark ideas in the mind of such a person and could influence the future course of the investigation in ways unpredictable by empirical rule.

In the Box (1966) article, I considered a simple but not unrealistic model, for which it was indeed found that the rate of progress for empirical feedback was maximized by ignoring all questions of significance. By contrast, for scientific feedback (which I believed had an important, although not exclusive, role) it was essential to have some idea of which effects could be distinguishable from noise, so that attempts to understand, explain, conjecture, and exploit were based on some moderately reproducible view of reality. It follows that if we are to use statistics not to *replace* but to *catalyze* the creativity of engineers and scientists, we need meaningful ways of detecting signals in noise such as are provided by significance tests and other less formal procedures in the same spirit, such as Daniel and Bayes plots.

A DATA-ANALYTIC APPROACH

The process of investigation, which is catalyzed by the use of statistical design and analysis, advances by a continual iteration between deduction and induction. The concept of scientific feedback is closely linked with the inductive phase in this iteration and with the *data-analytic* approach to statistics, the importance of which, following the leadership of J. W. Tukey, is now generally realized (see, e.g., Mosteller and Tukey, 1977; Tukey, 1977).

There is now more willingness to view the statistical investigator as a detective involved in an iterative and adaptive procedure in which deduction and induction alternate. Deduction and induction, respectively, are aided by two quite distinct statistical processes, *estimation* and *criticism* (see, e.g., Box, 1980). The tools of statistical criticism are studies of residuals and checks of fit, formal and informal. By contrast, the Taguchi philosophy seems to suffer from adherence to earlier, somewhat outmoded ideas in which the importance of data criticism as a generator of new ideas was not realized. The data-analytic view can be illustrated with the speedometer-cable data.

Two Rather Than Eight Effects Distinguishable from Noise

The reanalysis of Figure 7 suggests that only the factors *wire braid type* (E) and *wire diameter* (G) have effects distinguishable from noise [coating die type (K), liner OD (A), liner material (C), braiding tension (F), liner line speed (D), and liner tension (H), which were found significant by the biased ANOVA test, do not]. Since the sign of the G effect points to the choice of the *existing* wire diameter, the main conclusion from the experiment is that reduced shrinkage occurs when the type of wire braid is changed and that the effect is in location and not in dispersion. I believe that this simpler analysis would have provided a much more secure base from which the investigator could have reasoned. Using such an analysis he could, for example, have employed his special knowledge of the behavior of materials to consider possible reasons for the difference in shrinkage associated with the two types of braid. Unconfused by the (likely false) trails pointed to by the other six factors, he could then judge what should be looked at in the next stage of investigation. I have no special knowledge of the real technical and scientific aspects involved in this example, but it is easy to imagine an investigator who, having looked at the statistical analysis, might argue in this way: A major difference between the tested braids is that they are made with two very different types of plastic material, A and B. I know from my *technical knowledge* about these two particular types of plastic that their elasticities could be very differently affected by the high temperature to which they must be subjected during manufacture of the speedometer cable. This could account for the lesser shrinkage found using type B. *But if this is so, then I know of plastics of types C and D not previously tested having even better properties of this kind and at the next stage of the investigation I will include these as factors to be studied.*

A Discrepant Observation of Dispersion

Again looking at the dispersion analysis, the normal plot appears to follow not one, but two lines. Following Daniel (1959) this suggests the existence of a discordant run and points to run 12 for which the data (.54, .53, .53, .54) have a suspiciously *small* dispersion. The Bayes analysis of Box and Meyer (1987) tends to confirm this. Now small values of variance are, of course, desirable. If examination of the records excluded a copying error, inquiries ought now to be made to discover whether, for example, unusual process conditions might have accidentally occurred during this run. If so, a subsequent small experiment might be run to find out whether these unscheduled conditions could be reproduced and exploited to produce speedometer cables with smaller variation. In addition, a robust analysis could be conducted in which additional dispersion effects previously masked by the discordant value might show up. In this particular example, such analysis failed to show anything of further interest.

Two Components of Variance

There are two quite distinct components of variance in this example. The average estimated *within-run* variance (measuring variation of tests within four samples taken from the same cable) is .0013; the *between-run* variance (estimated from the normal plot) is .0170, about 13 times as big. This information could be valuable in thinking further about the process. The data imply that the process can be run very reproducibly *once a particular set of process conditions has been set up* but that the process conditions themselves may be hard to reproduce.

Notice also that in the identity $\frac{1}{4}\sum y^2 = \bar{y}^2 + \frac{3}{4}s^2$ the component $\bar{y}^2$ is on the average 117 times as large as the component $\frac{3}{4}s^2$. Thus (as is evident from inspection of the last three columns of Table 2) $\frac{1}{4}\sum y^2$ behaves like $\bar{y}^2$ slightly contaminated with noise from $\frac{3}{4}s^2$. Consequently, $SN_S \doteq -10\log_{10}\bar{y}^2 = -20\log_{10}\bar{y}$, and for these data an analysis of SN_S is essentially an analysis of the logged mean. This is the reason why, apart from differences in sign, the analyses of Figures 7(a) and 7(b) appear so similar.

Choice of Metric

Because these data cover a considerable range (from .07 to .62) and because we are interested in both location and dispersion effects, it is especially important to consider in what metric the analysis is best conducted. These issues were discussed earlier in this chapter, and a lambda plot of the kind discussed there is shown in Figure 8 for factors E, G, and $EG = B$. Lambda plots for effects of other factors did not suggest the existence of active effects other than E and G. The $EG (= B)$ interaction contrast is not particularly large for any value of λ, but it is of interest that it almost disappears when $\lambda = 0$.

TRAINING ENGINEERS AND SCIENTISTS IN STATISTICAL METHODS

Massive training programs for engineers and scientists are needed. But what should they be taught? The earlier success of the Japanese had sometimes prompted the simple answer that we should teach our engineers only Japanese techniques and, for problems involving experimental design and analysis, only "the Taguchi method."

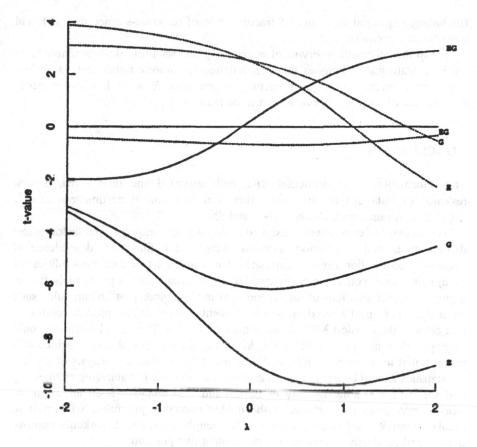

Figure 8. Speedometer example: $\cdots$, dispersion effects, and —, location effects, for various values of λ.

To answer the question at a deeper level, however, we need to consider some of the issues raised in the previous sections.

Taguchi has emphasized some extremely important concepts in quality engineering. Among these are (a) the need to achieve minimum variance at a desired quality value, (b) the need to design products insensitive to environmental disturbance, and (c) the need to design products insensitive to transmitted variation.

It is essential that we incorporate these ideas into the body of our own practice and teaching. When we consider the statistical means by which these engineering aims are to be achieved, however, we can and should substitute better and simpler methods of design and analysis when these exist. In addition, our engineers should be trained to exploit the sequential-iterative-interactive-adaptive nature of investigation rather than to attempt to answer all questions with one-shot experiments. They must also be taught to use the best of our home-grown products—exploratory data analysis (including, in particular, computer graphics, interactive analysis, and appropriate transformation techniques), sequential use of experimental design (including sequential assembly of fractions and of response-surface designs), and evolutionary operation.

To tap the enormous reservoir of engineering and scientific skills available to us we need statistical techniques that are designed to *catalyze* rather than to *replace* engineering inventiveness. Such techniques are easy to learn but the immense possibilities of such an approach wait to be realized.

CONCLUSIONS

The information in experimental data, both expected and unexpected, is best revealed by data-analytic methods rather than in terms of portmanteau criteria decided in advance such as SN_T, SN_S, and SN_L.

The analysis of experiments designed to discover how experimental factors affect dispersion as well as location is greatly simplified if functional dependence of dispersion on location can be eliminated. For a given set of data, possibilities for doing this may be conveniently reviewed by considering an appropriate analysis over a range of transformations of the response, as in the lambda plot. In particular, such an analysis is helpful when designed experiments are run to find process conditions that achieve the smallest MSE about some target value. The $\{SN_T\}$ criterion would be appropriate in an endeavor of this kind only for the special case in which σ is proportional to μ, but it would usually be less efficient than an analysis of log y.

Statistics should be introduced to engineers as a means of catalyzing engineering and scientific reasoning by way of design and data analysis. Such an approach, emphasizing graphical methods with suitable computer programs, will result in greater creativity and, if taught on a wide enough scale, could markedly improve quality and productivity and our overall competitive position.

ACKNOWLEDGMENTS

This research was sponsored by National Science Foundation Grant DMS-8420968, and by the Vilas Trust of the University of Wisconsin–Madison; it was aided by access to the Statistics Department research computer at the University of Wisconsin–Madison. My thanks are due to José Ramírez for help with the preparation and calculations for this article and to the referees, who made a number of valuable suggestions for improvement of an earlier draft.

PART F

Songs

.

I do not have a particularly good voice but I use to like to write and sing songs, particularly at Christmas parties and Gordon Conferences. Here are three of them.—George Box

F.1. There's No Theorem like Bayes Theorem
F.2. It's Distribution Free
F.3. I Am the Very Model of a Professor Statistical

CHAPTER F.1

There's No Theorem Like Bayes Theorem

Improving Almost Anything: Ideas and Essays, Revised Edition. By George Box and Friends
Copyright © 2006 John Wiley & Sons, Inc.

(To the tune of "There's No Business Like Show Business", by Irving Berlin)

VERSE (1)   The model, the data you can't wait to see
            The theta, beta, sigma, and the rho
            The Normal, the Poisson, the Cauchy, the t
            The need to specify what you don't know
            The likelihood for data you acquire
            The perspicacious choosing of the prior

REFRAIN   There's no theorem like Bayes theorem
            Like no theorem we know

            Everything about it is appealing
            Everything about it is a wow

            Let out all that a priori feeling
            You've been concealing right up to now

            There's no people like Bayes people
            All odd balls from the urn

            The other day you thought that you had got it straight
            Take my advice and don't celebrate

            A paradox by Lindley could arrive quite late
            Another Stone to unturn

REFRAIN   There's no theorem like Bayes theorem
            Like no theorem we know

            You can lose forever that perplexed look
            If you start to study it right now

            Even more enthralling than a sex book
            You'll find that textbook
            By Box and Tiao

            There's no dogma like Bayes dogma
            Its great knowing you're right

            We know of a fiducialist who knew the lot
            We thought at first he had hit the spot

            But after three more seminars we lost the plot
            We just could not see the light

REFRAIN   There's no theorem like Bayes theorem
            Like no theorem we know

            Fisher felt its use was quite restricted
            Except in making family plans for mice

            But there, he said, for pinning down a zygote
            I'd give it my vote
            And not think twice.

            There's no answers like Bayes answers
            Transparent, clear and precise

            Stein's conundrums you can solve without a blink
            Best estimators in half a wink

            You can even understand what makes 'em shrink
            Their properties are so nice

VERSE (2)   There's Raiffa and Schlaifer, Mosteller and Pratt
            There's Geisser, Zellner, Novick, Hill, and Tiao
            And these all are people who know what they're at
            They represent Statistics' finest flower
            And tho' on nothing else they could agree
            With us they'd join and sing in harmony.

REFRAIN   There's no theorem like Bayes theorem
            Like no theorem we know

            Just recall what Pearson said to Neyman
            Emerging from a region of type B

            "Its difficult explaining to the Lehmann
            I fear it lacks Bayes simplicity."

            There's no haters like Bayes haters
            They spit when they see a prior

            Be careful when you offer your posterior
            They'll try to kick it right through the door

            But turn the other cheek if it is not too sore
            Of error they may yet tire

REFRAIN   There's no theorem like Bayes theorem
            Like no theorem we know

            Critics carp at Bayes's hesitation
            Claiming that his doubts on what he'd done

            Led to late posthumous publication
            We will explain that to everyone.

            When Bayes got up to Heaven
            He asked for an interview

            Jehovah quickly told him he had got it right
            Bayes popped down earthwards at dead of night

            His spectre ceded Richard Price the copyright
            It's very strange but it's true!

CHAPTER F.2

It's Distribution Free

Improving Almost Anything: Ideas and Essays, Revised Edition. By George Box and Friends
Copyright © 2006 John Wiley & Sons, Inc.

(To be sung to the music of Gilbert & Sullivan's "When the Foeman Bares his Steel")

It's distribution free! Ta-ran-Ta-ra Ta-ran-Ta-ra
Due to Mann and Whit-en-nee Ta-ran-Ta-ra
And to them we give our thanks Ta-ran-Ta-ra Ta-ran-Ta-ra
As we fiddle with the ranks Ta-ran-Ta-ra
In complex calculations you need never be involved
A child of three could do it easily
But the thing we like the best
About this admirable test
It's distribution free!

When your data's rather skew Ta-ran-Ta-ra Ta-ran-Ta-ra
And you don't know what to do Ta-ran-Ta-ra
When decision is a must Ta-ran-Ta-ra Ta-ran-Ta-ra
And the calculator's bust Ta-ran-Ta-ra
And although you're pretty sure
            they dropped the beaker on the floor
And the data isn't all it ought to be
Your alternative hypothesis will feel a rosy glow
Chosen by a test that's distribution free!

Oh it's distribution free! Ta-ran-Ta-ra Ta-ran-Ta-ra
Well—I think it ought to be Ta-ran-Ta-ra
No problems need delay Ta-ran-Ta-ra Ta-ran-Ta-ra
We've assumed them all away Ta-ran-Ta-ra
We declared our independence in this country long ago
So we can all ignore the I.I.D.
But as they sound the bugle at the ending of the day
Be proud your test was distribution free!

When your alpha must be right Ta-ran-Ta-ra Ta-ran-Ta-ra
And your limits must be tight Ta-ran-Ta-ra
If with care you rank the ties Ta-ran-Ta-ra Ta-ran-Ta-ra
So that none can criticize Ta-ran-Ta-ra
Then whatever other errors you have made along the way
The first kind we will gladly guarantee
It is clearly evident
It's exactly five percent
If it's distribution free!

When you want a Ph.D. Ta-ran-Ta-ra Ta-ran-Ta-ra
Full of Mathematistry Ta-ran-Ta-ra
In this you should invest Ta-ran-Ta-ra Ta-ran-Ta-ra
A nonparametric test Ta-ran-Ta-ra

A thesis full of lemmas and of theorems by the score
Your committee all will welcome that with glee
For you their heads they'll bare
And create a special chair
If it's distribution free!

It's distribution free! Ta-ran-Ta-ra Ta-ran-Ta-ra
It's not like Student's t! Ta-ran-Ta-ra
No normality implied Ta-ran-Ta-ra Ta-ran-Ta-ra
Takes the Cauchy in its stride Ta-ran-Ta-ra
You could have a distribution like a stalagmetic mass
Inverted stalactitic don't you see?
And I very greatly doubt
That you'd be a fraction out
For it's distribution free!

When outliers abound Ta-ran-Ta ra Ta-ran-Ta-ra
And John Tukey can't be found Ta-ran-Ta-ra
Missing data everywhere Ta-ran-Ta-ra Ta-ran-Ta-ra
Here's a test beyond compare Ta-ran-Ta-ra
This procedure's not most powerful
        But we know how power corrupts
We rightly therefore shun this property
But we're going to mark the spot
Where this concept was begot
To be distribution free!

I Am the Very Model of a Professor Statistical

(To be sung to the music of Gilbert & Sullivan's "I am the Very Model of a Modern Major General")

I am the very model of a professor statistical
I understand the theory both exotical and mystical

The logic of my argument it is that matters most to me
My chance of making errors is exactly what it's s'posed to be

I relentlessly uncover any aberrant, contingency
I strangle it with rigor and I stifle it with stringency

I understand the different symbols be they Roman, Greek, or cuneiform
And every distribution from the Cauchy to the uniform

Chorus: And every distribution from the Canchy to the uniform
            And every distribution from the Cauchy to the uniform
            And every distribution from the Cauchy to the uni-uni-form

With derivation rigorous each lemma I can justify
My every estimator I am careful to robustify

In short in matters logical, mathematic, idealistical
I am the very model of a professor statistical

Chorus: In short in matters logical, mathematic, idealistical
            He is the very model of a professor statistical

I am the very model of a professor statistical
My art it is immaculate and thoroughly puristical

Judge me by my inner soul and not by my external face
Understand my mind's away in reproducing Hilbert space

With repetitious pleasantries my students all must learn to live
For my wit's one hypothesis to which there's no alternative

I never stoop to folly nor to action reprehensible
I always state assumptions whether ludicrous or sensible

Chorus: He always states assumptions whether ludicrous or sensible
            He always states assumptions whether ludicrous or sensible
            He always states assumptions whether ludicrous or sensi-sensi-ble

My manner it is modest and not the least hysterical
My errors they are normal both elliptical and spherical

In short in every aspect whether complex or realistical
I am the very model of a professor statistical.

Chorus: In short in every aspect whether complex or realistical
            He is the very model of a professor statistical

I am the very model of a professor statistical
I understand the theory both exotical and mystical

Tho' if all my expoundings in the discipline didactical
Could lead to a connection of the theory with the practical

If designs that I dub optimal with letters alphabetical
Were readily applicable and not only theoretical

If I didn't think consulting was a practice that was better missed
So I could tell a gymnast from a physical geneticist

Chorus: So he could tell a gymnast from a physical geneticist
            So he could tell a gymnast from a physical geneticist
            So he could tell a gymnast from a physical geneti-neti-cist

If decision theory argument could help me make a simple choice
At meetings of the faculty I feel I'd have a stronger voice

Why, then, in matters practical, applicable, heuristical
I'd be the very model of a professor statistical

Chorus: Why, then, in matters practical, applicable, heuristical
            He'd be the very model of a professor statistical

References

Abraham, B. and Ledolter, J. (1983), *Statistical Methods for Forecasting*. John Wiley & Sons, New York.

Adam, C. G. (1987), "Instrument panel process development," in *Fifth Symposium on Taguchi Methods*, pp. 93–106. Also reprinted in *Taguchi Methods. Case Studies from the U.S. and Europe*, 1989, G. Taguchi and Y. Wu (eds.). ASI Press, Dearborn, MI.

Adams, B. M. (1988), *Economically Optimal On-Line Quality Control Procedures*, unpublished Ph.D. thesis, University of Southwestern Louisiana, Dept. of Mathematics and Statistics.

Adams, B. M. and Woodall, W. I. (1989), "An analysis of Taguchi's on-line process-control procedure under a random-walk model," *Technometrics*, 31, 401–413.

Adams, B. M., Lowry, C., and Woodall, W. H. (1992), "The use (and misuse) of false alarms probabilities in control chart design," in H. J. Lenz, G. B. Wetherhill and P.-Th. Wilrich (eds.) *Frontiers in Quality on Control 4*, Physica-Verlag, Heidelberg, pp. 155–168.

Addelman, S. (1984), "Some two-level factorial plans with confounding," *Technometrics*, 6, 253–258.

Aitchison, J. and Aitken, C. G. G. (1976), "Multivariate binary discrimination by the kernel method," *Biometrika*, 413–420.

Alwan, L. C. and Roberts, H. V. (1988), "Time-series modeling for statistical process control," *Journal of Business and Economic Statistics*, 6, 87–95.

Altpeter, R. J., Box, G. E. P., and Kotnour, K. D. (1966), "A discrete predictor controller applied to sinusoidal perturbation adaptive optimization," *Instrument Society of American Transactions*, 5(3), 255–262.

Anderson, E. (1960), "A semigraphical method for the analysis of complex problems," *Technometrics*, 2, 387–391.

Anscombe, F. J. (1963), "Bayesian inference concerning many parameters with reference to supersaturated designs," *Bulletin of the International Statistics Institute*, 40, 721–733.

Aström, K. J. (1970), *Introduction to Stochastic Control Theory*. Academic Press, New York.

Aström, K. J. (1970), *Introduction to Stochastic Control*, Mathematics in Science and Engineering Series, Vol. 70. Academic Press, New York.

Aström, K. J. and Wittenmark, B. (1984), *Computer Controlled Systems: Theory and Design*. Prentice Hall, Englewood Cliffs, NJ.

Atkinson, A. C. (1972), "Planning experiments to detect inadequate regression models," *Biometrika*, **59**, 275–293.

Atkinson, A. C. (1973a), "Testing transformations to normality," *Journal of the Royal Statistical Society, Series B*, **35**, 473–479.

Atkinson, A. C. (1973b), "Multifactor second order designs for cuboidal regions," *Biometrika*, **60**, 15–19.

Atkinson, A. C. and Cox, D. R. (1974), "Planning experiments for discriminating between models" (with discussion), *Journal of the Royal Statistical Society, Series B*, **36**, 321–348.

Atkinson, A. C. and Fedorov, V. V. (1975a), "The design of experiments for discriminating between two rival models," *Biometrika*, **62**, 57–70.

Atkinson, A. C. and Fedorov, V. V. (1975b), "Optimal design: experiments for discriminating between several models," *Biometrika*, **62**, 289–303.

Atwood, C. L. (1969), "Optimal and efficient designs of experiments," *Annals of Mathematical Statistics*, **40**, 1570–1602.

Bacon, D. W. (1970), "Making the most of a one-shot experiment," *Industrial and Engineering Chemistry*, **62**(7), 27–34.

Bagshaw, M. and Johnson, R. (1977), "Sequential procedures for detecting parameter changes in a time-series model," *Journal of the American Statistical Association*, **72**, 593–597.

Bailey, S. P. and Box, G. E. P. (1980), "The duality of diagnostic checking and robustification in model building: some considerations and examples," *Technical Summary Report #2086*, Mathematics Research Center, University of Wisconsin–Madison.

Bailey, S. P., Chatto, K. A., Fellner, W. H., and Pfeifer, C. G. (1991), "Giving your response surface a robust workout," *Transactions* (45th Annual Quality Congress), American Society for Quality Control, Milwaukee, WI, pp. 645–652.

Baker, C. T. H. (1977), *The Numerical Treatment of Integral Equations*. Clarendon Press, Oxford.

Barker, T. (1986), "Quality engineering by design: Taguchi's philosophy," *Quality Progress*, December 1986, 32–37.

Barnard, G. A. (1946), "Sequential tests in industrial statistics," *Journal of the Royal Statistical Society, Series B*, **8**, 1–21.

Barnard, G. A. (1959), "Control charts and stochastic processes," *Journal of the Royal Statistical Society, Series B*, **21**, 239–271.

Barrios, E. (2004a) "Topics in Engineering Statistics", Ph.D. thesis, University of Wisconsin, Madison.

Barrios, E. (2004b) Bayesian Screening and Model Selection BSMD R-package. URL: http// cran.r-project.org/src/contrib/Descriptions/BSMD.html.

Bartlett, M. S. (1935), "Discussion of 'Complex experiments' by F. Yates," *Journal of the Royal Statistical Society, Series B*, **2**, 224–226.

Bartlett, M. S. (1936), "The square root transformation in analysis of variance," *Journal of the Royal Statistical Society, Series B*, **3**, 68–78.

Bartlett, M. S. and Kendall, D. G. (1946), "The statistical analysis of variance heterogeneity and the logarithmic transformation," *Journal of the Royal Statistical Society, Series B*, **8**, 128–150.

Basseville, M. and Nikiforov, I. V. (1993), *Detection of Abrupt Changes: Theory and Application*. Prentice Hall, Englewood Cliffs, NJ.

Baxley, R. V. Jr. (1990), "Discussion of 'Exponentially weighted moving average control schemes: properties and enhancements' by James M. Lucas and Michael S. Saccucci," *Technometrics*, **32**, 13–16.

Bender, C. M. and Orszag, S. A. (1984), *Advanced Mathematical Methods for Scientists and Engineers*. McGraw-Hill, New York.

Bergh, L. G. and MacGregor, J. F. (1987), "Constrained minimum variance controllers: internal model structure mid robustness properties," *Industrial and Engineering Chemistry Research*, **26**, 1158–1564.

Beveridge, W. I. B. (1950), *The Art of Scientific Investigation*. Vintage Books, New York.

Bible 1 Thessalonians 5:21.

BHHII, same as Box, Hunter, and Hunter (2005).

Bisgaard, S. (1988), *A Practical Aid for Experimenters*. Starlight Press, Madison, WI.

Bisgaard, S. (1989), "The quality detective: a case study," *Proceedings of the Royal Society of London*, 21–32.

Bisgaard, S. (1991), "Letter to the Editor," *The American Statistician.*"

Bisgaard, S. (1992), "The design and analysis of $2^{k-p} \times 2^{q-r}$ inner and outer array experiments," *Technical Report No. 90*, Center for Quality and Productivity Improvement, University of Wisconsin–Madison.

Bisgaard, S. (1993), "Quality quandaries: iterative analysis of data from two-level factorials," *Quality Engineering*, **6**(2), 319–330.

Bisgaard, S. (2000), "The design and analysis of $2^{k-p} \times 2^{q-r}$ split plot experiments," *Journal of Quality Technology*, **22**(1), 39–56.

Bisgaard, S. and Fuller, H. T. (1994), "Quality quandaries: analysis of factorial experiments with defects or defectives as the response," *Quality Engineering*, **7**(2), 429–443.

Bisgaard, S. and Steinberg, D. M. (1993), "The design and analysis of $2^{k-p} \times S$ prototype experiments," *Technical Report 90*, University of Wisconsin–Madison, Center for Quality and Productivity Improvement.

Blackeslee, T. R. (1980), *The Right Brain*. Anchor Press/Doubleday, Garden City, NY.

Bliss, C. I. (1935), "The calculation of the dosage–mortality curve," *Annals of Applied Biology*, **22**, 134–137.

Box, G. E. P. (1950), "Problems in the analysis of growth and wear curves," *Biometrics*, **6**, 362–389.

Box, G. E. P. (1952), "Multifactorial designs of first order," *Biometrika*, **39**, 49–57.

Box, G. E. P. (1954a), "The exploration and exploitation of response surfaces: some general considerations and examples," *Biometrics*, **10**(1), 16–60.

Box, G. E. P. (1954b), "Some theorems on quadratic forms applied in the study of analysis of variance problems: I. Effect of inequality of variance in the one way classification," *Annals of Mathematical Statistics*, **25**(2), 290–302.

Box, G. E. P. (1957a), "Evolutionary operation: a method for increasing industrial productivity," *Applied Statistics*, **VI**(2), 3–23.

Box, G. E. P. (1957b), "Integration of techniques in process development," in *Transactions 11th Annual Convention of American Society for Quality Control*, pp. 687–702.

Box, G. E. P. (1960), "Fitting empirical data," *Annals of the New York Academy of Sciences*, **86**(3), 792–816.

Box, G. E. P. (1966), "A simple system of evolutionary operation subject to empirical feedback," *Technometrics*, **8**(1), 19–26.

Box, G. E. P. (1968), "Experimental design: response surfaces," in *International Encyclopedia of the Social Sciences*. Macmillan, New York, pp. 254–259.

Box, G. E. P. (1976), "Science and statistics," *Journal of the American Statistical Association*, **71**(356), 791–799.

Box, G. E. P. (1979), "Robustness in the strategy of scientific model building," in *Robustness in Statistics*, Academic Press, New York, pp. 201–236.

Box, G. E. P. (1980), "Sampling and Bayes' inference in scientific modeling and robustness," *Journal of the Royal Statistical Society, Series A*, **143**(Pt. 4), 383–430.

Box, G. E. P. (1982), "Choice of response surface design and alphabetic optimality," *Utilitas Mathematica*, **21B**, 11–55. Also in *Proceedings of the Twenty-eighth Conference of the Design of Experiments in Army Research Development and Testing*, Report 83-2.

Box, G. E. P. (1983a), "An apology for ecumenism in statistics," in *Scientific Inference, Data Analysis, and Robustness*, G. E. P. Box, Tom Leonard, and Chien-Fu Wu (eds.). Academic Press, New York, pp. 51–84.

Box, G. E. P. (1983b), "Gwilym Jenkins, experimental design and the time series," *Journal Questio*, Madrid, Spain, **7**(4), 515–525.

Box, G. E. P. (1984), "The importance of practice in the development of statistics," *Technometrics*, **26**(1), 1–8.

Box, G. E. P. (1986a), "Discussion of 'Accumulation analysis,' a paper by V. N. Nair," *Technometrics*, **28**(4), 295–301.

Box, G. E. P. (1986b), "Studies in quality improvement: signal to noise ratios. Performance criteria and statistical analysis: Part I," *Technical Report 11*, Center for Quality and Productivity Improvement, University of Wisconsin–Madison.

Box, G. E. P. (1988), "Signal to noise ratios, performance criteria and transformations," *Technometrics*, **30**(1), 1–17.

Box, G. E. P. (1989a), "When Murphy speaks—listen," *Quality Progress*, **22**(10), 79–84.

Box, G. E. P. (1989b), "Quality improvement: an expanding domain for scientific method," *Philosophical Transactions of the Royal Society*, **A327**, 617–630.

Box, G. E. P. (1990), "Do interactions matter?" *Quality Engineering*, **2**(3), 365–369.

Box, G. E. P. (1991a), "Understanding exponential smoothing: a simple way of forecast sales and inventory," *Quality Engineering*, **3**(4), 561–566.

Box, G. E. P. (1991b), "Feedback control by manual adjustment," *Quality Engineering*, **4**(1), 143–151.

Box, G. E. P. (1991c), "Bounded adjustment charts," *Quality Engineering*, **4**(2), 331–338.

Box, G. E. P. (1992a), "Teaching engineers experimental design with a paper helicopter," *Quality Engineering*, **4**(3), 453–459.

Box, G. E. P. (1992b), "What can you find out from eight experimental runs?" *Quality Engineering*, **4**(4), 619–627.

Box, G. E. P. (1993a), "Sequential experimentation and sequential assembly of designs," *Quality Engineering,* **5**(2), 321–330.

Box, G. E. P. (1993b), "What can you find out from twelve experimental runs?" *Quality Engineering,* **5**(4), 663–668.

Box, G. E. P. (1993c), "What can you find out from sixteen experimental runs?" *Quality Engineering,* **5**(1), 167–178.

Box, G. E. P. (1993d), "Process adjustment and quality control," *Total Quality Management,* **4**(2), 215–227.

Box, G. E. P. (1993e), "How to get lucky," *Quality Engineering,* **5**(3), 517–524.

Box, G. E. P. (1994), "Statistics and quality improvement," *Journal of the Royal Statistical Society, Series A,* **157**(Pt. 2), 209–229.

Box, G. E. P. (1996), "Split plot experiments," *Quality Engineering,* **8**(3), 515–520.

Box, G. E. P. (1998), "Discussion on the paper 'The stochastic control of process capability' by N. D. Singpurwalla," *Test,* **7**, 33–38.

Box, G. E. P. (1999), "Statistics as a catalyst to learning by scientific method. Part II, A discussion," *Journal of Quality Technology* **31**(1), 16–72.

Box, G. E. P. and Behnken, D. W. (1960), "Some new three level designs for the study of quantitative variables," *Technometrics,* **2**(4), 455–475.

Box, G. E. P. and Bisgaard, S. (1987), "The scientific context of quality improvement," *Quality Progress,* 54–62.

Box, G. E. P. and Bisgaard, S. (1993), "What can you find out from 12 experimental runs?" *Quality Engineering,* **5**, 663–668.

Box, G. E. P. and Cox, D. R. (1964), "An analysis of transformations," *Journal of the Royal Statistical Society, Series B,* **26**(2), 211–252.

Box, G. E. P. and Draper, N. R. (1959), "A basis for the selection of a response surface design," *Journal of the American Statistical Association,* **54**, 622–654.

Box, G. E. P. and Draper, N. R. (1963), "The choice of a second order rotatable design," *Biometrika,* **50**(Pts. 3 and 4), 335–352.

Box, G. E. P. and Draper, N. R. (1965), "The Bayesian estimation of common parameters from several responses," *Biometrika,* **52**(Pts. 3 and 4), 355–365.

Box, G. E. P. and Draper, N. R. (1969), *Evolutionary Operation—A Statistical Method for Process Improvement.* John Wiley & Sons, New York.

Box, G. E. P. and Draper, N. R. (1975), "Robust designs," *Biometrika,* **62**(2), 347–352.

Box, G. E. P. and Draper, N. R. (1982), "Measures of lack of fit for response surface designs and predictor variable transformations," *Technometrics,* **24**(1), 1–8.

Box, G. E. P. and Draper, N. R. (1987), *Empirical Model-Building and Response Surfaces.* John Wiley & Sons, New York.

Box, G. E. P. and Fung, C. A. (1983), "Some considerations in estimating data transformations," *Technical Summary Report–Madison Research 2609,* University of Wisconsin Mathematics Center.

Box, G. E. P. and Fung, C. (1986), "Studies in quality improvement: minimizing transmitted variation by parameter design," CQPI Report No. 8.

Box, G. E. P. and Fung, C. A. (1994), "Is your robust design procedure robust?" *Quality Engineering,* **6**(3), 503–514.

Box, G. E. P. and Fung, C. (1995), "The importance of data transformation in designed experiments for life testing," *Quality Engineering,* **7**(3), 625–638.

Box, G. and Henson, T. L. (1969), "Model fitting and discrimination," *Technical Report No. 211*, Department of Statistics, University of Wisconsin, Madison, WI.

Box, G. and Henson, T. L. (1970), "Some aspects of mathematical modeling in chemical engineering," in *Proceedings of the Inaugural Conference of the Scientific Computation Centre and the Institute of Statistical Studies and Research*, Cairo University Press, Cairo, Egypt, p. 548.

Box, G. E. P. and Hill, W. J. (1967), "Discrimination among mechanistic models," *Technometrics*, **9**(1), 57–71.

Box, G. E. P. and Hill, W. J. (1974), "Correcting inhomogeneity of variance with power transformation weighting," *Technometrics*, **16**(3), 385–389.

Box, G. E. P. and Hunter, J. S. (1957), "Multifactor experimental designs for exploring response surfaces," *Annals of Mathematical Statistics*, **28**(1), 195–241.

Box, G. E. P. and Hunter, J. S. (1961a), "The 2^{k-p} fractional factorial designs, Pt. I," *Technometrics*, **3**(3), 311–351.

Box, G. E. P. and Hunter, J. S. (1961b), "The 2^{k-p} fractional factorial designs, Pt. II," *Technometrics*, **3**(4), 449–458.

Box, G. E. P., Hunter, W. G., Erjavec, J., and MacGregor, J. F. (1973), "Some problems associated with the analysis of multi-response data," *Technometrics*, **15**(1), 33–51.

Box, G. E. P., Hunter, W. G., and Hunter, J. S. (2005), *Statistics for Experimenters*. Second Edition, John Wiley & Sons, New York.

Box, G. E. P. and Jenkins, G. M. (1962), "Some statistical aspects of adaptive optimization and control," *Journal of the Royal Statistical Society, Series B*, **24**(2), 297–343.

Box, G. E. P. and Jenkins, G. M. (1963), "Further contributions to adaptive quality control: simultaneous estimation of dynamics: non-zero costs," in *Proceedings of the International Statistics Institute*, pp. 943–974.

Box, G. E. P. and Jenkins, G. M. (1966), "Models for prediction and control. VI: Diagnostic checking," *Technical Report No. 99*, University of Wisconsin–Madison, Dept. of Statistics.

Box, G. E. P. and Jenkins, G. M. (1968), "Discrete models for feedback and feedforward control," in *The Future Of Statistics*, D. G. Watts (ed.), Academic Press, New York, pp. 201–240.

Box, G. E. P. and Jenkins, G. M. (1970, 1976), *Time Series Analysis: Forecasting and Control*, Holden-Day, (1st and 2nd editions).

Box, G. E. P., Jenkins, G. M., and Bacon, D. W. (1968), "Models for forecasting seasonal and non-seasonal time series," in *Spectral Analysis of Time Series*, B. Harris (ed.). John Wiley & Sons, New York, pp. 271–311.

Box, G. E. P., Jenkins, G. M., and Reinsel, G. (1994), *Time Series Analysis: Forecasting and Control*, 3rd ed. Prentice-Hall, Englewood Cliffs, NJ.

Box, G. E. P., Jenkins, G. M., and MacGregor, J. F. (1974), "Some recent advances in forecasting and control, Pt. II," *Journal of the Royal Statistical Society, Series C*, **23**(2), 158–179.

Box, G. E. P., Joiner, L., Rohan, S., and Sensenbrenner, F. J. (1989), "Quality in the community—one city's experience," *Quality Progress*.

Box, G. E. P. and Jones, S. (1986), "Discussion of 'Testing in industrial experiments with ordered categorical data' by Vijayan N. Nair," *Technometrics*, **28**, 295–301.

Box, G. E. P. and Jones, S. (1990), "An investigation of the method of accumulation analysis," *Total Quality Management*, **1**(1), 101–113.

Box, G. E. P. and Jones, S. P. (1991), "Robust product designs, Part II: second order models," *Technical Report 63*, Center for Quality and Productivity Improvement, University of Wisconsin–Madison.

Box, G. E. P. and Jones, S. (1992a), "Designing products that are robust to the environment," *Journal of Total Quality Management*, 3(3), 265–282.

Box, G. E. P. and Jones, S. (1992b), "Split-plot designs for robust product experimentation," *Journal of Applied Statistics*, 19(1), 3–26.

Box, G. E. P. and Jones, S. (2000) "Split-plot designs for robust product and process experimentation," *Quality Engineering*, 3(1).

Box, G. E. P. and Kramer, T. (1992), "Statistical process control and automatic process control—a discussion," *Technometrics*, 34(3), 251–267.

Box, G. E. P. and Liu, P. (1999), "Statistics as a catalyst to learning by scientific method. Part I, An example," *Journal of Quality Technology*, 31(1), 1–15.

Box, G. E. P. and Lucas, H. L. (1959), "Design of experiments in non-linear situations," *Biometrika*, 46(Pts. 1 and 2), 77–90.

Box, G. E. P. and Luceño, A. (1995), "Discrete proportional-integral control with constrained adjustment," *The Statistician, JRSS Series D*, 44(4), 479–495.

Box, G. E. P. and Luceño, A. (1997), *Statistical Control by Monitoring and Feedback Adjustment*. John Wiley & Sons, New York.

Box, G. E. P. and MacGregor, J. F. (1974), "The analysis of closed-loop dynamic-stochastic systems," *Technometrics*, 16(3), 391–398.

Box, G. E. P. and MacGregor, J. F. (1976), "Parameter estimation with closed-loop operating data," Invited address—*136th Annual Meeting of the American Statistical Association*, Boston, August 1976. Also *Technometrics*, 18(4), 371–380.

Box, G. E. P. and Meyer, R. D. (1986a), "An analysis for unreplicated fractional factorials," *Technometrics*, 28(1), 11–18.

Box, G. E. P. and Meyer, R. D. (1986b), "Dispersion effects from fractional designs," *Technometrics*, 28(1), 19–27.

Box, G. E. P. and Meyer, R. D. (1987), "Analysis of unreplicated factorials allowing for possibly faulty observations," in *Design, Data, and Analysis*, Colin Mallows (ed.). John Wiley & Sons, New York, pp. 1–12. Also CQPI Report #3, February 1986.

Box, G. E. P. and Meyer, R. D. (1993), "Finding the active factors in fractionated screening experiments," *Journal of Quality Technology*, 25(2), 94–105.

Box, G. E. P. and Newbold, P. (1971), "Some comments on a paper of Coen, Gomme, and Kendall," *Journal of the Royal Statistical Society, Series A*, 134(2), 229–240.

Box, G. E. P. and Ramírez, J. (1986), "Studies in quality improvement: signal to noise ratios. Performance criteria and statistical analysis: Part II," *Technical Report 12*. Center for Quality and Productivity Improvement, University of Wisconsin–Madison.

Box, G. E. P. and Ramírez, J. (1992), "Cumulative score charts," *Quality and Reliability Engineering*, 8, 17–27.

Box, G. E. P. and Tiao, G. C. (1973), *Bayesian Inference in Statistical Analysis*. John Wiley & Sons, New York.

Box, G. E. P. and Tiao, G. C. (1968a), "A Bayesian approach to some outlier problems," *Biometrika*, 55(1), 119–130.

Box, G. E. P. and Tiao, G. C. (1968b), "Bayesian estimation for means in the random effect model," *Journal of the American Statistical Association*, 63, 174–181.

Box, G. E. P. and Tidwell, P. (1962), "Transformation of the Independent Variables," *Technometrics*, **4**, 531–550.

Box, G. E. P. and Tyssedal, J. (1996), "Projective properties of certain orthogonal arrays," *Biometrika*, **83**(4), 950–955. Also CQPI Report #116, May 1994.

Box, G. E. P. and Tyssedal, J. (1995), Projection Properties of the Sixteen Run Two-Level Orthogonal Arrays, *Technical Report No. 135*, Center for Quality and Productivity, University of Wisconsin, Madison, WI.

Box, G. E. P. and Wetz, J. (1973), "Criteria for judging adequacy of estimation by an approximating response function," *Technical Report No. 9*, Statistics Department, University of Wisconsin, Madison, WI.

Box, G. E. P. and Wilson, K. B. (1951), "On the experimental attainment of optimum conditions," *Journal of the Royal Statistical Society, Series B*, **XIII**(1), 1–45.

Box, G. E. P. and Youle, P. V. (1955), "The exploration and exploitation of response surfaces: an example of the link between the fitted surface and the basic mechanism of the system," *Biometrics*, **11**(3), 287–323.

Box, J. F. (1978), *Fisher, The Life of a Scientist*. John Wiley & Sons, New York.

Box, M. J. (1968), "The occurrence of replications in the optimal designs of experiments to estimate parameters in non-linear models," *Journal of the Royal Statistical Society, Series B*, **30**, 290–302.

Brook, D. and Evans, D. A. (1972), "An approach to the probability distribution of Cusum run length," *Biometrika*, **59**, 539–549.

Blackburn, K. and Lammers, J (1994), *The World Record Paper Airplane Book*. Workmon, New York.

Byrne, D. M. and Taguchi, S. (1986), "The Taguchi approach to parameter design," *ASQC Quality Congress Transactions*, 168–177.

Chaloner, K. (1984), "Optimal Bayesian experimental design for linear models," *The Annals of Statistics*, **12**, 283–300.

Chen, G. G. and Box, G. E. P. (1979), "Further study of robustification via a Bayesian approach," *Technical Summary Report No. 1998*, Mathematics Research Center, University of Wisconsin–Madison.

Chen, G. G. and Box, G. E. P. (1989), "The weighting pattern of a Bayesian estimator" in *Robust Regression*, Lawrence, K. D. and Arthur, J. L. (eds.), Marcel Dekker, New York.

Cheng, C. S. (1995), "Some projection properties of orthogonal arrays," *Annals of Statistics*, **23**, 1223–1233.

Chernoff, H. (1973), "The use of faces to represent points in k-dimensional space graphically," *Journal of the American Statistical Association*, **68**, 361–368.

Cochran, W. G. (1934), "The distribution of quadratic forms in a normal system, with applications to the analysis of covariance," *Proceedings of the Cambridge Philosophy Society*, **30**, 178–191.

Cochran, W. G. (1954), "Some methods for strengthening the common χ^2 tests," *Biometrics*, **10**, 417–451.

Cochran, W. G. (1950), "The comparison of percentages in matched samples," *Biometrika*, **37**, 256–266.

Cochran, W. G. (1973), "Experiments for nonlinear functions," *Journal of the American Statistical Association*, **68**, 771–781.

Cochran, W. G. (1978), "Laplace's ratio estimator," in *Contributions to Survey Sampling and Applied Statistics, Papers in Honor of H. O. Hartley,* H. A. David (ed). Academic Press, New York.

Cochran, W. G. and Cox, G. M. (1957), *Experimental Designs.* John Wiley & Sons, New York.

Coen, P. G., Gomme, E. D., and Kendall, M. G. (1969), "Lagged relationships in economic forecasting,", *Journal of the Royal Statistical Society, Series A,* **132,** 133–152.

Cook, R. D. (1977), "Detection of influential observations in linear regression," *Technometrics,* **19,** 15–18.

Covey-Crump, P. A. K. and Silvey, S. D. (1970), "Optimal regression designs with previous observations," *Biometrika,* **57,** 551–566.

Cox, D. R. (1958), *Planning of Experiments.* John Wiley & Sons, New York.

Cressie, N. (1988), "A graphical procedure for determining non-stationarity in time series," *Journal of the American Statistical Association,* **83,** 1108–1116.

Crouch, T. (1989), *The Bishops Boys: a Life of Wilbur and Orville Wright.* W. W. Norton, New York.

Crowder, S. V. (1986), *Kalman Filtering and Statistical Process Control,* unpublished Ph.D. dissertation, Iowa State University, Dept. of Statistics.

Crowder, S. V. (1987), "A simple method for studying run-length distributions of exponentially weighted moving average charts," *Technometrics,* **29,** 401–408.

Crowder, S. V., Hawkins, D. M., Reynolds, M. R. Jr., and Yashchin, E. (1997), "Process control and statistical inference," *Journal of Quality Technology,* **29,** 134–139.

Daniel, C. (1959), "Use of half-normal plots in interpreting factorial two-level experiments," *Technometrics,* **1,** 311–342.

Daniel, C. (1962), "Sequences of fractional replicates in the 2^{p-g} series," *Journal of the American Statistical Association,* **58,** 403–429.

Daniel, C. (1976), *Applications of Statistics to Industrial Experimentation.* John Wiley & Sons, New York.

Daniels, H. E. (1938), "Some problems of statistical interest in wool research," *Journal of the Royal Statistical Society, Series B,* **5,** 89–112.

Davies, O. L. (ed.) (1954), *The Design and Analysis of Industrial Experiments.* Oliver and Boyd, London.

Davies, O. L. (ed.) (1958), *Statistical Methods in Research and Production.* Hafner Publishing Co., New York.

DeBaun, R. M. (1956), "Block effects in the determination of optimum conditions," *Biometrics,* **12,** 20–22.

De Bono, E. (1967), *The Use of Lateral Thinking.* Penguin Books, London.

DeGroot, M. H. (1987), "A conversation with George Box," *Statistical Science,* **2**(3), 239–258.

Delozier, M. (1994), "Application of response surfaces in evaluating tool performance in metal cutting," manuscript.

Deming, W. E. (1950), *Some Theory of Sampling,* Wiley, New York.

Deming, W. E. (1975), "On probability as a basis for action," *Ain Star,* **29**(4), 146–152.

Deming, W. E. (1986), *Out of the Crisis.* MIT, Center for Advanced Engineering Study, Cambridge, MA.

Dixon, W. J. (1953), "Processing data for outliers," *Biometrics*, **9**, 74–89.

Dodge, H. F. (1969, 1970), "Notes on the evolution of acceptance sampling plans," *Journal of Quality Technology*, **1**, 77–88; **2**, 155–162; **3**, 225–232; **4**, 1–8.

Draper, N. R. (1963), "Ridge analysis of response surfaces," *Technometrics*, **4**, 469–479.

Draper, N. R. and Lawrence, W. E. (1967), "Sequential designs for spherical weight functions," *Technometrics*, **9**, 517–529.

Draper, N. R. and Guttman, I. (1992), "Treating bias as variance for experimental design purposes," *Annals of the Institute of Statistical Mathematics*, **44**, 659–671.

Draper, N. R. and Smith, H. (1998), *Applied Regression Analysis*, 3rd ed. John Wiley & Sons, New York.

Draper, N. R. and Stoneman, D. M. (1964), "Estimating missing values in unreplicated two-level factorial and fractional factorial designs," *Biometrics*, **20**(3), 443–458.

Dobzhansky, T. (1958), "Evolution at work," *Science*, 1091–1098.

DuMouchel, W. and Jones, B. (1994), "A simple Bayesian modification of D-optimal designs to reduce dependence on an assumed model," *Technometrics*, **36**, 37–47.

Duncan, A. J. (1974), *Quality Control and Industrial Statistics*. R. D. Irwin, Homewood, IL.

Eckes, G. (2001) *The Six Sigma Revolution*. John Wiley & Sons, New York.

Ewan, W. D. and Kemp, K. W. (1960), "Sampling inspection of continuous processes with no autocorrelation between successive results," *Biometrika*, **47**, 363–380.

Evans, J. W. (1979), "Computer augmentation of experimental designs to maximize X'X," *Technometrics*, **21**, 321–330.

Fearn, T. and Maris, P. I. (1991), "An application of Box–Jenkins methodology to the control of gluten addition in a flour mill," *Applied Statistics*, **40**, 477–484.

Fedorov, V. V. (1972), *Theory of Optimal Experiments*. Academic Press, New York.

Ferger, D. (1995), "Nonparametric tests for nonstationary change-point problems," *The Annals of Statistics*, **23**, 1848–1861.

Fieller, E. C. (1940), "The biological standardisation of insulin," *Journal of the Royal Statistical Society, Supplement*, **7**, 1–64.

Finney, D. J. (1945), "The fractional replication of factorial arrangements," *Annals of Eugenics*, **12**(4), 291–301.

Finney, D. J. (1951), "On the distribution of a variate whose logarithm is normally distributed," *Journal of the Royal Statistical Society, Supplement*, **7**, 155–161.

Finney, D. J. (1952), *Statistical Method in Biological Assay*. Charles Griffin & Co., London.

Fisher, R. A. (1921), "Studies in crop variation I. An examination of the yield of dressed grain from Broadbalk," *Journal of Agricultural Sciences*, **11**, 107–135.

Fisher, R. A. (1924), "Studies in crop variation III. The influence of rainfall on the yield of wheat at Rothamstead," *Philosophical Transactions of the Royal Society of London, Series B*, **213**, 89–142.

Fisher, R. A. (1925a), *Statistical Methods for Research Workers*, Oliver and Boyd, Edinburgh.

Fisher, R. A. (1925b), "Theory of statistical estimation," *Proceedings of the Cambridge Philosophical Society*, **22**, 700–725.

Fisher, R. A. (1926), "The arrangement of field experiments," *Journal of the Ministry of Agriculture*, **33**, 503–513.

Fisher, R. A. (1935), *The Design of Experiments*. Oliver and Boyd, Edinburgh and London.

Fisher, R. A. (1956), *Statistical Methods and Scientific Inference*. Oliver and Boyd, Edinburgh.

Fisher, R. A. (1963), *Statistical Methods for Research Workers*, 13th edition. Oliver and Boyd, Edinburgh.

Fisher, R. A. (1966), *The Design of Experiments*, 8th edition. Hafner Press, New York.

Fisher, R. A. and Mackenzie, W. A. (1923), "Studies in crop Variation II. The manurial response of different potato varieties," *Journal of Agricultural Science*, **13**, 311–320.

Franklin, N. L., Pinchbeck, P. H., and Popper, F. (1956), "A statistical approach to catalyst development, Part 1. The effect of process variables on the vapour phase oxidation of naphthalene," *Transactions of the Institute of Chemical Engineers*, **34**, 280–293.

Fries, A. and Hunter, W. G. (1980), "Minimum aberration $2^{1} - P$ designs," *Technometrics*, **22**, 601–608.

Fuller, F. T. (1986), "Eliminating complexity from work: improving productivity by enhancing quality," *Center for Quality and Productivity Report No. 17*, University of Wisconsin–Madison.

Fuller, H. and Bisgaard, S. (1996), "A comparison of dispersion effect identification methods for unreplicated two-level factorials," *Center for Quality and Productivity Report No. 132*, University of Wisconsin–Madison.

Fung, C. A. (1986), *Statistical Topics in Off-Line Quality Control*, Ph.D. thesis, University of Wisconsin–Madison.

Gaddum, J. H. (1933), *Reports on Biological Standards III. Methods of Biological Assay Depending on a Quantal Response*, Medical Research Council, Special Report Series, No. 183.

Gaddum, J. H. (1945), "Lognormal distributions," *Nature*, **156**, 463–466.

Galton, F. (1886), "Family likeness in stature," *Proceedings of the Royal Society of London*, **40**, 42–63.

Goel, A. L. and Wu, S. M. (1971), "Determination of A.R.L. and a contour nomogram for Cusum charts to control normal mean," *Technometrics*, **13**, 221–230.

Goldsmith, C. H. and Whitfield, H. W. (1961), "Average run lengths in cumulative chart quality control schemes," *Technometrics*, **3**, 11–20.

Gorman, J. W. and Toman, R. J. (1966), "Selection of variables for fitting equations to data,", *Technometrics*, **8**, 27–51.

Gosset, W. S. (1970), *Letters from W. S. Gosset to R. A. Fisher, 1915–1936*, 2nd edition. Privately circulated.

Graves, S. (1993), "Compensation and employment security: overlooked keys to total quality," *Report No. 104*, University of Wisconsin Center for Quality and Productivity Improvement.

Griffin, B. A., Westman, A. E. R., and Lloyd, B. H. (1989/90), "Analysis of variance," *Quality Engineering*, **2**(2), 195–226.

Grodon, L. and Pollak, M. (1995), "A robust surveillance scheme for stochastically ordered alternatives," *The Annals of Statistics*, **23**, 1350–1375.

Gunter, B. (1987), "A perspective on the Taguchi methods," *Quality Progress*, **20**(6), 44–52.

Hahn, G. J. (1995), "Deming's impact on Industrial Statistics: Source Deflections," *The American Statistician*, **49**, 336–341.

Hahn, G. J., Faltin, F., Tucker, W. T., Richards, S., and Vander Wiel, S. A. (1988), "ASPC: making SPC more proactive," unpublished paper presented at the Fourth

National Symposium on Statistics in Design and Process Control, Arizona State University.

Haldane, J. B. S. (1970), "Karl Pearson, 1857–1957," in *Studies in the History of Statistics and Probability*, E. S. Pearson and M. G. Kendall (eds.). Hafner Press, Darien, CT.

Hall, M. J. (1961), "Hadamard matrices of order 16," *Jet Propulsion Laboratory*, Summary 1, pp. 21–26.

Hamada, I. and Wu, C. F. J. (1986), "A critical look at accumulation analysis and related methods," *Report #20*, Center for Quality and Productivity Improvement, University of Wisconsin–Madison.

Hamada, I. and Wu, C. F. J. (1992), "Analysis of designed experiments with complex aliasing," *J. Qual. Tech.*, **24**(3), 115–173.

Harris, T. J. (1988), "Interfaces between statistical process control and engineering process control," unpublished manuscript.

Harry, M. (1994), *The Vision of Six Sigma. Roadmap for Breakthrough*. Sigma Publishing Company, pp. 60–64.

Harry, M. and R. Schroeder (2000). *Six Sigma – The Breakthrough Management Strategy Revolutionizing the World's Top Corporations*. Currency Doubleday, New York.

Hartley, H. O. and Jayatillake, K. S. E. (1973), "Estimation for linear models with unequal variances," *Journal of the American Statistical Association*, **68**, 189–192.

Hebble, T. L. and Mitchell, T. J. (1972), "Repairing response surface designs," *Technometrics*, **14**, 767–779.

Hedayat, A. and Wallis, W. D. (1978), "Hadamard matrices and their applications," *Annals of Statistics*, **6**, 1184–1238.

Hellstrand, C. (1989), "The necessity of modern quality improvement and some experiences with its implementation in the manufacture of rolling bearings," *Philosophical Transactions of the Royal Society, Industrial Quality Rel.*, 51–56.

Hellstrand, C. (1991), "This example is from: Experiences in applying experimental designs at SKF," presented at the 19th European Meeting of Statistics, Barcelona, September 1991. Used by permission.

Herzberg, A. M. (1979), "Are theoretical designs applicable?" *Operations Research Verfahren/ Methods of Operations Research*, **30**, 68–76.

Herzberg, A. M. (1981), "The robust design of experiments: a review," *Technical Summary Reports 2218*, Mathematics Research Center, University of Wisconsin–Madison.

Herzberg, A. M. and Andrews, D. F. (1976), "Some considerations in the optimal design of experiments in non-optimal situations," *Journal of the Royal Statistical Society, Series B*, **38**, 284–289.

Herzberg, A. M. and Cox, D. R. (1969), "Recent work on the design of experiments: a bibliography and a review," *Journal of the Royal Statistical Society, Series A*, **132**, 29–67.

Hill, W. J. and Hunter, W. G. (1966), "A review of response surface methodology: a literature survey," *Technometrics*, **8**, 571–590.

Hoerl, A. E. (1959), "Optimum solution of many variables equations," *Chemical Engineering Progress*, **55**, 69–78.

Hoerl, A. E. and Kennard, R. W. (1970), "Ridge regression: applications to non-orthogonal problems," *Technometrics*, **12**, 69–82.

Hoerl, R. W. (1985), "Ridge analysis 25 years later," *American Statistician*, **39**, 186–192.

Hoerl, R. W. (1998), "Six sigma and the future of the quality profession," *Quality Progress*, June, 35–42.

Holt, C. C. (1957), "Forecasting trends and seasonals by exponentially weighted moving averages," *O. N. R. Memorandum, No. 52*, Carnegie Institute of Technology.

Hsieh, P. I. and Goodwin, D. E. (1986), "Sheet molded compound process improvement," in *Fourth Symposium on Taguchi Methods*. American Supplier Institute, Dearborn, NJ, pp. 13–21.

Huber, P. J. (1964), "Robust estimation of a location parameter," *Annals of Mathematical Statistics*, **35**, 73–101.

Huber, P. J. (1981), *Robust Statistics*. John Wiley & Sons, New York.

Hunter, J. S. (1983), "The birth of a journal," *Technometrics*, **25**, 3–7.

Hunter, J. S. (1986), "The exponentially weighted moving average," *Journal of Quality Technology*, **18**, 203–210.

Hunter, J. S. (1987), "Signal to noise ratio depicted," *Quality Progress*, **20**(5), 7–8.

Hunter, J. S. (1989), "A one-point equivalent to the Shewhart chart with Western Electric rules," *Quality Engineering*, **2**(1), 13–19.

Hunter, W. G. (1975), "101 Ways to design an experiment," *Technical Report #413*, University of Wisconsin–Madison, Dept. of Statistics.

Hunter, W. G. (1977), "Some ideas about teaching design of experiments, with 25 examples of experiments conducted by students," *American Statistician*, **31**, 1.

Hunter, W. G., O'Neill, J., and Wallen, C. (1987), "Doing more with less in the public sector," *Quality Progress*.

International Mathematical and Statistical Libraries, Inc. (1979), The IMSL, Houston

Ishikawa, K. (1976), "Guide to quality control," Asian Productivity Organization, Tokyo. Available in U.S.A. from UNIPUB, White Plains, NY.

Ishikawa, K. (1982), *Guide to Quality Control*. Kraus International Publications, UNIPUB, White Plains, NY.

Jeffreys, H. (1932), "An alternative to the rejection of observations," *Proceedings of the Royal Society, Series A*, **137**, 78–87.

Johnson, N. L. (1961), "A simple theoretical approach to cumulative sum control chart," *Journal of the American Statistical Association*, **56**, 835–840.

Johnson, N. L. and Kotz, S. (1970), *Distributions in Statistics Continuous Univariate Distributions I*. Houghton Mifflin, Boston.

Joiner, B. L. (1994), *Forth Generation Management*. McGraw-Hill, New York.

Jones, E. R. and Mitchell, T. J. (1978), "Design criteria for detecting model inadequacy," *Biometrika*, **65**, 541–551.

Jones, S. P. (1990), *Designs for Minimizing the Effect of Environmental Variables*, Ph.D. thesis, University of Wisconsin–Madison.

Jowett, G. H. (1952), "The accuracy of systematic sampling from conveyor belts," *Applied Statistics*, **1**, 50–59.

Jowett, G. H. (1955), "The comparison of means of sets of observations from sections of independent stochastic series," *Journal of the Royal Statistical Society, Series B*, **17**, 208–227.

Juran, J. M. (1988), *Quality Control Handbook*, 4th edition. McGraw-Hill, New York.

Kackar, R. N. (1985), "Off-line quality control parameter design, and the Taguchi method (with discussion)," *Journal of Quality Technology*, **17**(4), 176–209.

Kempthorne, O. (1952), *The Design and Analysis of Experiments*. John Wiley & Sons, New York.

Khuri, A. I. and Cornell, J. A. (1996), *Response Surfaces: Designs and Analyses*, 2nd edition. Marcel Dekker, New York.

Kiefer, J. (1958), "On the nonrandomized optimality and the randomized nonoptimality of symmetrical designs," *Annals of Mathematical Statistics*, **29**, 675–699.

Kiefer, J. (1959), "Optimum experimental designs," *Journal of the Royal Statistical Society, Series B*, **21**, 272–319.

Kiefer, J. (1975), "Optimal design: variation in structure and performance under change of criterion," *Biometrika*, **62**, 277–288.

Kiefer, J. and Studden, W. J. (1976), "Optimal designs for large degree polynomial regression," *Annals of Statistics*, **4**, 1113–1123.

Kiefer, J. and Wolfowitz, J. (1959), "Optimum designs in regression problems," *Annals of Mathematical Statistics*, **30**, 271–294.

Kiefer, J. and Wolfowitz, J. (1960), "The equivalence of two extremum problems," *Canadian Journal of Mathematics*, **12**, 363–366.

Kim, K. J. and Lin, K. J. (1998), "Dual Response Surface Optimization: A Fuzzy modelling approach," *Journal of Quality Technology*, **30**, 1–11.

Kramer, T. (1989), *Process Control from an Economic Point of View*, unpublished Ph.D. dissertation, University of Wisconsin–Madison, Dept. of Statistics.

Kuhn, T. (1962), *The Structure of Scientific Revolutions*. University of Chicago Press, Chicago.

Kulahci, M. (2000), *Applications of Plackett and Burman Designs to Split Plot Designs*, Unpublished Thesis, University of Wisconsin–Madison.

Lai, T. L. (1995), "Sequential changepoint detection in quality control and dynamical systems," *Journal of the Royal Statistical Society, Series B*, **57**, 613–658.

Ledolter, K. L. (1997), "Dorian Shainin's Variables Search Procedure: A Critical Assessment", *Journal of Quality Technology*, **29**, 237–247.

Lenth, R. V. (1989), "Quick and easy analysis of unreplicated factorials," *Technometrics*, **31**, 469–473.

León, R. V., Shoemaker, A. C. and Kacker, R. N. (1987), "Performance measures independent of adjustment" (with discussion), *Technometrics*, **29**, 253–285.

Levine, D. and Tyson, L. D. (1990), "Participation, productivity, and the firm's environment," in *Paying for Productivity*, A. S. Blinder (ed.). Brookings Institute, Washington, DC, pp. 183–243.

Lin, K.-T. and Draper, N. R. (1992), "Projection properties of Plackett and Burman designs," *Technometrics*, **34**(4), 423–428.

Lin, D. K. (1993), "Another look at first-order saturated designs: the *p*-efficient designs," *Technometrics*, **35**, 284–294.

Lindley, D. V. (1965), *Introduction to Probability and Statistics from a Bayesian Viewpoint, Part 2, Inference*. Cambridge University Press, Cambridge, U.K.

Lindley, D. V. and Smith, A. F. M. (1972), "Bayes estimates for the linear model," *Journal of the Royal Statistical Society, Series B*, **34**, 1–41 (with discussion).

Lord Acton (1887), *Letter to Bishop Mandell Creighton.*

Lorden, G. (1971), "Procedures for reacting to a change in distribution," *Annals of Mathematical Statistics*, **6**, 1897–1908.

Lorenzen, T. J. and Vance, L. C. (1986), "The economic design of control charts: a unified approach," *Technometrics*, **28**, 3–10.

Lucas, J. M. (1976), "Which response surface design is best: a performance comparison of several types of quadratic response surface designs in symmetric regions," *Technometrics*, **18**, 411–417.

Lucas, J. M. (1996), "The 1996 W. J. Youden Address: System Change and Improvement: Guidelines for Action When the System Resists," *American Statistical Association Proceedings.*

Lucas, J. M. and Crosier, R. B. (1982), "Fast initial response for CUSUM quality-control schemes: give your CUSUM a head start," *Technometrics*, **24**, 199–206.

Luceño, A. (1993), "Performance of EWMA versus last observation for feedback control," *Communications in Statistics—Theory and Methods*, **22**, 241–255.

Luceño, A., Gonzalez, F. J., and Puig-Pey, J. (1996), "Computing optimal adjustment schemes for the general tool-wear problem," *Journal of Statistical Computation and Simulation*, **54**, 87–113.

Luceño, A., and Box, G. E. P. (2000), "Influence of sampling interval, design limit, and autocorrelation on the average run length in cusum charts, " *Journal of Applied Statistics*, **27**(2), 172–182.

MacGregor, J. F. (1972), *Topics in the Control of Linear Processes Subject to Stochastic Disturbances*, unpublished Ph.D. dissertation, University of Wisconsin–Madison, Dept. of Statistics.

MacGregor, J. F. (1987), "Interfaces between process control and on line statistical control," *Computing Systems Technology Division Communications*, **10**, 9–20.

MacGregor, J. F. (1988), "On line statistical process control," *Chemical Engineering Progress*, Oct. 1988, 21–31.

MacGregor, J. F. (1990), "A different view of the funnel experiment," *Journal of Quality Technology*, **22**, 255–259.

Mantel, N. (1963), "Chi-square tests with one degree of freedom; extensions of the Mantel–Haenszel procedure," *Journal of the American Statistical Association*, **58**, 690–700.

Margolin, B. H. (1968), "Orthogonal main-effect $2^n 3^m$ designs and two-factor interaction aliasing,", *Technometrics*, **10**(3), 559–573.

Margolin, B. H. (1969), "Results on Factorial Designs of Resolution IV for the 2^n and $2^n 3^m$ series," *Technometrics*, **11**, 431–444.

Mayr, O. (1970), *The Origins of Feedback Control.* MIT Press, Cambridge, MA.

Mead, R. and Pike, D. J. (1975), "A review of response surface methodology from a biometric viewpoint," *Biometrics*, **31**, 803–851.

Meyer, R. D. (1996) mdopt: FORTRAN programs to generate MD-optimal screening and follow-up designs, and analysis of data. Statlib. URL: http//lib.statlib.cmu.cdu.

Meyer, R. D. and Box, G. E. P. (1992). "Finding the Active Factors in Fractionated Screening Experiments:" Technical Report No 80. Center for Quality and Productivity.

Meyer, R. D., Steinberg, D. M. and Box, G. E. P. (1996) "Follow-up designs to resolve confounding in multifactor experiments," *Technometrics* **38** (9), 303–313.

Michaels, S. E. (1964), "The usefulness of experimental design" (with discussion), *Journal of Applied Statistics*, **13**(3), 221–235.

Mitchell, T. J. and Miller, F. L. Jr. (1970), "Use of design repair to construct designs for special linear models," *Mathematical Division Annual Progress Report* (ORNL-4661), Oak Ridge National Laboratory, Oak Ridge, TN.

Montgomery, D. C. (1991), *Introduction to Statistical Quality Control*, 2nd edition. John Wiley & Sons, New York.

Morrison, S. J. (1957), "The study of variability in engineering design," *Applied Statistics*, **6**(2), 133–138.

Mosteller, F. and Tukey, J. W. (1977), *Data Analysis and Regression, a Second Course in Statistics*. Addison-Wesley, Reading, MA.

Muth, J. F. (1960), "Optimal properties of exponentially weighted forecasts of time series with permanent and transitory components," *Journal of the American Statistical Association*, **55**, 299–306.

Myers, R. H. and Montgomery, D. C. (1995), *Response Surface Methodology: Process and Product Optimization Using Designed Experiments*. John Wiley & Sons, New York.

Nair, V. N. (1986), "Testing in industrial experiments with ordered categorical data," *Technometrics*, **28**, 283–291.

Nair, V. N. (1992) (ed.), "Taguchi's parameter design: a panel discussion," *Technometrics*, **34**(2), 127–161.

Nair, V. N. and Pregibon, D. (1986), "A data analysis strategy for quality engineering experiments," *AT&T Technical Journal*, **65**, 73–84.

Nair, V. N. and Pregibon, D. (1987), "Analysis of dispersion effects: When to log?" *Technometrics*.

Nair, V. N. and Pregibon, D. (1988), "Analyzing dispersion effects from replicated factorial experiments," *Technometrics*, **30**(3), 247–257.

Nalimov, V. V., Golikovo, T. I., and Mikeshina, N. G. (1970), "On practical use of the concept of D-optimality," *Technometrics*, **12**, 799–812.

Nelder, J. A. (1998), "The selection of terms in response surfaces models—How strong is the weak-heredity principle?" *The American Statistician*, **52**, 315–318.

Neyman, J. and Pearson, B. S. (1931), *Bull. Int. Acad. Cracovie, A, p. 460*.

Occam (1330), William of, *Opus Nonaginta Dierum*.

Ott, E. R. (1975), *Process Quality Control*. McGraw-Hill, New York.

O'Shaughnessy, A. W. E. (1939), "The music makers," in *The Oxford Book of English Verse*. Oxford University Press, Oxford.

Page, E. S. (1954), "Continuous inspection schemes," *Biometrika*, **41**, 100–115.

Page, E. S. (1957), "On problems in which a change in a parameter occurs at some unknown point," *Biometrika*, **44**, 248–252.

Page, E. S. (1961), "Cumulative sum charts," *Technometrics*, **3**, 1–9.

Paude, P., Neuman, R. and Cavanagh, R. (2001) *The Six Sigma Way*. McGraw-Hill. New York.

Parkinson, C. N. (1957), *Parkinson's Law and Other Studies in Administration*. Houghton Mifflin, Boston.

Parry-Jones, R. (1999), *Engineering for Corporate Success in the New Millennium*. Royal Academy of Engineering, Westminster, London.

Pearson, E. S. (1938), "Discussion of H. E. Daniels' article: some problems of statistical interest in wool research," *Journal of the Royal Statistical Society, Series B*, **5**, 89–112.

Phadke, M. S. (1982), "Quality engineering using design of experiments," *Proceedings of the Section on Statistical Education*, American Statistical Association, pp. 11–20.

Phadke, M. S., Kackar, R. N., Speeney, D. V., and Grieco, M. J. (1983), "Off-line quality control in integrated circuit fabrication using experimental design," *The Bell System Technical Journal*, **62**, 1273–1309.

Pinchbeck, P. H. (1957), "The kinetic implications of an empirically fitted yield surface for the vapour phase oxidation of naphthalene to phthalic anhydride," *Chemical Engineering Science*, **6**, 105–111.

Plackett, R. L. and Burman, J. P. (1946), "The design of optimum multifactorial experiments," *Biometrika*, **33**, 305–325 and 328–332.

Potter, V. R. (1990), "Getting to the year 3000: Can bioethics overcome evolution's fatal flaw?" *Perspectives in Biological Medicine*, **34**, 89–98.

Pukelshein, F. (1980), "On linear regression designs which maximize information," *Journal of Statistical Planning and Information*, **4**, 339–362.

Quinlan, J. (1985), "Product improvement by application of Taguchi methods," in *Third Supplier Symposium on Taguchi Methods*. American Supplier Institute, Inc., Dearborn, MI.

Raghavarao, D. (1971), *Construction and Combinatorial Problems in Design of Experiments*. John Wiley & Sons, New York.

Ramirez, J. G. (1989), *Sequential Methods in Statistical Process Monitoring*, unpublished Ph.D. dissertation, University of Wisconsin–Madison, Dept. of Statistics.

Rao, C. R. (1947), "Factorial experiments derivable from combinatorial arrangements of arrays," *Journal of the Royal Statistical Society, Series B*, **9**, 128–139.

Roberts, S. W. (1959), "Control chart test based on geometric moving averages," *Technometrics*, **1**(3), 239–250.

Roberts, S. W. (1966), "A comparison of some control chart procedures," *Technometrics*, **8**, 411–430.

Scholtes, P. R. (1988), *The Team Handbook*. Joiner Associates, Madison, WI.

Scientific Computing Associates, P.O. Box 625, Dekalb, IL 60115.

Segan, J. and Sanderson, A. C. (1980), "Detecting change in a time series," *IEEE Transactions on Information Theory*, 11–26, 249–254.

Seiden, E. (1954), "On the problem of construction of orthogonal arrays," *Annals of Mathematical Statistics*, **25**, 151–156.

Shewhart, W. A. (1931), *Economic Control of Quality of Manufactured Product*. D. Van Nostrand, New York. (Republished in 1981, with a dedication by W. Edwards Deming, by the American Society for Quality Control, Milwaukee, WI.)

Shewhart, W. A. (1939), *Statistical Method from the Viewpoint of Quality Control*. Graduate School Department of Agriculture, Washington, DC.

Shoemaker, A. C., Tsui, K. L., and Wu, C. F. J. (1991), "Economical experimentation methods for robust design," *Technometrics*, **33**(4), 415–427.

Siegmund, D. (1985), *Sequential Analysis: Tests and Confidence Intervals*. Springer-Verlag, New York.

Sloane, N. J. A. (2005) "A Library of Hadamard Matrices," URL: www.research.att.com/ ~njas/hadamard.

Silvey, S. D. and Titterington, D. M. (1973), "A geometrical approach to optimal design theory," *Biometrika*, **60**, 21–32.

Slutsky, E. (1927), "The summation of random causes as the source of cyclic processes" (in Russian), *Problems of Economic Conditions*, **3**. English translation in *Econometrica*, **5**, 105 (1937).

Smith, W. (1992), "Presentation at the case study conference, Center for Quality and Productivity Improvement, University of Wisconsin–Madison.

Snedecor, G. W. and Cochran, W. G. (1980), *Statistical Methods*, 7th edition. The Iowa State University Press, Ames.

Snee, R. and R. Hoerl (2003), *Leading Six Sigma*, Prentice-Hall, Upper Saddle River, NJ.

Springer, S. P. and Deutsch, G. (1981), *Left Brain, Right Brain*. W. H. Freeman, San Francisco.

Srivastava, M. S. and Wu, Y. (1991), "A second order approximation to Taguchi et al.'s on-line control procedure," *Communications in Statistics—Theory and Methods*, **20**, 2149–2168.

Srivastava, M. S. and Wu, Y. (1991), "Taguchi's on-line control procedures and some improvements," *Technical Report 9121*, University of Toronto, Dept. of Statistics.

Stallings, B. (1993), "The new international context of development," *Items, Social Research Council*, **47**, 1–11.

Steel, R. G. D. and Torrie, J. H. (1980), *Principles and Procedures of Statistics: A Biometrical Approach*, 2nd edition. McGraw-Hill, New York.

Stein, C. (1956), "Inadmissibility of the usual estimator of the mean of a multivariate normal distribution," in *Proceedings of the Third Berkeley Symposium* (Vol. 1). University of California Press, Berkeley, pp. 197–206.

Steinberg, D. M. (1985), "Model robust response surface designs: scaling two-level factorials," *Biometrika*, **72**, 513–526.

Stewart, W. E., Henson, T. L., and Box, G. E. P. (1996), "Model discrimination and criticism with single-response data," *AIChE Journal*, **42**, 3055–3062.

Stewart, W. E., Shon, Y., and Box, G. E. P. (1998), "Discrimination and goodness of fit of multiresponse mechanistic models," *AIChE Journal*, **44**, 1404–1412.

Taguchi, G. (1974), "A new statistical analysis for clinical data, the accumulating, in contrast with the chi-square test," *Saishin Igaku (The Newest Medicine)*, **29**, 806–813.

Taguchi, G. (1981), *On-Line Quality Control During Production*. Japanese Standard Association, Tokyo.

Taguchi, G. (1986), *Introduction to Quality Engineering: Designing Quality into Products and Processes*. Kraus International Publications, White Plains, NY.

Taguchi, G. (1987), *System of Experimental Design*, Vols. I and II. UNIPUB, Kraus International Publications, White Plains, NY.

Taguchi, G., Elsayed, E. A., and Hsiang, T. (1989), *Quality Engineering in Production Systems*. McGraw-Hill, New York.

Takeuchi, K. and Hirotsu, C. (1982), "The cumulative chi-squares method against ordered alternatives in two-way contingency tables, reports of statistical application research," *Japanese Union of Scientists and Engineers*, **29**, 1–13.

Taguchi, G. and Phadke, M. S. (1984), "Quality engineering through design optimization," in *Conference Record Vol. 3, IEEE Globecom Conference*. IEEE, New York, pp. 1106–1113.

Taguchi, G. and Wu, Y. (1980), *Introduction to Off-Line Quality Control*. Central Japan Quality Control Association, Nagoya, Japan.

Taylor, W. A. (1991), *Optimization and Variation Reduction for Quality*. McGraw-Hill, New York.

Theil, H. (1963), "On the use of incomplete prior information in regression analysis," *Journal of the American Statistical Association*, **58**, 401–414.

Thurow, L. (1992), *Head to Head*. William Morrow and Company, New York.

Tiao, G. C. and Ali, M. M. (1971), "Analysis of correlated random effects linear model with two random components," *Biometrika*, **58**, 37–52.

Tippett, L. H. C. (1935), "Some applications of statistical methods to the study of the variation of quality of cotton yarn," *Journal of the Royal Statistical Society, Series B*, **1**, 27–55.

Titterington, D. M. (1975), "Optimal design: some geometrical aspects of D-optimality," *Biometrika*, **62**, 313–320.

Tribus, M. and Szonyi, G. (1989), "An alternative view of the Taguchi approach," *Quality Progress*, **22**(5), 46–52.

Tukey, J. W. (1960), "A survey of sampling from contaminated distributions," in *Contributions to Probability and Statistics: Essays in Honor of Harold Hotelling*. Stanford University Press, Stanford, pp. 448–485.

Tukey, J. W. (1977), *Exploratory Data Analysis*. Addison-Wesley, Reading, MA.

Van Dobben de Bruyn, C. S. (1968), *Cumulative Sum Tests—Theory and Practice* Hafner Publishing, New York.

Vining, G. G. and Myers, R. H. (1990), "Combining Taguchi and response surface philosophies: a dual response approach," *Journal of Quality Technology*, **22**(1), 38–45.

Vander Wiel, S. A. and Tucker, W. T. (1989), *Algorithmic Statistical Process Control: Literature Review, Implementation, and Research Opportunities*. Management Science and Statistics Program, Corporate Research and Development, General Electric Company.

Wald, A. (1947), *Sequential Analysis*. John Wiley & Sons, New York.

Wallis, W. A. (1980), "The Statistical Research Group 1942–45", *Journal of the American Statistical Association*, **75**, 320–335.

Whittle, P. (1963), *Prediction and Regulation by Linear Least-Squares Methods*. English Universities Press, London.

Whittle, P. (1973), "Some general points in the theory of optimal experimental design," *Journal of the Royal Statistical Society, Series B*, **35**, 123–130.

Wilcoxon, F. (1949), *Some Rapid Approximate Statistical Procedures*. American Cyanamid Company, Stanford, CT.

Wilson, G. T. (1970), *Modelling Linear Systems for Multivariate Control*, unpublished Ph.D. thesis, Dept. of System Engineering, University of Lancaster, England.

Winters, P. R. (1960), "Forecasting sales by exponentially weighted moving averages," *Management Science*, **6**, 324–342.

Wold, H. (1954), *A Study in the Analysis of Stationary Time Series*. Almquist Wiksell Book Co., Uppsala, Sweden.

Wold, H. (1966), "Nonlinear estimation by iterative least squares procedures," in *Research Papers in Statistics, Festschrift for J. Neyman*, F. N. David (ed.). John Wiley & Sons, New York.

Woodall, W. H. and Adams, B. M. (1993), "The statistical design of Cusum charts," *Quality Engineering*, **5**, 559–570.

Wynn, H. P. (1972), "Results in the theory and construction of D optimum experimental designs," *Journal of the Royal Statistical Society, Series B*, **34**, 133–147 (with discussion, pp. 170–185).

Yates, F. (1933), "The analysis of replicated experiments when the field results are incomplete. Emp.," *Journal of Experimental Agriculture*, **1**, 129–142.

Yates, F. (1937), *The Design and Analysis of Factorial Experiments*. Imperial Bureau of Soil Science, Harpenden, U.K.

Yates, F. (1948), "The analysis of contingency tables with groupings based on quantitative characters," *Biometrika*, **35**, 176–181.

Yates, F. (1967), "A fresh look at the basic principles of the design and analysis of experiments," in *Proceedings of the 5th Berkeley Symposium on Mathematical Statistics and Probability*, vol. IV, pp. 777–790. University of California Press, Berkeley.

Yates, F. (1970), *Experimental Design: Selected Papers of Frank Yates*. Hafner Publishing, Darien, CT.

Yates, F. (1985), "Complex experiments," *Journal of the Royal Statistical Society, Series B*, **2**, 181–223 (discussion pp. 223–247).

Youden, W. J. (1937), "Use of incomplete block replications in estimating tobacco-mosaic virus," *Contributions from Boyce Thomson Institute*, **IX**, 91–98.

Yule, G. U. (1927), "On a method of investigating periodicities in disturbed series," *Philosophical Transactions of the Royal Society of London, Series A*, **226**, 207.

Biography

Born October 18, 1919 in Gravesend, England, George E. P. Box did undergraduate work in chemisty at London University. During World War II, he worked with experimenters designing and analyzing experiments at the Chemical Defense Experiment Station in Porton, England while serving in the British Army. After the war, he earned his Bachelors of Science in Mathematical Statistics at London University. He later received a Ph.D. in Mathematical Statistics and a Doctor of Science degree from that unversity in 1952 and 1961. He has received honorary doctorate degrees from the University of Rochester, Carnegie-Mellon University, the University Carlos III Madrid, the University of Waterloo, Canada, and the Conservatoire National des Arts et Métiers, Paris.

Box's career in statistics started at Imperial Chemical Industries in England as Head of the Statistical Techniques Research Section. On leave from ICI, he served as a Visiting Research Professor at the University of North Carolina. He later returned to the United States, as the Director of Statistical Techniques Research Group at Princeton University in 1957. He became professor and the founding chairman of the Department of Statistics at the University of Wisconsin at Madison in 1960. In the period 1965 to 1971, between being professor and chairman (until 1968) of the Department of Statistics at UW–Madison, he was a visiting Ford Foundation Professor at the Harvard Business School, visiting Professor at the University of Essex in Colchester, England and served in a joint appointment at the Engineering Experiment Station at UW–Madison. In 1971, he was inducted into newly created Ronald Aylmer Fisher Chair of Statistics at the University of Wisconsin–Madison. In 1980, he was appointed distinguished Vilas Research Professor of Mathematics and Statistics at UW–Madison. From 1986, he has served as Director and Research Director of the Center or Quality Productivity Improvement at the university. He co-founded the latter in 1983 with Professor William Hunter and it has been from this base that much of George Box's research work on quality improvement reported in this volume, began and developed into practice.

George Box has written more than 200 papers and has coauthored 9 books, in a long and distinguished career. He has been recognized extensively by his peers and international organizations in the sciences and has received more than 25 honors including election as a Fellow of the Royal Society, honorary membership of the American Quality Society (ASQ) the Shewhart Medal, the Deming Medal, and the Youden and Brumbaugh awards. He has also received the Wilks Memorial Medal from the American Statistical Association and U.S. Army, the Royal Statistical Society Guy medals in Silver and in Gold, the Guggenheim Fellow award, and the Byron Bird Award for Excellence in Engineering Research. He was elected President of ASA and of the Institute of Mathematical Statistics (IMS). He is a fellow of the American Academy of Arts and Sciences, the ASA, IMS and RSS.

Indeed Dr. Box has been a highly regarded and recognized contributor and leader in numerous professional organizations and sciences. His work has bridged statistics with engineering, the physical sciences, and quality and productivity improvement. This book research work focuses on the area of quality and productivity improvement. His clear and concise writings on the concepts and methodology of design of experiments, control, and robustness among other techniques have helped the practitioner of quality improvement methods drive major changes in industry and business. Those corporations that have had major quality and productivity efforts have extensively used the methods of George Box and his co-workers. Methodologies, such as fractional factorial designs, response surface methods, evolutionary operation, time series analysis, have become part of the arsenal used by engineers, business majors, scientists, technicians; and others to drive improvement in their processes by reducing variation and cost while increasing throughput. Although based in statistical rigor, the practical aspect of his work as illustrated by the material just presented, is what sets it apart. George thinks like the scientist, the engineer and the business person who need to use statistical thinking and the scientific method to drive results. Some of this comes from his time in industry and his consulting and teaching experiences at Ford Motor Company, Monsanto, Hewlett-Packard, Boeing, the World Bank, and Federal Reserve Board, plus many other organizations and businesses.

George Box lives with wife Claire in Oregon, Wisconsin. He is known for his Monday night "beer parties" where students, faculty, and visitors listen to and discuss new research ideas presented by a guest. The informal atmosphere has helped generate new research, as well as life-long friendships among those who attended. Just as in his home, he has led many stimulating research discussions at the Gordon Research Conferences on Statistics in Chemistry and Chemical Engineering over the last 50 years. To the enjoyment of all, he is well known for his stories, skits, and singing at the Thursday night talent party. No one forgets the new learning received through his or her contact with George Box, nor the fun time they had in the process.

Books and Articles Written by George Box from 1982 to 2005

Books

1. *The Collected Works of George E. P. Box*, Two Volumes, G. C. Tiao et al. (eds.). Wadsworth, Belmont, CA, 1985.
2. *Empirical Model-Building and Response Surfaces* (with N. R. Draper). John Wiley & Sons, New York, 1987
3. *Statistical Control by Monitoring and Feedback Adjustment* (with A. Luceño). John Wiley & Sons, New York, 1997.

Articles

1. "An apology for ecumenism in statistics" (1983), in *Scientific Inference, Data Analysis, and Robustness*, G. E. P. Box, Tom Leonard, and Chien-Fu Wu (eds.), Academic Press New York, pp. 51–84.
2. "Gwilym Jenkins, experimental design and the time series" (1983), *Questiio*, 7(4), 515–525.
3. "The importance of practice in the development of statistics" (1984), *Technometrics*, 26(1), 1–8.
4. H. Kanemasu,. " Constrained nonlinear least squares" (1984), in *Essays in Honor of Oscar Kempthorne, Contributions to Experimental Design Linear Models and Genetic Statistics*. Marcel Dekker, New York.
5. "Experimental design for product improvement" (1985), *Communications in Statistics — Theory Methods*, 14(11), 2605.
6. R. D. Meyer, "Some new ideas in the analysis of screening designs" (1985), *Journal of Research of the National Bureau of Standards*, 90(6), 495–502.
7. "Discussion" (1985), *Journal of Quality Technology*, 17(4), 189–190.

8. R. D. Meyer, "An analysis for unreplicated fractional factorials" (1986), *Technometrics*, **28**(1), 11–18.

9. R. D. Meyer. "Dispersion effects from fractional designs" (1986), *Technometrics*, **28**(1), 19–27.

10. S. Jones, "Discussion of 'Accumulation analysis' a paper by V. N. Nair" (1986), *Technometrics*, **28**(4), 295–301.

11. C. Fung, "Studies in quality improvement: minimizing transmitted variation by parameter design" (1986), CQPI Report No. 8.

12. S. Jones, "An investigation of the method of accumulation analysis" (1986), *Technometrics*, **28**(4), 295–301. Also *Total Quality Management*, **1**(1), 101–113 (1990).

13. "Anatomy of some time series models" (1984), *Statistics: An Appreciation*, H. A. David and H. T. David (eds.), Iowa State University Press, Ames, pp. 489–508.

14. R. D. Meyer, "Analysis of unreplicated factorials allowing for possibly faulty observations" (1987), *Design, Data, and Analysis*, Colin Mallows (ed.). Wiley, New York, pp. 1–12. Also CQPI Report #3, February 1986.

15. D. A. Pierce and Paul Newbold, "Estimating trend and growth rates in seasonal time series" (1987), *Journal of the American Statistical Association*, **82**(397), 276–282.

16. D. Peña, "Identifying simplifying structure in time series" (1987), *Journal of American Statistical Association*, **82**(399), 836–843.

17. "In Memoriam: William G. Hunter" (1987), *Technometrics*, **29**(3), 251–252.

18. C. A. Fung, "Discussion of 'Performance measures independent of adjustment" a paper by R. Léon, A. Shoemaker, and R. Kackar" (1987), *Technometrics*, **29**(3), 270.

19. R. Kackar, V. N. Nair, M. Phadke, A. Shoemaker, and C. F. Wu, "On quality practice in japan" AT&T Statistical Research Report No. 45, (1987); and *Quality Progress*, March 1988, pp. 37–41.

20. S. Bisgaard, "The scientific context of quality improvement" (1987), *Quality Progress*, 54–62.

21. "Statistical design and analysis in quality improvement: the signal to noise ratio and transformation" (1987), in *Proceedings of the 46th Session of ISI*, Tokyo, Japan.

22. R. D. Meyer, "Some aspects of statistical design in quality improvement" (1987), in *Proceedings from the Second International Tampere Conference in Statistics* held in Tampere, Finland.

23. M. DeGroot (ed.), "A conversation with George Box" (1987), *Statistical Science*, **2**(3), 239–258.

24. R. D. Meyer, "Identification of active factors in unreplicated fractional factorial experiments" (1987), CQPI Report No. 23, February.

25. C. A. Fung, "Minimizing transmitted variation by parameter design" (1986), CQPI Report No. 8.

26. S. Bisgaard, "Statistical tools for improving designs" (1988), *Mechanical Engineering*, **110**(1), 32–40.

27. "Signal to noise ratios, performance criteria and transformations" (1988), *Technometrics*, **30**(1), 1–17.

28. S. Bisgaard and C. A. Fung, "An explanation and critique of Taguchi's contributions to quality engineering" (1988), *Quality and Reliability Engineering International*, **4**(2), 123–131.

29. P. M. Berthouex and A. Darjatmoko, "Discrimination upset analysis" (1988), CQPI Report No. 30, May.

30. "Quality Improvement: an expanding domain for the application of the scientific method" (1989), *Proceedings of the Royal Society, Industrial Quality and Reliability*, 139–152.

31. "When Murphy speaks—listen" (1989). *Quality Progress*, **22**(10), 79–84.

32. L. Joiner, S. Rohan and F. J. Sensenbrenner, "Quality in the community—one city's experience" (1989), *Quality Progress*.

33. S. Jones, "An investigation of the method of accumulation analysis" (1990), *Total Quality Management*, *1*(1), 101–113.

34. *Opinion: One Important Idea* Vol. 2, No. 3. (1989), Springer-Verlag, New York, p. 8.

35. T. Kramer, "Industrial process control—a multifaceted problem" (1990), in *ASQC's 44th Annual Quality Congress Transactions*, San Francisco, CA, pp. 86–95.

36. "Do interactions matter?" (1990), *Quality Engineering*, **2**(3), 365–369.

37. "Must we randomize our experiment?" (1990), *Quality Engineering*, **2**(4), 497–502.

38. "Good quality costs less? How come?" (1990–91), *Quality Engineering*, **3**(1), 85–90.

39. "A simple way to deal with missing observations from designed experiments" (1990–91), *Quality Engineering*, **3**(2), 249–254.

40. "Commentary on 'Communications between statisticians and engineers/physical scientists' by Hoadley and Kettenring" (1990), *Technometrics*, **32**(3).

41. "Research and training for quality improvement" (1990), *Resources in Education* (RAE), May.

42. "Discussion of the paper 'The unity and diversity of probability' by Glen Schafer" (1990), *Statistical Science*.

43. "Discussion of the paper 'Application in business and economic statistics—some personal views' by Harry V. Roberts" (1990), *Statistical Science*.

44. G. G. Chen, "The weighting pattern of a Bayesian robust estimator" (1990), *Robust Regression: Analysis and Application*. Marcel Dekker, New York, pp. 3–21.

45. "Finding bad values in factorial designs" (1991), *Quality Engineering*, **3**(3), 405–410.

46. "Understanding exponential smoothing: a simple way of forecast sales and inventory" (1991), *Quality Engineering*, **3**(4), 561–566.

47. "Feedback control by manual adjustment" (1991), *Quality Engineering*, **4**(1), 143–151.

48. S. Jones. "Designing products that are robust to the environment" (1992), *Journal of Total Quality Management*, **3**(3), 265–282.

49. J. Ramírez, "Cumulative score charts" (1992), *Quality and Reliability Engineering*, **8**, 17–27.

50. S. Jones, "Split-plot designs for robust product experimentation" (1992), *Journal of Applied Statistics*, **19**(1), 3–26.

51. T. Kramer, "Statistical process control and automatic process control—a discussion" (1992), *Technometrics*, **34**(3), 251–267.

52. "Bounded adjustment charts" (1991–92), *Quality Engineering*, **4**(2), 331–338.

53. "Teaching engineers experimental design with a paper helicopter" (1992), *Quality Engineering*, **4**(3), 453–459.

54. "What can you find out from eight experimental runs?" (1992), *Quality Engineering*, **4**(4), 619–627.

55. Joanne Wendelberger, "Identification and estimation of sources of transmitted variation" (1992), in *Proceedings of the Section on Physical and Engineering of the American Statistical Association*, pp. 71–76.

56. "Taguchi's parameter design: a panel discussion" (1992), discussants B. Abraham, G. Box, R. Kacker, T. Lorenzen, J. Lucas, J. MacKay, R. Myers, J. Nelder, M. Phadke, J. Sacks, A. Shoemaker, S. Taguchi, K. Tsui, G. Vining, W. Welch, and J. Wu., *Technometrics*, **34**(2), 127–161.

57. "Quality improvement—the new industrial revolution" (1993), *International Statistical Review*, **61**(1), 3–19.

58. "Sequential experimentation and sequential assembly of designs" (1993), *Quality Engineering*, **5**(2), 321–330.

59. "What can you find out from twelve experimental runs?" (1993), *Quality Engineering*, **5**(4), 663–668.

60. "What can you find out from sixteen experimental runs?" (1993), *Quality Engineering*, **5**(1), 167–178.

61. R. D. Meyer, "Finding the active factors in fractionated screening experiments" (1993), *Journal of Quality Technology*, **25**(2), 94–105.

62. "Process adjustment and quality control" (1993), *Total Quality Management*, **4**(2), 215–227.

63. "Sequential experimentation and sequential assembly of designs" (1993), *Quality Engineering*, **5**(2), 321–330.

64. "How to get lucky" (1993), *Quality Engineering*, **5**(3), 517–524.

65. "What engineers need to learn about statistics" (1993), in *Proceedings of the first scientific meeting of the international association for statistical education, Perugia*, 23–24 August 1993.

66. "Changing management policy to improve quality and productivity" (1994), *Quality Engineering*, **6**(4), 719–724.

67. C. Fung, "Is your robust design procedure robust?" (1993–4), *Quality Engineering*, **6**(3), 503–514.

68. A. Luceño, "Selection of sampling interval and action limit for discrete feedback adjustment" (1994), *Technometrics*, **36**(4), 369–378.

69. "Statistics and quality improvement" (1994), *Journal of the Royal Statistical Society, Series A*, **157**(2), 209–229.

70. A. Luceño, "Discrete proportional-integral control with constrained adjustment" (1995), *The Statistician, JRSS Series D*, **44**, 479–495.

71. J. Tyssedal, "Projective properties of certain orthogonal arrays" (1996), *Biometrika*, **83**(4), 950–955. Also CQPI Report #116, May 1994.

72. C. Fung, "The importance of data transformation in designed experiments for life testing" (1995), *Quality Engineering*, **7**(3), 625–638.

73. "Total quality: its origins and its future" (1995), CQPI Report No. 123, January.

74. "Regression analysis applied to happenstance data" (1995), *Quality Engineering*, **7**(4), 841–846.

75. R. D. Meyer and D. M. Steinberg, "Follow-up designs to resolve confounding in multifactor experiments" (1996), *Technometrics*, **38**(4), 303–313.

76. "Split plot experiments" (1996), *Quality Engineering*, **8**(3), 515–520.

77. W. Stewart and T. Henson, "Model discrimination and criticism with single-response data" (1996), *AIChE Journal*, **42**(11), 3055–3062.

78. P. M. Berthouex, "Time series models for forecasting wastewater treatment plant performance" (1996), CQPI Report No. 139, February 1996.

79. A. Luceño, "The anatomy and robustness of discrete proportional-integral adjustment and its application to statistical process control" (1996), CQPI Report No. 143, April.

80. "Scientific statistics, teaching, learning and the computer" (1996), CQPI Report No. 146, June.

81. "Role of statistics in quality and productivity improvement" (1996), *Journal of Applied Statistics*, **23**(1), 3–20.

82. J. Tyssedal, "Projective properties of the sixteen run two-level orthogonal arrays" (1995), CQPI Report #135, December.

83. A. Luceño, "Discrete proportional-integral adjustment and statistical process control" (1997), *Journal of Quality Technology*, **29**(3), 248–260.

84. "Scientific method: the generation of knowledge and quality" (1997), *Quality Progress*, **30**(1), 47–50.

85. A. Luceño, "Quality quandaries-models, assumptions and robustness" (1998), *Quality Engineering*, **10**(3), 595–598.

86. P. Liu, "Product design with response surface methods" (1998), CQPI Report No. 150.

87. W. Stewart and Y. Shon, (1998), "Discrimination and goodness of fit of multi-responsive mechanistic Models" *AIChE Journal*, **44**(6), 1404–1412.

88. "Quality quandaries-multiple sources of variation: variance components" (1998). *Quality Engineering*, **11**(1), 171–174.

89. "Quality quandaries—use of Cusum statistics in the analysis of data and in process monitoring (1999), *Quality Engineering*, **11**(3), 495–498.

90. P. Liu, "Statistics as a catalyst to learning by scientific method: Part I, an example" (1999), *Journal of Quality Technology*, **31**(1), 1–15.

91. "Statistics as a catalyst to learning by scientific method: Part II, a discussion" (1999), *Journal of Quality Technology*, **31**, 16–72.

92. G. E. P. Box and A. Luceño, "Six Sigma, Process Drift, Capability Indices, and Feedback Adjustment" (1999), *Quality Engineering*, **12**(3), 297–302.

93. G. E. P. Box, "The Invention of the Composite Design" (1999–2000), *Quality Engineering*, **12**(1), 119–122.

94. G. E. P. Box, *Box in Quality and Discovery*. New York: Wiley, 2000. (Eds. C. Tiao, Soren Bisgaard, William J. Hill, Daniel Pea, Stephen M. Stigler).

95. G. E. P. Box and S. Jones, "Split Plots for Robust Product and Process Experimentation" (2001–01), *Quality Engineering*, **13**(1), 127–134.

96. G. E. P. Box and A. Luceño, "Influence of the Sampling Interval, Decision Limit, and Autocorreclation on the Average Run Length in Cusum Charts" (2000), *Journal of Applied Statistics*, **12**(2), 177–183.

97. G. E. P. Box, "Comparisons, Absolute Values, and How I got to go to the Folies Bergres" (2001), *Quality Engineering*, **14**(1), 167–169.

98. G. E. P. Box, "Statistics for Discovery" (2001), *Journal of Applied Statistics*, **28**(3–4), 285–299.

99. G. E. P. Box and I. Han, "Experimental Design When There Are One or More Factor Constraints" (2001), *Journal of Applied Statistics*, **28**(8), 973–989.

100. G. E. P. Box and A. Luceño, "Feedforward as a Supplement to Feedback Adjustment in Allowing for Feedstock Changes" Tech. Rep. 180, Center for Quality and Productivity Improvement, University of Wisconsin, June 2001.

101. G. E. P. Box and J. Tyssedal, "Sixteen run designs of high projectivity for factor screening" (2001), *Communications in Statistics, Part B*, **30**(2), 217–228. Also CQPI Report No. 116, May 1994.

102. G. E. P. Box, S. Graves, S. Bisgaard, J. Van Gilder, K. Marko, J. James, M. Seifer, M. Poublon, and F. Fondale, "Detecting Mal functions in Dynamic Systems," Tech. Rep. 2001-01-0363, Society of Automotive Engineers, Inc., 2001. 2001-01-0363.

103. M. Kulahci and G. E. P. Box, "Catalysis o Discovery and Development in Engineering and Industry" (2003), *Quality Engineering*, **15**(3), 509–513.

104. G. Box, S. Bisgaard, S. Graves, M. Kulahci, K. Marko, J. James, J. Van Gilder, T. Ting, H. Zatorski, and C. Wu, "Performance Evaluation of Dynamic Monitoring Systems: The Waterfall Chart" (2003), *Quality Engineering*, **16**(2), 183–191.

Index

Improving Almost Anything: Ideas and Essays, Revised Edition. By George Box and Friends
Copyright © 2006 John Wiley & Sons, Inc.

WILEY SERIES IN PROBABILITY AND STATISTICS
ESTABLISHED BY WALTER A. SHEWHART AND SAMUEL S. WILKS

The *Wiley Series in Probability and Statistics* is well established and authoritative. It covers many topics of current research interest in both pure and applied statistics and probability theory. Written by leading statisticians and institutions, the titles span both state-of-the-art developments in the field and classical methods.

Reflecting the wide range of current research in statistics, the series encompasses applied, methodological and theoretical statistics, ranging from applications and new techniques made possible by advances in computerized practice to rigorous treatment of theoretical approaches.

This series provides essential and invaluable reading for all statisticians, whether in academia, industry, government, or research.

*Now available in a lower priced paperback edition in the Wiley Classics Library.
†Now available in a lower priced paperback edition in the Wiley–Interscience Paperback Series.

*Now available in a lower priced paperback edition in the Wiley Classics Library.
†Now available in a lower priced paperback edition in the Wiley–Interscience Paperback Series.

*Now available in a lower priced paperback edition in the Wiley Classics Library.
†Now available in a lower priced paperback edition in the Wiley–Interscience Paperback Series.

*Now available in a lower priced paperback edition in the Wiley Classics Library.

†Now available in a lower priced paperback edition in the Wiley–Interscience Paperback Series.

---

*Now available in a lower priced paperback edition in the Wiley Classics Library.

†Now available in a lower priced paperback edition in the Wiley–Interscience Paperback Series.

*Now available in a lower priced paperback edition in the Wiley Classics Library.
†Now available in a lower priced paperback edition in the Wiley–Interscience Paperback Series.

OCHI · Applied Probability and Stochastic Processes in Engineering and Physical Sciences

OKABE, BOOTS, SUGIHARA, and CHIU · Spatial Tesselations: Concepts and Applications of Voronoi Diagrams, *Second Edition*

OLIVER and SMITH · Influence Diagrams, Belief Nets and Decision Analysis

PALTA · Quantitative Methods in Population Health: Extensions of Ordinary Regressions

PANKRATZ · Forecasting with Dynamic Regression Models

PANKRATZ · Forecasting with Univariate Box-Jenkins Models: Concepts and Cases

* PARZEN · Modern Probability Theory and Its Applications

PEÑA, TIAO, and TSAY · A Course in Time Series Analysis

PIANTADOSI · Clinical Trials: A Methodologic Perspective

PORT · Theoretical Probability for Applications

POURAHMADI · Foundations of Time Series Analysis and Prediction Theory

PRESS · Bayesian Statistics: Principles, Models, and Applications

PRESS · Subjective and Objective Bayesian Statistics, *Second Edition*

PRESS and TANUR · The Subjectivity of Scientists and the Bayesian Approach

PUKELSHEIM · Optimal Experimental Design

PURI, VILAPLANA, and WERTZ · New Perspectives in Theoretical and Applied Statistics

† PUTERMAN · Markov Decision Processes: Discrete Stochastic Dynamic Programming

QIU · Image Processing and Jump Regression Analysis

* RAO · Linear Statistical Inference and Its Applications, *Second Edition*

RAUSAND and HØYLAND · System Reliability Theory: Models, Statistical Methods, and Applications, *Second Edition*

RENCHER · Linear Models in Statistics

RENCHER · Methods of Multivariate Analysis, *Second Edition*

RENCHER · Multivariate Statistical Inference with Applications

* RIPLEY · Spatial Statistics

* RIPLEY · Stochastic Simulation

ROBINSON · Practical Strategies for Experimenting

ROHATGI and SALEH · An Introduction to Probability and Statistics, *Second Edition*

ROLSKI, SCHMIDLI, SCHMIDT, and TEUGELS · Stochastic Processes for Insurance and Finance

ROSENBERGER and LACHIN · Randomization in Clinical Trials: Theory and Practice

ROSS · Introduction to Probability and Statistics for Engineers and Scientists

ROSSI, ALLENBY, and McCULLOCH · Bayesian Statistics and Marketing

† ROUSSEEUW and LEROY · Robust Regression and Outlier Detection

* RUBIN · Multiple Imputation for Nonresponse in Surveys

RUBINSTEIN · Simulation and the Monte Carlo Method

RUBINSTEIN and MELAMED · Modern Simulation and Modeling

RYAN · Modern Regression Methods

RYAN · Statistical Methods for Quality Improvement, *Second Edition*

SALEH · Theory of Preliminary Test and Stein-Type Estimation with Applications

* SCHEFFE · The Analysis of Variance

SCHIMEK · Smoothing and Regression: Approaches, Computation, and Application

SCHOTT · Matrix Analysis for Statistics, *Second Edition*

SCHOUTENS · Levy Processes in Finance: Pricing Financial Derivatives

SCHUSS · Theory and Applications of Stochastic Differential Equations

SCOTT · Multivariate Density Estimation: Theory, Practice, and Visualization

† SEARLE · Linear Models for Unbalanced Data

† SEARLE · Matrix Algebra Useful for Statistics

† SEARLE, CASELLA, and McCULLOCH · Variance Components

SEARLE and WILLETT · Matrix Algebra for Applied Economics

SEBER and LEE · Linear Regression Analysis, *Second Edition*

*Now available in a lower priced paperback edition in the Wiley Classics Library.
†Now available in a lower priced paperback edition in the Wiley–Interscience Paperback Series.

*Now available in a lower priced paperback edition in the Wiley Classics Library.
†Now available in a lower priced paperback edition in the Wiley–Interscience Paperback Series.

WU and HAMADA · Experiments: Planning, Analysis, and Parameter Design Optimization

WU and ZHANG · Nonparametric Regression Methods for Longitudinal Data Analysis

YANG · The Construction Theory of Denumerable Markov Processes

ZELTERMAN · Discrete Distributions—Applications in the Health Sciences

* ZELLNER · An Introduction to Bayesian Inference in Econometrics

ZHOU, OBUCHOWSKI, and McCLISH · Statistical Methods in Diagnostic Medicine